TECHNICAL DICTIONARY OF
VACUUM PHYSICS AND VACUUM TECHNOLOGY

TECHNICAL DICTIONARY OF

VACUUM PHYSICS AND
VACUUM TECHNOLOGY

**ENGLISH
GERMAN
FRENCH
RUSSIAN**

EDITED BY
DIPL.-PHYS. KARL HURRLE
DR.-ING. F. M. JABLONSKI AND
DIPL.-PHYS. HERBERT ROTH

CONTAINING ABOUT
5,000 TECHNICAL TERMS

PERGAMON PRESS
OXFORD · NEW YORK · TORONTO · SYDNEY
BRAUNSCHWEIG

PERGAMON PRESS LTD.
Headington Hill Hall, Oxford

PERGAMON PRESS INC.
Maxwell House, Fairview Park, Elmsford, New York 10523

PERGAMON OF CANADA LTD.
207 Queen's Quay West, Toronto 1

PERGAMON PRESS (AUST.) PTY. LTD.
19a Boundary Street, Rushcutters Bay, N.S.W. 2011, Australia

VIEWEG & SOHN GMBH
Burgplatz 1, Braunschweig

Copyright 1972 VEB Verlag Technik Berlin
First Edition 1972
Library of Congress Catalog Card No. 71-187211
Printed in the German Democratic Republic
08 016957 0

PREFACE

In this age of scientific and technological revolution, the international demand for information has increased enormously and the study and evalutation of technical literature, books, learned and trade journals, published in many languages, has grown similarly. The continuous development of scientific and technical concepts leading to the introduction of new terms has increased the need for reliable specialized dictionaries to aid readers of foreign-language publications.

Hence, in the field of Vacuum Physics and Technology with its almost universal range of application (from space travel and electrical engineering to dehydration and preservation of foodstuffs) a need was recognized for a reference source to facilitate the study and interpretation of international scientific and technical achievements. We decided, therefore, to compile a multilingual specialist dictionary for this particular field of science and engineering.

This first edition will be of considerable help to scientists engaged in teaching or research, and to manufacturing engineers, technicians, students, etc. The terms were compiled, from international publications on Vacuum Physics and Technology and related fields.

We are well aware that the first edition of such a comprehensive work of reference may have faults and omissions and we should appreciate receiving critical comments, corrections and proposals for additions, which should be addressed to VEB Verlag Technik, 102 Berlin, Oranienburger Strasse 13/14, GDR.

The authors and publishers

DIRECTIONS FOR USE
HINWEISE FÜR DIE BENUTZUNG
DIRECTIVES POUR L'USAGE
О ПОЛЬЗОВАНИИ СЛОВАРЕМ

1. **Examples of alphabetization**
   Beispiele für die alphabetische Ordnung
   Exemples de la classification alphabétique
   Примеры для алфавитного порядка

ball and socket joint	connecting glass
ballasted pumping speed	connecting terminal
ballast tank	connect in series
ball bearing	connection
ball valve	connection flange
Leckanzeigevorrichtung	thermische Oxydation
Leck bekannter Größe	thermische Raumkammer
leckfrei	thermischer Ausdehnungskoeffizient
Leck in dünner Wand	thermisches Gleichgewicht
Leckprüfung	thermische Transpiration
charge d'huile	métal de transition
chargement	métallisation adhésive
charge par ressort / à	métallisé sous vide
charge spatiale	métallurgie du vide
chargeur mécanique	métal mixte
испаритель мощности	стекловолокно
испарительная камера	стекло для запайки
испарительное газопоглощение	стекло для защиты от теплоизлучения
испарительный нагрев	стеклоцемент
испаритель с падающей пленкой	стекольный припой

2. Meanings of signs and abbreviations
Bedeutung der Zeichen und Abkürzungen
Signification des signes et des abréviations
Значение знаков и аббревиатур

( ) absorptive capacity (power) = absorptive capacity *or* absorptive power

[ ] adiabatic[al] = adiabatic *or* adiabatical

/   vacuo/in = in vacuo
    Vakuum/im = im Vakuum
    vide/sous = sous vide
    вакууме/в = в вакууме

< > these brackets contain explanations
    diese Klammern enthalten Erläuterungen
    ces parenthèses contiennent des explications
    в эти скобки заключены пояснения

s.   = see

s. a. = see also

‹US› = American English

‹sl› = slang

ENGLISH

CONTENTS

1.	Vacuum physics
1.1	Matter in the gaseous state
1.1.1	Real and ideal gases
1.1.2	Laws of the gaseous state
1.2	Kinetic theory of gases
1.3	Flow of gas
2.	Vacuum engineering
2.1	Values and measuring units
2.2	Vacuum gauges
2.2.1	Total pressure measuring devices
2.2.2	Partial pressure measuring devices
2.3	Vacuum pumps and getters
2.3.1	Positive displacement pumps
2.3.2	Molecular drag pumps
2.3.3	Ejector pumps
2.3.4	Diffusion pumps
2.3.5	Getter (ion) pumps
2.3.6	Cryo pumps
2.3.7	Pump accessories
2.4	Baffles
2.5	Valves and stopcocks
2.5.1	Isolation valves
2.5.2	Air release valves
2.5.3	Gas inlet valves
2.5.4	Valve (interlocking) combinations
2.5.5	Stopcocks
2.6	Vacuum lines
2.7	Joints
2.7.1	Fixed joints
2.7.2	Demountable joints
2.8	Vacuum vessels
2.9	Feedthroughs
2.10	Windows
2.11	Leak detection
2.11.1	Leak detectors
2.11.2	Methods of leak detection
2.12	Vacuum systems
2.12.1	Sealed vacuum systems
2.12.2	Pumped vacuum systems
2.13	UHV-technology
3.	Vacuum applications
3.1	Vacuum drying
3.2	Vacuum degassing
3.3	Vacuum impregnation
3.4	Vacuum brazing
3.5	Vacuum packaging
3.6	Vacuum distillation
3.7	Vacuum sublimation
3.8	Vacuum condensation
3.9	Vacuum filtration
3.10	Vacuum coating technique
3.11	Optics coatings
3.12	Protective coatings
3.13	Metallizing
3.14	Thin film technology
3.15	Vacuum melting
3.16	Vacuum degassing
3.17	Vacuum casting
3.18	Vacuum annealing
3.19	Vacuum sintering
3.20	Freeze drying
3.21	Space simulation

A

	English	German	French	Russian
A 1	**abnormal glow discharge**	anormale Glimm-entladung *f*	décharge *f* luminescente anormale	аномальный тлеющий разряд
A 2	**abrasion**	Abrieb *m*	abrasion *f*	истирание
A 3	**abrasive grain**	Schleifkorn *n*	grain *m* abrasif	абразивное зерно
A 4	**abrasive hardness,** scratch hardness	Ritzhärte *f*, Abnutzungs-widerstand *m*	résistance *f* à la rayure, dureté *f* sclérométrique	твердость на истирание, твердость по царапанию, склерометрическая твердость
A 5	**absolute Fahrenheit scale,** Rankine temperature scale	absolute Temperaturskale *f* <gemessen in Grad Rankine>, °R	échelle *f* de la température absolue	абсолютная температурная шкала Фаренгейта, температурная шкала Ренкина
A 6	**absolute manometer**	absolutes Manometer *n*	manomètre *m* absolu	абсолютный манометр
A 7	**absolute pressure**	absoluter Druck *m*	pression *f* absolue	абсолютное давление
A 8	**absolute temperature**	absolute Temperatur *f*	température *f* absolue	абсолютная температура
A 9	**absolute temperature scale**	absolute Temperaturskale *f*	échelle *f* de température absolue	абсолютная температурная шкала
A 10	**absolute vacuum,** perfect vacuum	absolutes Vakuum *n*	vide *m* absolu	абсолютный вакуум, идеальный вакуум
A 11	**absolute vacuum gauge, absolute vacuummeter**	absolutes Vakuummeter *n*	vacuomètre *m* absolu	абсолютный вакуумметр
A 12	**absolute viscosity,** dynamic viscosity, coefficient of viscosity	dynamische Zähigkeit *f*, dynamische Viskosität *f*	viscosité *f* dynamique	динамическая вязкость
A 13	**absolute zero**	absoluter Nullpunkt *m*	zéro *m* absolu	абсолютный нуль
A 14	**absolute zero of temperature**	absoluter Nullpunkt *m* der Temperatur	zéro *m* absolu de la température	абсолютный нуль температуры
A 15	**absorb**	absorbieren	absorber	абсорбировать
A 16	**absorbate**	Absorptiv *n*	substance *f* absorbée	абсорбированное вещество, абсорбтив
A 17	**absorbent, absorbing agent**	Absorbens *n*, Absorptions-mittel *n*	absorbant *m*	абсорбент, поглощающее вещество, поглотитель
A 18	**absorbing power,** absorptive capacity (power), absorptivity, absorption capacity	Absorptionsvermögen *n*	pouvoir *m* absorbant (d'absorption), absorbabilité *f*, absorptivité *f*	абсорбционная способность, поглотительная способность
A 19	**absorbing trap,** absorption trap	Absorptionsfalle *f*	piège *m* à absorption	абсорбционная ловушка
A 20	**absorption**	Absorption *f*	absorption *f*	абсорбция
	absorption capacity	*s.* absorbing power		
A 21	**absorption factor**	Absorptionsfaktor *m*	facteur *m* d'absorption	коэффициент аккомодации при абсорбции
A 22	**absorption pump**	Absorptionspumpe *f*	pompe *f* à absorption	абсорбционный насос
A 23	**absorption refrigerator,** absorption-type refrigerator	Absorptionskälte-maschine *f*	réfrigérateur *m* à absorption	абсорбционный электрохолодильник
A 24	**absorption spectrum**	Absorptionsspektrum *n*	spectre *m* d'absorption	спектр поглощения, абсорбционный спектр
	absorption trap	*s.* absorbing trap		
A 25	**absorption-type refrigerating system**	Absorptionskältesatz *m*	système *m* frigorifique d'absorption	абсорбционный холодильный агрегат
	absorption-type refrigerator	*s.* absorption refrigerator		
	absorptive capacity (power), absorptivity	*s.* absorbing power		
A 26	**accelerating electrode**	Beschleunigungselektrode *f*	électrode *f* d'accélération, deuxième électrode	ускоряющий электрод
A 27	**accelerating field**	Beschleunigungsfeld *n*	champ *m* d'accélération	ускоряющее поле
A 28	**accelerating voltage**	Beschleunigungsspannung *f*	tension *f* accélératrice (d'accélération)	ускоряющее напряжение
A 29	**acceptor**	Akzeptor *m*	accepteur *m*	акцептор
A 30	**accessories**	Zubehör *n*	accessoires *mpl*	принадлежности
A 31	**accidental error**	zufälliger Fehler *m*	erreur *f* fortuite	случайная ошибка
A 32	**accommodation coefficient**	Akkommodations-koeffizient *m*, Akkommodationszahl *f*	coefficient *m* d'accommodation	коэффициент аккомодации
A 33	**accommodation coefficient for condensation,** condensation coefficient	Kondensationskoeffizient *m*, Kondensationszahl *f*	coefficient *m* de condensation	коэффициент аккомодации при конденсации
A 34	**accumulation method, accumulation technics** <leak detection>	Anreicherungsverfahren *n* <Lecksuche>	procédé *m* de concentration	метод накопления <в течеискании>
A 35	**accuracy**	Genauigkeit *f*	précision *f*	точность
A 36	**accuracy of adjustment**	Einstellgenauigkeit *f*	exactitude *f* d'ajustage	точность настройки
A 37	**acidproof, acid-resistant**	säurebeständig	résistant aux acides, anti-acide	кислотоустойчивый, кислотостойкий
A 38	**acting time**	Regelzeit *f*	temps *m* de réglage	продолжительность переходного процесса, время регулирования
A 39	**actinium**	Aktinium *n*	actinium *m*	актиний
A 40	**action cycle,** operation cycle	Arbeitstakt *m*	phase *f* (cycle *m*, période *f*, rythme *m*) de travail	рабочий такт

A 41	**activated alumina**	aktiviertes Aluminium-oxid n <Sorptions-mittel>	oxyde m d'aluminium activé	активированная окись алюминия <газопогло-титель>
A 42	**activated charcoal**	Aktivkohle f	charbon m actif (adsorbant)	активированный уголь
A 43	**activation analysis**	Aktivierungsanalyse f	analyse f d'activation	активационный анализ
A 44	**activation energy**	Aktivierungsenergie f	énergie f d'activation	энергия активации
A 45	**activation of oxide-coated cathodes**	Aktivierung f von Oxidkatoden	activation f des cathodes à oxydes	активировка оксидных катодов
A 46	**activator**	Aktivator m	activateur m	активатор
A 47	**active alloy process** <titanium or circonium hydride technique>	Aktivmetallverfahren n <Titan- oder Zirkon-_hydridverfahren>	procédé m du métal actif	технология с примене-нием гидридов титана или циркония
A 48	**active carbon trap**	Sorptionsfalle f mit Aktivkohle	piège m à sorption avec charbon actif	ловушка с активиро-ванным углем
A 49	**active getter**	Aktivgetter m	getter m actif	активный газопогло-титель (геттер)
A 50	**activity**	Aktivität f	activité f	активность
A 51	**actuating shaft**	Steuerwelle f	arbre m de commande	распределительный вал
A 52	**adapt**	anpassen	adapter	подгонять, приспосабли-вать
A 53	**adapter**, intermediate piece, transition piece	Zwischenstück n	raccord m intermédiaire, pièce f intermédiaire, adapteur m	промежуточный элемент, переходник
	adapter flange	s. adapting flange		
A 54	**adapter glass,** intermediate glass	Zwischenglas n	verre m intermédiaire	переходное стекло
A 55	**adapting flange,** adapter flange	Anpassungsflansch m	bride f d'adaptation	переходный фланец
A 56	**adatom**	Adatom n	adatome m	адатом
A 57	**addition,** admixture	Zusatz m, Zuschlag m	addition f, ajoutage m	добавка, примесь, присадка
A 58	**additional layer**	Zusatzschicht f	filet m additionnel (supplémentaire)	дополнительный слой
A 59	**adhere,** stick	stecken, kleben, haften	adhérer	склеивать, прилеплять, прилипать
A 60	**adherence,** adhesive power (strength), adhesion	Haftfestigkeit f	adhérence f	прилипаемость, адгезив-ная прочность, проч-ность сцепления
A 61	**adherence time,** dwell-ing time, residence time, retention time, stay period, sticking time	Verweilzeit f	durée f, temps m de con-tact, temps d'exposition	время удерживания
A 62	**adherent**	fest haftend, haftbeständig	adhésif	прилипший, прочно приставший
	adhesion	s. adherence		
A 63	**adhesive**	Kleber m, Klebemittel n	colle f, agglutinant m, matière f collante	склеивающее (клейкое) вещество, клей, адгезив
A 64	**adhesive coating**	haftfeste Bedampfung f	métallisation f adhésive, enduit m adhésif, traite-ment m adhésif par métallisation	приклеенное покрытие
A 65	**adhesive film**	Klebefolie f	lamelle f collante (adhésive)	адгезивный слой
A 66	**adhesiveness**	Adhäsionsvermögen n	adhésivité f	адгезионная способность
	adhesive power (strength)	s. adherence		
A 67	**adhesive wax**	Klebwachs n	cire f adhésive	прилипающий воск
A 68	**adiabatic[al]**	adiabatisch	adiabatique	адиабатический
A 69	**adiabatic exponent,** specific heat ratio	Adiabatenexponent m	indice m adiabatique	показатель степени в уравнении адиабаты, показатель адиабаты
A 70	**adipocerous**	fettwachsartig	adipocéreux	из трупного жира
A 71	**adjustable leak,** variable leak	einstellbares Leck n	fuite f réglable	регулируемая течь
A 72	**adjust to zero**	auf Null einstellen	amener au zéro, mettre à zéro	устанавливать на нуль
A 73	**admit air,** vent, break vacuum, aerate	belüften	respirer, aérer	аэрировать, впускать воздух, продувать
A 74	**admittance area,** nozzle (jet) clearance area, pump throat area, aperture (annular) gap	Diffusionsspaltfläche f	plan m de clivage de diffusion	площадь кольцевого за-зора, площадь поверх-ности диффузии
	admixture	s. addition		
A 75	**adsorbate**	Adsorbat n, adsorbiertes Gas n	gaz m adsorbé, adsorbat m	адсорбат
A 76	**adsorbed layer,** adsorption layer	Adsorptionsschicht f	couche f d'adsorption	адсорбирующий слой
A 77	**adsorbent,** adsorbing agent	Adsorbens n, Adsorp-tionsmittel n	adsorbant m	адсорбент, адсорбирую-щее вещество
A 78	**adsorption**	Adsorption f	adsorption f	адсорбция
A 79	**adsorption-desorption technique**	Adsorption-Desorptions-Technik f	technique f d'adsorption et de desorption	адсорбционно-десорб-ционная техника
A 80	**adsorption gas-chromatography**	Adsorptionsgas-chromatografie f	chromatographie f de gaz	адсорбционная газовая хроматография

	English	German	French	Russian
A 81	**adsorption isotherm**	Adsorptionsisotherme f	isotherme f d'adsorption	изотерма адсорбции
	adsorption layer, adsorbed layer	Adsorptionsschicht f	couche f d'adsorption	адсорбирующий слой
A 82	**adsorption power,** adsorptivity, adsorptive power	Adsorptionsvermögen n	pouvoir m adsorbant	адсорбционная способность
A 83	**adsorption pump**	Adsorptionspumpe f	pompe f à adsorption	адсорбционный насос
A 84	**adsorption spectrometer**	Adsorptionsspektrometer n	spectromètre m à adsorption	адсорбционный спектрометр
A 85	**adsorption time**	Adsorptionszeit f	temps m d'adsorption	продолжительность адсорбции
A 86	**adsorption trap,** molecular sieve baffle	Adsorptionsfalle f	piège m à adsorption, trappe f d'adsorption	адсорбционная ловушка
A 87	**adsorption vacuum gauge**	Adsorptionsvakuummeter n, Adsorptionsvakuumprüfgerät n	vacuomètre m à adsorption	адсорбционный вакуумметр
	adsorptive power, adsorptivity	s. adsorption power		
	aerate, admit air, vent, break vacuum	belüften	respirer, aérer	аэрировать, впускать воздух, продувать
A 88	**aerated plastics**	Schaumstoff m	matière f de mousse	пенопласт
A 89	**aeration,** airing, air inlet, venting	Belüftung f	aération f, entrée f d'air	аэрация, впуск воздуха, обдувание
A 90	**afterdrying,** final drying, secondary drying, subsequent drying	Nachtrocknung f, Endtrocknung f	dessiccation f secondaire (finale), séchage m final (secondaire)	досушивание, досушка, вторичная осушка, конечная (окончательная) сушка
A 91	**afterglow**	Nachglimmen n, Nachleuchten n	reluminescence f	послесвечение
A 92	**age hardening**	Alterungshärtung f	durcissement m de vieillissement	твердение при старении
A 93	**ag[e]ing of thoriated filaments**	Alterung f thorierter Katoden	vieillissement m des filaments thoriés	старение торированных катодов
A 94	**agitator**	Rühreinrichtung f	agitateur m	мешалка
A 95	**air admittance valve,** venting (air release) valve, airing (bleed, air inlet) valve	Belüftungsventil n, Flutventil n, Lufteinlaßventil n	soupape f à admettre de l'air, robinet m d'entrée d'air	вентиль пуска воздуха, напускной вентиль
A 96	**air ballasting,** gas ballasting	Gasballastbetrieb m	admission f du lest d'air	напуск балластного газа
A 97	**air blast**	Druckluftstrom m	flux m d'air comprimé	дутье, струя воздуха
A 98	**air bleeding**	Einblasen n von Luft	entrée f de l'air	продувка, выпуск (удаление) воздуха
A 99	**airborne dryer,** suspended particle dryer, supporting gas dryer	Schwebegastrockner m	sécheur m de gaz pendant	газовая сушилка суспендированного вещества
A 100	**air-cooled**	luftgekühlt	refroidi par air, refroidi à l'air	охлажденный воздухом
A 101	**air-escape valve,** air-relief valve	Entlüftungsventil n	soupape f de purge (sortie d'air)	выпускной воздушный вентиль
A 102	**air filter**	Luftfilter n	filtre m à air	воздушный фильтр
A 103	**air flap valve**	Luftklappe f	volet m d'aération, clapet m à air	воздушный клапан
A 104	**air furnace,** reverberating furnace	Flammofen m	four m à réverbère	пламенная печь, отражательная печь
A 105	**air-gas mixture**	Gas-Luft-Gemisch n	mélange m d'air et de gaz	газовоздушная смесь
A 106	**air impact,** air shock, incrush of air	Lufteinbruch m	entrée f d'air accidentelle, pénétration f d'air	прорыв воздуха, воздушный удар
	airing	s. aeration		
	airing valve	s. air admittance valve		
	air inlet, airing, aeration, venting	Belüftung f	aération f, entrée f d'air	аэрация, впуск воздуха, обдувание
	air inlet valve	s. air admittance valve		
A 107	**air liquefaction**	Luftverflüssigung f	liquéfaction f de l'air	сжижение воздуха
A 108	**air lock**	Luftschleuse f	sas m d'air	воздушный шлюз, шлюзовая камера
A 109	**air-operated valve**	druckluftbetätigtes Ventil n	soupape f à commande d'air comprimé	клапан с пневматическим приводом
A 110	**airproof,** hermetic[al]	hermetisch, luftdicht	hermétique	герметичный
A 111	**air refrigerating machine,** cold-air machine	Kaltluftmaschine f	machine f à air froid	воздушная холодильная машина
A 112	**air release plug,** venting screw	Entlüftungsschraube f	vis f d'évacuation d'air, vis de sortie d'air	пробка (винт) для выпуска воздуха
	air release valve	s. air admittance valve		
	air-relief valve	s. air-escape valve		
A 113	**air-relieving hole,** pumping hole	Entlüftungsloch n	foration f à évacuation de l'air, foration de désaération	отверстие для откачки воздуха
A 114	**air scrubber,** air washer	Luftwäscher m	laveur (scrubber) m d'air	воздухоочиститель
A 115	**air separation plant**	Luftzerlegungsanlage f, Luftspaltsäule f	décompositeur m d'air, colonne f de séparation d'air	установка для разделения воздуха
	air shock	s. air impact		
A 116	**air speed**	Saugvermögen n für Luft	débit m d'une pompe à air	скорость откачки воздуха

A 117	**air-to-air-plant**	Luft-zu-Luft-Anlage f, kontinuierliche Anlage f	installation f continue	установка непрерывного действия <с атмосферной загрузкой и выгрузкой>
	air washer	s. air scrubber		
A 118	**alembic,** still, distillation still	Destillationskolben m, Destillationsblase f	ampoule f à distiller	перегонная колба
A 119	**aligning pin**	Paßstift m, Fixierstift m	cheville (goupille) f d'ajustage	центрирующая (фиксирующая) шпилька
A 120	**aliphatic compound**	Fettverbindung f	composé m aliphatique	алифатическое соединение
A 121	**alkyl phenol resins**	Alkylphenolharze npl	résines fpl alkylphénoliques	алкилфенольные смолы
A 122	**all-glass ion gauge head**	Allglas-Ionisationsmeßzelle f	cellule f de mesure à ionisation en verre	стеклянный ионизационный манометр
A 123	**all-metal construction**	Ganzmetallausführung f	construction f tout-métal (entièrement métallique)	цельнометаллическая конструкция
A 124	**all-metal valve**	Ganzmetallventil n	robinet m tout-métal, vanne f entièrement métallique	цельнометаллический вентиль
A 125	**allowable error**	zulässiger Fehler m	erreur f admissible	допустимая ошибка
A 126	**alloy**	Legierung f	alliage m	сплав
A 127	**alloying addition**	Legierungszusatz m	alliage m d'addition	легирующая добавка
A 128	**alpha emitter**	Alpha-Strahler m	alpha-radiateur m	альфа-излучатель
A 129	**alpha particle**	Alpha-Teilchen n	particule f alpha	альфа-частица
A 130	**alphatron**	Alphatron n	alphatron m	радиоактивный (радиоизотопный) ионизационный манометр, альфатрон
A 131	**altitude chamber,** low-pressure chamber	Höhenkammer f, Unterdruckkammer f	chambre f de dépression	барокамера
A 132	**altitude simulator,** space simulator	Höhensimulator m, Raumsimulator m	simulateur m cosmique, chambre f spatiale	имитатор космического пространства
A 133	**alumina,** aluminium oxide	Aluminiumoxid n	oxyde m d'aluminium	глинозем, окись алюминия
A 134	**aluminium foil**	Aluminiumfolie f	feuille f d'aluminium	алюминиевая фольга
A 135	**aluminium hydroxide**	Tonerdehydrat n	hydrate m d'aluminium	гидроокись алюминия
	aluminium oxide, alumina	Aluminiumoxid n	oxyde m d'aluminium	глинозем, окись алюминия
A 136	**aluminizing**	Aluminiumbedampfung f	vaporisation f d'aluminium	алюминирование
A 137	**amalgamate**	amalgamieren	amalgamer	амальгамировать
A 138	**amalgamation**	Amalgamation f	amalgamation f	амальгамация
A 139	**ambient pressure,** environmental pressure	Umgebungsdruck m	pression f ambiante	давление окружающей среды
A 140	**ambient temperature**	Umgebungstemperatur f	température f de l'air ambiant, température ambiante	окружающая температура, температура окружающей среды
A 141	**ammonia leak detection**	Lecksuche f mit Ammoniak, Ammoniaklecksuche f	détection f de fuites par ammoniaque	течеискание с помощью аммиака
A 142	**analyzer tube**	Analysatorröhre f	tube m analyseur	анализаторная трубка
A 143	**analyzing film,** polaroid analyzer	Polaroidfolie f als Analysator	analyseur m de feuille polaroïdale	поляроидный анализатор
A 144	**angle of incidence**	Einfallswinkel m	angle m d'incidence	угол падения
A 145	**angle tee,** Y-branch	Abzweigung f	dérivation f, embranchement m, ramification f	тройник в виде буквы «Y»
A 146	**angle valve,** corner valve, right angle valve	Eckventil n	soupape f à passage angulaire, vanne f d'équerre	угловой вентиль
A 147	**Ångström unit**	Ångström-Einheit f	unité f Ångström	ангстрем
A 148	**angular distribution**	Winkelverteilung f	distribution f angulaire	угловое распределение
A 149	**angular spread**	Divergenz f, Winkelbereich m	divergence f	угловой сектор
A 150	**anneal**	glühen, anlassen, tempern	détremper, recuire, adoucir	отжигать, прокаливать
A 151	**anneal electrically**	elektrisch ausglühen	recuire par électricité	электрически отжигать (прокаливать)
A 152	**annealing furnace**	Glühofen m	four m à recuire	печь для отжига
A 153	**annealing of stainless steel**	Glühen n von rostfreiem Stahl	incandescence f d'acier inoxydable	отжиг нержавеющей стали
A 154	**annealing point,** A.P.	Kühlpunkt m, 15-Minuten-Entspannungstemperatur f, obere Entspannungstemperatur f	température f supérieure de desserrage	температура отжига, точка отжига
A 155	**annular burner,** ring burner	Ringbrenner m	brûleur m circulaire (annulaire)	кольцевая горелка
A 156	**annular clearance seal**	Ringspaltdichtung f	joint m annulaire	уплотнение кольцевого зазора
	annular gap	s. admittance area		
A 157	**annular jet nozzle,** inverted jet nozzle, umbrella jet	ringförmige Strahlumlenkdüse f	diffuseur m, diffuseur-déflecteur m, tuyère f à jet inversé	зонтичная ступень, зонтичное сопло, обращенное сопло Лаваля
A 158	**annular jet pump**	Ringdüsendampfstrahlpumpe f	éjecteur m à tuyère annulaire, éjecteur diffuseur	пароструйный насос с кольцевым соплом
A 159	**annular nozzle**	Ringdüse f	diffuseur m, tuyère f à jet inversé	кольцевое сопло

A 160	annular piston valve, annular sleeve valve	Ringkolbenschieber *m*	tiroir *m* à piston annulaire	камерная поршневая задвижка
A 161	annular seal	Ringanschmelzung *f*	scellement *m* annulaire	кольцевой спай
	annular sleeve valve	*s.* annular piston valve		
A 162	anode dark space	Anodendunkelraum *m*, anodischer Dunkelraum *m*	espace *m* obscur anodique	анодное темное пространство, анодная темная область
A 163	anode dissipation	Anodenverlustleistung *f*	dissipation *f* anodique	рассеяние [мощности] на аноде
A 164	anode distance	Anodenabstand *m*	distance *f* anodique	анодное расстояние
A 165	anode grid	Anodengitter *n*	grille *f* anodique	анодная сетка
A 166	anodic glow	Anodenglimmhaut *f*, anodisches Glimmlicht *n*	luminescence *f* anodique	анодное свечение
A 167	anodic oxidation	anodische Oxydation *f*	oxydation *f* anodique	анодирование
A 168	anodize	eloxieren	oxyder électrolytiquement, anodiser	анодировать
A 169	anticathode	Antikatode *f*	anticathode *f*	антикатод
A 170	anticreep barrier, oil creep barrier, antimigration	Kriechsperre *f*, Ölkriechsperre *f*	barrière *f* de fuites d'huile	барьер для защиты от проникновения масла по поверхности, антимиграционный барьер
A 171	antifriction bearing	Wälzlager *n*	palier *m* à roulement	подшипник качения
A 172	antifriction metal	Gleitlagermetall *n*	métal *m* de palier lisse	антифрикционный сплав
A 173	antiknock	klopffest	antidétonant	антидетонационный
	antimigration	*s.* anticreep barrier		
A 174	antimony	Antimon *n*	antimoine *m*	сурьма
A 175	anti-reflection, reflection reduction	Reflexionsverminderung *f*	abaissement *m* réfléchissant	ослабление отражения
A 176	anti-reflection coating, anti-reflex coating, bloomed coating	Antireflexschicht *f*, Antireflexbelag *m*, reflexmindernde Schicht *f*	couche *f* antiréfléchissante	антиотражательное покрытие
A 177	anti-vacuum	Gegenvakuum *n*	antidépression *f*	противоразрежение
A 178	anti-vibration mounting	Schwingungsdämpfer *m*	amortisseur *m* de vibrations	амортизационное крепление
	A. P.	*s.* atmospheric pressure		
	A. P.	*s.* annealing point		
	aperture gap	*s.* admittance area		
A 179	aperture impedance	Strömungswiderstand *m* einer Öffnung	résistance *f* au flux d'un orifice	сопротивление апертуры
A 180	apron	Düsenschirm *m*	écran *m* à éjecteur (injecteur, tuyère)	экран сопла
A 181	aquadag coating	Aquadag-Belag *m*, Graphitschicht *f*	enduit *m* aquadag, couche *f* de graphite	покрытие аквадагом
A 182	arcatom welding, atomic hydrogen arc welding	Arcatom-Schweißverfahren *m*, Arcatom-Schweißung *f*, Arcatom-Schweißen *n*	arcatome-soudage *f*, soudure *f* arcatome, soudure à l'hydrogène atomique	атомно-водородная дуговая сварка, атомно-дуговая сварка
A 183	arc discharge	Bogenentladung *f*	décharge *f* à l'arc	дуговой разряд
A 184	arc furnace	Lichtbogenofen *m*	four *m* à arc	дуговая печь
A 185	arc melting	Lichtbogenschmelzen *n*	fusion *f* à l'arc	плавка в дуговой печи
A 186	arc welding	Lichtbogenschweißung *f*	soudure *f* à l'arc	[электро]дуговая сварка
A 187	ardometer	Gesamtstrahlungspyrometer *n*, Ardometer *n*	pyromètre *m* à rayonnement total	пирометр суммарного излучения
A 188	argon arc welding process	Argonarc-Schweißverfahren *n*	procédé *m* de soudure argonarc	способ аргоно-дуговой сварки, способ дуговой сварки плавящимся электродом в среде аргона
A 189	argon instability	Argoninstabilität *f*	instabilité *f* d'argon	аргонная нестабильность
A 190	argon sputtering treatment	Argonreinigung *f* durch Zerstäuben	épuration *f* à l'argon par pulvérisation	очистка путем распыления в аргоне
A 191	argon stability	Argonstabilität *f*	stabilité *f* d'argon	аргонная стабильность
A 192	argon treatment	Argonbehandlung *f*	traitement *m* à l'argon	аргоновая обработка
A 193	artificial zeolite	künstlicher Zeolith *m* <Sorptionsmittel>	zéolite *f* artificielle	искусственный цеолит
A 194	asbestos	Asbest *m*	asbeste *m*, amiante *m*	асбест
A 195	asbestos board	Asbestpappe *f*	carton *m* d'amiante	асбестовый картон
A 196	asbestos cord	Asbestschnur *f*	ficelle *f* d'amiante, cordonnet *m* d'amiante	асбестовый корд
A 197	aspirator pump, water jet pump, water blast pump	Wasserstrahlpumpe *f*	pompe *f* à jet d'eau	водоструйный насос
A 198	Aston dark space	Astonscher Dunkelraum *m*	chambre *f* noire d'Aston	первая катодная темная область, астоново темное пространство
A 199	astrotorus baffle	Astrotorusbaffle *n*	baffle *m* astrotorus	ловушка с астероидальным профилем жалюзи
	atm	*s.* atmosphere		
A 200	atmosphere, standard atmosphere, normal atmosphere, atmospheric pressure <sea level, °C>, A.P., atm	normaler Atmosphärendruck *m* <760 Torr>, physikalische Atmosphäre *f*, Normdruck *m*, atm	pression *f* atmosphérique normale, atmosphère *f* physique	физическая атмосфера, атм
A 201	atmosphere gauge pressure, atmospheric excess pressure	Atmosphärenüberdruck *m*	surpression *f* atmosphérique	избыточное атмосферное давление

		atmospheric pressure, standard atmosphere, normal atmosphere, \<sea level, °C\>, atmosphere, A.P., atm	normaler Atmosphärendruck *m* \<760 Torr\>, physikalische Atmosphäre *f*, Normdruck *m*, **atm**	pression *f* atmosphérique normale, atmosphère *f* physique	физическая атмосфера, атм
A 202		atmospheric pressure	Atmosphärendruck *m*	pression *f* atmosphérique	атмосферное давление
A 203		atomically clean	atomar sauber	pur en état atomique	атомно-чистый
A 204		atomic beam	Atomstrahl *m*	rayon (jet) *m* atomique	атомный пучок
A 204		atomic hydrogen arc welding, arcatom welding	Arcatom-Schweißverfahren *n*, Arcatom-Schweißung *f*, Arcatom-Schweißen *n*	arcatome-soudage *f*, soudure *f* arcatome, soudure à l'hydrogène atomique	атомно-водородная дуговая сварка, атомнодуговая сварка
A 205		atomic number	Kernladungszahl *f*, Ordnungszahl *f*	nombre *m* ordinal	атомный номер, порядковый номер элемента, менделеевское число
A 206		atomic volume	Atomvolumen *n*	volume *m* atomique	атомный объем
A 207		atomization, atomizing spraying	Atomisieren *n*, Druckverdüsung *f*, Zerstäubung *f* \<Flüssigkeiten\>	pulvérisation *f*, atomisation *f*	распыление жидкости, атомизация
A 208		atomize, spray	zerstäuben \<Flüssigkeiten\>	pulvériser	распылять жидкость
		atomizing praying	*s.* atomization		
A 209		audible leak indicator, squealer, audio leak indicator	akustischer Leckanzeiger *m*	indicateur *m* acoustique de fuite	звуковой индикатор течи
A 210		audible signal, audio-alarm	akustisches Signal *n*	signal *m* sonore	акустический (звуковой) сигнал
		audio leak indicator	*s.* audible leak indicator		
A 211		austenitic stainless steel	austenitischer rostfreier Stahl *m*	acier *m* austénitique inoxydable	аустенитная нержавеющая сталь
A 212		automatic machine	Automat *m*	appareil *m* automatique	автомат
A 213		automatic pump	Pumpautomat *m*	manège (carrousel) *m* de pompage, banc *m* de pompage rotatoire	автоматический насос
A 214		automatic shutter	automatische Blende *f*	diaphragme *m* automatique, écran *m* automatique	автоматический затвор, автоматическая диафрагма
A 215		autoradiographic technique	Autoradiografie *f*	autoradiographie *f*	авторадиография, радиоавтография
A 216		auxiliary cathode	Hilfskatode *f*	cathode *f* auxiliaire	вспомогательный катод
A 217		auxiliary electrode	Hilfselektrode *f*	électrode *f* auxiliaire	вспомогательный электрод
A 218		auxiliary voltage, bias voltage	Vorspannung *f*	tension *f* de polarisation, tension auxiliaire	напряжение смещения
A 219		average excitation energy	mittlere Anregungsenergie *f*	énergie *f* moyenne d'excitation	средняя энергия возбуждения
A 220		average linear expansion coefficient	mittlerer linearer Ausdehnungskoeffizient *m*	moyen coefficient *m* linéaire de dilatation	средний коэффициент линейного расширения
A 221		average molecular velocity	mittlere Teilchengeschwindigkeit *f*, mittlere Molekulargeschwindigkeit *f*	vitesse *f* moyenne (des particules) moléculaire	средняя скорость молекул
A 222		average temperature, mean temperature	Durchschnittstemperatur *f*, mittlere Temperatur *f*	température *f* moyenne	средняя температура
A 223		Avogadro's law	Avogadrosches Gesetz *n*	loi *f* d'Avogadro	закон Авогадро
A 224		Avogadro's number \<number of molecules contained in one gramme-molecule\>	Loschmidtsche Zahl *f* \<6,025 · 10²³ Mol⁻¹\>	nombre *m* d'Avogadro	число Авогадро
A 225		axial-flow blower	Axialgebläse *n*	soufflerie *f* axiale	осевой вентилятор
A 226		axial-flow compressor	Axialverdichter *m*	compresseur *m* axial	осевой компрессор, нагнетатель
A 227		axial-flow pump	Axialpumpe *f*	pompe *f* à flux axial	осевой насос
A 228		axial gasket	Axialdichtring *m*	rondelle *f* axiale \<de joint\>, anneau *m* d'étanchéité axial	уплотнительное кольцо для оси
A 229		axial piston pump	Axialkolbenpumpe *f*	pompe *f* à piston axiale	осевой поршневой насос

B

		b	*s.* bar		
B 1		babbit metal, bearing metal	Lagermetall *n*	métal *m* antifriction	подшипниковый (антифрикционный) сплав, баббит, сплав для заливки подшипников
B 2		back diffusion	Rückdiffusion *f*	rétrodiffusion *f*, diffusion *f* de retour	обратная диффузия
B 3		backfilling	Rückfüllung *f*	recharge *f*	повторное наполнение, перенаполнение
B 4		back-flow of oil, oil suck-back, oil return	Ölrückfluß *m*	reflux (retour) *m* d'huile	обратный поток масла, обратное течение масла
B 5		background current, dark current	Untergrundstrom *m*	courant *m* de fond	фоновый ток
B 6		background gas	Untergrundgas *n*	sous-sol gaz *m*	фоновый газ
B 7		back ignition	Rückzündung *f*	allumage *m* de retour	обратное зажигание
B 8		backing line, fore-line	Vorvakuumleitung *f*	canalisation *f* pour vide primaire	форвакуумная магистраль

B 9	**backing line condenser**	vorvakuumseitiger Abscheider *m*	séparateur *m* sur le côté prévide	уловитель (ловушка) для конденсата на форвакуумной стороне
B 10	**backing-line connection,** fore-vacuum connection, fore-arm connection, outlet flange, fore-pump connection, exhaust fitting	Vorvakuumanschluß *m*	raccord *m* de vide préliminaire	соединение [насоса] с форвакуумом
B 11	**backing-line trap,** fore-line trap	Vorvakuumfalle *f*	piège *m* à (pour le) vide primaire	форвакуумная ловушка, маслосборник
B 12	**backing-line valve,** fore-pump valve, fore-line valve	Vorpumpenventil *n*, Vorvakuumventil *n*	soupape *f* [de pompe] à vide préliminaire	форвакуумный вентиль
B 13	**backing pressure,** fore pressure	Vorvakuumdruck *m*	pression *f* du vide préliminaire	выпускное давление
B 14	**backing pump,** fore pump	Vorvakuumpumpe *f*, Vorpumpe *f*	pompe *f* primaire (à prévide, à vide préliminaire, pour vide primaire)	насос предварительного разрежения, форвакуумный насос
B 15	**backing space,** backing volume	Vorvakuumraum *m*	espace *m* sous vide primaire, volume *m* à prévide	форвакуумное пространство ‹диффузионного насоса›
B 16	**backing-space leak detection technique**	Lecksuche *f* mit Anschluß des Lecksuchers an das Vorvakuum	détection *f* par l'espace primaire	поиски течеискателем, присоединенным к форвакууму
B 17	**backing-space technique**	Vorvakuumtechnik *f* ‹Anschluß des Lecksuchers zwischen Diffusionspumpe und Vorpumpe›	technique *f* du vide préliminaire	метод течеискания с присоединением течеискателя к форвакуумной магистрали
	backing volume	*s.* backing space		
B 18	**back migration**	Rückkriechen *n*, Rückverdampfung *f* infolge Kriechens	migration *f* en retour, reflux *m* par migration	испарение рабочей жидкости в систему
B 19	**back pressure valve,** non-return valve	Rückschlagventil *n*	valve *f* de rebondissement, soupape *f* de rebondissement (retenue)	обратный клапан, возвратный клапан
B 20	**back-seat gasket**	Doppelsitzdichtung *f*	joint *m* à double siège	двухседельное уплотнение
B 21	**back streaming**	Rückströmung *f* ‹Treibmitteldampf›	courant *m* inverse ‹de poudre›	обратный поток паров рабочей жидкости
B 22	**back-streaming rate**	Rückströmrate *f*	quote-part *f* de flux inverse, quote-part de reflux [par migration]	скорость обратного течения
B 23	**baffle**	Baffle *n*, Prallfläche *f*, Kondensationsfläche *f*, Dampfsperre *f*	chicane *f*, baffle *m*	отражатель, механическая ловушка
B 24	**baffle plate,** deflector, repeller	Prallplatte *f*	déflecteur *m*, écran *m*, chicane *f*	отражательная пластина
B 25	**baffle valve**	Baffleventil *n*	vanne *f* à baffle (écran)	тарельчатый вентиль
B 26	**bakeable**	ausheizbar	étuvable	прогреваемый
B 27	**bake cycle,** baking cycle	Ausheizzyklus *m*	cycle *m* d'étuvage, période *f* de dégazage	цикл прогрева
B 28	**bake-out control**	Ausheizsteuerung *f*	commande *f* d'étuvage, réglage *m* d'étuvage	регулировка процесса нагрева
B 29	**bake-out furnace,** bake-out oven	Ausheizofen *m*	four *m* d'étuvage, four de chauffe, étuve *f*	печь для нагрева
B 30	**bake-out jacket,** baking jacket	Ausheizmantel *m*	chemise *f* (manchon *m*) d'étuvage	нагревательная рубашка
	bake-out oven	*s.* bake-out furnace		
B 31	**bake-out table**	Ausheiztisch *m*	table *f* à étuver (sécher en chauffant)	стол для прогрева ‹при обезгаживании›
B 32	**bake-out temperature,** baking temperature	Ausheiztemperatur *f*	température *f* de chauffage	температура прогрева ‹при обезгаживании›
B 33	**baking**	Ausheizen *n* ‹mittels Ofens›	chauffage *m*	прогрев в печи
	baking cycle	*s.* bake cycle		
	baking jacket	*s.* bake-out jacket		
	baking temperature, bake-out temperature	Ausheiztemperatur *f*	température *f* de chauffage	температура прогрева ‹при обезгаживании›
B 34	**balanced method**	Nullabgleichverfahren *n*	méthode *f* de compensation	компенсационный метод
B 35	**balance pressure**	Ausgleichsdruck *m*	pression *f* de compensation	уравновешивающее давление
B 36	**ball and socket joint**	Kugelgelenk *n*	joint *m* à rotule (bille)	шаровой шлиф
B 37	**ballasted pumping speed**	Saugvermögen *n* bei Gasballast	débit *m* d'une pompe à lest d'air	быстрота действия насоса с подачей балласта
B 38	**ballast tank**	Vorvakuumkessel *m*, Vorvakuumbehälter *m*	réservoir *m* intermédiaire	балластный бак, форвакуумный бак
B 39	**ball bearing,** journal bearing	Kugellager *n*	roulement (palier) *m* à billes	шариковый подшипник
B 40	**ball check valve**	Kugelrückschlagventil *n*	robinet *m* (vanne *f*) sphérique de retenue	шаровой запорный клапан
B 41	**ball valve,** globe valve, spherical valve	Kugelventil *n*	soupape *f* à bille (boulet), soupape sphérique	вентиль со сферическим клапаном, шаровой клапан

B 42	**band coating,** strip coating, roll coating	Bandbedampfung f	vaporisation f à bande, métallisation f au déroulé	нанесение покрытия на ленту
B 43	**band spektrum**	Bandenspektrum n	spectre m cannelé	полосатый спектр
B 44	**bar,** b	Bar n, absolute Atmosphäre f	bar, atmosphère f absolue	бар
B 45	**bar drawing,** mandrel drawing	Stopfenzug m	traction f de bouchage	волочение на оправке
B 46	**bare gauge tube,** nude ion gauge, nude gauge, high-speed gauge	Einbausystem n, Eintauchsystem n ‹eines Ionisationsvakuummeters›	cellule f de mesure insérée, système m d'immersion	ионизационный манометр без оболочки, открытый ионизационный манометр
B 47	**bar feed lock**	Stangenvorschubschleuse f	écluse f d'avance à barre	шлюз для подачи прутка
B 48	**barium carbonate**	Bariumkarbonat n	carbonate m de baryum	карбонат бария, углекислый барий
B 49	**barium oxide**	Bariumoxid n	oxyde m de baryum	окись бария
B 50	**barometer**	Barometer n	baromètre m	барометр
B 51	**barometer reading,** barometric level	Barometerstand m	niveau m de baromètre	барометрическое давление
B 52	**barometric compensation**	Barometerkompensation f	compensation f barométrique	барометрическая компенсация
B 53	**barometric equation,** Boltzmann barometric equation	barometrische Höhenformel f	formule f barométrique [de Boltzmann]	барометрическая формула [Больцмана]
	barometric level	s. barometric reading		
B 54	**barrel of a pump**	Pumpenstiefel m	corps m de pompe	цилиндр насоса
B 55	**barrier layer,** depletion layer	Sperrschicht f	couche f d'arrêt, couche de barrage	запорный слой
B 56	**base coat of lacquer,** undercoat	Grundlackschicht f	couche f de laque d'accrochage, couche de fond de vernis	лаковая основа
B 57	**base lacquer,** prime lacquer	Grundlack m	laque f d'accrochage, couche f support de laque	грунтовочный лак
B 58	**base plane**	Basisfläche f	plaine f de base	плоскость базы (основания)
B 59	**base plate,** mounting plate	Grundplatte f, Montageplatte f	plaque f de base (montage), plaque-support f	опорная плита
B 60	**basic circuit**	Prinzipschaltung f	principe m schématisé de câblage	принципиальная схема
B 61	**basic test**	Grundlagenuntersuchung f	analyse (recherche) f fondamentale	фундаментальное исследование
B 62	**basket coil**	Korbspule f	bobine f clissée	корзиночная катушка
B 63	**batch operation,** batch process	Chargenbetrieb m	opération f de charge	циклический режим
B 64	**batchwise**	intermittierend	intermittent, discontinu	прерывистый
B 65	**bath metal**	Badmetall n	métal m de bain	[расплавленный] металл в ванне
B 66	**Bayard-Alpert gauge**	Bayard-Alpert-Röhre f	jauge f de Bayard-Alpert	манометр Байярда-Альперта
B 67	**Bayard-Alpert ionization gauge**	Glühkatodenionisationsvakuummeter n nach Bayard und Alpert	manomètre m d'ionisation à cathode chaude de Bayard et Alpert	ионизационный манометр Байярда-Альперта
B 68	**bayonet cap**	Bajonettsockel m	culot m à baïonnette	цоколь Свана, штифтовой цоколь
B 69	**bayonet fixing**	Bajonettverschluß m	fermeture f à baïonnette	штифтовой (байонетный) запор
B 70	**bayonet holder**	Bajonettfassung f	douille f à baïonnette	патрон Свана, штыковой (штифтовой) патрон
B 71	**bead,** flange	bördeln	border	отбортовывать
B 72	**beading machine**	Sickenmaschine f, Bördelmaschine f	machine f de suintement	кромкогибочный станок
B 73	**beam aperture**	Strahlapertur f	ouverture f du jet	апертура пучка, угловой раствор пучка
B 74	**beam current**	Strahlstrom m	courant m de faisceau électronique	ток пучка
B 75	**beam density**	Strahldichte f	densité f du jet (rayon)	плотность пучка (струи)
B 76	**beamed**	gestrahlt, gerichtet, gebündelt	lancé	направленный, лучевой
B 77	**beam generation**	Strahlerzeugung f	production f de jet	генерация пучка
B 78	**beam guide tube,** beam guiding tube	Strahlleitrohr n	tube m directeur du rayon, tube directeur du jet	лучевод
B 79	**beaming effect**	Richtstrahlwirkung f	effet m du faisceau lancé	действие направленного луча
B 80	**beam penetration**	Strahleindringtiefe f	enfoncement m d'un rayon	глубина проникновения луча
B 81	**beam power**	Strahlleistung f	puissance f de rayon	мощность луча
B 82	**beam pressure**	Strahldruck m	pression f du jet (rayon)	давление струи (пучка)
B 83	**beam splitter**	Strahlteiler m	diviseur m de rayon	устройство для расщепления пучка
B 84	**beam spot**	Strahlfleck m	spot m de rayons	пятно от луча
B 85	**bearing clearance**	Lagerspiel n	jeu m du palier	зазор в подшипнике
	bearing metal, babbit metal	Lagermetall n	métal m antifriction	подшипниковый (антифрикционный) сплав, баббит, сплав для заливки подшипников

B 86	bearing ring, supporting ring	Tragring *m*	anneau *m* portatif, bague-support *f*	опорное кольцо
B 87	beat frequency	Taktfrequenz *f*	fréquence *f* de phase	частота биений (повторения)
B 88	Beaufort scale	Beaufort-Skale *f*	échelle *f* Beaufort	шкала Бофорта
B 89	bell-and-spigot joint	Muffenverbindung *f*	assemblage *m* à manchons	соединение раструбом
B 90	bell jar	Rezipientenglocke *f*, Vakuumkammer *f*, Vakuumglocke *f*	chambre (cloche) *f* à vide, cloche du récipient	вакуумный колпак, вакуумная камера
B 91	bell-jar plant	Glockenanlage *f*	installation *f* de cloche	установка с вакуумным колпаком
B 92	bell-jar plate	Rezipiententeller *m*	plaque *f* du récipient	плита вакуумного колпака
B 93	bellows	Faltenbalg *m*, Federbalg *m*, Federungskörper *m*	soufflet *m*	сильфон
B 94	bellows-sealed high vacuum valve	Federbalghochvakuumventil *n*	robinet *m* piant expansible à vide poussé	высоковакуумный вентиль с сильфонным уплотнением
B 95/6	bellows-sealed valve, bellows-seal valve	Federbalgventil *n* Ventil *n* mit Federbalgdichtung	vanne *f* à soufflet, robinet-vanne *m* à soufflet	вентиль с сильфонным уплотнением, сильфонный вентиль
B 97	bellows-type null reading differential manometer	Federbalgdifferentialmanometer *n* mit Nullablesung	manomètre *m* différentiel à soufflet avec lecture de zéro	сильфонный дифференциальный манометр с нулевым отсчетом
B 98	bell-shaped valve	Glockenventil *n*	soupape *f* à cloche	чашечный вентиль
B 99	belt	Riemen *m*	courroie *f*	ремень
B 100	belt conveyor	Bandförderer *m*	convoyeur *m*, transporteur *m* à courroie, bande *f* transporteuse	ленточный конвейер
B 101	belt conveyor dryer, belt-type dryer	Bandtrockner *m*	sécheur *m* à bande	ленточная конвейерная сушилка
B 102	bend, elbow	Rohrbogen *m*, Rohrkrümmer *m*	coude *m*, raccord *m* coudé (angulaire)	колено трубы
B 103	bending radius	Krümmungsradius *m*	rayon *m* de courbure	радиус кривизны
B 104	bending stress fatigue limit	Dauerbiegefestigkeit *f*	résistance *f* continue à la flexion, endurance *f* à la flexion	предел усталости при переменном изгибе
B 105	Bernoulli formula, Bernoulli's equation	Bernoullische Gleichung *f*	équation *f* de Bernoulli	уравнение Бернулли
B 106	beryllium window	Berylliumfenster *n*	fenêtre *f* de glucinium	бериллиевое окно
B 107	beta particle	Beta-Teilchen *n*	particule *f* bêta	бета-частица
B 108	betatron	Betatron *n*	bétatron *m*	бетатрон
B 109	BET isotherm, Burnauer-Emmet-Teller isotherm	BET-Isotherme *f*, Burnauer-Emmet-Teller-Isotherme *f*	BET-isotherme *f*	БЭТ-изотерма, изотерма Бурнауера-Эммета-Теллера
B 110	bevel seat valve, inclined seat valve, slanting seat valve	Schrägsitzventil *n*	vanne *f* (robinet *m*) à siège oblique	вентиль с наклонным шпинделем
B 111	bias sputtering	Gegenfeldzerstäubung *f*	désintégration (dispersion, pulvérisation) *f* inverse	распыление под напряжением смещения
	bias voltage, auxiliary voltage	Vorspannung *f*	tension *f* de polarisation, tension auxiliaire	напряжение смещения
B 112	bifilar winding	Bifilarwicklung *f*	enroulement *m* bifilaire	бифилярная намотка (навивка)
B 113	billet	Knüppel *m*, Barren *m*	lingot *m*	заготовка, болванка, чушка, слиток
B 114	bilux bulb	Biluxlampe *f*	lampe *f* à deux filaments	лампа двойного света, лампа с двойной нитью
B 115	bimetallic contact	Bimetallkontakt *m*	contact *m* bimétallique	биметаллический контакт
B 116	bimetallic strip gauge, bimetal vacuum gauge	Bimetallvakuummeter *n*	manomètre *m* bimétallique, vacuomètre *m*	манометр с биметаллической полоской
B 117	bimetallic switch	Bimetallschalter *m*	interrupteur *m* bimétallique	биметаллический выключатель
B 118	bimetallic thermometer	Bimetallthermometer *n*	thermomètre *m* bimétallique	биметаллический термометр
B 119	bimetal pressure controller, bimetal pressure switch	Bimetalldruckschalter *m*	interrupteur *m* bimétallique à pression, manostat *m* bimétallique	биметаллический регулятор давления
	bimetal vacuum gauge	s. bimetallic strip gauge		
B 120	binary thermodiffusion factor	binärer Thermodiffusionsfaktor *m*	facteur *m* binaire de la thermodiffusion	термодиффузионный коэффициент для бинарной смеси
B 121	binding, wiring	Abbinden *n*	câblage *m*	схватывание, связывание
B 122	binding energy	Bindungsenergie *f*	énergie *f* de liaison (combinaison)	энергия связи
B 123	bivalent	zweiwertig	bivalent	бивалентный
B 124	black body	schwarzer Körper *m*	corps *m* noir	черное тело
B 125	black chromium plating	Schwarzverchromen *n*	chromage *m* noir	черное хромирование
B 126	black discharge	unsichtbare Entladung *f* <im Gasentladungsrohr>	décharge *f* obscure (noire), décharge non visible	темный разряд
B 127	black discharge, black vacuum	Druck *m* unter 10⁻³ Torr	pression *f* sous 10⁻³ Torr	«черный» («темный») вакуум

B 128	blank flange, blank-off flange (plate), cover plate, blind flange	Blindflansch m	bride f aveugle (obturatrice, d'obturation), joint m plein, fausse f bride	глухой фланец, заглушка
B 129	blank off	blindflanschen	obturer par bride, fermer par une fausse bride	заглушать ‹часть трубопровода›
	blank-off flange (plate)	s. blank flange		
B 130	blank-off pressure, limiting (ultimate, final) pressure, ultimate vacuum	Enddruck m	pression f finale	предельный вакуум, предельное (остаточное) давление
B 131	blank test	Blindversuch m	épreuve f (essai m) à blanc	слепая проба
B 132	blast cleaning	Putzstrahlen n	rayons mpl de nettoyage	пескоструйная очистка
B 133	Blears effect	Blears-Effekt m	effet m de Blears	эффект Блирса
B 134	bleed, blow off, release, valve	abblasen	souffler, laisser échapper, lâcher, évacuer	выпускать
B 135	bleeder, bleeder cock, drain cock, discharge tap	Ablaßhahn m	robinet m de vidange	спускной кран, сливной кран
B 136	bleed gas, flush gas, purge gas	Spülgas n	gaz m de balayage (rinçage)	газ для продувки, промывочный газ
	bleed valve, air admittance valve, venting valve, air release valve, airing valve, air inlet valve	Belüftungsventil n, Flutventil n, Lufteinlaßventil n	soupape f à admettre de l'air, robinet m d'entrée d'air	вентиль пуска воздуха, напускной вентиль
	blind flange, blank flange, blank-off flange, blank-off plate, cover plate	Blindflansch m	bride f aveugle (obturatrice, d'obturation), joint m plein, fausse f bride	глухой фланец, заглушка
B 137	blind hole	Sackloch n	trou m borgne	глухое отверстие
B 138	blister	Gußblase f	soufflure f	пузырь в металле, раковина
B 139	blistering, bubble formation	Blasenbildung f	formation f de bulles	образование пузырьков
B 140	Bloch boundary, Bloch wall	Bloch-Wand f, Blochsche Wand f	paroi f de Bloch	граница доменов
B 141	block cathode	Blockkatode f	cathode f bloc	блок-катод
B 142	blocking efficiency	Sperrwirkung f	effet m d'arrêt, effet m de blocage	запирающее (блокирующее) действие
B 143	block talc, natural steatite	Naturspeckstein m, Talk m	stéatite f naturelle, talc m	природный стеатит, тальк
B 144	bloom, slab	Bramme f, Walzblock m	brame f	сляб, листовой (плоский) слиток, болванка
B 145	bloomed	antireflexvergütet	trempé antiréflexe	просветленный ‹в оптике›, антиотражательный
	bloomed coating, anti-reflection coating, anti-reflex coating	Antireflexschicht f, Antireflexbelag m, reflexmindernde Schicht f	couche f antiréfléchissante	антиотражательное покрытие
B 146	bloom steel	Luppenstahl m	acier m en loupes	кричная сталь
B 147	blower	Gebläse n	soufflante f	воздуходувка
B 148	blown film	Blasfolie f	lamelle f de soufflage	выдутая пленка
	blow off, bleed, release, valve	abblasen	souffler, laisser échapper, lâcher, évacuer	выпускать
B 149	blow-off check valve, blow-off valve	Abblasventil n	soupape f de purge, soupape d'échappement d'évacuation	продувочный вентиль
B 150/1	blow pipe	Gebläsebrenner m	chalumeau m	паяльная горелка
B 152	body-centered	raumzentriert	centré	объемно-центрированный
B 153	body of valve, valve body	Ventilkörper m	corps m de soupape	корпус вентиля
B 154	boiler	Dampfgefäß n, Dampfkessel m	bouilleur m	котел
B 155	boiler	Siedegefäß n	bouilleur m, chaudière f	кипятильник, испаритель
B 156	boiler pressure	Treibmitteldampfdruck m im Siedegefäß	tension f de vapeur de fluide moteur dans la chaudière	давление рабочего пара в кипятильнике
B 157	boiling, ebullition	Sieden n	ébullition f	кипение
B 158	boiling delay	Siedeverzug m	retard m à l'ébullition	задержка кипения
B 159	boiling point	Siedepunkt m	point m d'ébullition	точка кипения
B 160	bolt cathode	Bolzenkatode f	cathode f de boulon	болт-катод
	Boltzmann barometric equation, barometric equation	barometrische Höhenformel f	formule f barométrique [de Boltzmann]	барометрическая формула [Больцмана]
B 161	Boltzmann constant, universal gas constant related to one molecule	Boltzmann-Konstante f	constante f universelle des gaz, constante de Boltzmann	постоянная Больцмана
B 162	bombarding	Beschießen n ‹Elektrodenentgasung durch Elektronen- oder Ionenbeschuß›	bombardement m ‹dégazage des électrodes par bombardement électronique ou ionique›	бомбардировка

B 163	**bomb gas,** tank gas	Flaschengas *n*	gaz *m* en bouteille	баллонный газ
B 164	**bonding**	Bindung *f*, Verbindung *f*, Kleben *n*	liaison *f*, assemblage *m*, jonction *f*, collage *m*	схватывание, соединение, сцепление, связь
B 165	**bond strength**	Bindungsstärke *f*, Bindungskraft *f*	force *f* de liaison	прочность (сила) сцепления
B 166	**bonnet-body seal**	Ventildeckel-Ventilkörper-Dichtung *f*	joint *m* de chapeau et de corps de la vanne	уплотнение между крышкой и корпусом вентиля
B 167	**bonnet gasket**	Ventildeckeldichtung *f*	joint *m* de couvercle de soupape	уплотнение крышки вентиля
B 168	**booster diffusion pump**	Diffusionspumpe *f* zwischen Haupt- und Vorpumpe	booster pompe *f* à diffusion	бустерно-диффузионный насос
B 169	**booster diode**	Spardiode *f*	diode *f* de récupération	бустерный диод
B 170	**booster pump,** intermediate pump	Booster-Pumpe *f*, Zwischenpumpe *f*	pompe *f* intermédiaire, pompe de booster	бустерный насос, вспомогательный насос
B 171	**booster-type diffusion pump**	Treibdampfpumpe *f*	pompe *f* à diffusion du type booster	бустерно-диффузионный насос
B 172	**borosilicate glass**	Borsilikatglas *n*	verre *m* de borosilicate	боросиликатное стекло
B 173	**bottom discharge**	Untenentleerung *f*	vidange *f* inférieure	донная разгрузка
B 174	**bottom pouring crucible**	Bodenabgußtiegel *m*	creuset *m* à épancher de fond	тигель с донным сливом
B 175	**bottom pouring method**	Bodengießmethode *f*	procédé *m* de coulée par le fond	донная разливка
B 176	**bottom pour ladle**	Stopfengußpfanne *f*	poche *f* à quenouille	разливочный ковш с донным стаканом
B 177	**bottom tapping**	Bodenabstich *m*	coulée *f* du sol	литье в почвенные формы
B 178	**boundary condition**	Randbedingung *f*	condition *f* extérieure (de bord)	краевое условие
B 179	**boundary friction**	Grenzflächenreibung *f*	friction *f* de l'interface (de surface limite)	поверхностное трение
B 180	**boundary layer**	Grenzschicht *f*	couche *f* limite	[по]граничный слой
B 181	**boundary-layer flow**	Grenzschichtströmung *f*	écoulement (courant) *m* de couche limite	течение в пограничном слое
B 182	**boundary lubrication**	Grenzflächenschmierung *f*	lubrification *f* de surfaces limites	смазка граничных поверхностей
B 183	**Bourdon [tube type vacuum] gauge**	Bourdon-Vakuummeter *n*, Bourdon-Rohr *n*, Federrohrvakuummeter *n*, Röhrenfedermanometer *n*	vacuomètre (tube) *m* de Bourdon, vacuomètre à tube élastique	манометр Бурдона
B 184	**box pump**	Kapselpumpe *f*	pompe *f* à palettes	капсульный насос
B 185	**box-type oven**	Kastenofen *m*	four *m* de forme rectangulaire	коробчатая печь
B 186	**Boyle's law**	Boyle-Mariottesches Gesetz *n*	loi *f* de Boyle-Mariotte	закон Бойля-Мариотта
B 187	**braze,** nard-solder	hartlöten	braser	паять твердым припоем
B 188	**brazing,** hard soldering	Hartlöten *n*	brasage *m*	пайка твердым припоем
B 189	**breakable glass seal**	Aufschlagventil *n*	vanne *f* à rupture	разрушаемый стеклянный спай, чертик
B 190	**breakdown strength,** disruptive strength, dielectric strength	Durchschlagfestigkeit *f*	résistance *f* au claquage, rigidité *f* diélectrique, résistance au percement disruptif	пробивная прочность, электрическая прочность
B 191	**breakdown voltage,** break-through	Durchschlagsspannung *f*	tension *f* au percement disruptif	пробивное напряжение
B 192	**breaking capacity**	Schaltleistung *f*	capacité *f* de rupture	разрывная мощность, отключающая способность
B 193	**breaking elongation,** fracture strain, total extension	Bruchdehnung *f*	dilatation *f* de rupture, allongement *m* de rupture	растяжение при разрыве, удлинение при разрыве
B 194	**breaking forepressure**	„Durchbruchs"-Vorvakuumdruck *m* <der dem zehnfachen An­saugdruck einer Diffu­sionspumpe bei norma­lem Vorvakuum ent­spricht>	pression *f* primaire de rupture <au-dessus de laquelle la pression d'admission d'une pompe à diffusion a augmenté de plus de dix fois>	выпускное давление срыва <при котором выпускное давление высоковакуумного на­соса возрастает более чем в 10 раз>
B 195	**breaking strength**	Bruchfestigkeit *f*	résistance *f* à la rupture (traction)	прочность на излом, предел излома
B 196	**break-off seal,** break seal	Trümmerventil *n*	soupape *f* à rupture, break-seal *m*	разрушаемая перегородка, клапан одноразового действия
B 197	**break-off tip** break seal	Abbrechspitze *f* s. break-off seal	pointe *f* de chute	разбиваемый кончик
	break-through, breakdown voltage	Durchschlagsspannung *f*	tension *f* au percement disruptif	пробивное напряжение
	break vacuum, admit air, vent, aerate	belüften	respirer, aérer	аэрировать, впускать воздух, продувать
B 198	**breathing film**	atmende Folie *f*	lamelle *f* respiratoire	респираторная пленка
B 199	**bridge,** shunt	überbrücken	shunter, dériver	шунт
B 200	**bridge breaker**	Brückenbrecher *m*	broyeur (concasseur) *m* de pont	прерыватель мостика

B 201	**bridge circuit**	Brückenschaltung f	couplage m (montage m, circuit m, connexion f) en pont	мостовая схема
B 202	**bright annealing**	Blankglühen n	recuit m brillant	светлый отжиг
B 203	**bright-drawing**	Blankziehen n	étirage m brillant	светлая протяжка, чистовое волочение
B 204	**brightness temperature**	Leuchttemperatur f	température f lumineuse	яркостная температура
B 205	**Brinell hardness**	Brinellhärte f	dureté f Brinell	твердость по Бринелю
B 206	**brittle lacquer**, brittle varnish	Reißlack m	vernis m givré	ломкий (растрескивающийся) лак
B 207	**brittleness**	Sprödigkeit f	fragilité f	хрупкость, ломкость
B 208	**brittle point**	Kaltbrüchigkeitstemperatur f	température f cassante à froid	температура хрупкости
	brittle varnish, brittle lacquer	Reißlack m	vernis m givré	ломкий (растрескивающийся) лак
B 209	**Brownian movement**	Brownsche Bewegung f	mouvement m brownien	броуновское движение
	brush discharge, bunch discharge	Büschelentladung f	aigrette f, décharge f en aigrette	кистевой разряд
B 210	**brush still**	Bürstendestillieranlage f	distillerie f des balais	щеточный перегонный аппарат
B 211	**bubble chamber**	Blasenkammer f	chambre f de bulles	пузырьковая камера
	bubble formation, blistering	Blasenbildung f	formation f de bulles	образование пузырьков
B 212	**buckling resistance**, cross breaking strength, resistance to lateral bending	Knickfestigkeit f	résistance f au flambage	прочность при продольном изгибе
B 213	**buff**, cloth brush	Schwabbelscheibe f	disque m en lisière de tissu	полировальный круг
B 214	**building block principle**, unit-assembly principle	Baukastenprinzip n	principe m des unités de montage pour l'agencement rapide <des appareillages à vide>, principe de boîte de construction	принцип сборки из готовых узлов
B 215	**build-up method**	Anreicherungsmethode f	méthode f de concentration, méthode d'enrichissement	метод обогащения
B 216	**bulb potential**	Röhrenwandpotential n	potentiel m de la paroi de tube	потенциал стенки
B 217	**bulb-sealing machine**	Kolbenanschmelzmaschine f	machine f à piston de fusion	запаечная машина
B 218	**bulk absorption**	Volumenabsorption f	absorption f volumétrique	объемное поглощение
B 219	**bulk conductivity**	Volumenleitfähigkeit f, Masseleitfähigkeit f	conductibilité f de volume (masse)	объемная проводимость
B 220	**bulk density**	Schüttdichte f	densité f de déchargement	объемная плотность
B 221	**bulk diffusion**	Volumendiffusion f	diffusion f volumétrique	объемная диффузия
B 222	**bulk getter**	Massivgetter m	getter m massif	нераспыляемый газопоглотитель
B 223	**bulk material**	massiver Stoff m	matière f massive	сплошной материал, целиковое вещество
B 224	**bumping**	stoßweises Sieden n	ébullition f instable, évaporation f pulsatoire	кипение рывками
B 225	**bunch discharge**, brush discharge	Büschelentladung f	aigrette f, décharge f en aigrette	кистевой разряд
B 226	**bunching**	periodische Elektronenzusammenballung f	agglomération f périodique des électrons	периодическое электронное группирование
	Burnauer-Emmet-Teller isotherm, BET isotherm	BET-Isotherme f, Burnauer-Emmet-Teller-Isotherme f	BET-isotherme f	БЭТ-изотерма, изотерма Бурнауера-Эммета Теллера
B 227	**burner**	Brenner m	brûleur m	горелка
B 228	**burner orifice**	Brennermund m	buse f du chalumeau, orifice m du brûleur	отверстие горелки
B 229	**burn in**	einbrennen <Katodenstrahl>	brûler	выжигать, обжигать, вжигать
B 230	**burning**, combustion	Verbrennung f	combustion f, brûlage m, grillage m	сгорание, горение, сжигание
B 231	**burn-out**	Durchbrennen n	brûlure f	перегорание, пережог
B 232	**bursting disk**, rupture disk	Berstscheibe f, Bruchplatte f	plaque f de rupture	предохранительная мембрана (диафрагма)
B 233	**butterfly valve**, quarter swing valve, throttle valve	Drosselventil n, Drosselklappe f	papillon m d'obturation, robinet m d'étranglement	дроссельный вентиль
B 234	**button crucible**	Knopfschmelztiegel m	creuset m à bouton	луночный тигель
B 235	**button melt**	Knopfschmelze f, Knopfprobe f	bouton-fusion m	проба, отлитая в виде лепешки
B 236	**butt seal**	Stumpfanschmelzung f	fondage m bout à bout	стыковая припайка, стыковой спай
B 237	**butt-welding**	Stumpfschweißen n	soudure f bout à bout	стыковая сварка
B 238	**bypass**	umleiten, umgehen	détourner, contourner	шунтировать, обходить
B 239	**bypass line**	Umgehungsleitung f, Umwegleitung f	conduite f by-pass (de dérivation)	байпасный трубопровод, байпасная магистраль
B 240	**bypass valve**, roughing valve	Grobpumpventil n, Ventil n in der Umgehungsleitung, Ventil in der Umwegleitung	soupape f dans la conduite de dérivation	байпасный вентиль

C

	CAB	s. controlled-atmosphere brazing		
C 1	**cabinet dryer,** drying cabinet	Schranktrockner m	sécheur m en armoire, chambre f à sécher, exsiccateur m	сушильный шкаф
C 2	**cable tank**	Kabelkessel m	chaudière f à câble	бак для кабеля
C 3	**cadmium tungstate**	Kadmiumwolframat n	tungstate m de cadmium	вольфрамат кадмия
C 4	**caesium,** cesium	Zäsium n	césium m	цезий
C 5	**calander**	kalandern	calandrer	каландрировать, прокатывать, сатинировать
C 6	**calcination**	Kalzinierung f	calcination f	кальцинирование
C 7	**calcium tungstate**	Kalziumwolframat n	tungstate m de calcium	вольфрамат кальция
C 8	**calibrate**	kalibrieren	calibrer	калибровать
C 9	**calibrated capillary method,** capillary method, capillary technique at constant pressure	Kapillarmethode f	procédé m capillaire	капиллярный метод
C 10	**calibrated conductance method,** two-gauge method	Leitwertmethode f	méthode f de conductance	метод определения быстроты действия насоса по известной пропускной способности, метод двух манометров
C 11	**calibrated leak,** sensitivity calibrator, standard (reference, test) leak	Eichleck n, Testleck n, Vergleichsleck n, Leck n bekannter Größe, Bezugsleck n	fuite f calibrée	стандартная течь, калиброванная течь, эталонная течь
C 12	**calibrated leak rating**	Eichleckdimensionierung f	dimensionnement m de fuite calibrée	калиброванное натекание
C 13	**calibration**	Eichung f, Kalibrierung f	calibrage m	калибровка, градуировка
C 14	**calibration curve**	Kalibrierungskurve f	courbe f d'étalonnage	градуировочная (калибровочная) кривая
C 15	**calibration gas**	Eichgas n	gaz m étalon	калибровочный газ
C 16/7	**calibrator**	Kalibriervorrichtung f	calibrateur m	калибратор
C 18	**canal rays**	Kanalstrahlen mpl	rayons mpl canaux	каналовые лучи
C 19	**capacitor film thickness monitor**	Kondensatorschichtdickenmeßgerät n	appareil m de mesure des couches minces à condensateur	емкостный измеритель толщины пленок
C 20	**capacity,** load, throughput	Förderleistung f, Saugleistung f, Fördermenge f, Durchsatz m, Durchsatzmenge f	flux m massique, débit m massique	скорость откачки
C 21	**cap-collar gasket,** L-ring shaft seal	Hutmanschette f	joint m à chapeau, embouti m à capuchon, joint d'arbre à section angulaire	колпачковая манжета, уплотняющее кольцо с L-образным сечением
C 22	**capillarity,** capillary attraction	Kapillarwirkung f	effet m capillaire	капиллярное притяжение, капиллярность
C 23	**capillary,** capillary pipe (tube)	Kapillarröhre f, Kapillare f	tube m capillaire, capillaire m	капилляр, капиллярная трубка
	capillary attraction	s. capillarity		
C 24	**capillary condensation**	Kapillarkondensation f	condensation f capillaire	капиллярная конденсация
C 25	**capillary crack,** hair-line crack	Haarriß m	fente (fissure) f capillaire	волосяная (капиллярная) трещина
C 26	**capillary depression**	Kapillardepression f	dépression f capillaire	капиллярная депрессия
C 27	**capillary drag,** sticking vacuum	Klebevakuum n	vide m adhésif (collant, d'adhésion du mercure)	вакуум прилипания
	capillary method	s. calibrated capillary method		
	capillary pipe	s. capillary		
	capillary technique at constant pressure	s. calibrated capillary method		
	capillary tube	s. capillary		
C 28	**cap jet**	Manteldüse f	injecteur m enveloppe, buse (tuyère) f cuirassée	сопло с обтекателем (оболочкой)
C 29	**cap nut,** coupling nut	Überwurfmutter f	écrou m à raccord (chapeau), écrou-chapeau m	накидная (стяжная) гайка
C 30	**capsula gauge, capsula type vacuum gauge**	Kapselfedervakuummeter n, Kapselmeßgerät n	vacuomètre m à capsule	капсульный пружинный вакуумметр
C 31	**capture coefficient, capture probability,** trapping coefficient (efficiency), sticking probability	Einfangkoeffizient m, Haftwahrscheinlichkeit f, Einfangwahrscheinlichkeit f	coefficient m d'arrêt, coefficient de blocage, probabilité f du pouvoir adhérent, probabilité d'adhérence, probabilité d'adhésion, probabilité de fixation (collage)	коэффициент захвата (прилипания), вероятность захвата (прилипания)
C 32	**carbon-arc evaporation**	Kohlebogenverdampfung f	évaporation f à l'arc de charbon	испарение в дуге с угольными электродами

C 33	carbon dioxide snow, dry ice, solidified carbon dioxide gas, solid carbon dioxide	Trockeneis *n*, feste Kohlensäure *f*, Kohlensäureschnee *m*	glace *f* sèche, carboglace *f*, anhydride *m* carbonique en neige, neige *f* carbonique, acide *m* carbonique solide	сухой лед, твердая углекислота, снег из твердой углекислоты
C 34	carbon evaporation	Kohlespitzenverdampfung *f*	évaporation *f* [de pointes] de charbon	испарение выступов углерода
C 35	carbonized filament	Glühdraht *m* mit Wolframkarbidbekleidung	filament *m* de carbide de tungstène	карбонизированный вольфрамовый подогреватель
C 36	carbon steel	unlegierter Stahl *m*	acier *m* carboné	углеродистая сталь
C 37	carburization, carburizing	Karburieren *n*, Aufkohlung *f*	carburation *f*	карбюрация
C 38	Carnot cycle	Carnotscher Kreisprozeß *m*	cycle *m* de Carnot	цикл Карно
C 39	Carnot efficiency	Carnotscher Wirkungsgrad *m*	rendement *m* de Carnot	коэффициент полезного действия цикла Карно
C 40	carousel	Karussell *n*	carrousel *m*	круговой конвейер, карусель
C 41	carrier gas	Fördergas *n*, Trägergas *n*	gaz *m* porteur	газ-носитель
C 42	carrier wire	Trägerdraht *m*	fil *m* porteur	несущий провод
C 43	cartridge heater, rod heater	Heizpatrone *f*	cartouche *f* chauffante, élément *m* chauffant en forme de barre	патронный нагреватель
C 44	cascade condenser	Kaskadenkühler *m*	réfrigérant (condenseur) *m* en cascade	каскадный конденсатор (охладитель)
C 45	cascade connection	Kaskadenschaltung *f*	montage *m* (connexion *f*) en cascade	каскадное соединение
C 46	cascade dryer, louvre dryer	Rieseltrockner *m*	installation *f* de séchage par ruissellement	каскадная сушилка
C 47	cascade electron	Kaskadenelektron *n*	électron *m* de cascade	каскадный электрон
C 48	cascade system	Kaskadenanordnung *f*	système *f* en cascade	каскадная установка
C 49	case-hardening	Einsatzhärten *n*	cémentation *f*	поверхностное твердение, поверхностная закалка
C 50	cast form	Formschluß *m*	forme *f* de coulée	скрепление формы
C 51	casting	Gießen *n*, Guß *m*	moulage *m*	литье, разливка, отливка
C 52	casting	Abguß *m*	pièce *f* coulée	отливка, разливка
C 53	casting ladle, chill, pouring ladle, foundry ladle	Gießpfanne *f*	cuvette (coupe, écuelle, poche) *f* de coulée	литейный (разливочный) ковш, кокиль, форма
C 54	casting resin	Gießharz *n*	résine *f* synthétique (à couler, à mouler)	смола для заливки (литья), литьевая смола
C 55	cast iron	Gußeisen *n*	fonte *f* [de moulage]	чугун
C 56	cast joint	Gußverbindung *f*	joint *m* de coulée	уплотнение заливкой
C 57	cast welding	Gießschweißen *n*	soudure *f* par coulage	сварка заливкой расплавленным металлом
C 58	catalyst, catalyzer	Katalysator *m*, Kontaktsubstanz *f*	catalyseur *m*	катализатор
C 59	cataphoresis	Kataphorese *f*	cataphorèse *f*	катафорез
C 60	cathetometer	Kathetometer *n*	cathétomètre *m*	катетометр
C 61	cathode activation	Katodenaktivierung *f*	activation *f* cathodique	катодная активировка
C 62	cathode ageing	Katodenalterung *f*	vieillissement *m* cathodique	старение катода
C 63	cathode breakdown, cathode conversion	Katodenumsetzung *f*	conversion *f* cathodique	активировка катода
C 64	cathode drop, cathode fall	Katodenfall *m*	descente *f* cathodique	катодное падение
C 65	cathode emission efficiency	Elektronenausbeute *f*	rendement *m* électronique	эффективность катода
	cathode fall	s. cathode drop		
C 66	cathode glow	Katodenglimmlicht *n*	luminescence *f* cathodique	катодное свечение
C 67	cathode heating time	Katodenanheizzeit *f*	temps *m* de chauffage de la cathode	время разогрева катода
C 68	cathode poisoning	Katodenvergiftung *f*	empoisonnement *m* de la cathode	отравление катода
	cathode ray, electron beam	Katodenstrahl *m*	rayon *m* cathodique	электронный луч
C 69	cathode-ray deflection	Katodenstrahlablenkung *f*	déviation *f* de rayons cathodiques	отклонение электронного луча
C 70	cathode-ray pencil	Katodenstrahlbündel *n*	faisceau *m* cathodique	электронный пучок
C 71	cathode sputtering, cathodic sputtering	Katodenzerstäubung *f*	pulvérisation *f* cathodique	катодное распыление
C 72	cathodic electro-cleaning	katodisches Reinigen *n*	épuration *f* cathodique	катодная очистка
C 73	cathodic etching, vacuum etching	katodisches Ätzen *n*, Glimmen *n*, Abglimmen *n*, Beglimmen *n*	gravure *f* cathodique	катодное травление
	cathodic sputtering	s. cathode sputtering		
C 74	cathodoluminescence	Katodolumineszenz *f*	cathodoluminescence *f*	катодолюминесценция
C 75	cavitation	Kavitation *f*	cavitation *f*	кавитация
C 76	cavity, hole	Loch *n*, Lunker *m*	trou *m*, retassure *f*	дырка
C 77	cavity	Hohlraum *m*	cavité *f*	полость, раковина
C 78	cellular glass, glass foam	Schaumglas *n*	verre *m* mousse	пеностекло
C 79	celluloid lacquer	Zaponlack *m*	laques *fpl* zapon	нитроцеллюлозный (цапоновый) лак

C 80	cement	Bindemittel *n*, Kleber *m*	liant *m*, fixateur *m*	связующее вещество, замазка
C 81	cemented socket joint, cemented spigot and socket joint	geklebte Steckmuffe *f*	manchon *m* douille collé, raccord *m* collé	приклеенный цоколь, посаженный на мастике цоколь
C 82	cementing plant	Verklebeanlage *f*	installation *f* à coller	установка для приклеивания
C 83	centigrade scale	Celsius-Skale *f*	échelle *f* centigrade (Celsius)	шкала Цельсия
C 84	centigrade temperature scale	Temperaturskale *f* nach Celsius	Celsius-graduation *f* de température	температурная шкала Цельсия
C 85	centre rod, tie rod	Verbindungsstab *m*, Zentralstab *m*	bâton *m* de connexion, bâton central	стержень для крепления и центровки
C 86	centrifugal airpump, centrifugal blower, rotary blower	Kreiselgebläse *n*, Fliehkraftgebläse *n*, Zentrifugalgebläse *n*	soufflante *f* centrifuge	центробежная воздуходувка
C 87	centrifugal casting	Schleuderguß *m*	coulée *f* centrifuge	центробежное литье, центробежная отливка
C 88	centrifugal compressor, radial flow compressor	Kreiselverdichter *m*, Radialverdichter *m*	compresseur *m* centrifuge	центробежный компрессор
C 89	centrifugal dryer	Trockenschleuder *f*, Zentrifugiertrockner *m*	essoreuse *f*	центробежная сушилка
C 90	centrifugal drying	Zentrifugaltrocknung *f*	dessiccation *f* centrifuge	центробежная сушка
C 91	centrifugal freeze dryer	Schleudergefriertrockner *m*	sécheur *m* de congélation à rotation	центробежная сублимационная сушилка
C 92	centrifugal freeze drying	Schleudergefriertrocknung *f*	séchage *m* de congélation à rotation	центробежная сублимационная сушка
C 93	centrifugal molecular distillation plant, centrifugal molecular still	Zentrifugal-Molekulardestillationsanlage *f*	installation *f* de distillation centrifuge moléculaire	центробежный молекулярный дистиллятор
C 94	centrifugal pump	Kreiselpumpe *f*	pompe *f* giratoire (centrifuge)	центробежный насос
C 95	centring device	Zentriervorrichtung *f*	dispositif *m* de centrage	центрирующее приспособление
C 96	centring ring	Zentrierring *m*	anneau *m* de centrage	центрирующее кольцо
C 97	ceramic crucible	Keramiktiegel *m*	creuset *m* céramique	керамический тигель
C 98	ceramic disk	Keramikscheibe *f*	disque *m* céramique	керамический диск, керамическая шайба
C 99	ceramic insulating material	Isolierkeramik *f*	céramique *f* isolante	керамический изоляционный материал
C 100	ceramic-metal terminal	Metallkeramikanschluß *m*	raccord *m* céramique-métal	металлокерамический ввод
C 101	ceramics	Keramik *f*	céramique *f*	керамика
C 102	ceramic seal	Keramikabschmelzung *f*	scellement *m* céramique	керамический спай
C 103	ceramic-to-metal seal, metal-ceramic seal	Keramik-Metall-Verbindung *f*, Metall-Keramik-Verbindung *f*	joint (scellement) *m* céramique-métal, scellement *m* métal-céramique	соединение (спай) металла с керамикой, металлокерамическое соединение
C 104	cesium, caesium	Zäsium *n*	césium *m*	цезий
	chamber furnace	Kammerofen *m*	four *m* à chambres	печь периодического действия, камерная печь
C 105	chamber test, envelope test, hood pressure test, overall test	Hüllentest *m*, Haubenlecksuchverfahren *n*, Haubenleckprüfung *f*	test *m* d'enveloppe, épreuve *f* d'étanchéité	проверка помещением объекта в оболочку
C 106	change from solid to liquid state	Übergang *m* vom festen in den flüssigen Aggregatzustand	transition *f* de l'état solide à l'état liquide	переход из твердого в жидкое состояние
C 107	change of state	Zustandsänderung *f*	changement *m* d'état	изменение состояния
C 108	channel iron	U-Eisen *n*	fer *m* en U	швеллерная сталь, сталь корытного профиля
C 109	characteristic peak throughput	Saugleistungsmaximum *n*	débit *m* massique maximum d'une pompe, puissance *f* maxima d'une pompe	максимальная производительность
C 110	charcoal trap	Kohlefalle *f*, Sorptionsfalle *f*	piège *m* à sorption de charbon	ловушка с активированным углем
C 111	charge carrier	Ladungsträger *m*	porteur *m* de charge	носитель заряда
C 112	charge carrier density	Ladungsträgerdichte *f*	densité *f* des porteurs de charge	плотность носителей заряда
C 113	charge distribution	Ladungsverteilung *f*	distribution *f* de charge	распределение зарядов
C 114	charging, feeding, loading	Beschickung *f*	chargement *m*	питание, загрузка
C 115	charging device, feeding device	Beschickungsvorrichtung *f*	appareil *m* d'alimentation, appareil de chargement, appareil à charger, dispositif *m* d'alimentation, chargeur *m* mécanique	загрузочное устройство, питающее устройство
C 116	charging tray, loading tray	Fülltablett *n*	plateau *m* de remplissage	загрузочный лоток
C 117	charging tray for sublimation	Sublimationshorde *f*	claie *f* à sublimation	загрузочная решетка для сублимации
C 118	charging valve	Chargierventil *n*	soupape *f* de chargement	питательный (загрузочный) клапан
C 119	Charles' law, Gay-Lussac's law	Gay-Lussacsches Gesetz *n*	loi *f* de Gay-Lussac	закон Гей-Люссака

C 120	**check crack,** shrinkage crack	Schrumpfriß *m*	fissure *f* de contraction, fente *f* par contraction	трещина при усадке, усадочная трещина
C 121	**check valve**	Kontrollventil *n*	soupape *f* de contrôle, robinet *m* de contrôle, vanne *f* de contrôle	контрольный (регулирующий) клапан
C 122	**chemically pure,** cp	chemisch rein	chimiquement pur	химически чистый
C 123	**chemisorption,** chemosorption	Chemisorption *f*	sorption *f* chimique	хемосорбция
C 124	**chemisorption bond**	Chemisorptionsbindung *f*	liaison *f* par sorption chimique	хемосорбционная связь
	chemosorption, chemisorption	Chemisorption *f*	sorption *f* chimique	хемосорбция
C 125	**chevron baffle**	Chevron-Baffle *n*, Rasterdampfsperre *f*	baffle *m* à chevron (écrans angulaires multiples)	шевронный отражатель
C 126	**chevron seal**	Chevron-Dichtung *f*	chevron-garniture *f* étanche	шевронное уплотнение
C 127	**chill, cool down, cool off**	abkühlen	refroidir, réfrigérer, refraîchir	охлаждать
	chill, casting ladle, pouring ladle, foundry ladle	Gießpfanne *f*	cuvette (coupe, écuelle, poche) *f* de coulée	литейный (разливочный) ковш, кокиль, форма
C 128	**chilled water**	Kaltwasser *n*, gekühltes Wasser *n*	eau *f* réfrigérée	охлажденная вода
C 129	**chilling,** cooling	Kühlung *f*, Abschreckung *f*	refroidissement *m*	охлаждение
C 130	**chilling time**	Kühlzeit *f*	temps *m* refroidissant (de réfrigération)	продолжительность охлаждения
C 131	**chip**	Span *m*	tournure *f*, copeau *m*	стружка, осколок
C 132	**chip-off,** flake-off, scale-off	abblättern ‹von Schichten›	effeuiller	отслаиваться, отпадать
C 133	**chloride of cobalt**	Chlorkobalt *n*	chlorure *m* de cobalt	хлорид кобальта
C 134	**chloride of platinum**	Chlorplatin *n*	chlorure *m* de platine	хлорид платины
C 135	**choke time**	Drosselzeit *f*	temps *m* d'étranglement	время дросселирования
C 136	**chrome-plated**	verchromt	chromé	хромированный
C 137	**chromium-plate**	verchromen	chromer	хромировать
C 138	**circular accelerator**	Zirkularbeschleuniger *m*, Kreisbeschleuniger *m*	accélérateur *m* circulaire	циклический ускоритель
C 139	**circular aperture**	Kreisblende *f*	diaphragme *m* circulaire	кольцевая диафрагма
C 140	**circular stem**	Kreuzquetschfuß *m*	socle *m* pressant en croix, tige *f* en croix	крестообразная ножка
C 141	**circulate by pumping,** pump over	umpumpen	transvaser (faire circuler) par pompage	перекачивать
C 142	**circulating heating thermostat**	Umlaufheizthermostat *m*	thermostat *m* chauffant par circulation	термостат с циркуляционным подогревом
C 143	**circulating line**	Umpumpleitung *f*	conduite *f* de circulation par pompage	циркуляционный трубопровод
C 144	**circulation degassing,** continuous degassing by the siphon method, siphon degassing	Umlaufentgasung *f*	dégazage *m* par circulation, dégagement *m* par circulation	циркуляционное (сифонное) обезгаживание
C 145	**circulation pump,** recirculating pump	Umwälzpumpe *f*	pompe *f* de circulation	циркуляционный насос, рециркуляционный насос
C 146	**circumferential speed,** peripheral speed (velocity)	Umfangsgeschwindigkeit *f*	vitesse *f* périphérique	окружная скорость
C 147	**clamp**	festklemmen	serrer	зажимать
C 148	**clamp flange**	Klammerflansch *m*	bride *f* à griffe	фланец, прожимаемый ‹к контрфланцу› скобами
C 149	**clamping paste**	Kontaktschmierstoff *m*	lubrifiant *m* de contact	контактная паста
C 150	**clamping ring,** locking ring	Klemmring *m*, Spannring *m*	anneau (collier) *m* de serrage	кольцевой зажим, хомут
C 151	**Clapeyron formula**	Clausius-Clapeyronsche Gleichung *f*	équation *f* de Clausius-Clapeyron	уравнение Клапейрона
C 152	**Clausing's correction factor, Clausing's factor,** transmission probability	Clausing-Faktor *m*	facteur *m* de Clausing	поправочный коэффициент Клаузинга
C 153	**cleaning,** purification	Reinigung *f*	purification *f*, épuration *f*, rectification *f*, nettoyage *m*	очистка
C 154	**clean-up**	Gasaufzehrung *f*, Getterung *f*	absorption *f* de gaz, effet *m* getter	газопоглощение
C 155	**clean-up pump**	Gasaufzehrungspumpe *f*	pompe *f* d'absorption de gaz, pompe d'effet getter, pompe de getterage, pompe de gettérisation	газопоглотительный насос
C 156	**clean-up time**	Waschzeit *f*, Aufzehrungszeit *f*	temps *m* de clean up, temps getter	время газопоглощения
C 157	**clean-up time,** recovery (settling, rest) time	Erholungszeit *f*, Wiederansprechzeit *f*	temps *m* régénérateur, temps de régénérabilité	время [для] восстановления, период восстановления
C 158	**clearance between the plates,** shelf spacing	Plattenabstand *m*	écartement *m* des plaques	расстояние между пластинами
C 159	**clearance seal**	Spaltdichtung *f*	joint *m* à fente	уплотнение зазора
C 160	**cleavability**	Spaltbarkeit *f*	fissilité *f*, désintégration *f*, clivage *m*	расщепляемость, раскалываемость

C 161	**cleavage face, cleavage plane**	Spaltfläche *f*	plan *m* de clivage	плоскость кливажа (спайности)
C 162	**cleavage product**	Spaltprodukt *n*	produit *m* de clivage (fission)	продукт расщепления
C 163	**cleave**	spalten ‹Kristalle›	cliver	раскалывать ‹кристалл›
C 164	**clip,** hose press clamp, pinchcock, squeezing cock	Quetschhahn *m*	pince *f* pressante	зажим для шланга
C 165	**closed-circuit cooling**	Kreislaufkühlung *f*	refroidissement *m* par circulation	циркуляционное охлаждение
C 166	**closed-end mercury manometer**	geschlossenes Quecksilbermanometer *n*	manomètre *m* clos à mercure	ртутный манометр с запаянным концом
C 167	**closed-loop water circuit**	Wasserkreislauf *m*	cycle *m* d'eau, circulation *f* d'eau	циркуляция воды
	cloth brush, buff	Schwabbelscheibe *f*	disque *m* en lisière de tissu	полировальный круг
C 168	**cloud of electrons,** electron cloud	Elektronenwolke *f*	nuage *m* d'électrons	электронное облако
C 169	**cloud point**	Trübungspunkt *m*	point *m* de trouble	точка помутнения
C 170	**cluster crystal**	Kristalldruse *f*	druse *f* de cristal	кристаллическая друза
C 171	**coarse pumping,** roughing	Grobevakuieren *n*, Vorpumpen *n* (des Rezipienten)	pompage *m* primaire, évacuation *f* préliminaire, prévidage *m*, évacuation grossière	откачка до предварительного разрежения
C 172	**coasting**	Vorwärmen *n*, Anwärmen *n* ‹einer Treibmittelpumpe›	préchauffage *m*, chauffage *m* préliminaire	предварительный подогрев ‹паромасляного насоса›
C 173	**coat**	überziehen, beschichten, bedecken, auftragen	couvrir	покрыть
C 174	**coating by vapour decomposition,** pyrolytic plating	pyrolytische Beschichtung *f*	revêtement *m* pyrolytique	пиролитическое нанесение
C 175	**coating getter**	Deckgetter *m*, Schichtgetter *m*	getter *m* superficiel, couche-getter *f*	слойный газопоглотитель (геттер)
C 176	**coating layer**	Aufdampfschicht *f*	couche *f* de métallisation sous vide, couche d'évaporation	напыленный слой
C 177	**coating technique**	Beschichtungstechnik *f*	technique *f* de métallisation	технология напыления
C 178	**coat of synthetic resin**	Gießharzmantel *m*	gaine *f* en résine synthétique	покрытие синтетической смолой
C 179	**coaxial leadthrough**	koaxiale Durchführung *f*	traversée *f* coaxiale	коаксиальный ввод
C 180	**coefficient of expansion,** dilatation coefficient	Ausdehnungskoeffizient *m*	coefficient *m* de dilatation	коэффициент расширения
C 181	**coefficient of performance,** figure of merit, quality factor	Leistungsziffer *f*, Gütegrad *m*, Gütezahl *f*	coefficient (facteur) *m* de puissance	коэффициент мощности
	coefficient of viscosity, dynamic viscosity, absolute viscosity	dynamische Zähigkeit *f*, dynamische Viskosität *f*	viscosité *f* dynamique	динамическая вязкость
C 182	**coherence length**	Kohärenzlänge *f*	longueur *f* de cohérence	длина когерентности
C 183	**cohesion**	Kohäsion *f*	cohésion *f*	когезия
C 184	**coil condenser,** tubular cooler	Röhrenkühler *m*	radiateur (réfrigérant) *m* tubulaire	змеевиковый конденсатор (охладитель)
C 185	**coiled-coil [filament]**	Doppelwendel *f*	filament *m* à double bobinage	биспиральная нить накала
C 186	**coiled-coil heater**	Doppelwendelheizer *m*	chauffeur *m* à doubles spirales	биспиральный подогреватель
C 187	**coiled cooling pipe,** worm	Kühlschlange *f*	serpentin *m* réfrigérant (refroidisseur, condenseur)	охлаждающий змеевик
C 188	**coined gasket seal**	Dichtung *f* mit abgesetztem Rechteckring	joint *m* avec un anneau rectangulaire (à angle droit)	уплотнение с рельефной прокладкой
	cold air machine, air refrigerating machine	Kaltluftmaschine *f*	machine *f* à air froid	воздушная холодильная машина
C 189	**cold-air refrigerating machine**	Kaltgaskältemaschine *f* ‹mit Luft als Kältemittel›	installation *f* frigorifique à air froid	газовая холодильная машина ‹с температурой воздуха − 80 °C и ниже›
C 190	**cold boiler,** vacuum cooking appliance	Vakuumkocher *m*, Vakuumkochapparat *m*	marmite *f* à vide	вакуумный кипятильник (бойлер)
C 191	**cold cap**	Kühlkappe *f*, Düsenhut *m*	voûte *f* de véhicule du froid	холодный колпак
C 192	**cold cap** ‹buffle›, cooled cover, vapourcatching cone cup	Düsenhutdampfsperre *f*	baffle *m* à chapeau, baffle-chapeau *m*	охлаждаемый колпачковый отражатель, охлаждаемая коническая насадка
C 193	**cold-cathode**	Kaltkatode *f*	cathode *f* froide	холодный катод
C 194	**cold-cathode gas discharge**	Kaltkatodengasentladung *f*	décharge *f* des cathodes froides dans un gaz	газовый разряд с холодным катодом
C 195	**cold-cathode inverted magnetron gauge**	umgekehrtes Magnetronvakuummeter *n*	magnétron-vacuomètre *m* inverse	инверсно-магнетронный манометр с холодным катодом
C 196	**cold-cathode ionization gauge,** Philips ionization gauge, Penning gauge	Kaltkatodenvakuummeter *n*, PhilipsVakuummeter *n*, Penning-Vakuummeter *n*	manomètre *m* à ionisation à cathode froide	[магнитный] электроразрядный манометр с холодным катодом, манометр Пеннинга

C 197	cold-cathode ion pump, penning-type pump, sputter-ion pump	Ionenzerstäuberpumpe f	pompe f ionique à cathode froide, penning-pompe f, pompe ionique de pulvérisation	[магнитный] электроразрядный насос, ионный насос с холодным катодом, насос Пеннинга, ионно-распылительный насос
C 198	cold-cathode ion source	Kaltkatodenionenquelle f	source f d'ions à cathode froide	ионный источник с холодным катодом
C 199	cold-cathode magnetron gauge, magnetron vacuum gauge	Magnetronvakuummeter n	manomètre m magnétron	магнетронный вакуумметр
C 200	cold extrusion	Kaltfließpressen n	pressage m d'écoulement froid	холодная прессовка, выдавливание
C 201	cold finger, cooling finger, thimble trap	Kühlfinger m	doigt m refroidisseur, sonde f de refroidissement	пальцевой холодильник
C 202	cold flow	Kaltfließen n ‹unter Druck bei Zimmertemperatur›	écoulement m froid, fluage m froid	хладотекучесть
C 203	cold gilding	Kaltvergoldung f	dorure f à froid	холодное золочение
C 204	cold heading	Kaltstauchen n	refoulement m à froid	холодное осаживание, холодная обжимка
C 205	cold mirror	Kaltlichtspiegel m	miroir m froid	холодное зеркало
C 206	cold mould furnace, consumable-electrode vacuum arc furnace	Vakuum-Lichtbogenofen m mit selbstverzehrender Elektrode	four m à arc sous vide avec une électrode consommable	вакуумная дуговая печь с расходуемым электродом, вакуумная дуговая печь с холодным кристаллизатором
C 207	cold pressure welding	Kaltpreßschweißen n	soudage m à froid par pression	холодная сварка давлением
C 208	cold-rolled	kaltgewalzt, kaltverformt	laminé à froid	холоднокатанный
C 209	cold rolling	Kaltwalzen n	laminage m à froid	холодная прокатка
C 210	cold-setting	kaltabbindend, kalthärtend	tremper à froid	холодное схватывание
C 211	cold stamping	Kaltprägen n	estampage m à froid	холодное штампование
C 212	cold-strained wire	kaltgereckter Draht m	fil m étiré à froid	холоднотянутая проволока
C 213	cold trap, refrigerated trap	Kühlfalle f	piège m refroidi (à froidi, cryogénique, réfrigérant, de refroidissement)	охлаждаемая ловушка
C 214	cold-trap efficiency	Wirksamkeit f der Kühlfalle	efficacité f du piège refroidi	эффективность охлаждаемой ловушки, защитная способность охлаждаемой ловушки
C 215	cold vulcanization	Kaltvulkanisation f	vulcanisation f à froid	холодная вулканизация
C 216	cold-wall high-temperature internal-element vacuum brazing furnace	Kaltwandvakuumhochtemperaturlötofen m mit Innenheizer	four m à soudure sous vide à haute température de paroi froide avec chauffeur intérieur	высокотемпературная вакуумная печь с холодными стенками и внутренним нагревателем
C 217	cold welding, diffusion welding	Kaltschweißen n	soudage m à froid	холодная сварка, диффузионная сварка
C 218	cold-weld pinch-off device	Vorrichtung f zum Kaltpreßschweißen und Abklemmen	dispositif m à soudure froide sous pression et à déserrage	устройство для отпайки методом холодной сварки
C 219	cold working	Kaltverformung f	déformation f à froid	холодная обработка ‹давлением›
C 220	collecting side, inlet (intake, suction, vacuum) side	Saugseite f, Vakuumseite f	côté m vide (d'aspiration)	собирающая сторона, сторона впуска
C 221	collision cross-section	Stoßquerschnitt m	section f droite du choc	поперечное сечение соударений
C 222	collision frequency	Stoßhäufigkeit f, Volumenstoßhäufigkeit f, mittlere Gesamtstoßzahl f pro Zeit und Volumeneinheit	fréquence f de collision	частота соударений
C 223	collision frequency per molecule, collision probability (rate)	Stoßzahl f ‹pro Zeiteinheit›, Stoßwahrscheinlichkeit f	nombre m (fréquence f, taux m) de collision	число соударений
C 224	collodion	Kollodium n	collodion m	коллодий
C 225	colloidal	kolloidal	colloïdal	коллоидный
C 226	colophony, rosin	Kolophonium n	colophane f	канифоль
C 227	colour temperature	Farbtemperatur f	température f de couleur	цветовая температура
C 228	column structure	Säulenstruktur f	structure f de colonne	колонная конструкция (структура)
	combustion, burning	Verbrennung f	combustion f, brûlage m, grillage m	сгорание, горение, сжигание
C 229	compact pumping set, packaged pumping system, pumping unit	betriebsfertiger Pumpstand m	poste (bâti, banc, stand) m de pompage	откачной стенд
C 230	comparison tube ‹McLeod›	Vergleichskapillare f ‹McLeod›	capillaire m de comparaison	сравнительный капилляр ‹компрессионного манометра›
C 231	compensating body	Ausgleichskörper m	corps m d'équilibrage, élément m compensateur	компенсирующий элемент

C 232	**complete fusion,** melt flux	Schmelzfluß *m*	matière *f* fusée, flux *m* de matière en fusion	расплав, плавка, флюс
C 233	**component chamber**	Komponentenkammer *f*	chambre *f* de composante	компонентная камера
C 234	**composite metal,** ply metal	Verbundmetall *n*	métal *m* mixte	металлокерамическая композиция, плакированный металл
C 235	**compound,** series single stage, multi stage	mehrstufig	à plusieurs étages	многоступенчатый
C 236	**compound glass,** multi-layer glass	Verbundglas *n*	verre *m* compound (à vitre collé, à couches multiples)	составное стекло
C 237	**compound mechanical pump**	mehrstufige mechanische Pumpe*f*, Duplexpumpe*f*	pompe *f* mécanique à étages	многоступенчатый механический насос
C 238	**compression**	Verdichtung *f*	condensation *f*, compression *f*	компрессия
C 239	**compression chamber**	Kompressionsraum *m*	chambre *f* de compression	камера сжатия
C 240	**compression factor** <McLeod>	Kompressionsfaktor *m* <Mc Leod>	facteur *m* de compression	коэффициент компрессии <компрессионного манометра>
C 241	**compression gauge,** compression-type vacuum gauge	Kompressionsvakuummeter *n*	vacuomètre *m* à compression	компрессионный манометр
C 242	**compression limit,** pressure limit, maximum compression pressure	Verdichtungsenddruck *m*	compression *f* finie de sortie	предельное давление сжатия
C 243	**compression ratio,** pressure ratio	Druckverhältnis *n*, Kompressionsverhältnis *n*	rapport *m* de compression, taux *m* de pression	степень сжатия, степень компрессии
C 244	**compression refrigerating system**	Verdichterkälteanlage *f*	installation *f* frigorifique de compression	компрессорная холодильная установка
C 245	**compression seal,** gasket seal, crush seal	Preßdichtung *f*	joint *m* à compression, raccord *m* à écrasement, joint à écrasement	нажимное уплотнение
C 246	**compression tube** <McLeod>	Kompressionskapillare *f* <McLeod>	capillaire *m* de compression	компрессионный капилляр <компрессионного манометра>
C 247	**compression-type glass-to-metal seal**	Druckglaseinschmelzung *f*	joint *m* à compression en verre	напряженный спай стекла с металлом
C 248	**compression-type refrigerating unit**	Verdichterkältemaschine *f*	réfrigérateur *m* de compression, machine *f* (groupe *m*) frigorifique de compression, machine à froid de compression	компрессорная холодильная машина
	compression-type vacuum gauge, compression gauge	Kompressionsvakuummeter *n*	vacuomètre *m* à compression	компрессионный манометр
C 249	**compressive strength**	Druckfestigkeit *f*	résistance *f* à la pression	предел прочности при сжатии
C 250	**compressor**	Verdichter *m*, Kompressor *m*	compresseur *m*	компрессор
C 251	**concentration gradient**	Dichtegradient *m*, Konzentrationsgefälle *n*	gradient *m* (chute *f*) de concentration	градиент концентрации, перепад концентрации
C 252	**concentration of ions**	Ionenkonzentration *f*	concentration *f* ionique	концентрация ионов
C 253	**concentration rate**	Konzentrationsverhältnis *n*, Verhältnis *n* von Partial- zu Totaldruck	débit *m* de concentration	относительная концентрация
C 254	**concentrator** <induction heating coil>	Konzentrator *m* <Induktionsspule>	concentrateur *m* <bobine à induction>	концентратор <катушки для индукционного нагрева>
C 255	**condensable gas**	kondensierbares Gas *n*	gaz *m* condensables	конденсирующийся газ, конденсируемый газ
C 256	**condensation**	Kondensation *f*	condensation *f*	конденсация
	condensation coefficient, accomodation coefficient for condensation	Kondensationskoeffizient *m*	coefficient *m* de condensation	коэффициент аккомодаций при конденсации
C 257	**condensation pump**	Kondensationspumpe *f*	pompe *f* à condensation	конденсационный насос
C 258	**condensation rate**	Kondensationsrate *f*, Kondensationsgeschwindigkeit *f*, spezifische Kondensationsgeschwindigkeit	taux *m* de condensation	скорость конденсации
C 259	**condensation water**	Kondenswasser *n*	eau *f* condensée (de condensation)	конденсационная вода
C 260	**condenser,** liquefier	Kondensor *m*, Kondensatabscheider *m*, Kondensator *m*	séparateur *m* de condensat, condensateur *m*	конденсатор
C 261	**condenser capacity,** condensing power, condenser output	Kondensatorkapazität *f*, Kondensatorleistung *f*	capacité *f* du condensateur	производительность конденсации
	condenser output	s. condenser capacity		
	condensing power, condenser capacity, condenser output	Kondensatorkapazität *f*, Kondensatorleistung *f*	capacité *f* du condensateur	производительность конденсации

C 262	conductance	Leitwert *m*	conductance *f*	проводимость
C 263	cone seal, conical seal	Kegeldichtung *f*, Schliffdichtung *f*	joint *m* conique	коническое уплотнение
C 264	cone shadowing	Kegelbedampfung *f*	amortissement *m* conique, métallisation *f* conique	конусное экранирование
C 265	cone shadowing	Kegelbeschattung *f*	ombragement *m* (protection *f*) conique	конусное затенение
C 266	confining grid	Sperrgitter *n*	grille *f* d'arrêt	ограничивающая (барьерная) сетка
C 267	confining liquid, packing fluid	Sperrflüssigkeit *f*	liquide *m* obturateur	жидкость для [перекрытия] затвора, уплотняющая жидкость
	conical seal, cone seal	Kegeldichtung *f*, Schliffdichtung *f*	joint *m* conique	коническое уплотнение
C 268	conjugated double-linkage	Zwillingsdoppel- verbindung *f*	double liaison *f* conjuguée	сопряженное двойное соединение
C 269	connecting flange, connection flange	Anschlußflansch *m*, Verbindungsflansch *m*	bride *f* d'assemblage	соединительный фланец
C 270	connecting glass	Verbindungsglas *n*	verre *m* de raccordement (jonction), verre pour assemblage	соединительное стекло
C 271	connecting terminal	Anschlußklemme *f*	griffe *f* de raccordement	соединительный зажим
C 272	connect in series	hintereinanderschalten	coupler (monter) en série	включать последовательно
C 273	connection connection flange	Anschluß *m*, Verbindung *f* *s.* connecting flange	raccord *m*	соединение
C 274	connector, coupling, fitting	Verbindungsstück *n*	raccord *m*, pièce *f* de rac- cordement	соединительная часть
C 275	constancy	Konstanz *f*	constance *f*	постоянство, стабиль- ность, устойчивость
C 276	constant pressure method	Methode *f* des konstanten Druckes	méthode *f* de la pression constante	метод постоянного давления
C 277	constant volume method	Methode *f* des konstanten Volumens	méthode *f* du volume constant	метод постоянного объема
C 278	constitution cement	Konstitutionskitt *m*	mastic *m* de constitution	структурный цемент
C 279	constructional element, structural member	Bauteil *n*	élément *m* de construction	структурный элемент
C 280	consumable electrode, consutrode, melting electrode	Abschmelzelektrode *f*, selbstverzehrende Elektrode *f*	électrode *f* consommable, électrode fusible [pour soudage à l'arc]	расходуемый электрод [плавильной печи]
C 281	consumable-electrode vacuum-arc evaporating	Vakuumbogenverdamp- fung *f* mit selbstver- zehrender Katode	évaporation *f* à arc sous vide avec électrode con- sommable	вакуумно-дуговое испа- рение с расходуемым электродом
	consumable-electrode vacuum arc furnace, cold mould furnace	Vakuum-Lichtbogen- ofen *m* mit selbstver- zehrender Elektrode	four *m* à arc sous vide avec une électrode cosom- mable	вакуумная дуговая печь с расходуемым элек- тродом, вакуумная ду- говая печь с холодным кристаллизатором
	consutrode	*s.* consumable electrode		
C 282	consutrode melting, pool melting	Badschmelzen *n*, Schmelzen *n* mit flüssi- ger Katode	fusion *f* à cathode liquide	плавка с переплав- ляемым электродом
C 283	contact angle, wettability angle	Randwinkel *m*, Benetzungswinkel *m*	angle *m* mouillant	краевой угол, угол смачивания
C 284	contact condenser, direct-contact condenser	Mischkondensator *m*	condensateur (conden- seur) *m* à mélange	смесительный (смеши- вающий) конденсатор
C 285	contact discontinuity	Kontaktunstetigkeit *f*	discontinuation *f* de contact	нарушение контакта
C 286	contact dryer	Kontakttrockner *m*	sécheur *m* de contact	контактная сушилка
C 287	contact drying	Kontakttrocknung *f*	séchage *m* de contact	контактная сушка
C 288	contact freezer	Kontaktgefrieranlage *f*	installation *f* de congéla- tion par contact	контактный морозиль- ный аппарат
C 289	contact freezing	Kontaktgefrieren *n*	congélation *f* par contact	контактное заморажива- ние
C 290	contact gettering	Kontaktgetterung *f*	sorption *f* à l'aide de couches-getter	контактное газопогло- щение
C 291	contact potential	Kontaktpotential *n*	potentiel *m* de contact	контактный потенциал
C 292	contact process	Kontaktverfahren *n*	procédé *m* avec masse catalytique	контактный способ <катализа>
C 293	container, tank, vessel	Rezipient *m*, Behälter *m*	récipient *m*	контейнер, вместилище
C 294	continuity equation	Kontinuitätsgleichung *f*	équation *f* de continuité	уравнение непрерыв- ности
C 295	continuous annealing furnace	Durchlaufglühofen *m*	four *m* continu à recuire	печь отжига непрерыв- ного действия
C 296	continuous casting method	Stranggußverfahren *n*	méthode *f* de fonte à file	метод непрерывной разливки
	continuous degassing by the siphon method, circulation degassing, siphon degassing	Umlaufentgasung *f*	dégazage *m* par circula- tion, dégagement *m* par circulation	циркуляционное (сифонное) обезгаживание
C 297	continuous flow, continuum flow	Strömungskontinuum *n*	continu *m* du courant	непрерывность потока
C 298	continuous-line recorder, strip recorder	Linienschreiber *m*	enregistreur *m* en trait continu	ленточный самописец

C 299	**continuous operation**	Dauerbetrieb m	marche f continue	длительная (продолжительная) работа
C 300	**continuous output**	Dauerleistung f	puissance f continue	мощность при непрерывной работе, номинальная мощность
C 301	**continuous pusher-type furnace,** tunnel furnace	Tunnelofen m, Durchsatzofen m	four m tunnel	туннельная печь, печь с проталкиванием деталей
C 302	**continuous sintering furnace**	Durchlaufsinterofen m	four m de frittage de passage continu	спекательная печь непрерывного действия
C 303	**continuous strip vacuum heat treatment furnace**	Vakuumdurchlaufbandglühofen m	four m à recuire pendant la coulée à bandes sous vide	вакуумная конвейерная печь
C 304	**continuous vacuum metallizer**	Vakuumdurchlaufmetallbedampfer m	permanent vaporisateur m métallique à vide	установка для непрерывной металлизации испарением в вакууме
C 305	**continuous wave magnetron**	Dauerstrichmagnetron n	magnétron m à fonctionnement continu	магнетрон непрерывного режима
C 306	**continuum**	Kontinuum n	spectre m continu	континуум
	continuum flow, continuous flow	Strömungskontinuum n	continu m de courant	непрерывность потока
C 307	**control desk,** switch desk	Schaltpult n, Steuerpult n	pupitre m de commande	пульт управления
C 308	**controlled-atmosphere brazing,** CAB	Schutzgaslöten n	brasage m (soudure f) en atmosphère gazeuse protectrice	пайка в защитной атмосфере
C 309	**controlled-atmosphere furnace**	Schutzgasofen m	four m de gaz protecteur, four de gaz de protection	печь с защитной атмосферой
C 310	**controlled gas-ballast valve**	steuerbares Gasballastventil n	soupape (vanne) f de lest de gaz réglable	регулируемый газобалластный вентиль
C 311	**controller**	Regelgerät n	appareil m de réglage (commande)	управляющее устройство
C 312	**control panel**	Schaltschrank m, Bedienungspult n, Schalttafel f	châssis m de distribution, armoire f de commande, caisson m de commande	распределительный щит
C 313	**control valve,** regulating valve	Regelventil n	soupape (vanne) f de réglage	регулирующий вентиль
C 314	**convection,** heat convection	Konvektion f	convection f	конвекция
C 315	**conversion factor**	Umrechnungsfaktor m, Umformfaktor m	facteur m de conversion	коэффициент пересчета, переводный множитель
C 316	**conveying chute**	Förderrutsche f	couloir m de transport	желоб для транспортировки, качающийся транспортер
C 317	**coolant,** refrigerant, cooling agent	Kühlmittel n, Kühlflüssigkeit f, Kältemittel n	réfrigérant m, agent m frigorifique, frigorigène m, moyen m réfrigérant	охладитель, охлаждающий агент, охлаждающая среда, холодильный агент, хладоагент
C 318	**coolant consumption**	Kühlmittelverbrauch m	consommation f de réfrigérant	расход хладоагента
	cool down, chill, cool off	abkühlen	refroidir, réfrigérer, refraîchir	охлаждать
C 319	**cool-down time**	Abkühlzeit f	temps m de refroidissement	время охлаждения
C 320	**cooled by hydrogen**	wasserstoffgekühlt	refroidi par l'hydrogène	охлаждаемый водородом
	cooled cover	s. cold cap <baffle>		
	cooling, chilling	Kühlung f, Abschreckung f	refroidissement m	охлаждение
	cooling agent, coolant, refrigerant	Kühlmittel n, Kühlflüssigkeit f, Kältemittel n	réfrigérant m, agent m frigorifique, frigorigène m, moyen m réfrigérant	охладитель, охлаждающий агент, охлаждающая среда, холодильный агент, хладоагент
C 321	**cooling brine**	Kühlsole f	saumure f frigorigène, eau f saline de refroidissement	холодильный рассол
C 322	**cooling by means of circulating water**	Wasserumlaufkühlung f	refroidissement m par circulation d'eau	охлаждение водяной циркуляцией
C 323	**cooling coil**	Kühlspirale f	hélice f de refroidissement	охлаждающий змеевик
C 324	**cooling fin,** cooling flange, cooling rib	Kühlrippe f	ailette f de refroidissement	охлаждающее ребро
	cooling finger, cold finger, thimble trap	Kühlfinger m	doigt m refroidisseur, sonde f de refroidissement	пальцевой холодильник
	cooling flange	s. cooling fin		
C 325	**cooling jacket**	Kühlmantel m	chemise f de refroidissement, enveloppe f réfrigérante	охлаждающая рубашка
C 326	**cooling load,** refrigeration load, heat load, refrigeration, duty	Kältebedarf m	besoin m frigorifique	расход холода
C 327	**cooling of grid**	Gitterkühlung f	refroidissement m de grille	охлаждение сетки

C 328	cooling rate	Abkühlungs-geschwindigkeit f	vitesse f de refroidissement	скорость (интенсивность) охлаждения
	cooling rib	s. cooling fin		
C 329	cooling system	Kühlaggregat n	machine f frigorifique, ré-frigérateur m, machine réfrigérante	холодильная система
C 330	cooling-water con-nection, water-cooling connection	Kühlwasseranschluß m	raccord m de l'eau de refroidissement	соединение охлаждаю-щей воды
C 331	cooling-water flow safety switch	Kühlwasserkontroll-schalter m	commutateur m de con-trôle d'eau de refroidis-sement, interrupteur m de sûreté d'eau de re-froidissement	предохранительный выключатель водяного охлаждения
C 332	cooling-water jacket	Kühlwassermantel m	chemise f [à circulation] d'eau de refroidisse-ment, enveloppe f à l'eau de refroidisse-ment	охлаждающая водяная рубашка
C 333	cooling-water pump	Kühlwasserpumpe f	pompe f d'eau de refroidissement	охлаждающий водяной насос
	cool off, chill, cool down	abkühlen	refroidir, réfrigérer, refraîchir	охлаждать
C 334	copper, copper plate	verkupfern	cuivrer	омеднять, покрывать медью
C 335	copperclad wire, copper-coated nickel-iron alloy wire	Kupfermanteldraht m	fil m enrobé de cuivre, fil d'enveloppe de cuivre	электрод в медной оболочке
C 336	copper crucible	Kupfertiegel m	creuset m de cuivre	медный тигель
C 337	copper crystallizer	Kupferkristallisator m	cristallerie f de cuivre	медный кристаллизатор
C 338	copper-foil gasket	Kupferfoliendichtung f	joint m de feuilles minces de cuivre	прокладка из медной фольги
C 339	copper-foil trap	Kupferfolienfalle f	piège m à feuille de cuivre	ловушка с медной фольгой
	copper plate, copper	verkupfern	cuivrer	омеднять, покрывать медью
C 340	copper-plating by immersion	Tauchverkupferung f	cuivrage m par immersion	омеднение погружением
C 341	copper-to-glass seal, housekeeper seal	Glas-Kupfer-Verschmelzung f	cellement m verre-cuivre, joint m verre-cuivre	спай стекла с медью
C 342	copper tubing seal, pinch-off seal	Abquetschdichtung f <durch Kaltpreß-schweißen>	joint m à compression	отсоединение <методом холодной сварки>
C 343	core wire, PG-wire	Seelendraht m, PG-Draht m	fil m d'âme	провод сердечника
C 344	corner gold ring seal	„corner"-Golddraht-dichtung f, Winkel-Golddrahtdichtung f, Ecken-Golddraht-dichtung f	joint m de fil en or	угловое уплотнение с кольцевой золотой прокладкой
C 345	corner joint, corner weld	Ecknaht f	joint m d'angle	угловое соединение
	corner valve, angle valve, right angle valve	Eckventil n	soupape f à passage angulaire, vanne f d'équerre	угловой вентиль
	corner weld, corner joint	Ecknaht f	joint m d'angle	угловое соединение
C 346	corpuscle, particle	Masseteilchen n	corpuscule m, particule f, élément m de masse	корпускула
C 347	corpuscular beam	Korpuskularstrahl m	rayon m corpusculaire	корпускулярный пучок
C 348	corrosion-proof, corrosion-resistant, stainless	korrosionsbeständig	inoxydable, non corrosif, résistant à la corrosion	нержавеющий
C 349	corrosion-protective finish	korrosionshindernde Schicht f	couche f résistant à la corrosion	антикоррозионное покрытие
	corrosion-resistant	s. corrosion-proof		
C 350	corrugated diaphragm	gewellte Membran f	membrane f ondulée, diaphragme m ondulé	гофрированная мембрана
C 351	corrugated tube	Wellrohr n	tube m ondulé	гофрированная трубка
C 352	corrugation	Rillung f, Riffelung f	cannelage m	волнистость, рифление
C 353	cosine law distribution	Lambertsche Kosinus-verteilung f	distribution f cosinusoï-dale de Lambert	закон Ламберта, коси-нусоидальный закон распределения
C 354	cosine law of emission, Lambert's law	Lambertsches Kosinus-gesetz n, Lambertsches Gesetz n	loi f cosinusoïdale de Lambert, loi de Lambert	закон Ламберта
C 355	Coulomb repelling force	Coulombsche Abstoßungskraft f	force f répulsive de Coulomb	кулоновская сила отталкивания
C 356	countercurrent exchanger	Gegenstromaustauscher m	échangeur m à contre-courant	теплообменник с встреч-ным током
C 357	counter flange, mating flange	Gegenflansch m	contre-bride f	контрфланец
C 358	counter flow	Gegenstrom m	contre-courant m	противоток
	coupling, connector, fitting	Verbindungsstück n	raccord m, pièce f de raccordement	соединительная часть
	coupling nut, cap nut	Überwurfmutter f	écrou m à raccord (cha-peau), écrou-chapeau m	накидная (стяжная) гайка
C 359	coverage density	Belegungsdichte f	densité f de revêtement	плотность покрытия

C 360	**coverage rate,** degree of coverage	Bedeckungsgrad *m*	degré *m* de couverture, degré d'absorption	степень покрытия
	cover plate, blank flange, blank-off flange, blank-off plate, blind flange	Blindflansch *m*	bride *f* aveugle, bride d'obturation, joint *m* plein, fausse *f* bride, bride obturatrice	глухой фланец, заглушка
	cp	*s.* chemically pure		
C 361	**crack,** flaw, fissure	Sprung *m*, Riß *m*	fente *f*, fissure *f*, rupture *f*, crevasse *f*, fêlure *f*	трещина, щель
C 362	**creation of vacuum**	Vakuumerzeugung *f*	production *f* de vide	создание вакуума
C 363	**creep barrier**	Kriechbarriere *f*	barrière-écran *f*, barrière *f* de grimpement	преграда против миграции
C 364	**creep behaviour**	Zeitstandsverhalten *n*	comportement *m* de fluage	режим текучести
C 365	**creeping intensity**	Dehnungsgeschwindigkeit *f*	vitesse *f* de dilatation	скорость растяжения
C 366	**creeping path**	Kriechweg *m*	ligne *f* de fuite	путь утечки
C 367	**creeping strength,** resistance to creepage, creep resistance	Kriechfestigkeit *f*, Dauerstandfestigkeit *f*	résistance *f* au fluage	сопротивление ползучести, предел ползучести
C 368	**creep limit**	Zeitdehngrenze *f*, Zeitstauchgrenze *f*	limite *f* d'allongement	предел ползучести
	creep resistance	*s.* creeping strength		
C 369	**creep rupture strength**	Zeitstandfestigkeit *f*	stabilité *f* temporaire	предел долговременной прочности
C 370	**critical backing pressure, critical forepressure,** limiting forepressure,	Vorvakuumgrenzdruck *m*, Grenzdruck *m* der Vorvakuumbeständigkeit	pression *f* limite d'amorçage	наибольшее выпускное (форвакуумное) давление
C 371	**critical point**	kritischer Punkt *m*	point *m* critique	критическая точка
C 372	**critical pressure**	kritischer Druck *m*	pression *f* critique	критическое давление
C 373	**critical temperature**	kritische Temperatur *f*	température *f* critique	критическая температура
C 374	**critical wavelength**	Grenzwellenlänge *f*	longueur *f* limite d'onde	критическая длина волны
C 375	**Crookes' dark space**	Crookesscher Dunkelraum *m*, Hittorfscher Dunkelraum, Katodendunkelraum *m*	chambre *f* noire de Crookes (Hittorf), chambre noire cathodique	круксово темное пространство, второе катодное темное пространство
C 376	**Crookes' tube**	Crookessche Röhre *f*	tube *m* de Crookes	трубка Крукса
C 377	**cross,** cross-piece	Kreuzstück *n*	pièce *f* (tuyau *m*) en croix	крестовина
	cross breaking strength, buckling resistance, resistance to lateral bending	Knickfestigkeit *f*	résistance *f* au flambage	прочность при продольном изгибе
C 378	**cross fire**	Kreuzfeuerbrenner *m*	brûleur *m* en croix	горелка со скрещивающимися огнями
C 379	**cross-flow blower**	Querstromgebläse *n*	soufflerie *f* de courant transversal	вентилятор (компрессор) с поперечным потоком
C 380	**cross-linkage,** interlacing	Vernetzung *f*	réticulation *f*	сшивание, установление поперечной связи
C 381	**cross-over**	engster Strahlquerschnitt *m*	profil *m* du rayon très étroit	кроссовер
C 382	**cross-over forepressure,** take-hold pressure	Vorvakuumdruck *m* <bei dem der Ansaugdruck gleich dem Vorvakuumdruck ist>	pression *f* du vide préliminaire	противодавление полного срыва <при котором впускное давление становится равным выпускному>
	cross-piece, cross	Kreuzstück *n*	pièce *f* (tuyau *m*) en croix	крестовина
C 383	**crucible**	Schmelztiegel *m*	creuset *m*	тигель
C 384	**crucible evaporation**	Tiegelverdampfung *f*	évaporation *f* de creuset	тигельное испарение
C 385	**crucible-free zone melting**	tiegelfreies Zonenschmelzen *n*	fusion *f* en zones sans creuset	бестигельная зонная плавка
C 386	**crucible furnace**	Tiegelofen *m*	four *m* à creuset	тигельная печь
C 387	**crucible heater**	Tiegelheizer *m*	chauffeur *m* à creuset	подогреватель тигля
C 388	**crucible holder**	Tiegelhalter *m*	porte-creuset *m*	держатель тигля
C 389	**crucible reaction**	Tiegelreaktion *f*	réaction *f* de creuset	реакция в тигле
C 390	**crude metal**	Rohmetall *n*	métal *m* brut	необработанный (черновой) металл
	crush seal	*s.* compression seal		
C 391	**cryodrying,** freezedrying, lyophilization, sublimation from the frozen state	Gefriertrocknung *f*, Lyophilisation *f*	lyophilisation *f*, dessiccation *f* par congélation, cryodessiccation *f*	сублимационная сушка из замороженного состояния, лиофилизация
C 392	**cryogenerator,** gasrefrigerating machine	Gaskältemaschine *f*	machine *f* frigorifique à gaz	криогенная установка
C 393	**cryogenetic pump,** cryopump	Kryopumpe *f*	pompe *f* cryostatique, (cryogénique)	криогенный насос
C 394	**cryogenetic pumping,** cryopumping	Kryopumpen *n*	cryopompage *m*, pompage *m* cryogénique	криогенная откачка
C 395	**cryogenetic trapping,** cryotrapping	Kryotrapping *n*	cryotrapping *m*, cryofixation *f*, captage *m* cryogénique	криозахват, «сопутствующая» откачка, «спутная» откачка
C 396	**cryogetter pump**	Kryogetterpumpe *f*	pompe *f* à cryo-getter	криогеттерный (криосорбционный) насос
	cryopump	*s.* cryogenetic pump		
	cryopumping	*s.* cryogenetic pumping		

C 397	**cryopumping surface,** cryosurface	Kryofläche *f*	surface *f* cryogénique, cryosurface *f*	криопанель
C 398	**cryosorption pump**	Kryosorptionspumpe *f*	pompe *f* cryostatique à sorption	криосорбционный насос
C 399	**cryostat principle**	Kryostatenprinzip *n*	principe *m* de cryostat	принцип «криостата»
C 400	**cryosublimation trap**	Kryosublimationsfalle *f*	trappe *f* cryogène de sublimation, baffle *m* cryogénique de sublimation	криосублимационная ловушка
	cryosurface, cryo-pumping surface	Kryofläche *f*	surface *f* cryogénique, cryosurface *f*	криопанель
	cryotrapping	*s.* cryogenetic trapping		
C 401	**cryotrapping effect**	Kryotrappingeffekt *m*	effet *m* de cryotrapping, effet de cryofixation	эффект криозахвата <поглощение некон-денсируемых газов вместе с легко конден-сирующимся газом>
C 402	**crystal diode test set**	Diodenmeßgerät *n*	appareil *m* de mesure pour diodes	измеритель параметров полупроводниковых диодов
C 403	**crystal face**	Kristallfläche *f*	facette *f*	грань кристалла
C 404	**crystal growing furnace,** crystal pulling machine	Kristallziehofen *m*, Kristallziehanlage *f*	four *m* (installation *f*) à étirer des cristaux, four de tirage de cristaux, installation de tirage de cristaux	печь для выращивания кристаллов, установка для выращивания кристаллов
C 405	**crystal imperfection**	Kristallstörstelle *f*, Kristallstörung *f*, Kristallfehler *m*	dérangement *m* cristallin	несовершенство кри-сталла, нарушение в кристалле
C 406	**crystal nucleus,** seed crystal	Kristallisationskeim *m*, Kristallisationskern *m*, Impfling *m*, Kristall-keim *m*	germe *m* cristallin, amorce *f* de cristalli-sation	затравочный кристалл, зародыш кристалла
C 407	**crystal offsetting**	Kristallversetzung *f*	dislocation *f* de cristal	дислокация в кристалле
C 408	**crystal plane,** grate plane, net plane	Netzebene *f*	plaine *f* de réseau	плоскость [кристалли-ческой] решетки
C 409	**crystal pulling**	Kristallziehen *n*	tirage *m* de cristaux	вытягивание кристалла
	crystal pulling machine, crystal growing furnace	Kristallziehofen *m*, Kristallziehanlage *f*	four *m* (installation *f*) à étirer des cristaux, four de tirage de cristaux, installation de tirage de cristaux	печь для выращивания кристаллов, установка для выращивания кристаллов
C 410	**crystal transducer,** high-frequency quartz crystal, oscillating quartz crystal, quartz crystal oscillator, vibrating quartz crystal	Schwingquarz *m*	quartz *m* oscillatoire (de résonance)	кристалл кварцевого генератора, генери-рующий кварцевый кристалл, генерирую-щий кварц
C 411	**crystal wafer**	Kristallscheibchen *n*, Kristallplättchen *n*	plaquette *f* cristalline	кристаллическая пластина
C 412	**crystal whisker**	Haarkristall *m*	cristal *m* capillaire	волосяной кристалл, вискер
C 413	**cup spring,** plate spring	Tellerfeder *f*	rondelle *f* élastique, ressort *m* en disque	дисковая (тарельчатая) пружина
C 414	**cure**	Aushärtung *f*	trempe *f*	твердение, отверждение
C 415	**Curie point**	Curie-Punkt *m*	point *m* de Curie	точка Кюри
C 416	**Curie temperature**	Curie-Temperatur *f*	température *f* de Curie	температура Кюри
C 417	**current feed-through,** current lead-in, current lead-through	Stromdurchführung *f*	traversée *f* de courant, sortie *f* de courant, passage *m* de courant	токоввод
C 418	**current-limiting**	strombegrenzend	limitant le courant	токоограничивающий
C 419	**cut-off relay,** overload relay	Überstromrelais *n*	relais *m* à excès de courant, relais de surintensité	реле максимального тока, максимальное реле, реле перегрузки
C 420	**cut-off valve,** isolation valve, seal-off valve	Absperrventil *n*, Verschlußventil *n*	vanne *f* d'arrêt, soupape *f* d'arrêt, soupape de fer-meture, vanne obtura-trice, robinet *m* obtura-teur	запорный вентиль, изо-ляционный вентиль, запорный клапан
C 421	**cut-off wheel,** cutting wheel	Trennscheibe *f*	disque *m* intermédiaire	отделительный (раз-делительный) диск
C 422	**cutting edge**	Schneide *f*	tranchant *m*, taillant *m*	режущая кромка, лезвие
	cutting wheel, cut-off wheel	Trennscheibe *f*	disque *m* intermédiaire	отделительный (раз-делительный) диск
C 423	**cyclic batch still**	Durchlaufchargen-destillationsanlage *f*	poste *m* de distillation à cycle intermittent	перегонная установка с циклической загруз-кой
C 424	**cyclic timer**	Programmschalter *m*	commandeur *m* à programme	программный переклю-чатель (регулятор)
C 425	**cycling time**	Zykluszeit *f*, Taktzeit *f*	temps *m* de cycle (répétition)	период повторения
C 426	**cycloidally focused mass spectrometer,** cycloidal spectrom-eter	Zykloiden-Massen-spektrometer *n*	spectromètre *m* de masse cycloïdale	циклоидальный масс-спектрометр
C 427	**cyclotron**	Zyklotron *n*	cyclotron *m*	циклотрон
C 428	**cyclotron frequency**	Zyklotronfrequenz *f*	fréquence *f* de cyclotron	циклотронная частота
C 429	**cyclotron resonance**	Zyklotronresonanz *f*	résonance *f* du cyclotron	циклотронный резонанс

C 430	cyclotron resonance frequency	Zyklotronresonanzfrequenz *f*	fréquence *f* de résonance de cyclotron	циклотронная резонансная частота
C 431	cylinder dryer, drum dryer, rotary dryer, roller dryer	Zylindertrockner *m*, Walzentrockner *m*, Trommeltrockner *m*, Röhrentrockner *m*	sécheur *m* cylindrique (tournant, tubulaire)	[вращающаяся] барабанная сушилка, сушильный барабан, вальцовая сушилка
C 432	cylindrical diaphragm, rolling diaphragm	Rollmembran *f*	soufflet *m*, membrane *f* cylindrique	цилиндрическая мембрана, сильфон
C 433	Czochralski technique	Czochralski-Verfahren *n*	procédé *m* de Czochralski	технология Чохральского

D

D 1	Dalton's law	Daltonsches Gesetz *n*	loi *f* de Dalton	закон Дальтона
D 2	damping capillary	Dämpfungskapillare *f*	capillaire *f* à l'atténuation	демпфирующий капилляр
D 3	damping diode	Dämpfungsdiode *f*	diode *f* d'amortissement	демпфирующий диод
	dark current, background current	Untergrundstrom *m*	courant *m* de fond	фоновый ток
D 4	dark discharge	Dunkelentladung *f*, Townsend-Entladung *f*	décharge *f* obscure (de Townsend)	темный (таунсендовский) разряд
D 5	dead-beat	aperiodisch	apériodique	апериодический, затухающий
D 6	dead space, waste space	schädlicher Raum *m*	espace *m* mort (nuisible)	мертвое (вредное) пространство
D 7	deaerating, venting, deairing, ventilation	Entlüftung *f*	ventilation *f*, évacuation *f* de l'air, respiration *f*, désaération *f*	выпуск воздуха, удаление воздуха, деаэрация, деаэрирование
D 8	deaerating vessel, deaeration vessel	Luftabscheidegefäß *n*, Entgasungsgefäß *n*	récipient *m* de dégagement, vase *m* de dégazage, séparateur *m* d'air, appareil *m* à séparer l'air	деаэрационная камера, воздухоотделитель
	deairing	s. deaerating		
D 9	decant, pour-off	dekantieren	décanter	декантировать
D 10	decantation	Dekantieren *n*	décantation *f*	переливание, сцеживание, декантация
D 11	decarbonize the tungsten filament	den Wolframfaden entkarbonisieren	décarboniser le filament de tungstène	декарбонировать вольфрамовый катод
D 12	decarburize	entkohlen	décarburer	обезуглероживать
D 13	decay constant	Zerfallskonstante *f*	constante *f* radioactive (de désintégration)	радиоактивная постоянная
D 14	deceleration, delay, retardation, time lag	Verzögerung *f*	retard *m*, retardation *f*, ralentissement *m*, temporisation *f*	задержка, запаздывание
D 15	decoating	Entfernen *n* einer Aufdampfschicht	élimination *f* de couches métallisées	удаление напыленного слоя
D 16	decomposing, decomposition	Zersetzung *f*	décomposition *f*	разложение
D 17	decomposition hazard	Zersetzungsanfälligkeit *f*	susceptibilité *f* de décomposition	склонность к разложению
D 18	decomposition occurring at the electrodes	an den Elektroden auftretende Zersetzungen *fpl*	décompositions *fpl* apparaissantes aux électrodes	разложение, происходящее на электродах
D 19	decompressor, expansion engine	Expansionsmaschine *f*	machine *f* à expansion	детандер, расширительная машина
D 20	decorative coating	Schmucküberzug *m*	revêtement *m* ornemental (d'ornement)	декоративное покрытие
D 21	decrease in the number of ions, depletion of ions	Ionenverarmung *f*	appauvrissement *m* d'ions, diminution *f* de concentration d'ions	уменьшение числа ионов, обеднение ионами
D 22	decrease of tube vacuum	Erniedrigung *f* des Röhrenvakuums	diminution *f* du vide d'un tube	ухудшение вакуума в лампе
D 23	decrement gauge, decrement viscosity gauge, viscosity manometer, viscosity vacuum gauge, viscosity type of gauge, friction type vacuum gauge	Reibungsvakuummeter *n*	manomètre *m* à amortissement (viscosité), jauge *f* à vide à frottement, vacuomètre *m* d'adhérence, vacuomètre de frottement, tube *m* gradué à viscosité	динамический (вязкостный) манометр
D 24	deep cooling	Tiefkühlung *f*	refroidissement *m* à basse température	низкотемпературное охлаждение, глубокое охлаждение
D 25	deep drawn vacuum forming, vacuum deep drawing	Tiefziehansaugen *n*	pompage *m* d'emboutissage	вакуумная формовка с глубокой вытяжкой
D 26	deep-freeze cabinet, deep-freezing chest	Tiefkühltruhe *f*	réfrigérateur *m* à basse température, armoire *f* frigorifique, congélateur *m* à basse température, coffre *m* congélateur	холодильный шкаф для глубокого замораживания
D 27	deep freezing	Tiefgefrieren *n*	congélation *f* à basse température	низкотемпературное замораживание, глубокое замораживание
	deep-freezing chest	s. deep-freeze cabinet		
D 28	deep weld effect	Tiefschweißeffekt *m*	effet *m* de soudage à basse température	эффект глубокого проплавления

D 29	deep welding	Tiefschweißen *n*	soudure *f* à basse température	сварка с глубоким проплавлением
D 30	defect site	Störstellenplatz *m*	endroit *m* d'impuretés	место дефектов
D 31	deflatable bag moulding, vacuum bag moulding	Vakuumgummisackverfahren *n*	procédé *m* par le sac en caoutchouc	вакуумная формовка в резиновую форму
D 32	deflecting magnet	Ablenkmagnet *m*	aimant *m* de déviation	отклоняющий магнит
	deflector, baffle plate, repeller	Prallplatte *f*	déflecteur *m*, écran *m*, chicane *f*	отражательная пластина
D 33	deformation point	Deformationstemperatur *f*	température *f* de déformation	точка деформации, температура размягчения
D 34	defrost, thaw, thaw-off	entfrosten, auftauen, abtauen	dégeler, dégivrer	размораживать, оттаивать
D 35	defroster	Enteisungsanlage *f*	installation *f* de dégivrage	антиобледенитель, разморахиватель
D 36	defrosting, thawing	Auftauen *n*, Abtauen *n*	dégel *m*, dégivrage *m*	оттаивание
D 37	degas <US>, degasify	entgasen	dégazer	дегазировать, удалять газы
D 38	degas controller	Entgasungsregler *m*	régulateur *m* de dégazage	регулятор обезгаживания
D 39	degasification, degassing, outgassing	Entgasung *f*, Entgasen *n*	dégazage *m*, dégazation *f*	обезгаживание
	degasify	s. degas <US>		
	degassing	s. degasification		
D 40	degassing bell	Entgasungsglocke *f*	cloche *f* de dégazage	колпак для обезгаживания
D 41	degassing by bombarding	Entgasung *f* durch Bombardement	dégazage *m* par bombardement	обезгаживание путем бомбардировки
D 42	degassing column	Entgasungskolonne *f*	colonne *f* de dégazage	колонка для обезгаживания
D 43	degassing gun, hot air gun	Heißluftdusche *f*	séchoir *m* électrique	ручной теплоэлектровентилятор
D 44	degassing plant	Entgasungsanlage *f*	installation *f* de dégazage	установка для обезгаживания
D 45	degassing rate	Entgasungsrate *f*	quote-part *f* de dégazage	скорость обезгаживания
D 46	degassing tower	Entgasungsturm *m*	tour *f* de dégazage	колонна для обезгаживания
D 47	degrease	entfetten	dégraisser	обезжиривать, удалять смазку
D 48	degree Celsius, degree centigrade	Grad *m* Celsius	degré *m* Celsius (centésimal)	градус Цельсия
	degree of coverage, coverage rate	Bedeckungsgrad *m*	degré *m* de couverture, degré d'adsorption	степень покрытия
D 49	degree of dissociation	Dissoziationsgrad *m*	degré *m* de dissociation	степень диссоциации
D 50	degree of purity	Reinheitsgrad *m*	degré *m* de pureté	степень чистоты
D 51	degree of working	Bearbeitungsgrad *m*	degré *m* d'usinage	степень обработки
D 52	degree Rankine	Grad *m* Rankine	degré *m* Rankine	градус Ренкина
D 53	degree scale, tuning scale	Einstellskale *f*	échelle *f* graduée, cadran *m* de réglage	шкала настройки, градуированная шкала
D 54	dehumidifying, dehydration	Trocknung *f*, Entfeuchtung *f*, Feuchtigkeitsentzug *m*	déshydratation *f*, séchage *m*, dessiccation *f*	обезвоживание, дегидратация
D 55	dehydrating agent, desiccant, drying agent	Trockenmittel *n*	agent *m* siccatif (de séchage), desséchant *m*, dessiccateur *m*	осушитель <вещество>
	dehydration	s. dehumidifying		
D 56	de-ice	enteisen	déglacer	удалять лед
	delay	s. deceleration		
D 57	delayed condensation	verzögerte Kondensation *f*	condensation *f* retardée	задержанная конденсация
D 58	delay time	Verzögerungszeit *f*, Verzugzeit *f*	temps *m* de retardement	время запаздывания, время задержки
D 59	delivery screw, discharging screw, drain plug	Ablaßschraube *f*, Ablaßstopfen *m*	vis *f* de décharge, bouchon *m* de décharge	спускной винт, спускная пробка
D 60	demagnetization	Entmagnetisierung *f*	désaimantation *f*	размагничивание
D 61	demountable high-vacuum system	demontierbare Hochvakuumanlage *f*	installation *f* démontable à vide poussé	разборная вакуумная система
D 62	demountable joint	lösbare Verbindung *f*	connexion *f* détachable	разборное соединение
D 63	density of gas, gas density	Gasdichte *f*	densité *f* de gaz	плотность газа
D 64	deoxidant, deoxidizer	Desoxydationsmittel *n*	désoxydant *m*	восстановитель, раскислитель
	depletion layer, barrier layer	Sperrschicht *f*	couche *f* d'arrêt, couche de barrage	запорный слой
	depletion of ions	s. decrease in the number of ions		
D 65	deposition chamber	Bedampfungskammer *f*	chambre *f* de métallisation	камера для напыления
D 66	deposition mask	Bedampfungsmaske *f*	masque *m* d'évaporation	напыленная маска
D 67	deposition rate	Aufdampfrate *f*	taux *m* (vitesse *f*) de déposition par vaporisation, taux de métallisation sous vide, taux d'évaporation	скорость напыления, скорость осаждения
D 68	deposition technology	Beschichtungstechnik *f*, Aufdampftechnik *f*	technologie *f* de déposition par vaporisation, technologie de métallisation sous vide	технология нанесения пленки испарением
D 69	depression constant	Depressionskonstante *f*	constante *f* de la dépression	постоянная депрессии

	English	German	French	Russian
D 70	depth of penetration, penetrating depth	Eindringtiefe *f*	profondeur *f* de pénétration	глубина проникновения
D 71	descale	entzundern	décalaminer	удалять окалину
	desiccant	*s.* dehydrating agent		
D 72	desiccation	Austrocknung *f*	dessèchement *m*	высушивание
D 73	desorption	Desorption *f*	désorption *f*	десорбция
D 74	desorption spectrometer	Desorptionsspektrometer *n*	spectromètre *m* à désorption	десорбционный спектрометр
D 75	desorption spectrometry	Desorptionsspektrometrie *f*	spectrométrie *f* de désorption	десорбционная спектрометрия
D 76	destillate pump	Destillatpumpe *f*	pompe *f* de distillat	дистиллятный насос
D 77	detection limit	Nachweisgrenze *f*	limite *f* de décèlement, limite d'indication	предел обнаружения
D 78	detection sensitivity limit	Nachweisempfindlichkeitsgrenze *f*	limite *f* de la sensibilité décelable	порог чувствительности обнаружения
D 79	deviation, drift	Abweichung *f*, Drift *f*	déviation *f*, aberration *f*	дрейф
D 80	deviation system	Ablenksystem *n*	système *m* de déviation	отклоняющая система
D 81	device for removing metal rods	Strangabzugsvorrichtung *f*	installation *f* d'échappement à filer	устройство для удаления металлического прутка
D 82	device for tilting of the crucible	Tiegelkippvorrichtung *f*	appareil *m* basculant à creuset	механизм для опрокидывания тигля
D 83	devitrification	Entglasen *n*, Entglasung *f*	dévitrification *f*	расстекловывание
D 84	devitrify	entglasen	dévitrifier	расстекловываться
D 85	Dewar flask, Dewar vessel	Dewar-Gefäß *n*	vase *m* Dewar	сосуд Дьюара
D 86	dewaxing device	Entwachsungseinrichtung *f*	installation *f* à décirer	устройство для депарафинизации
D 87	dew point	Taupunkt *m*	point *m* de rosée	точка росы
D 88	dial barometer	Zeigerbarometer *n*	baromètre *m* à aiguille	стрелочный барометр
D 89	dial gauge, pointer manometer	Zeigervakuummeter *n*	manomètre *m* à aiguille, manomètre à cadran	стрелочный манометр, стрелочный вакуумметр
D 90	dial thermometer	Zeigerthermometer *n*	thermomètre *m* à aiguille	стрелочный термометр
D 91	dialyse	dialysieren	dialyser	диализировать
D 92	diaphragm	Membran *f*	membrane *f*, diaphragme *m*	диафрагма, мембрана
D 93	diaphragm box	Membrandose *f*	boîte *f* à membrane	мембранная коробка
D 94	diaphragm gland	Spaltdichtung *f*	joint *m* de fente	диафрагменный сальник
D 95	diaphragm manometer	Membranmanometer *n*	jauge *f* du diaphragme, manomètre *m* à membrane	мембранный манометр
D 96	diaphragm pressure controller, diaphragm switch	Membrandruckschalter *m*, Membranschalter *m*	pressostat *m* à membrane, interrupteur *m* manométrique à membrane	мембранный регулятор давления
D 97	diaphragm pump	Membranpumpe *f*	pompe *f* à membrane	мембранный насос
	diaphragm switch	*s.* diaphragm pressure controller		
D 98	diaphragm vacuum gauge, membrane vacuum manometer	Membranvakuummeßgerät *n*, Membranvakuummeter *n*	vacuomètre *m* à membrane, manomètre *m* à membrane	мембранный вакуумметр
D 99	diaphragm valve	Membranventil *n*	robinet *m* à membrane	мембранный вентиль
D 100	die	Gesenk *n*, Stempel *m*, Zieheisen *n*, Ziehdüse *f*, Schnittwerkzeug *n*	filière *f*	фильера
D 101	die casting, injection moulding	Spritzgießen *n*	moulage *m* par injection	литье под давлением
D 102	dielectric constant	Dielektrizitätskonstante *f*	constante *f* diélectrique	диэлектрическая постоянная
D 103	dielectric drying	dielektrische Trocknung *f*	dessiccation *f* diélectrique	диэлектрическая сушка
D 104	dielectric heating	dielektrische Erwärmung *f*	chauffage *m* diélectrique	диэлектрический нагрев
D 105	dielectric high-frequency heating	dielektrische Hochfrequenzerwärmung *f*	échauffement *m* diélectrique à haute fréquence	высокочастотный диэлектрический нагрев
	dielectric strength, disruptive strength, breakdown strength	Durchschlagfestigkeit *f*	résistance *f* au claquage, rigidité *f* diélectrique, résistance au percement disruptif	пробивная прочность, электрическая прочность
D 106	differential collision cross-section	differentieller Stoßquerschnitt *m*	section *f* droite différentielle de choc	дифференциальное эффективное сечение соударений
D 107	differential gear	Differentialgetriebe *n*	engrenage *m* différentiel	дифференциальная передача
D 108	differential getter pump	Differentialgetterpumpe *f*	pompe *f* différentielle à getter	дифференциальный сорбционный насос
D 109	differential interference manometer	interferometrisches Differentialvakuummeter *n*, Interferenzvakuummeter *n*	vacuomètre *m* différentiel d'interférence	дифференциальный интерферометрический манометр
D 110	differential ionization	differentielle Ionisierung *f*	ionisation *f* partielle	дифференциальная ионизация
D 111	differential ionization coefficient	spezifische Ionisierung *f*	ionisation *f* spécifique	линейный коэффициент ионизации
D 112	differential leak detection	Druckdifferenzlecksuchmethode *f*	méthode *f* de détection de fuites par différence de pression	метод течеискания по разности давлений
D 113	differential leak detector	Differentiallecksuchgerät *n*, Druckdifferenzlecksucher *m*	détecteur *m* différentiel de fuites	дифференциальный течеискатель

D 114	**differentially pumped lock**	Druckstufenschleuse *f*	écluse *f* à étage de pression	шлюз с дифференциальной откачкой
D 115	**differential manometer**	Differentialmanometer *n*, Differenzdruckvakuummeter *n*	manomètre *m* différentiel	дифференциальный манометр
D 116	**differential micromanometer**	Differentialmikromanometer *n*	micromanomètre *m* différentiel	дифференциальный микроманометр
D 117	**differential Pirani leak detector**	Differential-Pirani-Lecksuchgerät *n*	Pirani-détecteur *m* différentiel de fuites	дифференциальный течеискатель Пирани
D 118	**differential pressure**	Differenzdruck *m*	pression *f* différentielle	разность давлений
D 119	**differential pressure measurement**	Differenzdruckmessung *f*	mesure *f* de pression différentielle	измерение дифференциального давления
D 120	**differential pumping groove**	evakuierter Zwischenkanal *m* ‹einer Doppeldichtung›	conduit *m* vidé de transmission	дифференциально откачиваемая канавка ‹двойного уплотнения›
D 121	**differential pumping stage**	Differentialpumpstufe *f*	étage *m* de pompage différentiel	ступень с дифференциальной откачкой
D 122	**differential pumping system**	Druckstufensystem *n*	système *m* d'étage de pression	откачная система с дифференциальной откачкой
D 123	**diffraction of low-energy electrons, low-voltage electron diffraction, low-energy electron diffraction, LEED**	Beugung *f* langsamer Elektronen	diffraction *f* des électron lents (à basse énergie)	дифракция медленных электронов
D 124	**diffraction plane**	Beugungsebene *f*	plaine *f* de diffraction	плоскость дифракции
D 125	**diffuse**	diffus	diffus	диффузно, рассеянно
D 126	**diffuser**	Diffusor *m*, Staudüse *f*	diffuseur *m*	диффузор
D 127	**diffuser throat, throat of diffuser**	Diffusorhals *m* ‹engste Stelle einer Venturi-Düse›	col *m* de diffuseur	горловина диффузора
D 128	**diffusion**	Diffusion *f*	diffusion *f*	диффузия
D 129	**diffusion activity**	Diffusionsaktivität *f*	activité *f* de diffusion	диффузионная активность
D 130	**diffusion coefficient, diffusion constant**	Diffusionskoeffizient *m*, Diffusionskonstante *f*	coefficient *m* de diffusion	коэффициент диффузии
D 131	**diffusion-ejector pump, vapour booster pump**	Diffusionsejektorpumpe *f* ‹Pumpe mit Strahl- und Diffusionsstufen›	éjecteur *m* à diffusion	диффузионный эжекторный насос ‹насос с диффузионной и эжекторной ступенями›
D 132	**diffusion evaporation**	Diffusionsverdampfung *f*	vaporisation *f* de diffusion	диффузионное испарение
D 133	**diffusion membrane**	Diffusionsfenster *n*	fenêtre *f* de diffusion	диффузионная мембрана
D 134	**diffusion nozzle**	Diffusionsdüse *f*	diffuseur *m*	диффузионное сопло
D 135	**diffusion pump, vapour pump**	Diffusionspumpe *f*	pompe *f* à diffusion	диффузионный насос
D 136	**diffusion pump oil**	Diffusionspumpenöl *n*	huile *f* de pompe à diffusion	масло для диффузионного насоса
D 137	**diffusion rate**	Diffusionsgeschwindigkeit *f*	vitesse *f* de diffusion	скорость диффузии
D 138	**diffusion seal**	Diffusionsverbindung *f*	joint *m* de diffusion	диффузионное соединение
D 139	**diffusion stage**	Diffusionsstufe *f*	étage *m* de diffusion	диффузионная ступень
	diffusion welding, cold welding	Kaltschweißen *n*	soudage *m* à froid	холодная сварка, диффузионная сварка
	dilatation coefficient, coefficient of expansion	Ausdehnungskoeffizient *m*	coefficient *m* de dilatation	коэффициент расширения
D 140	**diode-pentode**	Diode-Pentode *f*	diode-pentode *f*	диод-пентод
D 141	**diode sputtering**	Diodenzerstäubung *f*	pulvérisation *f* de diodes	диодное распыление
D 142	**diode sputter-ion pump, two-electrode sputter-ion pump, diode type pump**	Ionenzerstäuberpumpe *f* vom Diodentyp	pompe *f* ionique à getter par pulvérisation cathodique	диодный магнитно-электроразрядный насос
D 143	**diode-triode**	Diode-Triode *f*	diode-triode *f*	диод-триод
	diode type pump	s. diode sputter-ion pump		
D 144	**dip brazing**	Tauch-Hartlöten *n*	brasage *m* par immersion	пайка погружением ‹твердым припоем›
D 145	**dip coating**	Beschichten *n* durch Eintauchen	trempage *m*, immersion *f*	покрытие погружением
D 146	**dip gettering**	Tauchgettern *n*	gettérisation *f* par immersion	нанесение газопоглотителя маканием
D 147	**dip ignition**	Tauchzündung *f*	allumage *m* par immersion	зажигание погружением
D 148	**dip soldering**	Tauchlöten *n*	soudure *f* par immersion	пайка погружением
	direct-contact condenser, contact condenser	Mischkondensator *m*	condenseur (condensateur) *m* à mélange	смешивающий (смесительный) конденсатор
D 149	**direct expansion refrigeration**	direkte Verdampfung *f* ‹Kühlverfahren›	vaporisation *f* directe ‹méthode de réfrigération›	охлаждение непосредственным испарением
D 150	**direct insert gauge, direct insertion gauge head**	Eintauchmeßröhre *f*	burette *f* d'immersion	иммерсионный датчик манометра, иммерсионный манометр
D 151	**directional effect**	Richtungseffekt *m*, Richtwirkung *f*	effet *m* directif	эффект направленности
D 152	**direction cosine**	Richtungskosinus *m*	cosinus *m* de direction	косинус [угла] направления

D 153	direction focusing	Richtungsfokussierung f	mise f au point de direction	фокусировка по направлению
D 154	direction of propagation	Fortpflanzungsrichtung f	direction f de propagation	направление распространения
D 155	directly heated	direkt geheizt	à chauffage direct	прямонакальный
D 156	dirt trap, scale trap	Schmutzfänger m	piège m à impuretés, collecteur m d'impuretés	грязеуловитель
D 157	disappearing filament pyrometer	Helligkeitspyrometer n, Glühfadenpyrometer n	pyromètre m à filament	оптический монохроматический пирометр, яркостный пирометр
D 158	discharge cleaning, glow discharge cleaning, ionic bombardment cleaning, ionic cleaning	Abglimmen n, Glimmreinigung f	nettoyage m par bombardement ionique, décharge f par effluves	очистка разрядом, очистка тлеющим разрядом, ионная очистка
D 159	discharge coefficient	Ausflußziffer f, Austrittszahl f	nombre m d'écoulement, coefficient m (nombre) de sortie	коэффициент расхода (истечения)
D 160	discharge current	Entladungsstrom m	courant m de décharge	разрядный ток
D 161	discharge device, discharge lock	Ausschleusvorrichtung f	dispositif m de décharge	устройство для выгрузки (шлюзования)
D 162	discharge [mist] filter, exhaust filter	Auspufffilter n ‹Ölfilter auf der Auspuffseite einer mechanischen Vakuumpumpe›	filtre m d'échappement	выхлопной фильтр ‹на выходе механического насоса›
D 163	discharge gap, discharge path	Entladungsstrecke f	distance f de décharge	разрядный промежуток
D 164	discharge line discharge lock discharge path	Druckleitung f s. discharge device s. discharge gap	conduite f sous pression	выпускной трубопровод
D 165	discharge pipe, exhaust pipe, outlet pipe	Auspuffleitung f ‹rotierende Pumpe›	tube m d'échappement, conduite f d'échappement	выхлопная труба, выпускной патрубок, выпускная труба
D 166	discharge port, exhaust port, exhaust, outlet	Auspuff m, Auspufföffnung f	échappement m, orifice m de refoulement, orifice d'échappement, orifice d'expulsion	выхлопное окно, выпускной патрубок
D 167	discharge pressure, outlet pressure, exhaust pressure, head pressure	Verdichtungsdruck m, Auspuffdruck m, Ausstoßdruck m	pression f de sortie, pression de refoulement, pression du vide préliminaire, pression d'échappement, pression finale	давление сжатия, выпускное давление, давление на выхлопе
D 168	discharge suppressor	Entladungssperre f	obstacle m à décharge, chicane f à décharge	экран для предотвращения возникновения разряда, экран для подавления разряда
D 169	discharge tap, bleeder, bleeder cock, drain cock discharge tube	Ablaßhahn m Entladungsrohr n	robinet m de vidange tube m à décharge	спускной кран, сливной кран разрядная трубка, разрядная лампа
D 170	discharge tube method of leak detection	Undichtigkeitsnachweis m mittels Entladungsrohres ‹Geißler-Rohr›	indicateur m de fuite par tube à décharge ‹tube de Geißler›	метод течеискания с помощью разрядной трубки
D 171	discharge valve, exhaust valve	Auspuffventil n	soupape f de refoulement, soupape d'échappement	выпускной клапан
D 172	discharging screw, delivery screw, drain plug	Ablaßschraube f, Ablaßstopfen m	vis f de refoulement, bouchon m de décharge	спускной винт, спускная пробка
D 173	disintegration time	Zerfallsgeschwindigkeit f	vitesse f de désintégration	скорость распада
D 173	disk cathode	Scheibenkatode f	cathode f à disque	дисковый катод
D 174	disk gasket, valve plate gasket	Ventiltellerdichtung f, Scheibendichtung f	joint m du disque de vanne, joint de l'obturateur d'un robinet	уплотнение тарелки вентиля (затвора)
D 175	disk seal	Scheibenanschmelzung f, Stumpfanglasung f	refondage m de disque	дисковый впай
D 176	disk-spring column, saucer spring	Tellerfedersäule f	colonne f de rondelle élastique	колонка с дисковыми пружинами
D 177	disk valve, poppet valve, plate valve	Tellerventil n, Plattenventil	soupape f à disque	тарельчатый вентиль
D 178	dislocating density	Versetzungsdichte f	densité f de la dislocation	плотность дислокаций
D 179	dislocation	Versetzung f	dislocation f	дислокация, нарушение кристаллической решётки
D 180	dispenser cathode	Vorratskatode f	cathode f de réserve	диспенсерный катод, распределительный катод
D 181	dispersal getter	Dispersionsgetter m	getter m de dispersion	дисперсионный газопоглотитель
D 182	dispersal gettering	Verdampfungsgetterung f	effet m getter d'évaporation	распылительное (испарительное) газопоглощение
D 183	disperse	dispergieren	disperser, répartir	диспергировать, эмульгировать
D 184	dispersion, scattering	Zerstreuung f, Streuung f, Dispersion f	dispersion f, diffusion f	дисперсия, рассеяние
D 185	displacement, displacement capacity	Förderleistung f, Saugvermögen n ‹einer mechanischen Pumpe›	puissance f d'une pompe, capacité f d'une pompe, capacité de pompage	производительность насоса

D 186	**displacement current**	Verschiebungsstrom *m*	courant *m* de décalage (déplacement)	ток смещения
D 187	**displacement manometer**	Verdrängermanometer *n*	manomètre *m* déplaceur	манометр с перемещающимся чувствительным элементом
D 188	**displacement piston,** displacing piston	Verdrängerkolben *m*	piston *m* déplaceur (de déplacement)	вытесняющий поршень, перемещающийся поршень
D 189	**displacement pump** displacing piston	Verdrängerpumpe *f* *s.* displacement piston	pompe *f* déplaceur	объёмный насос
D 190	**disruptive breakdown range**	Temperaturbereich *m* des rein elektrischen Durchschlages	zone *f* de température du percement disruptif	температурная область разрушающего пробоя
	disruptive strength, dielectric strength, breakdown strength	Durchschlagfestigkeit *f*	résistance *f* au claquage, rigidité *f* diélectrique, résistance au percement disruptif	пробивная прочность, электрическая прочность
D 191	**dissipation factor**	dielektrischer Verlustfaktor *m*	facteur *m* de pertes diélectriques	коэффициент [диэлектрических] потерь
D 192	**dissociation pressure**	Dissoziationsdruck *m*	pression *f* de dissolution (dissociation)	упругость диссоциации, давление диссоциации
D 193	**distance cathode,** distant cathode, remote cathode	Fernkatode *f*	cathode *f* à distance	удалённый катод
D 194	**distillation apparatus,** distilling apparatus, distiller	Destillierapparat *m*, Destillationsapparat *m*	installation *f* de distillation, distillateur *m*, appareil *m* distillatoire, alambic *m*	дистиллятор, перегонный (дистилляционный) аппарат
	distillation still, alembic, still	Destillationskolben *m*, Destillationsblase *f*	ampoule *f* à distiller	перегонная колба
D 195	**distilled water** distiller distilling apparatus	destilliertes Wasser *n* *s.* distillation apparatus *s.* distillation apparatus	eau *f* distillée	дистиллированная вода
D 196	**distilling plant**	Destillationsanlage *f*	distillerie *f*, installation *f* de distillation	дистилляционная установка
D 197	**distilling vessel**	Destillierbehälter *m*	réservoir *m* distillatoire cucurbite *f*	перегонный сосуд
D 198	**distributing chute**	Vibrationsrinne *f*	cannelure *f* de vibration	распределительный жёлоб, вибрационный жёлоб
D 199	**distributor,** manifold	Verteilerrohr *n*, Verteilerstück *n*, Verteiler *m*	distributeur *m*, répartiteur *m*, tuyau *m* collecteur	распределитель, загрузочное распределительное устройство, распределительная магистраль
D 200	**donor**	Donator *m*	donneur *m*	донор
D 201	**dosing equipment**	Dosiervorrichtung *f*	appareil *m* réglable	дозирующее устройство
D 202	**dosing pump**	Dosierpumpe *f*	pompe *f* de dosage	дозировочный насос
D 203	**dosing valve,** leak valve, variable leak valve	Gaseinlaßventil *n*, Dosierventil *n*	soupape *f* réglable, vanne *f* de dosage, fuite *f* réglable	дозировочный вентиль
D 204	**dot recorder,** single-point recorder	Punktschreiber *m*	enregistreur *m* en pointillé	точечный самописец
D 205	**double acting pump,** sucking and forcing pump, lift and force pump, suction and pressure pump	Saug- und Druckpumpe *f*	pompe *f* aspirante et refoulante	всасывающий и нагнетательный насос, насос двойного действия
D 206	**double bell-jar plant**	Doppelglockenanlage *f*	arrangement *m* à double cloche	установка с двойным колпаком
D 207	**double bell-jar principle**	Doppelglockenprinzip *n*	méthode *f* de la double enceinte, principe *m* de la cloche double	метод откачки сосуда под вакуумным колпаком
D 208	**double chamber method**	Doppelglockenmethode *f*	méthode *f* de double cloche	метод двойного колпака
D 209	**double chamber system**	doppelwandiges System *n*, Doppelglockenanordnung *f*	système *m* à double paroi	вакуумная система с двойным колпаком
D 210	**double-corner gold-seal flange**	Flansch *m* mit zwei konzentrischen Golddraht-Eckdichtungen	bride *f* avec deux joints d'équerre de fil d'or	фланец с угловым уплотнителем из двух концентрических золотых проволок
D 211	**double-focusing mass spectrograph**	doppelfokussierender Massenspektrograf *m*	spectrographe *m* de masse à double focalisation	масс-спектрограф с двойной фокусировкой
D 212	**double helix**	Gegenwendel *f*	hélice *f* double	двойная спираль со встречной намоткой
D 213	**double knife edge seal**	Doppelschneidendichtung *f*	joint *m* à double couteau	уплотнение с двойными ножевидными выступами, уплотнение с двухсторонними острыми кромками
D 214	**double O-ring seal**	O-Ring-Doppeldichtung *f*	anneau *m* à section circulaire, joint *m* torique [à section circulaire]	двойное уплотнение с круглой кольцевой прокладкой
D 215	**double quartz-thread pendulum**	Doppelfadenpendel *n* <Reibungsvakuummeter>	pendule *m* à deux fils	маятник с двойной кварцевой нитью <динамический манометр>

D 216	**double seal**	Doppeleinschmelzung *f*	double fusion *f*	двойной спай
D 217	**double-V butt weld**	X-Naht *f*	joint *m* en X, soudure *f* en X	сварное соединение встык с двусторонним скосом кромок, сварной стыковой шов в форме X
D 218	**double-walled ball jar**	Doppelwandglocke *f*	cloche *f* à double paroi	вакуумный колпак с двойными стенками
D 219	**double-walled vacuum chamber**	doppelwandige Vakuumkammer *f*	chambre *f* à vide à double paroi	вакуумная камера с двойными стенками
D 220	**doughnut,** evacuated doughnut chamber	kreisförmige Hochvakuumröhre *f* des Betatrons	tube *m* circulaire à vide poussé	тороидальная вакуумная камера [бетатрона]
D 221	**dovetail groove**	Schwalbenschwanznut *f*	encoche *f* en queue d'aronde	канавка в форме ласточкиного хвоста
D 222	**down evaporation,** inverted evaporation	Abwärtsverdampfung *f*	vaporisation *f* descendante	испарение вниз
D 223	**downstream**	stromab	courant descendant	в направлении течения
D 224	**downstream channel**	Abzugskanal *m*	canal *m* d'échappement, hotte *f* d'évacuation	выпускной канал, отводящий канал
D 225	**down-time**	Totzeit *f*, Ausfallzeit *f*, Leerlaufzeit *f*	temps *m* mort	«мертвое» время, время простоя
D 226	**drag**	Luftwiderstand *m*	résistance *f* de l'air	сопротивление воздуха, лобовое сопротивление
D 227	**drag balance**	Luftwiderstandswaage *f*	balance *f* de résistance de l'air	весы для измерения лобового сопротивления
D 228	**drain cock**	Entleerungshahn *m*	robinet *m* d'évacuation, robinet de vidage	спускной кран
	drain cock, bleeder, bleeder cock, discharge tap	Ablaßhahn *m*	robinet *m* de vidage	спускной кран, сливной кран
D 229	**draining,** purge	Entleerung *f*	vidange *f*, évacuation *f*	продувка
D 230	**draining board**	Abtropfbrett *n*	planche *f* d'égouttage	сушильная доска, сточная доска
D 231	**draining rack,** drying frame	Trockengestell *n*, Abtropfgestell *n*	support *m* de séchage, bâti *m* de séchage, châssis *m* de séchage	сушильная решетка
	drain plug	*s.* delivery screw		
D 232	**drain valve,** purger	Entleerungsventil *n*	soupape *f* de vidange	выпускной вентиль, спускной вентиль
D 233	**drape and vacuum forming**	Vakuumstreckformverfahren *n*	procédé *m* de drapage sous vide	позитивно-вакуумное формование
D 234	**drape forming**	Streckformen *n*	drapage *m*	позитивное формование
D 235	**drawing pump**	Saugpumpe *f*	pompe *f* aspirante (élévatoire)	всасывающий насос
D 236	**drawn metal**	gezogenes Metall *n*	métal *m* étiré	тянутый металл
D 237	**draw off,** suck off	absaugen	évacuer, pomper	отсасывать
D 238	**draw-out field**	Ziehfeld *n*	champ *m* de tirage	очищающее (вытягивающее) поле
D 239	**drier,** siccative	Trockenstoff *m*, Sikkativ *n*	siccatif *m*	сиккатив, обезвоживатель
	drift	*s.* deviation		
D 240	**drip feed lubricator**	Tropföler *m*	graisseur *m* de gouttes	капельная масленка
D 241	**drip melt**	Abtropfschmelze *f*	fusion *f* d'égouttage	капельная плавка
D 242	**drip tray**	Tropfschale *f*, Abtauwanne *f*	cuvette *f* de décongélation, capsule *f* de décongélation, cuvette d'égouttage	лоток для капель
D 243	**drive pinion**	Antriebsritzel *n*	pignon *m*	ведущая шестерня
D 244	**drive pulley**	Antriebsrolle *f*	rouleau *m* moteur	ведущий шкив
D 245	**drop in**	hinzutropfen	ajouter goutte à goutte	закапывать, вводить каплю за каплей
D 246	**droplet growth**	Tröpfchenwachstum *n*	croissance *f* de gouttelettes	капельный рост
D 247	**drop point**	Tropfpunkt *m*	point *m* de goutte	температура выпадения капель
D 248	**drop-tube**	Tropfendüse *f*	tuyère *f* à gouttes	пипетка
	drum dryer, rotary dryer, cylinder dryer, roller dryer	Trommeltrockner *m*, Röhrentrockner *m*, Zylindertrockner *m*, Walzentrockner *m*	sécheur *m* tournant (tubulaire, cylindrique)	сушильный барабан, [вращающаяся] барабанная сушилка, вальцовая сушилка
D 249	**drum freeze dryer**	Trommelgefriertrockner *m*	installation *f* (appareil *m*) de lyophilisation à tambour	роторная сублимационная сушилка
D 250	**dry box,** glove box	Handschuhkasten *m*	boîte *f* à gants	коробка для перчаток
D 251	**dryer**	Trockner *m*	sécheur *m*, séchoir *m*	сушилка
D 252	**dry film**	wasserabstoßende Schicht *f*	couche *f* hydrofuge	гидрофобный слой
D 253	**dry film lubricant,** dry lubricant	Trockenschmiermittel *n*	lubrifiant *m* sec	сухая смазка
D 254	**dry friction**	Trockenreibung *f*	friction *f* à sec	сухое трение
	dry ice, solidified carbon dioxide gas, carbon dioxide snow, solid carbon dioxide	Trockeneis *n*, feste Kohlensäure *f*, Kohlensäureschnee *m*	glace *f* sèche, carboglace *f*, anhydride *m* carbonique en neige, neige *f* carbonique, acide *m* carbonique solide	сухой лед, твердая углекислота, снег из твердой углекислоты
	drying agent	*s.* dehydrating agent		
D 255	**drying by sublimation**	Sublimationstrocknung *f*	dessication *f* par sublimation	сублимационная сушка

	drying cabinet, cabinet dryer	Schranktrockner m	sécheur m en armoire, chambre f à sécher, exsiccateur m	сушильный шкаф
D 256	drying chamber	Trockenkammer f	étuve f, chambre f de dessiccation (séchage)	сушильная камера
	drying frame	s. draining rack		
D 257	drying manifold	Trockenrechen m	râteau m sec, collecteur m pour la dessiccation	сушильная решетка, сушильный коллектор
D 258	drying material with ice nucleus inside	Trockenprodukt n mit Eiskern	produit m sec avec noyau de glace, desséchant m avec noyau de glace	сухой продукт с ледовым ядром
D 259	drying plant	Trocknungsanlage f	installation f de séchage (dessiccation)	сушильная установка
D 260	drying rack for ampoules	Rechen m für Ampullen- trocknung	collecteur m de tubes pour la dessiccation des ampoules	решетка для сушки ампул
D 261	drying rate, drying speed	Trocknungsrate f, Trocknungs- geschwindigkeit f	vitesse f de dessiccation (séchage)	скорость сушки
D 262	drying time	Trockenzeit f	temps m de séchage (dessiccation)	время сушки
	dry lubricant	s. dry film lubricant		
D 263	dry pumping system	Pumpanlage f mit Sorptionspumpe	installation f de pompage avec une pompe à sorption	откачная установка с сорбционным насосом, установка для без- масляной откачки
D 264	dry seal	Trockendichtung f	joint m sec	сухое уплотнение
D 265	dry vapour, superheated vapour	überhitzter Dampf m	vapeur f surchauffée	перегретый пар, сухой пар
D 266	duct, line, pipe, tubing	Rohrleitung f	canalisation f, tuyau m, conduite f	трубопровод
D 267	ductility	Duktilität f	ductilité f	пластичность, дуктиль- ность
D 268	dull silvering	Metallversilberung f	argenture f matte	тусклое серебрение, матовое серебрение
D 269	dumet seal	Verschmelzung f zwischen Dumet-Draht und Weichglas	Dumet-fusion f	спай стекла с прово- локой «Думет»
D 270	duoplasmatron ion source	Duoplasmatron- Ionenquelle f	source f d'ions de duo- plasmatron, canon m ionique de duo- plasmatron	дуоплазматронный ионный источник
D 271	Dushman viscosity manometer	Dushmansches Reibungsvakuum- meter n	manomètre m à amor- tissement de Dushman, manomètre à viscosité de Dushman	динамический манометр Дэшмана
D 272	dust filter	Staubfilter n	filtre m à poussière	фильтр для улавливания пыли
D 273	dust separator	Staubabscheider m	séparateur m de poussière, appareil m de dépous- siérage	пылеуловитель
	dwelling time, adherence time, resi- dence time, retention time, stay period, sticking time	Verweilzeit f	durée f, temps m de con- tact, temps d'exposition	время удерживания (выдержки, контакта, пребывания)
D 274	dye film	Farbschicht f	couche f colorée	слой краски
D 275	dynamical flow of gases, dynamic gas flow, gas dynamical flow	gasdynamische Strömung f	écoulement m (flux m) dynamique des gaz, courant m gazeux dynamique	газодинамическое течение
D 276	dynamic breakdown forepressure, normal breakdown fore- pressure	„normaler Durchbruch"- Vorvakuumdruck m	pression f normale du débit du vide prélimi- naire	нормальное выпускное давление срыва
D 277	dynamic forepressure, normal forepressure, operating forepressure	Vorvakuumdruck m ‹den eine Vorpumpe am Vorvakuumstutzen einer Treibmittelpumpe erzeugt›	pression f du vide préliminaire	выпускное (форваку умное) давление
D 278	dynamic gas flow	s. dynamical flow of gases		
	dynamic leak checking	dynamische Leckprüfung f	détection f dynamique de fuites	динамическая проверка на течи
D 279	dynamic pressure, stagnation pressure, velocity pressure, velocity head	Staudruck m, dynamischer Druck m	pression f dynamique	скоростной напор, дина- мическое давление
D 280	dynamic pressure stage	dynamische Druckstufe f	étage m de pression dynamique, degré m de pression dynamique	ступень с динамическим давлением
D 281	dynamic vacuum system, kinetic vacuum system, pumped vacuum system	dynamisches Vakuum- system n, dynamische Vakuumanlage f	système m dynamique à (du) vide	динамическая вакуум- ная система
	dynamic viscosity, coefficient of viscosity, absolute viscosity	dynamische Zähigkeit f, dynamische Viskosität f	viscosité f dynamique	динамическая вязкость
D 282	dyne	Dyn n	dyne f	дина

E

	English	German	French	Russian
E 1	**earth, earthing**	Erdung f	mise f à la terre	заземление
E 2	**earth potential**	Erdpotential n	potentiel m de terre	потенциал земли
E 3	**ebonite,** hard rubber	Hartgummi m, Ebonit n	ébonite f, caoutchouc m durci	эбонит, твердая резина
	ebullition, boiling	Sieden n	ébullition f	кипение
E 4	**ebullition**	Aufwallen n, Aufsprudeln n	effervescence f	вскипание, бурное кипение
E 5	**eddy current heating,** hysteresis heating	Wirbelstromheizung f	échauffement m par courants de Foucault	нагрев вихревыми токами
E 6	**edge dislocation**	Stufenversetzung f	déplacement m à étages	краевая дислокация
E 7	**edge joint,** edge weld	Stirnstoß m	choc m frontal	торцевое соединение
E 8	**edge seal,** push-in seal	Schneidenanglasung f, Innen- und Außen-anglasung f	scellement m de verre à pénétration	лезвенный спай
	edge weld	s. edge joint		
E 9	**edge welding**	Kantenschweißung f	soudage m d'arêtes	сварка встык, сварка кромок
E 10	**edgewise pad filter**	Kantenfilter n, Hochkantfilter n	filtre m debout	ребристый фильтр
E 11	**effective atomic number**	effektive Ordnungszahl f	nombre m ordinal effectif	эффективный атомный номер
E 12	**effective compression ratio**	effektives Kompressions-verhältnis n	relation f réelle de compression	эффективная степень сжатия
E 13	**effective dimensions**	Nutzraum m	volume m effectif, volume utile	эффективный объем, эффективное про-странство
E 14	**effective evaporation rate,** net evaporation rate	effektive Verdampfungs-rate f	quote-part f effective d'évaporation	эффективная скорость испарения
E 15	**effective speed,** operational speed, true speed, net speed	wirksames Saugvermögen n	débit m utile, débit effectif	эффективная скорость откачки, быстрота разрежения объекта
E 16	**efficiency**	Wirkungsgrad m, Ausbeute f	débit m, vitesse f d'aspira-tion, rendement m	коэффициент полезного действия
E 17	**effusion**	Effusion f	effusion f	эффузия
E 18	**effusion law**	Effusionsgesetz n, Grahamsches Gesetz n	loi f d'effusion	закон эффузии
E 19	**ejection**	Strahlverdichtung f	concentration f de rayon	эжекция, сжатие струи
E 20	**ejector,** ejector pump, steam-jet pump, vapour-jet pump	Dampfstrahlpumpe f, Ejektorpumpe f, Saugstrahlpumpe f	pompe f à jet de vapeur, éjecteur m [à vapeur], pompe à éjecteur	пароструйный насос, эжектор, эжекторный насос
E 21	**ejector nozzle,** nozzle, jet nozzle	Strahldüse f	tuyère f, buse f, diffuseur m, éjecteur m	эжектор, струйное сопло
	ejector pump	s. ejector		
E 22	**ejector stage**	Ejektorstufe f, Dampfstrahlstufe f	étage m d'éjecteur, étage d'éjection de vapeur	эжекторная ступень
E 23	**elasticity number**	Elastizitätszahl f	nombre m d'élasticité	коэффициент упругости
E 24	**elastic limit**	Elastizitätsgrenze f, Streckgrenze f	limite f de résistance	предел упругости
E 25	**elastic ratio**	Streckgrenzenverhältnis n	proportion f de la limite d'allongement	предела текучести
E 26	**elastomer**	Elastomer n	élastomère m	эластомер
E 27	**elastomeric sealant**	Elastomerdichtung f	joint m élastomérique	эластомерический уплотнитель
	elbow, bend	Rohrbogen m, Rohrkrümmer m	coude m, raccord m coudé (angulaire)	колено трубы
E 28	**elbow trap**	Winkelfalle f	trappe f angulaire	ловушка с коленом, угловая ловушка
E 29	**electrical clean-up**	elektrische Gasaufzehrung f	durcissement m électrique	электрическое погло-щение газа, жестчение
E 30	**electrical double layer**	elektrische Doppelschicht f	double surface f électrique	электрический двойной слой
E 31	**electrical mass filter,** mass filter	Massenfilter n, elektrisches Massenfilter	filtre m [électrique] de masse	электрический фильтр масс
E 32	**electrical resistance strain gauge,** resistance strain gauge	Dehnungsmeßstreifen m	bande f de mesure de dilatation	тензодатчик омического сопротивления
E 33	**electrical resistivity**	spezifischer elektrischer Widerstand m	résistivité f électrique	удельное электрическое сопротивление
E 34	**electric contact thermometer**	Kontaktthermometer n	thermomètre m à contacts électriques	контактный термометр
E 35	**electric discharge gettering**	Getterung f mittels ionisierender Gas-entladung	sorption f par getter de décharge ionisante	электроразрядное газопоглощение
E 36	**electric puncture strength**	elektrische Durchschlags-festigkeit f	résistance f électrique au percement	электрическая пробив-ная прочность
E 37	**electrode advance**	Elektrodenvorschub m	avance f d'électrode	подача электрода
E 38	**electrode bias**	Elektrodenvorspannung f	tension f de repos	смещение на электроде, напряжение смещения на электроде
E 39	**electrode burn-off**	Elektrodenabbrand m	usure f des électrodes	выгорание электрода, износ электрода
E 40	**electrode collar,** feed-through collar, feed-through ring	Durchführungsring m	collier m d'accès, cylindre-support m de traversées	проходное кольцо, проходная втулка

E 41	electrode holder	Elektrodenhalter m	support m d'électrode, porte-électrode m	держатель электрода
E 42	electro-erosion	Elektroerosion f	électro-érosion f	электроэрозия
E 43	electroforming	elektroerosive Bearbeitung f	usinage m électro-érosif	электроэрозионная обработка
E 44	electroluminescent pressure gauge	Elektrolumineszenzmanometer n	manomètre m d'électroluminescence	люминесцентный манометр
E 45	electrolysis	Elektrolyse f	électrolyse f	электролиз
E 46	electrolytic polishing	elektrolytisches Polieren n	polissage m (polissure f) électrolytique	электролитическая полировка
E 47	electrolytic silver	Elektrolytsilber n	argent m électrolytique	электролитическое серебро
E 48	electromagnetically operated	elektromagnetisch betätigt	à commande électromagnétique	электромагнитно управляемый, с электромагнитным приводом
E 49	electrometer tube	Elektrometerröhre f	tube m électrométrique	электрометрическая лампа
E 50	electron avalanche	Elektronenlawine f	avalanche f des électrons	электронная лавина
E 51	electron beam, cathode ray	Katodenstrahl m	rayon m cathodique	электронный луч
E 52	electron beam	Elektronenstrahl m	faisceau m électronique, rayon m électronique	электронный луч
E 53	electron beam annealing	Elektronenstrahlglühen n	incandescence f par bombardement électronique	электроннолучевой отжиг
E 54	electron beam annealing plant	Elektronenstrahlglühanlage f	installation f à incandescence à bombardement électronique	установка для электроннолучевого нагрева (отжига)
E 55	electron beam balancing	Elektronenstrahlabgleich m	équilibrage m de faisceau (rayon) électronique	уравновешивание электронным лучом
E 56	electron beam brazing	Elektronenstrahllöten n	soudage m (brasage m, soudure f) par bombardement électronique	электроннолучевая пайка
E 57	electron beam casting furnace	Elektronenstrahlgießofen m	four m de coulée à bombardement électronique	литейная печь с электроннолучевым нагревом
E 58	electron beam creation	Elektronenstrahlerzeugung f	production f des rayons d'électrons	выработка (создание) с помощью электронных лучей
E 59	electron beam current density	Elektronenstrahlstromdichte f	densité f de courant du faisceau électronique	плотность тока электронного луча
E 60	electron beam cutting	Elektronenstrahlschneiden n	coupure f par rayons d'électrons, coupure par rayon électronique	электроннолучевая резка
E 61	electron beam drilling	Elektronenstrahlbohren n	perçage m par rayons d'électrons	электроннолучевое сверление
E 62	electron beam drilling machine	Elektronenstrahlbohrgerät n	machine f à percer de faisceau électronique	машина для электроннолучевого сверления
E 63	electron beam evaporation	Elektronenstrahlverdampfung f	vaporisation f par faisceau d'électrons, vaporisation par bombardement électronique	электроннолучевое испарение
E 64	electron beam evaporation equipment (plant)	Elektronenstrahlverdampfungseinrichtung f, Elektronenstrahlverdampfungsanlage f	installation f de vaporisation de rayon électronique	установка для электроннолучевого испарения
E 65	electron beam evaporator	Elektronenstrahlverdampfer m	évaporateur m par rayons d'électrons, évaporateur par bombardement électronique	электроннолучевой испаритель
E 66	electron beam furnace, electron bombardment furnace	Elektronenstrahlofen m	four m à bombardement électronique	электроннолучевая печь, печь с нагревом электронной бомбардировкой
E 67	electron beam generator	Elektronenstrahlerzeuger m	générateur m des jets d'électrons	электроннолучевой генератор
E 68	electron beam heating	Elektronenstrahlheizung f	chauffage m par rayons électroniques, chauffage par rayons d'électrons	электроннолучевой нагрев
E 69	electron beam machine	Elektronenstrahlgerät n	appareil m à rayons d'électrons	электроннолучевая машина
E 70	electron beam machining	Elektronenstrahlbearbeitung f	usinage m de faisceau électronique	электроннолучевая обработка
E 71	electron beam melting	Elektronenstrahlschmelzen n	fusion f par bombardement électronique	электроннолучевая плавка
E 72	electron beam multiple-chamber furnace	Elektronenstrahlmehrkammerofen m	four m à bombardement électronique à plusieurs chambres	многокамерная электроннолучевая печь
E 73	electron beam multiple gun melting furnace	Elektronenstrahl-Mehrkanonen-Schmelzofen m	four m de fusion à canons multiples par faisceaux électroniques	электроннолучевая плавильная печь с несколькими пушками
E 74	electron beam polymerization	Elektronenstrahlpolymerisation f	polymérisation f à bombardement électronique	полимеризация электронным лучом
E 75	electron beam recording	Elektronenstrahlschreiben n	enregistrement m par rayon électronique	запись электронным лучом
E 76	electron beam sintering	Elektronenstrahlsintern n	frittage m par bombardement électronique	электроннолучевое синтерирование (спекание)

E 77	**electron beam steel strip evaporation plant**	Elektronenstrahl-Stahlbandbedampfungsanlage f	appareil m pour la métallisation de bande en acier par faisceaux électroniques	установка для электроннолучевого напыления на стальную ленту
E 78	**electron beam technology**	Elektronenstrahltechnologie f	technologie f des rayons d'électrons	электроннолучевая технология
E 79	**electron beam welding**	Elektronenstrahlschweißen n	soudage m par bombardement électronique	электроннолучевая сварка
E 80	**electron-bombarded vapour source**	elektronenstoßgeheizte Verdampferquelle f	source f de vaporisation bombardée par électrons	испаритель, нагреваемый электронной бомбардировкой
E 81	**electron bombardment**	Elektronenbeschuß m	bombardement m électronique (cathodique)	электронная бомбардировка
E 82	**electron bombardment evaporation source**	Elektronenstoßverdampferquelle f	source f de vaporisation à choc d'électrons	источник испаряющегося вещества с электронным нагревом
	electron bombardment furnace	s. electron beam furnace		
E 83	**electron bombardment ion source**	Elektronenstoßionenquelle f	source f d'ions par choc électronique	источник ионов, создаваемых электронной бомбардировкой
E 84	**electron capture,** electron trapping	Elektroneneinfang m, Elektronenanlagerung f	capture f des électrons, attachement m des électrons	электронный захват, захват электрона
E 85	**electron cloud,** cloud of electrons	Elektronenwolke f	nuage m d'électrons	электронное облако
E 86	**electron collision**	Elektronenstoß m	choc m électronique	соударение с электроном
E 87	**electron emission,** electronic emission	Elektronenemission f	émission f électronique	электронная эмиссия
E 88	**electron gas**	Elektronengas n	gaz m électronique	электронный газ
E 89	**electron gun**	Elektronenkanone f, Elektronenstrahler m	canon m électronique (à électrons)	электронная пушка
E 90	**electronic desorption**	Desorption f durch Elektronen	désorption f électronique	десорбция под действием электронов
	electronic emission	s. electron emission		
E 91	**electronic excitation**	Elektronenanregung f	impulsion f électronique, excitation f électronique	электронное возбуждение
E 92	**electronic valve,** thermionic valve	Elektronenröhre f	tube m électronique	электронная лампа, радиолампа
E 93	**electron impact**	Elektronenstoß m	choc m par électrons	электронный удар
E 94	**electron impact heating**	Elektronenstoßheizung f	chauffage m par choc d'électrons	нагрев электронной бомбардировкой
E 95	**electron impact ionization**	Elektronenstoßionisation f	ionisation f des électrons par choc	ионизация электронным ударом
E 96	**electron lens**	Elektronenlinse f	lentille f électronique	электронная линза
E 97	**electron micrograph**	elektronenoptische Abbildung f	image f d'optique électronique	электроннооптическое изображение
E 98	**electron mirror microscope**	Elektronenspiegelmikroskop n	microscope m électronique à miroir	зеркальный электронный микроскоп
E 99	**electron mobility**	Elektronenbeweglichkeit f	mobilité f des électrons	подвижность электронов
E 100	**electron optics**	Elektronenoptik f	optique f électronique	электронная оптика
E 101	**electron orbit**	Elektronenbahn f	orbite f d'électrons	электронная орбита
E 102	**electron path length**	Elektronenbahnlänge f	longueur f du trajet des électrons, chemin m parcouru des électrons	длина пробега электрона
E 103	**electron pressure**	Elektronendruck m	pression f des électrons	электронное давление
E 104	**electron probe**	Elektronen[strahl]sonde f	sonde f électronique	электронный зонд, электроннолучевой зонд
E 105	**electron recombination**	Elektronenrekombination f	recombinaison f électronique	электронная рекомбинация
E 106	**electron repeller**	Gegenfeldelektrode f für Elektronen	électrode f inverse pour électrons	отражатель электронов
E 107	**electron sheath**	Elektronenhülle f	orbite (enveloppe) f électronique, cortège m électronique planétaire	электронная оболочка у поверхности электрода
E 108	**electron shell**	Elektronenschale f	couche f électronique	электронная оболочка ‹атома или иона›
E 109	**electron strip beam**	Elektronenflachstrahl m	jet m plat d'électrons	ленточный электронный поток
	electron trapping	s. electron capture		
E 110	**electron tube manufacture**	Elektronenröhrenherstellung f	fabrication f des tubes électroniques	производство электронных ламп
E 111	**electron tunnelling**	Elektronendurchtunnelung f	effet m tunnel des électrons	туннельный эффект для электронов
E 112	**electropercussive welding,** percussion welding	Stoßschweißung f	soudure f à électro-percussion	ударная сварка
E 113	**electroplating**	Elektroplattieren n, elektrolytisches Metallisieren n, Galvanostegie f	placage m électrolytique	гальваностегия, гальванопокрытие
E 114	**electropneumatically operated**	elektropneumatisch betätigt	à commande électropneumatique	электропневматически управляемый, с электропневматическим приводом

E 115	**electropolish**	elektrolytisch polieren	polir électrolytiquement, vernir électrolyti- quement	электролитически полировать
E 116	**electropolishing**	Elektropolieren n	polissage m électriquement	электрополировка
E 117	**electrostatic shield**	elektrostatische Abschirmung f	protection f électrostatique	электростатический экран
E 118	**electrovalency**	Elektrovalenz f	électrovalence f	электрическая валентность
E 119	**eliquation,** liquation, segregation	Seigerung f	liquation f, ségrégation f	ликвация, сегрегация, зейгерование
E 120	**emanation**	Emanation f, Ausfluß m, Ausströmung f	émanation f, écoulement m, échappement m	эманация, выделение, истечение
E 121	**embed**	einbetten	insérer, noyer	вводить, заделывать
E 122	**embedding**	Einbetten n	action f de noyer	заделка <во что-либо>
E 123	**embrittlement**	Versprödung f	tendance f à être cassant	охрупчивание
E 124	**emission current density**	Emissionsstromdichte f	densité f de courant d'émission	плотность эмиссионного тока
E 125	**emission efficiency**	Emissionsmaß n <Katode>	efficacité f d'émission	эффективность эмиссии
E 126	**emission surface,** emitting surface	Emissionsfläche f	face f d'émission	эмиттирующая поверхность
E 127	**emission yield**	Emissionsausbeute f	rendement m d'émission	эмиссионный выход
E 128	**emissive power, emissivity**	Emissionsvermögen n	capacité f d'émission	эмиссионная способность
	emitting surface	s. emission surface		
E 129	**emulsify**	emulgieren	émulsionner	эмульгировать, образо- вывать эмульсию
E 130	**encapsulate**	[ein]kapseln	capsuler	заключать в капсулу
E 131	**encapsulation**	Einkapselung f, Ver- kapselung f, Umhüllung f, Hülle f	cuirasse f, blindage m	капсуляция, заключение в капсулу
E 132	**end block**	Endmaß n	mesure f finale	контрольная плитка
E 133	**end box,** terminal box	Kabelendverschluß m	boîte f d'extrémité du câble	присоединительная коробка, кабельный бокс
E 134	**end grid**	Endgitter n	grille f finale, grid m final	граничная сетка
E 135	**endothermic**	endotherm	endothermique	эндотермический
E 136	**end tube**	Endröhre f	tube m final	выходная лампа
E 137	**endurance test,** fatigue test	Dauerversuch m, Dauerprüfung f	essai m permanent	испытание на усталость, длительное испытание
E 138	**energy gap**	Energiesprung m	discontinuité f d'énergie	энергетическая зона, разрыв в энергети- ческой кривой
E 139	**energy-level diagram**	Termschema n, Energieniveaudia- gramm n	schéma m de termes	диаграмма энергети- ческих уровней
E 140	**energy spread**	Energiestreuung f	fuite f d'énergie	разброс энергий
E 141	**enthalpy,** heat content, total heat	Enthalpie f	enthalpie f	энтальпия, теплосо- держание
E 142	**enthalpy-controlled flow**	reibungsfreie Strömung f <bei hohen Drücken>	courant m sans frottement (friction)	течение, определяемое энтальпией, течение без трения <при высо- ком давлении>
E 143	**entropy**	Entropie f	entropie f	энтропия
E 144	**entry lock**	Eingangsschleuse f	écluse f d'entrée	входной шлюз
E 145	**envelope replica,** sheath replica	Hüllabdruck m	impression f envelop- pante, reproduction f enveloppante, épreuve f enveloppante	отпечаток оболочки
	envelope test, chamber test, hood pressure test, overall test	Hüllentest m, Hauben- lecksuchverfahren n, Haubenleckprüfung f	test m d'enveloppe, épreu- ve f d'étanchéité	проверка помещением объекта в оболочку
	environmental pressure, ambient pressure	Umgebungsdruck m	pression f ambiante	давление окружающей среды
E 146	**epitaxy**	Epitaxie f	epitaxie f	эпитаксия, структурное соответствие
E 147	**epoxy resin**	Epoxydharz n	résine f époxydique	эпоксидная смола
E 148	**equalizing valve**	Ausgleichsventil n	robinet m compensateur	выравнивающий вентиль
E 149	**equation for ideal gases,** general gas law	universelle Gasgleichung f	équation f universelle des gaz, équation des gaz idéaux	уравнение состояния идеального газа, уни- версальное газовое уравнение
E 150	**eraser test**	Abriebtest m	test m de déchets	испытание на истирание
E 151	**erg**	Erg n	erg m	эрг
E 152	**etched copper mask**	geätzte Kupfermaske f	masque m de cuivre corrodé	травленая медная маска
E 153	**etching by ion bombardment,** ion etching	Ionenätzen n	corrosion f ionique	ионное травление
E 154	**etching ink**	Glastinte f	encre f à verre	чернила для стекла
E 155	**etching powder**	Mattierpulver n	poudre f à dépolir	порошок для матирова- ния
E 156	**etching salt,** frosting salt	Mattiersalz n	sel m à ternir	соль для матирования

E 157	eutectic point	eutektischer Punkt *m*	point *m* eutectique	эвтектическая точка
E 158	evacuate, exhaust, pump out, pump down	evakuieren, auspumpen, abpumpen	évacuer, vider, faire le vide	откачивать, разрежать
	evacuated doughnut chamber, doughnut	kreisförmige Hoch-vakuumröhre *f* des Betatrons	tube *m* circulaire à vide poussé	тороидальная вакуумная камера [бетатрона]
E 159	evacuated fluidized layer sublimation	Vakuumfließbett-sublimation *f*	sublimation *f* de lit sous vide	вакуумная сублимация в псевдоожиженном слое, вакуумная субли-мация в движущемся слое
E 160	evacuating, exhausting	Auspuffprozeß *m*, Evakuierungsprozeß *m*	procédé *m* de pompage	откачка, удаление газа, вакуумирование
E 161	evacuation port, vacuum port, vacuum tubulation	Vakuumstutzen *m*	raccord *m* pour le vide, tubulaire *f* de vide, tu-bulaire d'évacuation	впускной патрубок
E 162	evaporant	Verdampfungsgut *n*	matière *f* d'évaporation	испаряемое вещество
E 163	evaporant ion source	Ionenstrom *m*, bedingt durch den Dampf-druck des Verdamp-fungsgutes	flux *m* ionique, courant *m* d'ions	испарительный ионный источник
E 164	evaporate, vaporize	verdampfen	s'évaporer, se vaporiser	испарять
E 165	evaporation, vaporization	Verdampfung *f*, Ver-dampfungsprozeß *m*, Verdunstung *f*	évaporation *f*, vapori-sation *f*	испарение
E 166	evaporation analysis	Verdampfungsanalyse *f*	analyse *f* de l'évaporation	испарительный анализ
E 167	evaporation cathode	Aufdampfkatode *f*	cathode *f* de métallisation sous vide	напыленный катод
E 168	evaporation chamber	Verdampfungskammer *f*	chambre *f* d'évaporation	испарительная камера
E 169	evaporation characteristic	Verdampfungs-charakteristik *f*	caractéristique *f* d'évaporation	характеристика испарения
E 170	evaporation coating, vacuum coating, evaporative coating	Aufdampfen *n*, Vakuum-bedampfung *f*	métallisation *f* sous vide	вакуумное напыление, нанесение в вакууме
E 171	evaporation coefficient ⟨ratio of actual to maximum rate of evaporation⟩	Verdampfungskoeffizient *m*, Transmissionsfaktor *m* ⟨Verhältnis von tat-sächlicher zu maxima-ler Verdampfungsrate⟩	facteur *m* de transmission, coefficient *m* d'évapora-tion ⟨taux du rapport effectif au rapport maximum de l'éva-poration⟩	коэффициент испарения ⟨отношение истинной скорости испарения к максимальной⟩
E 172	evaporation control	Verdampfungssteuerung *f*	réglage *m* par vaporisation	регулировка (контроль) испарения
E 173	evaporation controller	Verdampfungsregler *m*	régulateur *m* de vapori-sation	регулятор испарения
E 174	evaporation cooling	Verdampfungskühlung *f*	refroidissement *m* par vaporisation	охлаждение испарением
E 175	evaporation drying	Verdampfungs-trocknung *f*	séchage *m* par vaporisa-tion, séchage par ébullition	сушка испарением
E 176	evaporation heat	Verdampfungswärme *f*	chaleur *f* d'évaporation	теплота испарения
E 177	evaporation-ion pump, evaporator-ion pump	Ionenverdampferpumpe *f*	pompe *f* ionique par éva-poration (vaporisation)	испарительный ионный насос
E 178	evaporation loss, evaporative loss	Verdampfungsverlust *m*	perte *f* d'évaporation, perte par évaporation	потери на испарение
E 179	evaporation mask	Verdampfungsmaske *f*	masque *m* du vapori-sation	маска для испарения
E 180	evaporation rate, vaporization rate	Verdampfungsge-schwindigkeit *f*, Verdampfungsrate *f*	vitesse (rapidité) *f* de vaporisation, taux *m* d'évaporation	скорость испарения (парообразования)
E 181	evaporation rate control system	Verdampfungsraten-regelsystem *n*	système *m* de réglage du taux d'évaporation	система управления скоростью испарения
E 182	evaporation rate meter	Verdampfungsraten-meßgerät *n*	instrument *m* de mesure du taux d'évaporation	измеритель скорости испарения
E 183	evaporation source shutter	Blende *f* über der Ver-dampferquelle	écran *m* de la source d'évaporation	экран (заслонка) для источника испарения
E 184	evaporation source turret	Verdampferkarussell *n*	manège *m* tournant d'évaporation	карусель с испари-телями
E 185	evaporation synthesis	Verdampfungssynthese *f*	synthèse *f* de vapori-sation	испарительный синтез
E 186	evaporation technique	Verdampfungstechnik *f*	technique *f* d'évaporation	технология испарения
E 187	evaporation temper-ature, vaporization temperature	Verdampfungs-temperatur *f*	température *f* d'évapora-tion	температура испарения
	evaporative coating, vacuum coating, evaporation coating	Aufdampfen *n*, Vakuum-bedampfung *f*	métallisation *f* sous vide	вакуумное напыление, нанесение в вакууме
E 188	evaporative cooling	Verdunstungskühlung *f*	refroidissement *m* par évaporation	испарительное охлаждение
E 189	evaporative drying	Verdunstungstrocknung *f*	séchage *m* par évaporation	испарительная сушка
	evaporative loss, evaporation loss	Verdampfungsverlust *m*	perte *f* à l'évaporation, perte par évaporation	потери на испарение
E 190	evaporator, vaporizer	Verdampfer *m*	évaporateur *m*, vapori-sateur *m*	испаритель, выпарной аппарат
E 191	evaporator boat	Verdampferschiffchen *n*	nacelle *f* (creuset *m*) d'éva-poration	испарительная лодочка
	evaporator-ion pump	*s.* evaporation-ion pump		

E 192	**evaporator pump**	Verdampferpumpe *f*	pompe *f* à évaporation	испарительный насос
E 193	**evaporator source**	Verdampferquelle *f*	source *f* d'évaporation	источник испаряющего-ся вещества
E 194	**excitation**	Anregung *f*	excitation *f*, impulsion *f*	возбуждение
E 195	**excitation frequency**	Anregungsfrequenz *f*	fréquence *f* à l'excitation	частота возбуждения
E 196	**excitation voltage**	Anregungsspannung *f*	voltage *m* d'excitation	потенциал возбуждения
	exhaust	*s.* evacuate		выпускной патрубок
	exhaust, discharge port, exhaust port, outlet	Auspuff *m*, Auspuff-öffnung *f*	échappement *m*, orifice *m* de refoulement, orifice d'échappement, orifice d'expulsion	выхлопное окно, выпускной патрубок
E 197	**exhauster**	Absaugeanlage *f*, Exhaustor *m*	ventilateur *m* aspirant (d'extraction)	эксгаустер
	exhaust filter, discharge [mist] filter	Auspufffilter *n* <Ölfilter auf der Auspuffseite einer mechanischen Vakuumpumpe>	filtre *m* d'échappement	выхлопной фильтр <на выходе механического насоса>
	exhaust fitting, fore-vacuum connection, fore-arm connection, backing-line connection, outlet flange, fore-pump connection	Vorvakuumanschluß *m*	raccord *m* de vide préli-minaire	соединение [насоса] с форвакуумом
E 198	**exhaust gas turbine**	Abgasturbine *f*	turbine *f* d'échappement	турбина, работающая на отработавших газах
	exhausting	*s.* evacuating		
E 199	**exhausting power,** suction capacity, throughput	Saugleistung *f*	puissance *f* d'une pompe, puissance d'admission, débit *m* massique d'une pompe, rendement *m* d'aspiration, volume *m* aspiré	производительность по всасыванию
	exhaust line, discharge pipe	Abgasleitung *f*, Aus-puffleitung *f*	conduite *f* d'échappement	трубопровод для уходящих газов
	exhaust pipe, outlet pipe, discharge pipe	Auspuffleitung *f* <rotierende Pumpe>	tube *m* d'échappement, conduite *f* d'échappe-ment	выхлопная труба, вы-пускной патрубок, выпускная труба
	exhaust port, discharge port, exhaust, outlet	Auspuff *m*, Auspuff-öffnung *f*	échappement *m*, orifice *m* de refoulement, orifice d'échappement, orifice d'expulsion	выхлопное окно, выпус-кной патрубок
	exhaust pressure, outlet pressure, discharge pressure, head pressure	Verdichtungsdruck *m*, Auspuffdruck *m*, Ausstoßdruck *m*	pression *f* de sortie, pres-sion de refoulement, pression du vide préli-minaire, pression d'échappement, pression finale	давление сжатия, вы-пускное давление, давление на выхлопе
E 200	**exhaust steam**	Abdampf *m*	vapeur *f* d'échappement, vapeur épuisée	мятый пар, отработав-ший пар, выхлопной пар
E 201	**exhaust stem,** exhaust tubulation, pumping stem, pump-out tubulation	Pumpstengel *m*, Vakuumstutzen *m*	queusot *m*, tubulure *f* de pompage	откачной патрубок, штенгель
E 202	**exhaust system**	Absaugvorrichtung *f*, Pumpautomat *m*	installation *f* d'aspiration, banc *m* de pompage	откачная система
	exhaust tubulation	*s.* exhaust stem		
	exhaust valve, discharge valve	Auspuffventil *n*	soupape *f* de refoulement, soupape d'échappement	выпускной клапан
E 203	**exit aperture**	Austrittsblende *f*	diaphragme *m* de sortie	выпускная диафрагма, выходная диафрагма
E 204	**exoelectron emission**	Exoelektronenemission *f*	émission *f* exoélec-tronique	экзоэлектронная эмиссия
E 205	**exothermic**	exotherm	exothermique	экзотермический
E 206	**expand**	expandieren	détendre	расширить, растягивать
E 207	**expanding nozzle**	Expansionsdüse *f*	buse *f* d'expansion	расширяющееся сопло
E 208	**expansion chamber**	Expansionsraum *m*	chambre *f* de détente	камера расширения
	expansion engine, decompressor	Expansionsmaschine *f*	machine *f* à expansion	детандер, расширитель-ная машина
E 209	**expansion joint**	Dehnungsfuge *f*	joint *m* de dilatation	компенсационное соединение
E 210	**expansion rate**	Expansionsverhältnis *n*	proportion *f* de dilatation, proportion d'expansion	коэффициент расшире-ния
E 211	**expansion refrigeration**	Expansionskühlung *f*	refroidissement *m* par expansion	охлаждение расшире-нием
E 212	**expansion slide valve**	Expansionsschieber *m*	coulisse *f* à expansion	расширительный золот-ник
E 213	**expansion valve**	Expansionsventil *n*	clapet *m* de dilatation, clapet d'expansion	расширительный клапан, отсекающий клапан
E 214	**explosive evaporation,** explosive volatilization	Explosivverdampfung *f*	vaporisation *f* explosive	взрывное испарение
E 215	**explosive mixture**	Explosionsgemisch *n*	mélange *m* explosif	взрывчатая смесь, гремучая смесь
E 216	**explosive volatilization**	*s.* explosive evaporation		
	exposure time	Bestrahlungszeit *f*	temps *m* d'exposition à la lumière, temps d'irra-diation	время облучения
E 217	**exterior coating**	Schutzüberzug *m*	enveloppe *f* protectrice	защитное покрытие

E 218	external condenser	Außenkondensator *m*	condensateur *m* extérieur	наружный конденсатор
E 219	external diameter, outer diameter, OD, outside diameter	Außendurchmesser *m*	diamètre *m* extérieur	наружный диаметр
E 220	external field, separate field	Fremdfeld *n*	champ *m* étrangeur (perturbateur)	постороннее поле, внешнее поле
E 221	external thread, male thread	Außengewinde *n*	filet *m*, filetage *m*	наружная резьба
E 222	extract	extrahieren	extraire	экстрагировать, извлекать
E 223	extractant	Extraktionsmittel *n*	extracteur *m*	экстрагирующий агент
E 224	extractor gauge	Ionisationsvakuum-meter *n* mit Extraktor	manomètre *m* à ionisation avec extractor	ионизационный мано-метр с экстрактором
E 225	extrapolate	extrapolieren	extrapoler	экстраполировать
E 226	extrude	strangpressen	filer, former un boudin	выдавливать
E 227	extruder	Extruder *m*	extrudeuse *f*, presse *f* à ex-trusion	экструдер
E 228	extrusion	Strangpressen *n*	extrusion *f*	выдавливание, непре-рывное прессование
E 229	extrusion billet	Strangpreßrohling *m*	ébauche *f* pour presse à filet	заготовка для выдавли-вания
E 230	extrusion melting, ingot melting	Strangschmelzen *n*	fusion *f* par extrusion	плавка с вытягиванием слитка

F

F 1	fabrication metallurgy	Schmelz-, Legierungs- und Gießverfahren *npl*, bildsame Formgebung *f* durch Schmieden, Walzen, Pressen und Ziehen	procédé *m* de fusion, d'alliage et de coulage, formation *f* par for-geage, laminage, pres-sion et tirage	металлургические спосо-бы производства
F 2	face-centered	flächenzentriert	surface centrée	поверхностно центрированный
F 3	face-cubic-centered	kubisch-flächenzentriert	cubique-plaine centrée	кубически гранецентри-рованный
F 4	face seal, slide ring seal, slip-ring-seal	Gleitringdichtung *f*, Schleifringdichtung *f*	joint *m* à anneaux glis-sants, garniture *f* étanche à anneau glissant	скользящее кольцевое уплотнение, уплотне-ние со скользящим кольцом
F 5	Fahrenheit scale	Fahrenheit-Skale *f*	échelle *f* Fahrenheit	шкала Фаренгейта
F 6	Fahrenheit temperature scale	Temperaturskale *f* nach Fahrenheit	Fahrenheit-graduation *f* de température	температурная шкала Фаренгейта
F 7	fail-safe, reliable	betriebssicher	de fonctionnement sûr, sûr, éprouvé	надежный, безопасный
F 8	falling-film evapora-tor, thin-film evap-orator	Dünnschichtverdampfer *m*	vaporisateur *m* à couche fine	испаритель с падающей пленкой
F 9	falling film molecular still	Molekulardestillations-anlage *f* mit fallendem Film	distillerie *f* moléculaire à film tombant	молекулярный дистил-лятор с падающей пленкой
F 10	fan	Ventilator *m*	ventilateur *m*	вентилятор
F 11	Faraday cage	Faradayscher Käfig *m*	cage *f* de Faraday	клетка Фарадея
F 12	Faraday dark space	Faradayscher Dunkel-raum *m*	espace *m* noir de Faraday	фарадеева темная область, фарадеево темное пространство
F 13	fast-acting valve, quick-acting valve, rapid-action valve	Schnellschlußventil *n*	soupape *f* à fermeture instantanée, robinet *m* à fermeture rapide	быстродействующий вентиль (клапан)
F 14	fatigue	Ermüdung *f*	fatigue *f*	усталость
F 15	fatigue crack	Ermüdungsbruch *m*, Dauerbruch *m*	rupture *f* de fatigue	усталостное растрески-вание
F 16	fatigue limit, fatigue strength	Dauerfestigkeit *f*, Dauer-schwingfestigkeit *f*	résistance *f* continue (à la fatigue)	предел усталости (выносливости)
F 17	fatigue ratio	Dauerfestigkeits-verhältnis *n*	proportion *f* de résistance continue (à la fatigue)	коэффициент предела усталости (выносли-вости)
	fatigue strength	*s.* fatigue limit		
	fatigue test, endurance test	Dauerversuch *m*, Dauerprüfung *f*	essai *m* permanent	испытание на усталость, длительное испытание
F 18	faulty operation	Fehlbedienung *f*	opération *f* erronée	ошибочное управление
F 19	feather key	Federkeil *m*	coin *m* élastique	направляющая (призма-тическая) шпонка
F 20	feather valve	Federventil *n*	soupape *f* à ressort	подпружиненный вен-тиль
F 21	feed control	Vorschubregelung *f*	réglage *m* d'avance	регулирование подачи (загрузки)
	feeding, loading, charg-ing	Beschickung *f*	chargement *m*	питание, загрузка
	feeding device, charging device	Beschickungsvorrichtung *f*	appareil *m* d'alimentation, appareil de chargement, appareil à charger, dispositif *m* d'alimen-tation, chargeur *m* mécanique	загрузочное устройство, питающее устройство
F 22	feed of material	Materialzuführung *f*, Materialzugabe *f*	amenée *f* de matériaux	подача материала

F 23	feed pump	Speisepumpe f	pompe f alimentaire (d'alimentation)	питательный насос
F 24	feed pump	Förderpumpe f	pompe f d'alimentation	подающий насос
	feed-through collar (ring), electrode collar	Durchführungsring m	collier m d'accès, cylindre-support m de traversées	проходное кольцо, проходная втулка
F 25	feed tube lift pipe, lift tube, siphon	Heber m	siphon m	сифон
F 26	female thread, internal thread	Innengewinde n	taraudage m [de vis intérieur]	внутренняя резьба
F 27	Fermi edge	Fermi-Kante f	bord m de Fermi	граница Ферми
F 28	Fermi level	Fermi-Niveau n	niveau m de Fermi	уровень Ферми
F 29	ferrite-switching core	Ferritschaltkern m	commutateur m par noyau de ferrite	ферритовый сердечник с прямоугольной петлей
F 30	festoon dryer, loop dryer	Schleifentrockner m, Girlandentrockner m, Hängetrockner m	sécheur m suspendu (à boucles)	фестонная сушилка, подвесная сушилка
F 31	fibre glass cloth	Glasfasertuch n	trap m (toile f) de fibre de verre	ткань из стекловолокна
F 32	fibre texture	Fasertextur f	texture f fibreuse	волокнистая текстура
F 33	field-emission adsorption spectrometry	Feldemissions-Adsorptionsspektrometrie f	spectrométrie f par adsorption d'emission de champ	адсорбционная спектрометрия, основанная на электростатической эмиссии
F 34	field-emission electron microscope	Feldelektronenmikroskop n	microscope m à émission électronique	электростатический электронный микроскоп
F 35	field-emission ion microscope, field-ion microscope	Feldionenmikroskop n	microscope m à émission ionique de champ	электронно-ионный микроскоп
F 36	field-emission microscope	Feldemissionsmikroskop n	microscope m à émission de champ	злекуронно-автоэмиссионный микроскоп
F 37	field evaporation	Feldverdampfung f	évaporation f de champ	испарение в поле
F 38	field-ionization	Feldionisation f, Feldionisierung f, Ionisierung f im starken elektrischen Feld	émission f ionique de champ	ионизация под действием электрического поля высокой напряженности
F 39	field-ionization [vacuum] gauge	Feldionisationsvakuummeter n	vacuomètre m à émission ionique de champ	автоионизационный вакуумметр, ионизационный манометр с автоэлектронным катодом
	field-ion microscope, field-emission ion microscope	Feldionenmikroskop n	microscope m à émission ionique de champ	злектронно-ионный микоскоп
	figure of merit, coefficient of performance, quality factor	Leistungsziffer f, Gütegrad m, Gütezahl f	coefficient (facteur) m de puissance	коэффициент мощности
F 40	filament, hot filament, hot wire	Heizdraht m, Glühdraht m, Glühfaden m	filament m, filet m, fil m chaud (thermique)	нить накала, провод накала
F 41	filament burn-out	Durchbrennen n des Glühfadens	calcination f du filament	перегорание нити накала
F 42	filament life	Glühfadenlebensdauer f	vie f de filament, durée f de vie de filament, durée de service de filament	долговечность нити накала
F 43	filament temperature	Glühfadentemperatur f	température f de filament	температура нити накала
F 44	filler	Füllmittel n	masse (matière) f de remplissage	наполнитель
F 45	filler metal	Zusatzwerkstoff m, Elektrodenmetall n	matière f à faire le plein	присадочный металл
F 46	filler rod	Schweißstab m	barre (baguette) f de soudure	сварочный (присадочный) пруток
F 47	fillet weld	Kehlnaht f	joint m d'angle	угловой сварной шов
F 48	filling pressure	Fülldruck m	pression f de remplissage (charge)	давление наполняющего газа
F 49	filling unit	Abfüllvorrichtung f	appareil m à vider (soutirer)	устройство для наполнения, дозирующее устройство
F 50	film-coating	Befilmen n	traitement m par film	покрытие пленкой
F 51	film coefficient of heat transfer, surface coefficient of heat transfer, surface film conductance	Wärmeübergangszahl f	coefficient m d'écoulement thermique, nombre m de la transmission thermique, coefficient de transmission thermique	коэффициент теплоотдачи (теплопередачи)
F 52	film composition	Schichtzusammensetzung f	composition f de couche	состав пленки
F 53	film continuity	Schichtkontinuität f	continuité f de couches	целостность (непрерывность) пленки
F 54	film lattice	Schichtgitter n	grille f à couche	слоистая решетка
F 55	film purity	Schichtreinheit f	pureté f de couche	чистота слоя
F 56	film thickness measuring unit, film-thickness meter, film-thickness monitor	Schichtdickenmeßgerät n	appareil m de mesure de l'épaisseur de couche, instrument m de mesure à épaisseur de couche, instrument de mesure d'épaisseur de couche	измеритель толщины пленки
F 57	filter	filtern	filtrer	фильтровать

F 58	**filter by means of vacuum**	abnutschen	filtrer par aspiration	фильтровать с помощью вакуума
F 59	**filtering**	Aussiebung f	filtrage m	фильтрование
F 60	**filter out**	ausfiltern	éliminer par filtration	отфильтровывать
F 61	**filtration**	Filtration f	filtration f	фильтрование, фильтрация
	final drying, afterdrying, secondary drying, subsequent drying	Nachtrocknung f, Endtrocknung f	dessiccation f secondaire (finale), séchage m final (secondaire)	досушивание, досушка, вторичная осушка, конечная (окончательная) сушка
	final pressure, blank-off (limiting, ultimate) pressure, ultimate vacuum	Enddruck m	pression f finale	предельный вакуум, предельное (остаточное) давление
F 62	**fine etching**	Feinbeizen n	décapage m final	чистовое травление
F 63	**fine pressure,** head (inlet, intake, suction) pressure	Ansaugdruck m	pression f d'aspiration, pression d'admission	впускное давление
F 64	**fine vacuum,** medium vacuum	Feinvakuum n <$1 \cdots 10^{-3}$ Torr>	vide m moyen (intermédiaire)	средний вакуум
F 65	**finish,** surface finish, surface quality	Oberflächenbeschaffenheit f, Aussehen n, Oberflächengüte f	fini m de la surface, qualité f de surface, constitution f de la surface	чистота (качество, состояние) поверхности, отделка
F 66	**finished size**	Fertigmaß n	mesure f finie	окончательный размер, размер после доводки
F 67	**first melt**	Erstschmelze f	fonte f primaire	первая (первичная) плавка
	fissure	$s.$ flaw		
	fitting, connector, coupling	Verbindungsstück n	raccord m, pièce f de raccordement	соединительная часть
F 68	**fixed joint**	feste (unlösbare) Verbindung f	jonction f (raccordement m) fixe	неразъемное соединение
F 69	**fixed point**	Fixpunkt m	point m fixe	реперная точка температурной шкалы
	flake-off, chip-off, scale-off	abblättern <von Schichten>	effeuiller	отслаиваться, отпадать
F 70	**flame-spraying**	Flammspritzen n	injection f à flammes	газопламенное напыление, газопламенная металлизация
F 71	**flame welding**	Flammenschweißen n	soudage m par flammes	газовая сварка
	flange, bead	bördeln	border	отбортовывать
F 72	**flange**	Flansch m	bride f, collet m, rebord m	фланец
F 73	**flanged joint,** flange joint, flange termination	Flanschverbindung f, Flanschanschluß m	joint m de bride, joint à brides	фланцевое соединение, соединение с отбортовкой кромок
F 74	**flanged pipe**	Bördelrohr n	tuyau m rabattu	труба с фланцем (бортиком)
F 75	**flange gasket,** flange seal	Flanschdichtung f	garniture f de joint, joint m de bride	фланцевое уплотнение
F 76	**flange joint**	$s.$ flanged joint		
	flange nick	Flanschkerbe f	encoche f à brides	шейка фланца
	flange seal, flange gasket	Flanschdichtung f	garniture f de joint, joint m de bride	фланцевое уплотнение
	flange termination	$s.$ flange joint		
F 77	**flap valve**	Klappventil n	vanne f à charnière, vanne (robinet m) à clapet articulé	шарнирный (откидной, створчатый, клапанный) вентиль
F 78	**flared joint**	Bördelverbindung f	joint m rabattu	соединение с отбортовкой
F 79	**flaring machine**	Tellerdrehmaschine f	machine f tournante à disque	станок с вращающимся диском (кругом)
F 80	**flash-butt welding**	Abbrennschweißen n, elektrisches Stumpfschweißen n	soudure f par rapprochement, soudure électrique bout à bout	контактная стыковая сварка
F 81	**flash cooler**	Vakuumkühler m	condenseur m à vide	вакуумный охладитель
F 82	**flash dryer**	Rohrtrockner m, Schnelltrockner m	sécheur m par tuyau (action rapide)	установка для сушки распылением (методом сброса давления)
F 83	**flash drying plant**	Stromtrocknungsanlage f	installation f de séchage par courant	установка для сушки мгновенным распылением
F 84	**flash evaporation**	Stoßverdampfung f	évaporation f de poussée	испарение вспышкой, ударное испарение
F 85	**flash-filament method**	Druckimpulsvakuummessung f, Flashfilament-Methode f	méthode f du filament chauffé	метод вспышки
F 86	**flash-filament technique**	Flash-filament-Technik f, Impulsdesorption f	technique f du filament chauffé	импульсная десорбция
F 87	**flash heat**	Verdampfungsglühen n, Hochglühen n zum Verdampfen	évaporation f par recuit, calcination f haute	испарительный нагрев
F 88	**flashing**	blitzartiges Hochheizen n	désorption f thermique pulsée	мгновенное испарение
F 89	**flashover**	Funkenüberschlag m	éclatement m, jaillissement m	пробой
F 90	**flashover strength**	Überschlagsfestigkeit f	résistance f de conturnement (jaillissement)	поверхностная электрическая прочность

F 91	**flash point**	Flammpunkt m	point m d'inflammation	точка (температура) воспламенения
F 92	**flash welding**	Abschmelzschweißung f	soudage m par fonte	стыковая сварка оплавлением
F 93	**flat-butt weld,** square-butt weld	zweiseitige Vollnaht f	pleine soudure f bilatérale, plein joint m bilatéral	стыковой сварной шов без скоса кромок
F 94	**flat film,** flat sheet	Flachfolie f	lamelle f plate	плоская пленка, плоская тонкая пластина
F 95	**flat gasket**	Dichtungsring m aus Plattenmaterial, Flachdichtung f	joint m plat	плоская прокладка, плоское уплотнительное кольцо
F 96	**flat-press seal**	Quetschfußeinschmelzung f	réfondage m à tige (socle pressant)	запайка плоской ножки
	flat sheet, flat film	Flachfolie f	lamelle f plate	плоская пленка, плоская тонкая пластина
F 97	**flat stem**	Flachquetschfuß m	plate tige f pressante	плоская ножка
F 98	**flatten**	flachdrücken, plattdrücken	aplatir	плющить, выравнивать
	flaw, crack, fissure	Riß m, Sprung m	fissure f, rupture f, crevasse f, fente f, fêlure f	щель, трещина
F 99	**flexible joint**	bewegliche Kupplung f	connexion f flexible	подвижное (гибкое) соединение
F 100	**flexible metallic tube,** metal hose	Metallschlauch m	tuyau m métallique flexible, tuyau flexible en métal	металлический шланг
F 101	**flex seal**	Lötung f mit flexiblem Ausgleichsteil	soudure f avec compensation flexible, brasure f avec compensation flexible	пайка с гибкими переходными частями
F 102	**flexural rigidity**	Biegesteifigkeit f	résistance f à la raideur	жесткость при изгибе
F 103	**flexural strength,** maximum surface stress in bend, transverse strength, modulus of rupture	Biegefestigkeit f	résistance f à la flexion	предел прочности при изгибе, прочность при изгибе
F 104	**flicker noise**	Funkelrauschen n	bruit m de scintillation	шум фликкер-эффекта
F 105	**flighted dryer**	Rieseltrocknung f	dessication f par ruissellement	оросительная сушилка
F 106	**flip-flop circuit,** multi-vibrator	Flip-Flop-Schaltung f	montage (circuit) m flip-flop, multivibrateur m	мультивибратор
F 107	**floating potential**	Schwebepotential n	potentiel m oscillant	плавающий потенциал
F 108	**floating-zone melting**	Schwebezonenschmelzverfahren n, vertikales Zonenschmelzen n	fusion f de zone verticale, fusion à zone flottante	вертикальный способ зонной плавки
F 109	**float switch**	Schwimmerschalter m	interrupteur m flotteur	поплавковый выключатель
F 110	**float valve**	Schwimmerventil n	soupape f flotteur, robinet-flotteur m	поплавковый клапан (вентиль)
F 111	**float zone,** fusion zone	Schmelzzone f	zone f de fusion	зона плавления
F 112	**flocculation**	Flockung f, Flockenbildung f	floculation f	флоккуляция, выпадение хлопьями
F 113	**flock point**	Flockpunkt m	point m de floculation	температура помутнения
F 114	**flooded system**	geflutetes System n	système m noyé	заполненная система
F 115	**flooding**	Fluten n	fluctuation f	затопление
F 116	**flood lubrication,** pressure lubrication	Druckumlaufschmierung f	graissage m central sous pression, graissage par circulation sous pression	централизованная (циркуляционная) смазка под давлением
F 117	**flotation**	Flotation f, Schwimmaufbereitung f	flottation f	флотация
F 118	**flow,** stream	Strömung f, Strom m	écoulement m	поток, течение
F 119	**flow back,** reflux	zurückströmen, zurückfließen	refluer	течь обратно
F 120	**flow impedance,** flow resistance	Strömungswiderstand m	impédance f, résistance f à l'écoulement	сопротивление течению газа
F 121	**flow in**	einströmen	affluer	втекать
F 122	**flow meter**	Strömungsmesser m, Durchflußmesser m	débitmètre m, appareil m de contrôle du courant	расходомер
F 123	**flow of gas,** gas flow	Gasstrom m, Gasströmung f	flux (courant) m de gaz, flux (courant) gazeux	газовый поток, поток газа
F 124	**flow out**	ausströmen	se dégager, s'échapper	вытекать
F 125	**flow path**	Strömungsweg m	trajet m du courant	траектория (путь) потока
F 126	**flow point**	Fließtemperatur f, Fließpunkt m	température f continue	температура размягчения, точка (температура) текучести
F 127	**flow rate,** flow speed, rate of flow	Strömungsgeschwindigkeit f	vitesse f d'écoulement	скорость течения, скорость потока
F 128	**flow rate gain**	Durchflußmengenverstärker m	amplificateur m du débit volumétrique	усилитель расхода
	flow resistance	s. flow impedance		
	flow speed, flow rate, rate of flow	Strömungsgeschwindigkeit f	vitesse f d'écoulement	скорость течения, скорость потока
F 129	**flow switch**	Strömungsschalter m	interrupteur m de flux	реле расхода (потока)
F 130	**flow type pump**	Strömungspumpe f	pompe f d'écoulement	насос, в котором используется течение (поток) вещества

F 131	fluid flow	Flüssigkeitsströmung *f*	coulée *f* liquide	течение (поток) жидкости
F 132	fluid-free	treibmittelfrei	sans fluide	без рабочей жидкости
F 133	fluid-free vacuum	treibmittelfreies Vakuum *n*	vide *m* sans agent de propulsion, vide sans poudre propulsive, vide sans fluide	безмасляный вакуум
F 134	fluidity	Fließverhalten *n*, Fluidität *f*	fluidité *f*	текучесть
F 135	fluid jet pump	Flüssigkeitsstrahlpumpe *f*	trompe *f* à vide, pompe *f* à jet de liquide	жидкоструйный насос
F 136	fluid-ring pump, liquid ring pump	Flüssigkeitsringpumpe *f*	pompe *f* à anneau de liquide	жидкостнокольцевой насос
F 137	fluorescent coating	Fluoreszenzschicht *f*	couche *f* fluorescente	флюоресцентное покрытие
F 138	fluorescent screen	Fluoreszenzschirm *m*	écran *m* fluorescent	люминесцентный (флюоресцирующий) экран
F 139	fluoride resistent	fluoridfest	résistant aux fluorures	стойкий по отношению к солям фтора
F 140	flush, scavenge	ausschwemmen, ausspülen <mit Gas>	balayer	промывать, вымывать, смывать <струей>
	flush gas, bleed gas, purge gas	Spülgas *n*	gaz *m* de balayage (rinçage)	газ для продувки, промывочный газ
F 141	flushing	Spülen *n*, Spülung *f*	balayage *m*, lavage *m*	промывка газом
F 142	flying-spot recorder	Lichtpunktschreiber *m*	enregistreur *m* à point lumineux	самописец со световым лучом
F 143	foam, froth	schäumen	écumer, mousser	пениться
F 144	foam, froth	Schaum *m*	écume *f*, mousse *f*	пена, накипь
F 145	focal plane	Brennebene *f*	plaine *f* focale	фокальная плоскость
F 146	focusing	Fokussierung *f*	concentration *f* du faisceau, focalisation *f*, mise *f* au point focal	фокусировка
F 147	foil gasket	Foliendichtung *f*	joint *m* par feuilles minces, joint de lamelles	прокладка из фольги
F 148	foil window	Folienfenster *n*	fenêtre *f* de lamelles	окно из фольги
F 149	fold	falzen	plier	сгибать, фальцевать
F 150	foolproof	narrensicher, unempfindlich gegen Fehlbedienung	non interchangeable	защищенный от неосторожного обращения, не требующий квалифицированного обслуживания
F 151	foolproofness, operational reliability	Betriebssicherheit *f*	sécurité *f* de marche, sécurité de service, sécurité d'exploitation	эксплуатационная надежность, безопасность в работе
F 152	forced convection	erzwungene Konvektion *f*	convection *f* forcée	принудительная конвекция
F 153	forced lubrication, pressure lubrication	Druckschmierung *f*	graissage *m* sous pression, graissage forcé	принудительная смазка, смазка под давлением
F 154	forcing screw	Abdrückschraube *f*	vis *f* à repousser	отжимный винт
F 155	fore-arm, side arm	Vorvakuumstutzen *m*	raccord *m* au vide primaire	форвакуумный патрубок
	fore-arm connection, fore-vacuum connection, backing-line connection, outlet flange, fore-pump connection, exhaust fitting	Vorvakuumanschluß *m*	raccord *m* de vide préliminaire	соединение [насоса] с форвакуумом
F 156	foreign atom	Fremdatom *n*	atome *m* étranger, atome d'autre origine	примесный атом
F 157	foreign layer	Fremdschicht *f*	couche *f* étrangère	пленка примеси
	fore-line, backing line	Vorvakuumleitung *f*	canalisation *f* pour vide primaire	форвакуумная магистраль
	fore-line trap, backing-line trap	Vorvakuumfalle *f*	piège *m* à vide primaire, piège pour le vide primaire	форвакуумная ловушка, маслосборник
	fore-line valve, backing-line valve	Vorvakuumventil *n*	robinet *m* primaire (de vide primaire)	форвакуумный вентиль
	fore-line valve	*s.* fore-pump valve		
	forepressure, backing pressure	Vorvakuumdruck *m*	pression *f* du vide préliminaire	выпускное давление
F 158	forepressure tolerance, tolerable forepressure, maximum forepressure, maximum backing pressure	„zulässiger" Vorvakuumdruck *m* <der einem um 10% größeren Ansaugdruck als bei normalem Vorvakuum entspricht>	pression *f* tolérée du vide préliminaire	максимально-допустимое выпускное давление <при превышении которого впускное давление возрастает более, чем на 10%>
F 159	forepressure side	Vorvakuumseite *f*	côté *m* du vide préliminaire, côté prévide	форвакуумная сторона
F 160	fore-pump, pre-pump	vorpumpen	prévider, faire le vide primaire	предварительно откачивать
	fore pump, backing pump	Vorvakuumpumpe *f*, Vorpumpe *f*	pompe *f* primaire (à prévide, à vide préliminaire, pour vide primaire)	насос предварительного разрежения, форвакуумный насос

	fore-pump connection, fore-vacuum connection, fore-arm connection, backing-line connection, outlet flange, exhaust fitting	Vorvakuumanschluß *m*	raccord *m* de vide préliminaire	соединение [насоса] с форвакуумом
F 161	fore-pumping time, roughing time	Vorpumpzeit *f*, Grobpumpzeit *f*	temps *m* de pompage primaire, temps de prévidage	время предварительной откачки, время достижения форвакуума
	fore-pump valve, backing-line valve, fore-line valve	Vorpumpenventil *n*, Vorvakuumventil *n*	soupape *f* de pompe à vide préliminaire	форвакуумный вентиль
F 162	fore vacuum, primary vacuum	Vorvakuum *n*	vide *m* préliminaire	форвакуум, предварительный вакуум
	fore-vacuum connection, fore-arm connection, backing-line connection, outlet flange, fore-pump connection, exhaust fitting	Vorvakuumanschluß *m*	raccord *m* de vide préliminaire	соединение [насоса] с форвакуумом
F 163	fore-vacuum subassembly	Vorvakuumbauteil *n*	partie *f* constituante du vide préliminaire, pièce *f* détachée du vide préliminaire	компонент форвакуумной системы
F 164	fore-vacuum vessel	Vorvakuumbehälter *m*	réservoir *m* intermédiaire, recipient *m* du vide préliminaire, réserve *m* de vide primaire	форвакуумный баллон
F 165	forgeability	Schmiedbarkeit *f*	forgeabilité *f*	ковкость
F 166	forgeable alloy, wrought alloy	Knetlegierung *f*	alliage *m* à pétrir, alliage de pétrissage, alliage baratté	ковкий сплав, прокованный сплав, пластичный сплав
F 167	forged flange	geschmiedeter Flansch *m*	bride *f* forgée	кованный фланец
F 168	formation of dust	Staubentwicklung *f*	dégagement *m* (formation *f*) de poussière	пылеобразование
F 169	forming, potting	Verformung *f*	déformation *f*, formage *m*, façonnage *m*	деформация
F 170	forming gas, mixed gas <mixture of hydrogen and nitrogen>	Formiergas *n* <Wasserstoff-Stickstoff-Gemisch>	gaz *m* mixte <mélange de hydrogène et d'azote>	формиргаз <смесь водорода с азотом>
F 171	forming voltage	Formierspannung *f*	voltage *m* de formation, tension *f* formée (activée)	напряжение формовки формирующее напряжение
F 172	form persistance	Formbeständigkeit *f*	persistance *f* de forme	постоянство формы
	foundry ladle, casting ladle, chill, pouring ladle	Gießpfanne *f*	cuvette (coupe, écuelle, poche) *f* de coulée	литейный (разливочный) ковш, кокиль , форма
F 173	fractional-adsorption gas analysis	Gasanalyse *f* durch fraktionierte Adsorption	analyse *f* des gaz par adsorption fractionnée	анализ газов способом фракционной адсорбции
F 174	fractional-condensation gas analysis	Gasanalyse *f* durch fraktionierte Kondensation	analyse *f* des gaz par condensation fractionnée	анализ газов способом фракционной конденсации
F 175	fractional-desorption gas analysis	Gasanalyse *f* durch fraktionierte Desorption	analyse *f* des gaz par désorption fractionnée	анализ газов способом фракционной десорбции
F 176	fractional distillation	fraktionierte Destillation *f*	distillation *f* fractionnée	фракционированная дистилляция, дробная перегонка
F 177	fractional-evaporation gas analysis	Gasanalyse *f* durch fraktionierte Verdampfung	analyse *f* des gaz par évaporation fractionnée	анализ газов способом фракционного испарения
F 178	fractionating brush	Fraktionierbürste *f*	brosse *f* tournante (fractionnante)	фракционирующая щетка
F 179	fractionating column, separation column	Trennsäule *f*, Fraktionierkolonne *f*	colonne *f* de séparation (fractionnement)	разделительная колонна
F 180	fractionating pump, purifying pump, self-fractionating pump	fraktionierende Pumpe *f*	pompe *f* fractionnante (à rectification)	разгоночный насос, фракционирующий насос
	fracture strain, breaking elongation, total extension	Bruchdehnung *f*	dilatation *f* de rupture, allongement *m* de rupture	растяжение при разрыве, удлинение при разрыве
F 181	freckle segregation	Fleckseigerung *f*	ressuage *m* tacheté	пятнистая ликвация (сегрегация)
F 182	free blowing	Vakuumformen *n*	moulage *m* du vide	вакуумное формование
F 183	freedom from vibration	Schwingungsfreiheit *f*	liberté *f* d'oscillations	свобода от колебаний (вибраций)
F 184	free energy	freie Energie *f*	énergie *f* libre	свободная энергия
F 185	free expansion	freie Expansion *f*	expansion *f* libre	свободное расширение
F 186	free-falling film evaporator	Freifallverdampfer *m*	évaporateur *m* à chute libre	испаритель со свободно падающей пленкой
F 187	free-flow valve	Freiflußventil *n*	vanne *f* (robinet *m*) à flux libre	прямопролетный вентиль
F 188	free-levitation method, levitation melting	Schwebeschmelzen *n*	fusion *f* par lévitation	плавка во взвешенном состоянии
F 189	free-path distillation	Freiwegdestillation *f*	distillation *f* à libre parcours	дистилляция со свободным пробегом

F 190	**freeze**	gefrieren	congeler	замораживать
F 191	**freeze concentration**	Gefrierkonzentration f, Konzentrieren n durch Gefrieren	concentration f par congélation	концентрирование вымораживанием
F 192	**freeze-dried product**	gefriergetrocknetes Gut n	produit m lyophilisé, marchandise f séchée par congélation, produit séché par congélation	высушенный сублимацией продукт
F 193	**freeze-dry,** lyophilize, lyophile	gefriertrocknen	sécher par congélation, lyophiliser	сушить сублимационным способом
	freeze dryer	s. freeze drying plant		
	freeze-drying, cryodrying, lyophilization, sublimation from the frozen state	Gefriertrocknung f, Lyophilisation f	lyophilisation f, des iccation f par congélation, cryodessiccation f	сублимационная сушка из замороженного состояния, лиофилизация
F 194	**freeze-drying chamber**	Gefriertrocknungskammer f	chambre f lyophilisation, chambre de séchage par congélation	камера для сублимационной сушки
F 195	**freeze-drying plant,** freeze dryer	Gefriertrocknungsanlage f	installation f de lyophilisation	установка для сублимационной сушки
F 196	**freeze-drying rate**	Gefriertrocknungsgeschwindigkeit f	vitesse f du séchage par congélation	скорость сублимационной сушки
F 197	**freeze etching**	Gefrierätzung f	gravure f de congélation	сублимационная очистка ‹с замораживанием›
F 198	**freeze-etch technique**	Gefrierätztechnik f	technique f de cryodécapage	технология сублимационной очистки поверхности ‹из замороженного состояния›
F 199	**freezing**	Gefrieren n	congélation f	замораживание
F 200	**freezing capacity**	Gefrierleistung f	capacité (puissance) f de congélation	производительность по вымораживанию, производительность морозильного аппарата
F 201	**freezing device**	Einfriervorrichtung f	appareil m de congélation	замораживающее устройство
F 202	**freezing mixture**	Kältemischung f	mélange m frigorifique	охлаждающая смесь
F 203	**freezing point**	Gefrierpunkt m	point m de congélation	точка (температура) замерзания
F 204	**freezing point depression**	Gefrierpunktserniedrigung f	abaissement m du point de congélation	понижение точки (температуры) замерзания
F 205	**freezing rate**	Einfriergeschwindigkeit f	vitesse f de congélation	скорость замораживания
F 206	**freezing time**	Gefrierzeit f	temps m de congélation	продолжительность замораживания
F 207	**freezing tunnel,** tunnel freezer	Gefriertunnel m	tunnel m de congélation	туннельный морозильный аппарат, морозильный туннель
F 208	**freezing zone**	Erstarrungsbereich m	zone f de solidification	диапазон температур затвердевания
F 209	**frequency shift**	Frequenzverschiebung f	décalage m de fréquence	сдвиг по частоте, частотный сдвиг
F 210	**friction**	Reibung f	frottement m, friction f	трение
F 211	**friction brake**	Reibungsbremse f	frein m à friction	фрикционный тормоз
F 212	**friction-controlled flow**	Reibungsströmung f	frottement m d'écoulement	течение при наличии трения
	friction-type vacuum gauge, decrement gauge, decrement viscosity gauge, viscosity manometer, viscosity vacuum gauge, viscosity type of gauge	Reibungsvakuummeter n	manomètre m à amortissement (viscosité), jauge f à vide à frottement, vacuomètre m d'adhérence, vacuomètre de frottement, tube m gradué à viscosité	динамический (вязкостный) манометр
F 213	**friction welding,** spin welding	Reibungsschweißen n	soudure f par frottement	сварка трением
F 214	**fringe field**	Randfeld n, Streufeld n	champ m de dispersion	поле рассеяния
F 215	**fringing**	Streuflußbildung f	structure f de flux de dispersion	краевой поток
F 216	**frit**	Fritte f	fritte f	фритта, продукт обжига
F 217	**front surface mirror**	Oberflächenspiegel m	miroir m superficiel (de surface), miroir réfléchissant à la surface	поверхностное зеркало, зеркальная поверхность
F 218	**frost**	mattieren	ternir, dépolir	наводить «мороз», матировать
F 219	**frosting**	Reifbildung f	formation f de givre	образование инея
	frosting salt, etching salt	Mattiersalz n	sel m à ternir	соль для матирования
	froth	s. foam		
	froth, foam	Schaum m	écume f, mousse f	пена, накипь
F 220	**fuel element**	Brennstoffelement n	élément m combustible	топливный элемент
F 221	**fuel gas**	Brenngas n	fuel-gaz m	горючий (топливный) газ
F 222	**fugacity**	Fugazität f, Flüchtigkeit f	fugacité f	фугитивность, летучесть, абсолютная активность
F 223	**full annealing**	Ausglühen n	recuit m	полный отжиг
F 224	**fundamental constant**	Grundgröße f	quantité f fondamentale	основная постоянная (величина)
F 225	**fundamental frequency**	Eigenfrequenz f, Grundfrequenz f	fréquence f propre	основная частота (гармоническая)

F 226	**furnace brazing**	Hartlöten n mittels Ofens, Ofenlöten n	brasage m par four	пайка в печи <твердым припоем>
F 227	**furnace capacity**	Ofenkapazität f	capacité f d'un four	емкость печи
F 228	**furnace ladle**	Ofenpfanne f	poêle f de four	ковш печи
F 229	**furnace tank**	Ofenkessel m	chaudière f de four	корпус печи
F 230	**furnace throughput**	Ofendurchsatz m	fournée f	пропускная способность печи, производительность печи
F 231	**fused seal**, fusion joint	Verschmelzung f	fusion f	спай
F 232	**fused silica**, quartz glass	Quarzglas n	verre m quartzeux	кварцевое стекло
F 233	**fusibility**	Schmelzbarkeit f	fusibilité f	плавкость
F 234	**fusible**	schmelzbar	fusible	плавкий
F 235	**fusible plug**	Schmelzeinsatz m	lame f (élément m, fil m) fusible	плавкая вставка, плавкая пробка
F 236	**fusing time**	Abschmelzdauer f	durée f de fusion	продолжительность плавки
F 237	**fusion**, melting	Schmelzen n, Schmelzprozeß m	fusion f, fondage m	плавка, плавление, расплавление
F 238	**fusion analyzer**	Heißextraktionsanalysenanlage f	analyseur m d'extraction à chaude	установка для анализа методом горячей экстракции
	fusion joint, fused seal	Verschmelzung f	fusion f	спай
F 239	**fusion point**, melting point	Schmelzpunkt m	point m de fusion	точка плавления
F 240	**fusion welding**	Schmelzschweißen n, Abschmelzschweißen n, Schmelzschweißung f	soudage m par fusion	сварка плавлением
	fusion zone, float zone	Schmelzzone f	zone f de fusion	зона плавления

G

G 1	**Gaede gas ballast pump**	Gasballastpumpe f nach Gaede	pompe f à injection d'air de Gaede	газобалластный насос Геде
G 2	**Gaede mercury rotary pump**	Rotationsquecksilberpumpe f nach Gaede	pompe f Gaede, trompe f rotative à mercure de Gaede	ртутный вращательный насос Геде
G 3	**Gaede's mol vacuummeter**	Gaedesches Molekularvakuummeter n	vacuomètre m moléculaire de Gaede	мольвакуумметр Геде
G 4	**gamma rays**, γ rays	Gammastrahlen mpl	rayons mpl gamma, rayons γ	гамма-лучи
G 5	**gas access**	Gaszutritt m	accès m du gaz	доступ (приток) газа
G 6	**gas admittance valve**, gas inlet valve	Gaseinlaßventil n	soupape f (robinet m) d'admission des gaz	вентиль для впуска газа
G 7	**gas analysis**	Gasanalyse f	analyse f des gaz	газовый анализ
G 8	**gas at rest**	ruhendes Gas n	gaz m permanent	покоящийся газ, неподвижный газ
G 9	**gas backstreaming**	Gasrückströmung f	flux m inverse des gaz, reflux m des gaz	обратный поток газа
G 10	**gas ballast**, vented exhaust	Gasballast m	lest m de gaz, lest (injection f) d'air	газовый балласт
	gas ballasting, air ballasting	Gasballastbetrieb m	admission f du lest d'air	напуск балластного газа
G 11	**gas ballast pump**, vented-exhaust [mechanical] pump	Gasballastpumpe f	pompe f à injection d'air, pompe ballast à gaz	газобалластный насос
G 12	**gas-ballast roughing-holding pump**	Grobvakuum- und Haltepumpe f mit Gasballasteinrichtung	pompe f d'entretien avec dispositif de lest d'air, pompe de maintien avec dispositif de lest d'air, pompe pour maintenir le vide avec dispositif de lest d'air	газобалластный насос для создания и поддержания предварительного разрежения
G 13	**gas ballast valve**	Gasballastventil n	robinet m à injection d'air, robinet de lest d'air, soupape f de lest d'air	клапан для напуска балластного газа, газобалластный клапан, газобалластное устройство
G 14	**gas bleed flange**, purge gas flange	Spülgasflansch m	bride f de gaz de balayage	фланец для подсоединения промывочного газа
G 15	**gas bleed valve**	Spülgasventil n	robinet m (soupape f) de rinçage à gaz	газовыпускной вентиль
G 16	**gas-bubble pump**	Gasblasenpumpe f, Mammutpumpe f	pompe f mammouth, émulseur m à air comprimé	маммут-насос, газопузырьковый насос
G 17	**gas burst**	Gasausbruch m	échappée f de gaz, dégagement m spontané (instantané) de gaz	выброс газа
G 18	**gas chromatography**	Gaschromatografie f	chromatographie f gazeuse	газовая хроматография
G 19	**gas composition**	Gaszusammensetzung f	composition f de gaz	газовый состав
G 20	**gas content**	Gasgehalt m	teneur f en gaz, teneur gazeuse	газовое содержание, газосодержание
	gas density, density of gas	Gasdichte f	densité f de gaz	плотность газа
G 21	**gas discharge colour method**	Lecksuche f mit Geißlerrohr	détection f de fuites avec le tube de Geissler	метод обнаружения течи по свечению газового разряда

G 22	gas dosing	Gasdosierung f	dosage m gazeux	дозирование газа
G 23	gas dosing leak	Gasdosierleck n	fuite f de dosage de gaz	течь для дозировки газа
	gas dynamical flow, dynamical flow of gases, dynamic gas flow	gasdynamische Strömung f	écoulement m (flux m) dynamique des gaz, courant m gazeux dynamique	газодинамическое течение
G 24	gaseous	gasförmig	gazeux	газообразный
G 25	gaseous adsorption	Gasadsorption f	adsorption f gazeuse	адсорбция газа
G 26	gaseous discharge	Gasentladung f	décharge f dans un gaz	электрический ток (разряд) в газе, газовый разряд
G 27	gaseous electronics	Gaselektronik f	électronique f de gaz	газовая электроника
G 28	gaseous ion	Gasion n	ion m de gaz	газовый ион
G 29	gas equation	Gleichung f des idealen Gases	équation f du gaz parfait (idéal)	уравнение идеального газа
G 30	gas evolution	Gasentwicklung f	dégagement m gazeux	выделение газа
G 31	gas factor	Gasfaktor m	facteur m de gaz	коэффициент, учитывающий род газа
G 32	gas-filling valve	Gaseinfüllventil n	robinet m de remplissage des gaz	вентиль для наполнения газом
G 33	gas film	Gasschicht f, Gashaut f	couche f de gaz	газовая пленка
	gas flow, flow of gas	Gasstrom m, Gasströmung f	flux (courant) m de gaz, flux (courant) gazeux	газовый поток, поток газа
G 34	gas-flow adjusting flange	Gasführungsflansch m	bride f à conduite de gaz, rebord m à guidage de gaz	фланец для подвода потока газа
G 35	gas flowmeter	Gasströmungsmesser m	mesureur m du courant gazeux	газовый расходомер
G 36	gas focusing	Gasfokussierung f	focalisation f des gaz	газовая фокусировка
G 37	gasfree high purity copper, GFPH copper	GFPH-Kupfer n	cuivre m GFPH, cuivre sans gaz et de grande pureté	медь особо чистая с низким содержанием газов, медь GFPH
G 37a	gas-fusion welding	Gasschmelzschweißung f	soudure f autogène oxyacétylénique	газовая сварка
G 38	gas generation	Gaserzeugung f	génération (production) f de gaz	получение газа, генерация газа
G 39	gas-impermeable	gasundurchlässig	imperméable aux gaz	газонепроницаемый
G 40	gas injection	Gasinjektion f	injection f de gaz	инжекция газа
	gas inlet valve, gas admittance valve	Gaseinlaßventil n	soupape f (robinet m) d'admission des gaz	вентиль для впуска газа
G 41	gas-ion constant, gauge sensitivity factor, gauge constant	Gasionenkonstante f, Röhrenkonstante f	constante f des ions de gaz	постоянная газовых ионов
G 42	gas ion current	Ionenstrom m, bedingt durch den Restgasdruck	courant m ionique	ионный ток, обусловленный остаточными газами
G 43	gas jet	Gasstrahl m	jet m de gaz	газовая струя
G 44	gas jet pump	Gasstrahlpumpe f	éjecteur m à gaz	газовый эжекторный насос, газовый эжектор
G 45	gasket, seal, sealing, sealing-in, packing, sealing element	Dichtung f, Dichtungselement n, Einschmelzstelle f, Abdichtung f, Einschmelzung f, Verschmelzung f	joint m, joint étanche (scellé, d'étanchéité), scellement m, étanchement m, jonction f étanche, point m de scellement, élément m d'étanchéité	уплотнение, соединение, уплотняющий элемент, спай, место спая, запайка, впайка
G 46	gasket groove	Dichtungsnut f	rainure f pour joint étanche	канавка для уплотнения
G 47	gasketless	dichtungslos	sans joint	беспрокладочный
G 48	gasket ring, packing ring, sealing ring	Dichtungsring m	anneau m d'étanchéité, rondelle f de joint	уплотнительное (уплотняющее, прокладочное) кольцо, прокладка
G 49	gasket scratch	Kratzer m auf der Dichtung	gratture f sur le joint	риска на прокладке
	gasket seal, crush seal, compression seal	Preßdichtung f	joint m à compression, raccord m à écrasement, joint à écrasement	нажимное уплотнение
G 50	gas kinetics	Gaskinetik f	cinétique f des gaz	газовая кинетика
G 51	gas permeability	Gasdurchlässigkeit f	perméabilité f aux gaz	газопроницаемость
G 52	gas permeation method	Nachschwitzmethode f	méthode f de perméation	метод, основанный на разнице в газопроницаемости
G 53	gas plating	Aufwachsverfahren n	revêtement m gazeux, dépôt m gazeux	способ наращивания <разложением газообразного соединения>
G 54	gas purging	Entgasen n	dégagement m (émission f) de gaz, dégazage m	выпуск газа
G 55	gas purification	Gasreinigung f	épuration f des gaz	очистка газа
	gas-refrigerating machine, cryo-generator	Gaskältemaschine f	machine f frigorifique à gaz	криогенная установка
G 56	gas separation plant	Gaszerlegungsanlage f	installation f à séparer des gaz	установка для разделения газов
G 57	gas-solid interaction	Wechselwirkung f Gas—Festkörper	action f réciproque de gaz et solide	взаимодействие газа с твердым телом
G 58	gas solubility	Gaslöslichkeit f	solubilité f des gaz	растворимость газа
G 59	gas source	Gasquelle f	source f de gaz	источник газа

G 60	**gas supply**	Gasvorrat *m*	dépôt *m* de gaz, réserve *f* gazeuse	запас газа
G 61	**gas tight**	gasdicht	étanche aux gaz, imperméable aux gaz	газоплотный
G 62	**gas tightness**	Gasdichtheit *f*	étanchéité *f* aux gaz, imperméabilité *f* aux gaz	газоплотность
G 63	**gas volumeter**	Gasvolumeter *n*	volumètre *m* à gaz	газовый волюметр, измеритель объема газа
G 64	**gate valve,** slide valve, vacuum gate valve	Torventil *n*, Schieberventil *n*, Vakuumschieber *m*	robinet *m* [à tiroir], vanne *f* à tiroir (coulisse), soupape *f* à passage direct à vide	задвижка, заслонка, вакуумная задвижка
G 65	**gauge**	Lehre *f*	jauge *f*, calibre *m*	измеритель, измерительный прибор
G 66	**gauge,** gaugehead, head, sensing head	Meßröhre *f*, Meßkopf *m*, Meßgerät *n*	burette *f*, tube *m* de mesure, cellule *f* de mesure, tête *f* de mesure, tête manométrique	датчик манометра
G 67	**gauge constant**	Manometerkonstante *f* <McLeod>, Röhrenkonstante *f*	constante *f* de manomètre	постоянная манометра
	gauge constant s. gas-ion constant **gaugehead,** head, sensing head, gauge	Meßröhre *f*, Meßkopf *m*	burette *f*, tube *m* de mesure, cellule *f* de mesure, tête *f* de mesure, tête manométrique	датчик манометра
G 68	**gauge sensitivity,** sensitivity of a gauge	Röhrenempfindlichkeit *f*	sensibilité *f* d'un tube électronique	чувствительность манометра
	gauge sensitivity factor s. gas-ion constant **Gay-Lussac's law,** Charles' law	Gay-Lussacsches Gesetz *n*	loi *f* de Gay-Lussac	закон Гей-Люссака
G 69	**gear pump**	Zahnradpumpe *f*	pompe *f* à engrenages	зубчатый насос, шестеренчатый насос
G 70	**Geiger counter, Geiger counter tube, Geiger-Müller counter, Geiger-Müller tube**	Geiger-Müller-Zählrohr *n*, Geiger-Zähler *m*	compteur *m* d'après Geiger-Müller, tube *m* compteur Geiger-Müller	счетчик Гейгера-Мюллера
G 71	**Geissler's tube**	Geißlerrohr *n*	tube *m* de Geissler	трубка Гейслера
	general gas law, equation for ideal gases	universelle Gasgleichung *f*	équation *f* universelle des gaz, équation des gaz idéaux	универсальное газовое уравнение, уравнение состояния идеального газа
G 72	**germanium**	Germanium *n*	germanium *m*	германий
G 73	**getter**	Getter *m*, Gettermaterial *n*	getter *m*	газопоглотитель, геттер
G 74	**getter-ion pump,** ion-getter pump	Ionengetterpumpe *f*	pompe *f* getter-ionique, pompe ionique à getter	геттеро-ионный насос, сорбционно-ионный насос
G 75	**getter plate**	Getterplättchen *n*	palette *f* du getter	пластина газопоглотителя
G 76	**getter pump**	Getterpumpe *f*	pompe *f* [à] getter, pompe à sorption par getter	газопоглотительный насос
	GFPH copper, gasfree high purity copper	GFPH-Kupfer *n*	cuivre *m* GFPH, cuivre sans gaz et de grande pureté	медь особо чистая с низким содержанием газов, медь GFPH
G 77	**ghost line**	Geisterlinie *f*	ligne *f* fantôme	линия фантома
G 78	**gimbal mounting**	Kardanaufhängung *f*	suspension *f* de cardan	карданный подвес
G 79	**gland packing,** stuffing-box packing	Stopfbüchsenpackung *f*	garniture *f* de presse-étoupe	набивка сальника
G 80	**glass bead**	Glasperle *f*	perle *f* de verre	стеклянная бусинка
G 81	**glass bell jar**	Glasglocke *f*	cloche *f* de verre	стеклянный колпак
G 82	**glass bellows**	Glasbalg *m*	peau *f* de verre	стеклянный сильфон
G 83	**glass bulb blowing machine**	Kolbenblasmaschine *f*	machine *f* à piston de soufflage	машина для выдувания стеклянных колб
G 84	**glass cement**	Glaskitt *m*	mastic *m* à verre	стеклоцемент
G 85	**glass-ceramics** <"pyroceram">	auskristallisierte Gläser *npl* <"Pyroceram">	verres *mpl* cristallisés	ситаллы, пирокерамы
G 86	**glass envelope**	Glasmantel *m*	enveloppe *f* de verre	стеклянная оболочка
G 87	**glass fibre**	Glasfaser *f*	fibre *f* de verre	стекловолокно
G 88	**glass fibre reinforced**	glasfaserverstärkt	renforcé par fibre de verre	армированный стекловолокном
G 89	**glass filter**	Glasfilter *n*	filtre *m* de verre	стеклянный фильтр
	glass foam, cellular glass	Schaumglas *n*	verre *m* mousse	пеностекло
G 90	**glass for sealing-in molybdenum**	Molybdänglas *n*	verre *m* molybdique	молибденовое стекло
G 91	**glass for sealing-in platinum**	Platinglas *n*	verre *m* de platine	платиновое стекло
G 92	**glass for sealing-in tungsten**	Wolfram-Einschmelzglas *n*	verre *m* se soudant bien au tungstène	вольфрамовое стекло
G 93	**glass ground joint**	Glasschliff *m*	rodage *m* en verre	стеклянный шлиф
G 94	**glass stem**	Glas[quetsch]fuß *m*	socle *m* à compression en verre, tige *f* de verre	стеклянная ножка
G 95	**glass termination**	Glasanschluß *m*	jonction *f* (connexion *f*, raccord *m*) de verre	стеклянный наконечник
G 96	**glass-to-metal seal**	Metall-Glasverschmelzung *f*, Metallanglasung *f*	scellement *m* verre-métal, joint *m* verre-métal	спай стекла с металлом

G 97	**glass tube**	Glasrohr *n*, Glasröhre *f*	tube *m* de verre	стеклянная трубка
G 98	**glaze, glazing**	Glasur *f*	glaçure *f*	глазурь
	globe valve, ball valve, spherical valve	Kugelventil *n*	soupape *f* à bille (boulet), soupape sphérique	вентиль со сферическим клапаном, шаровой клапан
G 99	**glove box,** dry box	Handschuhkasten *m*	boîte *f* à gants	коробка для перчаток
	glow discharge	Glimmentladung *f*	décharge *f* luminescente	тлеющий разряд, тихий разряд
	glow discharge clean-ing, discharge clean-ing, ionic bombard-ment cleaning, ionic cleaning	Abglimmen *n*, Glimm-reinigung *f*	nettoyage *m* par bombar-dement ionique, dé-charge *f* par effluves	очистка разрядом, очи-стка тлеющим раз-рядом, ионная очистка
G 100	**glow discharge device**	Glimmeinrichtung *f*	installation *f* de combus-tion lente	установка, использующая тлеющий разряд
G 101	**glow discharge tube**	Glimmentladungsrohr *n*	tube *m* de décharge lumineuse	лампа тлеющего разряда
G 102	**glow discharge unit**	Abglimmeinrichtung *f*	installation *f* de nettoyage par bombardement ionique	установка очистки тлеющим разрядом
G 103	**gold wire gasket, gold wire seal**	Golddrahtdichtung *f*	joint *m* de fil en or	уплотнение с золотой проволокой
G 104	**graded**	abgestuft, stufenweise	graduel	сортированный, градуированный
G 105	**graded glass seal tubulation, graded seal,** multiple-graded glass seals	Schachtelhalm *m*, Über-gangsglasrohr *n*	tube *m* en verre de transi-tion	стеклянный переход, соединение, состоящее из нескольких переход-ных стекол
G 106	**grain boundary**	Korngrenze *f*	limite *f* de grain	граница зерна
G 107	**grain boundary diffusion, grain boundary scattering**	Korngrenzenstreuung *f*, Korngrenzendiffusion *f*	dispersion *f* de limites de grains	разброс границ зерен
G 108	**grain growth**	Kornwachstum *n*, Kristallwachstum *n*	croissance *f* des grains	рост зерна
G 109	**grain size**	Korngröße *f*	grosseur *f* des grains, granulation *f*	размер зерна
G 110	**gram-atom**	Grammatom *n*	atome-gramme *m*	грамм-атом
G 111	**gramme molecule,** mole	Grammolekül *n*, Mol *n*	mole *f*	грамм-молекула, моль
G 112	**granule**	Granulat *n*	granule *m*	гранула
G 113	**graphite coating**	Graphitüberzug *m*	couche *f* graphitique	графитовое покрытие
G 114	**graphite crucible**	Graphittiegel *m*	creuset *m* graphitique	графитовый тигель
G 115	**graphite electrode**	Graphitelektrode *f*	électrode *f* en graphite	графитовый электрод
	grate plane, crystal plane, net plane	Netzebene *f*	plaine *f* de réseau	плоскость [кристалли-ческой] решетки
G 116	**gravity**	Schwerkraft *f*, Schwere *f*	pesanteur *f*, gravité *f*	сила тяжести, гравита-ция
G 117	**gravity drain plug**	Schwerkraftölablaß-schraube *f*	vis *f* de vidange d'huile en champ de gravitation	гравитационная масло-спускная пробка
G 118	**grazing incidence**	streifender Einfall *m*, streifende Inzidenz *f*	incidence *f* touchante	касательное падение
G 119	**green fodder dryer**	Grünfuttertrockner *m*	sécheur *m* à fourrage vert	сушилка для зеленого корма
G 120	**grey body**	grauer Körper *m*	corps *m* gris	серое тело
G 121	**grey filter**	Neutralglasfilter *n*	filtre *m* gris (neutre)	серый (нейтральный) фильтр
G 122	**grid end covering**	Gitterendabdeckung *f*	couverture *f* finale de grille	оконечное покрытие сетки
G 123	**grid pitch**	Gittersteigung *f*	montée *f* de grille	шаг сетки
G 124	**grinding hardness**	Schleifhärte *f*	dureté *f* de polissage	сопротивление шлифованию
G 125	**gripping device**	Einspannvorrichtung *f*	dispositif *m* de serrage	зажимное приспособле-ние
G 126	**gross density**	Rohdichte *f*	densité *f* brute	брутто-плотность
G 127	**ground-glass joint**	Glasschliffverbindung *f*	joint *m* de rodage en verre	стеклянное шлифовое соединение
G 128	**ground-in ball and socket joint**	Kugelschliff *m*	rodage *m* sphérique	шаровое шлифовое сое-динение, шаровой шлиф
G 129	**ground joint**	Schliffverbindung *f*	joint *m* rodé, raccord *m* à rodage	пришлифованное (притертое) соеди-нение
G 130	**group of pumps,** pump set	Pumpsatz *m*	groupe (agrégat) *m* de pompage	насосный (откачной) агрегат
G 131	**growth rate,** rate of growth	Wachstumsgeschwindig-keit *f*, Wachstumsrate *f*	vitesse *f* de croissance	скорость роста
G 132	**grub screw**	Madenschraube *f*	vis *f* pointeau (sans tête)	винт без головки, по-тайной винт
G 133	**guaranteed vacuum**	garantiertes Vakuum *n*	vide *m* garanti	гарантированный вакуум
G 134	**guard ring**	Schutzring *m*	bague *f* protectrice	охранное кольцо
G 135	**guard-ring electrode assembly**	Schutzringelektroden-system *n*	système *m* d'électrodes avec anneau protecteur	электродная система с охранными коль-цами
G 136	**guard vacuum**	Schutzvakuum *n*, Zwischenvakuum *n*, Vakuummantel *m*	vide *m* de protection, espace *m* vide intermé-diaire, espace évacué in-termédiaire	защитный вакуум

G 137	guide bushing	Führungslager n	support-guide m	направляющая опора
G 138	guide electrode	Leitelektrode f	électrode f conductrice	направляющий электрод
G 139	guide pin	Führungsstift m	fiche f de guidage	направляющий штифт
G 140	guiding rod	Führungsstange f	guidon m, tige f de guidage	направляющий стержень
G 141	gun lock	[Elektronen-]Strahlerschleuse f	écluse f du radiateur [électronique]	шлюз для [пропускания] электронного луча
G 142	gutta-percha	Guttapercha f	gutta-percha f	гуттаперча

H

	hair-line crack, capillary crack	Haarriß m	fente (fissure) f capillaire	волосяная трещина, капиллярная трещина
H 1	hairpin cathode, hairpin-shaped cathode	Haarnadelkatode f	cathode f en U	шпилькообразный катод
H 2	half life <period>	Halbwertszeit f	demi-vie f	период полураспада
H 3	halid, halogen	Halogen n	halogène m	галоид
H 4	halide detector	Halogenanzeigegerät n	détecteur m de fuites aux halogènes	галоидный индикатор (детектор)
H 5	halide diode	Halogen-Diode f	diode f halogène	галоидный диод
H 6	halide diode detector head	Halogen-Dioden-Meßzelle f	détecteur m de fuites de diodes halogènes	галоидный диодный детектор
H 7	halide leak detector, halogen leak detector, halogen sensitive leak detector	Halogenlecksuchgerät n, Halogenlecksucher m	détecteur m de fuites aux halogènes, détecteur de fuites sensible aux halogènes	галоидный течеискатель
H 8	halide torch	Suchlampe f	lampe f de recherche	галоидная лампа (горелка)
	halogen, halid	Halogen n	halogène m	галоид
	halogen leak detector	s. halide leak detector		
	halogen sensitive leak detector	s. halide leak detector		
H 9	halogen small sniffer	Halogenkleinschnüffler m	petit détecteur m de fuites aux halogènes	малый щуп галоидного течеискателя
H 10	hand-operated valve, manually operated valve	handbetätigtes Ventil n	valve f à commande manuelle, valve actionnée à la main	вентиль с ручным приводом
H 11	hand torch	Handgebläse n	soufflante f à commande manuelle	ручная горелка
H 12	harden	härten	tremper	упрочнять
H 13	hardening, strengthening	Verfestigung f	solidification f, durcissement m, consolidation f	упрочнение, наклеп
H 14	hard facing, hard surfacing	Auftragschweißen n	apport m par soudure	покрытие (наплавка) твердым слоем
H 15	hard glass to stainless steel sealing	Verschmelzung f zwischen rostfreiem Stahl und Hartglas	fusion f d'acier inoxydable et verre dur	спай нержавеющей стали с тугоплавким стеклом
H 16	hardness	Härte f	dureté f	твердость
H 17	hardness test	Härteprüfung f	essai m de dureté	испытание на твердость
	hard rubber, ebonite	Hartgummi m, Ebonit n	ébonite f, caoutchouc m durci	эбонит, твердая резина
	hard-solder, braze	hartlöten	braser	паять твердым припоем
H 18	hard-solder flux	Flußmittel n für Hartlot	fondant m à braser	флюс для пайки твердым припоем
	hard soldering, brazing	Hartlöten n	brasage m	пайка твердым припоем
H 19	hard-sphere model	Starrkugelmodell n	modèle m de sphère rigide	модель твердой (абсолютно упругой) сферы
	hard surfacing	s. hard facing		
	head, gaugehead, sensing head, gauge	Meßröhre f, Meßkopf m	burette f, tube m de mesure, cellule f de mesure, tête f de mesure, tête manométrique	датчик манометра
H 20	head	Pumpenkopf m	tête f de pompe	головка насоса
H 21	header	Sammelstück n	morceau m collecteur	коллектор-сборник
	head pressure, outlet pressure, discharge pressure, exhaust pressure	Verdichtungsdruck m, Auspuffdruck m, Ausstoßdruck m	pression f de sortie, pression de refoulement, pression du vide préliminaire, pression d'échappement, pression finale	давление сжатия, выпускное давление, давление на выхлопе
	head pressure, inlet pressure, intake pressure, fine pressure, suction pressure	Ansaugdruck m	pression f d'aspiration, pression d'admission	впускное давление
H 22	heatable container	ausheizbarer Rezipient m	récipient m étuvable	прогреваемый контейнер
H 23	heat conduction	Wärmeleitung f	conduction f thermique (de chaleur)	теплопередача
H 24	heat conduction, thermal conduction	Wärmeleitung f	conductibilité f thermique	теплопроводность <явление>
H 25	heat-conduction gauge, thermal gauge, thermal conductivity vacuum gauge	Wärmeleitungsvakuummeter n	jauge f thermique, vacuomètre m à conductibilité thermique, manomètre m thermique	тепловой манометр, теплоэлектрический манометр

H 26	**heat conduction power,** thermal conductance	Wärmeleitvermögen n, Wärmeleitfähigkeit f	conductibilité f thermique, conductivité f calorifique	теплопроводность, коэффициент теплопроводности
	heat content, enthalpy, total heat	Enthalpie f	enthalpie f	энтальпия, теплосодержание
	heat convection, convection	Konvektion f	convection f	конвекция
H 27	**heat crack,** hot tear, pull	Wärmeriß m	fente f thermique	термическая трещина, тепловой разрыв
H 28	**heat dissipation**	Wärmeverteilung f	dissipation (distribution) f de chaleur	рассеяние тепла
H 29	**heat distortion point**	Wärmefestigkeitsgrenze f	limite f de rupture thermique	предел термостойкости
H 30	**heated electron evaporation rate sensing device**	Verdampfungsratenmeßanordnung f mit geheizten Elektroden	disposition f de l'appareillage des taux de vaporisation avec électrodes chauffées	устройство для измерения скорости испарения с подогревными электродами
H 31	**heater capacity, heater input,** heater loading (power, wattage)	Heizleistung f	consommation f de courant de chauffage, rendement m calorifique, puissance f de chauffage absorbé	мощность подогревателя, мощность накала
H 32	**heater insert**	Heizereinschub m	réchaud m sur tiroir	подогревательная вставка
	heater loading	s. heater capacity		
	heater power	s. heater capacity		
	heater wattage	s. heater capacity		
H 33	**heat exchange**	Wärmeaustausch m	échange m thermique	теплообмен
H 34	**heat exchanger**	Wärmeaustauscher m	échangeur m thermique	теплообменник
H 35	**heating jacket**	Heizmantel m	chemise f chauffante (de chauffage)	тепловая рубашка
H 36	**heat insulation,** thermal insulation	Wärmeisolation f	isolation f thermique (calorifuge)	теплоизоляция
	heat load, refrigeration load, refrigeration duty, cooling load	Kältebedarf m	besoin m frigorifique	расход холода
H 37	**heat of desorption**	Desorptionswärme f	chaleur f de désorption	теплота десорбции
H 38	**heat of fusion**	Schmelzwärme f	chaleur f de fusion	теплота плавления
H 39	**heat of mixing**	Mischungswärme f	chaleur f de mélange	теплота смешения
H 40	**heat of solution**	Lösungswärme f	chaleur f de dissolution	теплота растворения
H 41	**heat of sublimation,** sublimation heat	Sublimationswärme f	chaleur f de sublimation	теплота сублимации
H 42	**heat pump**	Wärmepumpe f	pompe f à (de) chaleur	тепловой насос
H 43	**heat-reflecting glass**	Wärmeschutzglas n	protection (isolation) f thermique de verre	теплозащитное стекло, теплоотражающее стекло
H 44	**heat reflecting interference filter**	Interferenz-Wärmereflexionsfilter n	filtre m d'interférence à réflexion de chaleur	теплоотражающий интерференционный фильтр
H 45	**heat reflector**	Wärmereflektor m	réflecteur m thermique	тепловой отражатель
H 46	**heat shield,** radiation shield	Strahlungsschirm m, Strahlungsschutzschirm m	écran m protecteur contre le rayonnement, écran contre la radiation	тепловой экран, экран для защиты от излучения
H 47	**heat-shock resistance,** resistance to thermal shock	Temperaturwechselbeständigkeit f, Abschreckfestigkeit f	résistance f à choc de température, résistance de trempe	устойчивость к термоудару
H 48	**heat stagnation**	Wärmestau m	accumulation f (refoulement m) thermique	скопление (накопление) тепла
H 49	**heat transfer,** heat transmission, thermal transmission	Wärmeübergang m, Wärmeübertragung f	écoulement m thermique, transmission f de chaleur, transmission calorifique (thermique), conduction f thermique	теплопередача, перенос тепла
H 50	**heat transfer coefficient**	Wärmeübergangszahl f	coefficient m de transmission thermique	коэффициент теплопередачи
H 51	**heat transition**	Wärmedurchgang m	passage m de chaleur	прохождение тепла
	heat transmission	s. heat transfer		
H 52	**heat treatment**	Wärmebehandlung f	traitement m thermique	термообработка
H 53	**heavy current lead-in,** power feedthrough, high-current feedthrough	Hochstromdurchführung f	traversée f à courant de haute intensité, passage m à courant fort	сильноточный ввод
	HEED	s. high energy electron diffraction		
H 54	**heliarc welding process,** inert-gas-shielded nonconsumable-electrode arc welding method ‹Heliarc›	Heliarc-Verfahren n, Wolfram-Inertgas-Schweißen n mit Helium als Schutzgas	procédé m Heliarc	способ гелиево-дуговой сварки, способ дуговой сварки в гелиевой среде
H 55	**helical**	schraubenförmig	hélicoïde	винтовой, спиральный
H 56	**helical evaporator filament,** helical filament evaporator	Verdampferwendel f	filament m d'évaporation hélicoïdal	спиральный испаритель
H 57	**helical filament**	Heizwendel f	spirale f chauffante (du filament)	спираль накала
	helical filament evaporator	s. helical evaporator filament		

H 58	**helium gas refrigeration plant**	Heliumgaskälteanlage f	réfrigérateur m à gaz d'hélium	гелиевая холодильная установка
H 59	**helium leak detector**	Heliumlecksuchgerät n	détecteur m de fuites par hélium	гелиевый течеискатель
H 60	**helium permeation**	Heliumdurchlässigkeit f	pénétration f d'hélium	проникновение гелия
H 61	**helium permeation through glass**	Heliumdurchlässigkeit f von Glas	perméabilité f d'hélium de verre	проникновение гелия через стекло
H 62	**helium recovery plant**	Heliumrückgewinnungsanlage f	installation f à récupération pour l'hélium	гелиевая рекуперационная установка
H 63	**helix pitch error**	Wendelsteigungsfehler m	défaut m de la descente hélicoïdale	отклонение шага намотки
H 64	**Helmholtz coil**	Helmholtz-Spule f	bobine f de Helmholtz	катушка Гельмгольца
	hermetic[al], airproof	hermetisch, luftdicht	hermétique	герметический
H 65	**hermetically sealed**	hermetisch abgedichtet, luftdicht abgeschlossen	étoupé (fermé) hermétiquement, hermétiquement clos	герметично заделанный
H 66	**hermetically sealed electrode**	luftdicht eingeschmolzene Elektrode f	électrode f scellée étanche à l'air	герметично запаянный электрод
H 67	**hermetical seal, hermetical sealing**	hermetischer Verschluß m, hermetische Abdichtung f	fermeture f hermétique	герметичное уплотнение
H 68	**high-accuracy measurement**	Präzisionsmessung f	mesure f de précision	прецизионное измерения
H 69	**high alumina ceramics**	Keramiken fpl mit hohem Aluminiumoxidgehalt	céramiques fpl à teneur haut d'oxyde d'aluminium	высокоглиноземистая керамика
H 70	**high-boiling**	hochsiedend	bouilli à température élevée	кипящий при высокой температуре
H 71	**high conductance low loss liquid nitrogen trap**	Hochleistungstiefkühlfalle f	piège m à basse température à haut rendement	ловушка с высокой пропускной способностью, охлаждаемая жидким азотом
	high-current feedthrough	s. heavy current lead-in		
H 72	**high energy electron diffraction, HEED**	Beugung f schneller Elektronen	diffraction f des électrons à grande vitesse	дифракция быстрых электронов
H 73	**high-frequency ion source**	Hochfrequenzionenquelle f	source f d'ions à haute fréquence	высокочастотный ионный источник
H 74	**high-frequency mass spectrometer**	Hochfrequenzmassenspektrometer n	spectromètre m de masse à haute fréquence	высокочастотный масс-спектрометр
	high-frequency quartz crystal, crystal transducer, oscillating quartz crystal, quartz crystal oscillator, vibrating quartz crystal	Schwingquarz m	quartz m oscillatoire (de résonance)	кристалл кварцевого генератора, генерирующий кварцевый кристалл, генерирующий кварц
H 75	**high-frequency vacuum tester**, spark-coil detector	Hochfrequenzvakuumprüfer m	appareil m à haute fréquence pour le contrôle du vide, contrôleur m à vide à haute fréquence	искровой течеискатель
H 76	**high perveance electron beam**	Elektronenstrahl m hoher Perveanz	rayon m électronique de haute perveance, faisceau m électronique de haute perveance	электронный пучок с высоким первеансом
H 77	**high-pressure ionization gauge**	Ionisationsvakuummeter n für hohe Drücke	indicateur m à ionisation de haute pression, vacuomètre m d'ionisation à haute pression	ионизационный манометр для высоких давлений
H 78	**high silica glass**	quarzähnliches Glas n	verre m semblable à silice	стекло с большим содержанием двуокиси кремния, стекло «викор»
H 79	**high-speed diffusion pump**	Hochleistungsdiffusionspumpe f	pompe f à diffusion à haut débit	высокопроизводительный диффузионный насос
	high speed gauge, nude ion gauge, nude gauge, bare gauge tube	Einbausystem n, Eintauchsystem n <eines Ionisationsvakuummeters>	cellule f de mesure insérée, système m d'immersion	ионизационный манометр без оболочки, открытый ионизационный манометр
H 80	**high-speed pumping unit**	Hochleistungspumpstand m	installation f de pompes à grand débit	высокопроизводительный насосный агрегат
H 81	**high vacuum**	Hochvakuum n	vide m poussé	высокий вакуум
H 82	**high vacuum coater, high vacuum coating plant**	Hochvakuumaufdampfanlage f	installation f poste de métallisation sous vide, installation de métallisation sous vide	вакуумный испаритель
H 83	**high-vacuum column still**	Hochvakuumdestillationskolonne f	colonne f de distillation sous vide poussé	высоковакуумная перегонная колонна
H 84	**high-vacuum connection**, high-vacuum fitting, high-vacuum joint	Hochvakuumverbindung f	assemblage m à vide poussé, raccord m à vide poussé	высоковакуумное соединение
H 85	**high-vacuum cut-off**	Hochvakuumsperre f	fermeture f de vide poussé	высоковакуумная отсечка
H 86	**high-vacuum diagnostic tube**	Hochvakuumdiagnostikröhre f	tube m diagnostique sous vide poussé	высоковакуумная диагностическая лампа

	high-vacuum fitting	*s.* high-vacuum connection		
H 87	high vacuum fractional distillation	fraktionierte Hochvakuumdestillation *f*	distillation *f* fractionnée sous vide poussé	фракционированная перегонка под высоким вакуумом
H 88	high-vacuum fusion process, high-vacuum melting method	Hochvakuumschmelzverfahren *n*	fusion *f* (fondage *m*) sous vide poussé	процесс плавки в высоком вакууме
H 89	high-vacuum growing	Hochvakuumzüchtung *f* ‹von Kristallen›	élève *f* sous vide poussé	выращивание в высоком вакууме
H 90	high vacuum installation, high vacuum plant	Hochvakuumanlage *f*	installation *f* de vide poussé	высоковакуумная установка
	high-vacuum joint	*s.* high-vacuum connection		
	high-vacuum melting method	*s.* high-vacuum fusion process		
	high vacuum plant	*s.* high vacuum installation		
H 91	high-vacuum precision casting	Hochvakuumpräzisionsguß *m*	coulée *f* de précision sous vide poussé	высоковакуумное прецизионное литье
H 92	high-vacuum pump	Hochvakuumpumpe *f*	pompe *f* à vide poussé	высоковакуумный насос
H 93	high-vacuum rectifier	Hochvakuumgleichrichter *m*	redresseur *m* à vide poussé	кенотронный выпрямитель
H 94	high-vacuum sintering furnace	Hochvakuumsinterofen *m*	four *m* à recuire à vide poussé, four de frittage à vide poussé	высоковакуумная печь для спекания
H 95	high-vacuum stopcock	Hochvakuumhahn *m*	robinet *m* pour vide poussé	высоковакуумный кран
H 96	high-vacuum treatment tube	Hochvakuumtherapieröhre *f*	tube *m* de thérapie sous vide poussé	вакуумная лампа для терапии
H 97	high-vacuum valve	Hochvakuumventil *n*	soupape *f* à vide poussé	высоковакуумный вентиль
H 98	high-vacuum wax impregnating plant	Hochvakuumanlage *f* für Wachsimprägnierung	installation *f* à vide poussé à imprégnation de cire	высоковакуумная установка для пропитки воском
H 99	high-voltage welding	Hochspannungsschweißen *n*	soudure *f* à haute tension	высоковольтная сварка
H 100	Ho coefficient, Ho factor, speed factor	Ho-Faktor *m*	coefficient *m* Ho	коэффициент Хо
H 101	holding pump	Haltepumpe *f*	pompe *f* d'entretien, pompe de maintien, pompe pour maintenir le vide	поддерживающий насос
H 102	holding time	Entspannungstemperatur *f*, Haltezeit *f*, Kühldauer *f*	temps *m* de maintien, durée *f* de non-contamination	продолжительность охлаждения ‹в печи›
H 103	holding vacuum, take-hold pressure	Haltevakuum *n*	vide *m* d'entretien, vide à maintenir	давление, созданное поддерживающим насосом, поддерживающий вакуум
H 104	hole, cavity	Loch *n*, Lunker *m*	trou *m*, retassure *f*	дырка
H 105	hole capture	Löchereinfang *m*	blocage *m* des trous	захват дыркой
H 106	hollow beam	Hohlstrahl *m*	jet *m* creux	полый пучок
H 107	hollow cathode	Hohlkatode *f*	cathode *f* évidée	полый катод
H 108	hollow plunger	Sperrschieber *m*	tiroir *m* de blocage	полый плунжер
H 108	homogeneity	Homogenität *f*	homogénéité *f*	гомогенность, однородность
H 109	homogenization	Homogenisierung *f*	homogénéisation *f*	гомогенизация
H 110	homogenize	homogenisieren	homogénéiser	гомогенизировать
H 111	homopolar	homöopolar	homéopolaire	гомеополярный
H 112	honeycomb	Wabe *f*	alvéole *f*	ячейка
H 113	honeycomb structure	Wabenstruktur *f*	structure *f* en nid d'abeille, configuration *f* en nid d'abeille, structure croisée, configuration croisée	ячеистая структура, сотовая структура
	hood pressure test, chamber test, envelope test, overall test	Hüllentest *m*, Haubenlecksuchverfahren *n*, Haubenleckprüfung *f*	test *m* d'enveloppe, épreuve *f* d'étanchéité	проверка помещением объекта в оболочку
H 114	hood technique, hood testing	Umhüllungstechnik *f*, Hüllenmethode *f*	technique *f* d'enveloppement, méthode *f* d'enveloppe	метод проверки помещением объекта в оболочку
H 115	hoop drop relay	Fallbügelregler *m*	régulateur *m* à étrier mobile	регулятор с падающей дужкой
H 116	hoop stress	Ringspannung *f*	effort *m* annulaire	тангенциальное (кольцевое) напряжение
H 117	horizontal box furnace	horizontaler Kammerofen *m*	four *m* à chambres horizontales	горизонтальная камерная печь
H 118	hose	Schlauch *m*	tuyau *m* flexible, outre *f*	шланг
H 119	hose clamp, hose clip	Schlauchklemme *f*	pince *f* (collier *m*) pour tuyaux	зажим для шланга
H 120	hose nipple, hose nozzle	Schlauchtülle *f*, Schlauchwelle *f*	olive *f*, raccord *m* pour tuyaux	оливка, наконечник для шланга
	hose press clamp, clip, pinchcock, squeezing cock	Quetschhahn *m*	pince *f* pressante	зажим для шланга

	hot air gun, degassing gun	Heißluftdusche f	séchoir m électrique	ручной теплоэлектровентилятор
H 121	hot cathode, thermionic cathode	Glühkatode f	cathode f chaude (chauffée, incandescente, à incandescence)	термоэлектронный катод, термокатод, накаленный катод
H 122	hot-cathode ionization gauge, hot-filament ionization gauge, hot-wire ionization gauge	Glühkatodenionisationsvakuummeter n	manomètre m d'ionisation à cathode chaude	ионизационный манометр с накаленным катодом
H 123	hot-cathode magnetron ionization gauge	Magnetron-Ionisationsvakuummeter n mit Glühkatode, Lafferty-Ionisationsvakuummeter n	magnétron-vacuomètre m à ionisation avec cathode chauffée, Lafferty-vacuomètre m à ionisation	магнетронный ионизационный манометр с накаленным катодом
H 124	hot-cathode magnetron ionization gauge with photocurrent suppressor	Magnetronvakuummeterröhre f mit heißer Katode und Unterdrückung des Foto-Elektronenstromes	manomètre m à magnétron à cathode chaude et suppression du courant photoélectrique	магнетронный ионизационный манометр с накаленным катодом с подавлением фототока
H 125	hot-cathode vacuum tube, thermionic tube	Glühkatodenröhre f	tube m à cathode chaude, tube thermionique	электронная лампа с накаленным катодом
H 126	hot-cathode valve	Glühkatodenventil n	soupape f à cathode incandescente	кенотрон
H 127	hot cooling	Siedekühlung f	refroidissement m à l'ébullition	охлаждение кипением
H 128	hot-dip process	Tauchveredelung f	raffinage m par immersion	очистка погружением в нагретом состоянии
H 129	hot extraction gas analysis, vacuum fusion gas analysis	Heißextraktionsgasanalyse f	analyse f d'extraction à chaude	экстракционный анализ газов путем вакуумного расплавления
H 130	hot extraction process, vacuum fusion gas extraction	Heißextraktionsverfahren n	procédé m d'extraction à chaude	процесс горячей экстракции
	hot filament, filament, hot wire	Heizdraht m, Glühdraht m, Glühfaden m	filament m, filet m, fil m chaud (thermique)	нить накала, провод накала
	hot-filament ionization gauge	s. hot-cathode ionization gauge		
H 131	hot-gas defrosting	Abtauen n mit Heißgas	dégel m par gaz chaud	оттаивание горячим газом
H 132	hot-plate magnetic stirrer	Heizplattenmagnetrührer m	agitateur m magnétique à plaque de chauffage	магнитная мешалка с нагретой пластиной
H 133	hot-platinum halogen detector	Platin-Halogen-Lecksuchgerät n	détecteur m de fuites au platine-halogène	галоидный течеискатель с накаленным платиновым анодом
H 134	hot pressing	Heißpressen n	serrage m chaud	горячая прессовка
H 135	hot rolling	Heißwalzen n, Warmwalzen n	laminage m à chaud	горячая прокатка
H 136	hot shortness	Warmbrüchigkeit f	fragilité f à chaud	красноломкость
H 137	hot strength	Warmfestigkeit f	résistance f au rouge	теплостойкость, термостойкость
	hot tear, heat crack, pull	Wärmeriß m	fente f thermique	термическая трещина, тепловой разрыв
H 138	hot tensile strength	Warmzerreißfestigkeit f	résistance f à la rupture chaude	сопротивление разрыву в нагретом состоянии
H 139	hot vulcanization	Heißvulkanisation f	vulcanisation f à chaud	горячая вулканизация
	hot wire, hot filament, filament	Heizdraht m, Glühdraht m, Glühfaden m	filament m, filet m, fil m chaud (thermique)	нить накала, провод накала
	hot-wire ionization gauge	s. hot-cathode ionization gauge		
H 140	hot wire scissors	Glühdrahtzange f	pince f au filament	клещи для горячей проволоки
H 141	hot wire welding	Trennahtschweißen n	soudure f par couture	роликовая сварка через проволочную прокладку
H 142	hourly throughput	Stundendurchsatz m	débit m horaire	часовая производительность
	housekeeper seal, copper-to-glass seal	Glas-Kupfer-Verschmelzung f	scellement m verre-cuivre, joint m verre-cuivre	спай стекла с медью
H 143	Houston ionization gauge	Houstonsches Ionisationsvakuummeter n	manomètre m à ionisation de Houston	ионизационный манометр Хаустона
H 144	humidity, moisture	Feuchtigkeit f	humidité f	влажность
H 145	hydraulic radius	hydraulischer Radius m	rayon m hydraulique	гидравлический радиус
H 146	hydride process	Hydridverfahren n <Metall-Keramik-Verbindung>	hydrogénation f	гидридный процесс <пайки металла с керамикой>
H 147	hydrocarbon	Kohlenwasserstoff m	hydrocarbure m	углеводород
H 148	hydrocarbon-free vacuum	kohlenwasserstofffreies Vakuum n	vide m sans hydrocarbures	вакуум без остаточного давления углеводородов
H 149	hydrogen brazing	Hartlöten n unter Wasserstoffatmosphäre	brasure f sous hydrogène, soudure f forte sous hydrogène	пайка твердым припоем в водороде
H 150	hydrolytic classification	hydrolytische Klassifikation f	classification f hydrolytique	гидролитическая классификация
	hysteresis heating, eddy current heating	Wirbelstromheizung f	échauffement m par courants de Foucault	нагрев вихревыми токами

I

I 1	ice condenser	Eiskondensator *m*	condensateur *m* à glace	конденсатор со льдом
I 2	ice formation	Eisbildung *f*	formation *f* de glace	образование льда
I 3	ice nucleus	Eiskern *m*	nucléus *m* de glace, noyau *m* de glace	ледяное ядро, мутная серцевина в блоке льда
I 4	iconoscope	Ikonoskop *n*	iconoscope *m*	иконоскоп
I 5	ideal crystal, perfect crystal	Idealkristall *m*	cristal *m* idéal	идеальный кристалл
I 6	ideal gas, perfect gas	ideales Gas *n*	gaz *m* idéal (parfait)	идеальный газ
I 7	ideal gas constant, universal gas constant	universelle Gaskonstante *f*	constante *f* universelle des gaz	универсальная газовая постоянная
I 8	ideal pump, perfect pump	ideale Pumpe *f* ‹Pumpe mit konstantem Saug-vermögen über den gesamten Druck-bereich›	pompe *f* idéale	идеальный насос ‹с независящей от давле-ния быстротой действия›
I 9	ignition	Entflammung *f*	inflammation *f*, ignition *f*	зажигание
I 10	ignition quality	Glutfestigkeit *f*	résistance *f* à l'incandes-cence	стойкость к пережогу
I 11	ignition voltage	Zündspannung *f*	tension *f* d'ignition (d'amorçage, d'allu-mage)	напряжение зажигания, напряжение возни-кновения разряда
I 12	immerseable finger	Eintauchkühlfinger *m*	doigt *m* de refroidissement à plonger, doigt refroi-disseur d'immersion	погружаемый пальцевой холодильник
I 13	immersion freezer	Tauchgefrieranlage *f*	installation *f* frigorifique par immersion	установка для замора-живания погружением
I 14	impact ionization	Stoßionisation *f*	ionisation *f* par choc	ударная ионизация
I 15	impact probability, probability of collision	Stoßwahrscheinlichkeit *f*	probabilité *f* du choc, taux *m* de collision	вероятность соударения, вероятность столкно-вения
I 16	impact strength	Schlagbiegefestigkeit *f*, Schlagzähigkeit *f*, Stoßfestigkeit *f*	résilience *f*, résistance *f* au choc	ударная вязкость, сопро-тивление удару
I 17	impact strength testing	Schlagbiegeversuch *m*	essai *m* de flexion au choc	испытание на удар при изгибе
I 18	impeller, rotary piston	Drehkolben *m*	rotor *m*, piston *m* rotatif	поворотный плунжер, ротор
I 19	imperfect crystal	Realkristall *m*	cristal *m* réel	реальный (несовершен-ный) кристалл
I 20	imperfect gas, real gas	reales Gas *n*	gaz *m* réel	реальный (неидеаль-ный) газ
I 21	impingement rate, rate of incidence, number of collisions on a wall	Flächenstoßhäufigkeit *f*, Stoßzahlverhältnis *n*, mittlere Wandstoßzahl *f* pro Zeit- und Flächeneinheit, mittlere spezifische Wandstoßrate *f*	fréquence *f* du choc de surface, nombre *m* de collisions avec une paroi, fréquence de collision, taux *m* d'inci-dence	частота соударений с по-верхностью
I 22	implode	implodieren	imploder	взрываться внутрь
I 23	implosion	Implosion *f*	implosion *f*	взрыв, вызванный на-ружным давлением, раздавливание, имплозия
I 24	implosion guard, protection against implosion	Implosionsschutz *m*	protection *f* contre l'implosion	защита от имплозии
I 25	impregnate	imprägnieren	imprégner	импрегнировать
I 26	impregnated cathode	imprägnierte Katode *f*	cathode *f* imprégnée	импрегнированный катод
I 27	impregnating installa-tion, impregnation plant	Imprägnieranlage *f*	installation *f* d'imprégna-tion	установка для пропитки, пропиточная установ-ка
I 28	impregnation	Imprägnieren *n*, Imprägnierung *f*	imprégnation *f*	пропитка, импрегниро-вание
I 29	impulse	Impuls *m*	impulsion *f*	импульс
I 30	impulse transfer	Impulsübertragung *f*	transfert *m* d'impulsion	передача импульса
I 31	impulse voltage, surge voltage	Stoßspannung *f*	tension *f* de choc, surten-sion *f* impulsionnelle	импульсное напряжение
I 32	impurity	Verunreinigung *f*	impureté *f*	загрязнение, примесь
I 33	Inbus key	Inbus-Schlüssel *m*	Inbus-clé *f*, Inbus-clef *f*	ключ Инбуса
I 34	incandescent filament lamp	Glühlampe *f*	lampe *f* à incandescence	лампа накаливания
	inclined seat valve, slanting seat valve, bevel seat valve	Schrägsitzventil *n*	vanne *f* (robinet *m*) à siège oblique	вентиль с наклонным шпинделем
I 35	increase of gas pressure	Gasdruckanstieg *m*	accroissement *m* de pression gazeuse	повышение давления газа
I 36	increase of tube vacuum	Erhöhung *f* des Röhrenvakuums	accroissement *m* du vide d'un tube	улучшение вакуума в лампе
	incrush of air, air impact, air shock	Lufteinbruch *m*	entrée *f* d'air accidentelle, pénétration *f* d'air	прорыв воздуха, воздуш-ный удар
I 37	incrustation	Krustenbildung *f*	incrustation *f*	образование корки (налета)
I 38	indentation hardness	Eindringhärte *f*	dureté *f* de pénétration	твердость на вдавлива-ние, сопротивление вдавливанию

I 39	**independent of voltage**	spannungsunabhängig	indépendant de la tension	независимый от напряжения
I 40	**indirectly heated**	indirekt geheizt	à chauffage indirect	с косвенным подогревом, с косвенным накалом
I 41	**indium wire seal**	Indium-Drahtdichtung *f*	joint *m* métallique d'indium	индиевое проволочное уплотнение
I 42	**induction coupled plasma torch,** induction plasma torch	Induktionsplasmabrenner *m*	brûleur *m* d'induction à plasma	индукционная плазменная горелка
I 43	**induction-heated melting furnace**	Induktionsschmelzofen *m*	four *m* de fusion par induction	индукционная плавильная печь
I 44	**induction heating,** radio-frequency heating, rf heating	Induktionsheizung *f*, induktive Wärmebehandlung *f*	chauffage *m* par induction	индукционный нагрев, высокочастотный нагрев
I 45	**induction melting**	Induktionsschmelzen *n*	fusion *f* par induction	индукционная плавка
	induction plasma torch	*s.* induction coupled plasma torch		
I 46	**induction stirring**	induktives Rühren *n*	agitation *f* inductive	индукционное перемешивание
I 47	**inert gas,** noble gas, rare gas	Edelgas *n*	gaz *m* inerte, gaz rare	инертный газ, редкий газ, благородный газ
I 48	**inert gas consumable-electrode arc welding,** sigma welding	Metall-Inertgas-Schweißen *n*	soudage *m* gaz inerte-métal	дуговая сварка в среде инертного газа с расходуемым электродом
I 49	**inert-gas shielded-arc welding**	Inertgasschweißung *f*	soudure *f* sous gaz inerte	дуговая сварка в среде инертного газа
I 50	**inert-gas-shielded consumable-electrode process (Aircomatic)**	Aircomatic-Verfahren *n*	procédé *m* « Aircomatic »	способ Айркоматик, сварка плавящимся электродом в аргоновой среде
	inert-gas-shielded nonconsumable-electrode arc welding method, heliarc welding process <Heliarc>	Heliarc-Verfahren *n*, Wolfram-Inertgas-Schweißen *n* mit Helium als Schutzgas	procédé *m* Heliarc	способ гелиево-дуговой сварки, способ дуговой сварки в гелиевой среде
I 51	**inflection point**	Knickpunkttemperatur *f*	point *m* de pliage de la température	точка перегиба температурной кривой
I 52	**inflow rate**	Einlaßrate *f*	taux *m* d'entrée	скорость всасывания
I 53	**infra-red absorption**	Infrarotabsorption *f*	absorption *f* infrarouge	абсорбция инфракрасных лучей
I 54	**infra-red adsorption spectroscopy**	Infrarotadsorptionsspektroskopie *f*	spectroscopie *f* d'adsorption infrarouge	адсорбционная спектроскопия в инфракрасных лучах
I 55	**infra-red heating**	Infrarotheizung *f*	chauffage *m* infrarouge	инфракрасный нагрев, нагрев инфракрасными лучами
I 56	**infra-red leak detector**	Infrarotlecksuchgerät *n*	détecteur *m* de fuites d'un récipient par rayons infrarouges	инфракрасный течеискатель
I 57	**ingot**	Schmelzblock *m*	barre *f* (lingot *m*, saumon *m*) fusible	слиток
I 58	**ingot lock**	Schmelzblockschleuse *f*	écluse *f* à lingot fusible	шлюз для слитка
	ingot melting, extrusion melting	Strangschmelzen *n*	fusion *f* par extrusion	плавка с вытягиванием слитка
I 59	**ingot mould**	Blockform *f*, Barrenform *f*	lingotière *f*	изложница
I 60	**ingot puller**	Blockabzug *m*	échappement *m* de lingot	приспособление для вытягивания слитка
I 61	**ingot support**	Kokillentisch *m*	support (appui) *m* de lingotière	основание изложницы
I 62	**ingot weight**	Schmelzblockgewicht *n*	poids *m* du lingot	вес слитка
I 63	**ingot withdrawing device**	Blockabzugsvorrichtung *f*	appareil *m* d'échappement de lingots, dispositif *m* d'échappement de lingots	устройство для извлечения слитка
I 64	**inhomogeneous**	inhomogen	hétérogène, non homogène	неоднородный
I 65	**initial pressure**	Anfangsdruck *m*, Startdruck *m*	pression *f* initiale	начальное давление
I 66	**initial susceptibility**	Anfangssuszeptibilität *f*	susceptibilité *f* initiale	начальная чувствительность
I 67	**injection condenser,** jet condenser	Einspritzkondensator *m*	condenseur *m* par injection, condenseur à jet, injecteur *m* condenseur	конденсатор смешения с охлаждением впрыском
	injection moulding, die casting	Spritzgießen *n*	moulage *m* par injection	литье под давлением
I 68	**inleakage rate,** leakage rate, leak rate	Leckrate *f*, Undichtigkeit *f*	défaut *m* d'étanchéité, débit *m* d'une fuite, taux *m* de fuite	натекание, скорость натекания
I 69	**inlet area,** intake area, throat, throat area	Ansaugquerschnitt *m*	section *f* d'aspiration	площадь сечения горловины сопла
I 70	**inlet flange,** intake flange, inlet-pipe connection	Eintrittsstutzen *m*	manchon *m* d'entrée	впускной фланец, входной патрубок
	inlet-pipe connection	*s.* inlet flange		

	inlet pressure, intake pressure, fine pressure, head pressure, suction pressure	Ansaugdruck *m*	pression *f* d'aspiration, pression d'admission	впускное давление
	inlet side, collecting side, intake side, suction side, vacuum side	Saugseite *f*, Vakuumseite *f*	côté *m* vide, côté d'aspiration	собирающая сторона, сторона впуска
I 71	inline [exhaust] system	Pumpstraße *f*	train *m* de pompage	подвижная откачная система
I 72	inline valve, straight-through valve, through-way valve, straight-way valve	Durchgangsventil *n*	soupape *f* droite, soupape de traversée, vanne *f* à passage droit	проходной вентиль
I 73	inner diameter	Innendurchmesser *m*	diamètre *m* intérieur	внутренний диаметр
I 74	insertion ion gauge, nude ion source	Eintauchionenquelle *f*	source *f* ionique d'immersion	открытый ионный источник
I 75	inside frosted	innenmattiert	dépoli (satiné) intérieurement	матированный изнутри
I 76	inside lap seal	Innenzylinderlötung *f*	soudure (brasure) *f* de cylindre intérieur	внутреннее паяное цилиндрическое соединение
I 77	inside tubular seal	Innenanglasung *f*	scellement *m* de verre intérieur	внутренний трубчатый спай, спай с внутренней трубкой
I 78	inspection glass, observation window, sight glass, sight port, sight window, viewing glass, viewing port, viewing window	Beobachtungsfenster *n*, Schauglas *n*, Einblickfenster *n*	fenêtre *f* de contrôle, regard *m*, regard de contrôle, fenêtre d'observation, fenêtre de coup d'œil, hublot *m*	контрольное (смотровое) стекло, глазок, смотровое отверстие (окно), наблюдательное окно
I 79	inspection window for oil level	Kontrollfenster *n* für den Ölstand	fenêtre *f* de contrôle à niveau d'huile	контрольное окно для наблюдения за уровнем масла
I 80	inspiration, sucking, suction	Saugen *n*, Ansaugen *n*, Einsaugen *n*	succion *f*, aspiration *f*	всасывание, засасывание
I 81	installation, plant, plant equipment	Anlage *f*	installation *f*, appareillage *m*	установка, устройство
I 82	instantaneous annealing point	Schnellentspannungstemperatur *f*	température *f* de détente instantanée	температура быстрого отжига
I 83	insulant, insulating material	Isolierstoff *m*, Isoliermaterial *n*	isolant *m*, matière *f* isolante	изоляционный материал
I 84	insulated metal sheathed wire	Rohrdraht *m*	fil *m* sous tube	трубчатый (бронированный) провод
I 85	insulating compound insulating material	Isoliermasse *f* *s.* insulant	masse *f* isolante	изолирующий состав
I 86	insulating power	Isolierfähigkeit *f*	pouvoir *m* isolant	изоляционная способность
	intake area, inlet area, throat, throat area	Ansaugquerschnitt *m*	section *f* d'aspiration	площадь сечения горловины сопла
	intake flange	*s.* inlet flange		
I 87	intake line, suction line	Saugleitung *f*	conduite (canalisation) *f* d'aspiration, tuyau *m* de pompage	впускной (всасывающий) трубопровод
I 88	intake port, suction port, pump inlet	Ansaugöffnung *f*	fente *f* d'aspiration, ouverture *f* d'aspiration	впускной патрубок
	intake pressure, inlet pressure, fine pressure, head pressure, suction pressure	Ansaugdruck *m*	pression *f* d'aspiration, pression d'admission	впускное давление
	intake side	*s.* inlet side		
I 89	integral leakage, total leakage, total leaks	Gesamtundichtigkeit *f*, Leckrate *f*	fuite *f* totale	общее натекание
I 90	integrated electronics	Festkörpermikroelektronik *f*	micro-électronique *f* des corps solides	интегральная электроника, микроэлектроника
I 91	interaction	Wechselwirkung *f*	réaction *f* réciproque	взаимодействие
I 92	interchangeable, replaceable	austauschbar	échangeable	взаимозаменяемый
I 93	interchangeable vaporizer system	Wechselverdampfer *m*	évaporateur *m* interchangeable	сменный испаритель
I 94	interconnecting technique	Verbindungstechnik *f*	technique *f* de jonction (raccordement)	техника соединений
I 95	intercrystalline fracture	interkristalliner Bruch *m*	fraction *f* intercristalline	межкристаллический излом, межкристаллическая трещина
I 96	interface	Grenzfläche *f*	surface *f* limite	граничная поверхность, поверхность раздела
I 97	interface energy, interface surface energy, interfacial energy	Grenzflächenenergie *f*	énergie *f* interfaciale (de surface limite)	поверхностная энергия
I 98	interfacial film	Grenzflächenfilm *m*	film *m* interfacial (de surface limite)	граничная пленка
I 99	interfacial layer, intermediate layer, interlayer	Zwischenschicht *f*	couche *f* intermédiaire, lit *m* intercalé	прослойка
I 100	interference coating, interference layer interlayer	Interferenzschicht *f*	couche *f* d'interférence	интерференционная пленка

I 101	**interference film system**	Interferenzschichtsystem *n*	système *m* des couches interférentielles (d'interférence)	система с интерференционными пленками
I 102	**interference filter**	Interferenzfilter *n*	filtre *m* d'interférence	интерференционный фильтр
	interference layer	*s.* interference coating		
I 103	**interference tube**	Interferenzröhre *f*	tube *m* à l'interférence	интерференционная лампа
I 104	**interferometric manometer**	interferometrisches Vakuummeter *n*	vacuomètre *m* interférentiel	интерферометрический манометр
I 105	**interferometric oil manometer**	interferometrisches Ölvakuummeter *n*	vacuomètre *m* interférométrique à huile	интерферометрический масляный манометр
I 106	**interior**	Innenraum *m*	espace *m* intérieur	внутреннее пространство
	interlacing, cross-linkage	Vernetzung *f*	réticulation *f*	сшивание, установление поперечной связи
	interlayer	*s.* interfacial layer		
I 107	**interlock**	verriegeln	bloquer	блокировать
I 108	**interlock, inter-locking,** locking	Verriegelung *f*, Verblok-ken *n*, Verriegeln *n*, Verblockung *f*	verrouillage *m*, blocage *m*	блокировка, запирание
I 109	**intermediate con-denser,** interstage condenser	Zwischenkondensator *m*, Zwischenkondensor *m*	condensateur *m* intermé-diaire, condenseur *m* in-terposé (intermédiaire)	промежуточный кон-денсатор
I 110	**intermediate flow**	Übergangsströmung *f* ‹Gleitströmung›	étendue *f* (domaine *m*) transitoire	переходная область течения ‹течение со скольжением›
	intermediate glass, adapter glass	Zwischenglas *n*	verre *m* intermédiaire	переходное стекло
	intermediate layer	*s.* interfacial layer		
	intermediate piece, adapter, transition piece	Zwischenstück *n*	raccord *m* intermédiaire, pièce *f* intermédiaire, adapteur *m*	промежуточный элемент, переходник
	intermediate pump, booster pump	Booster-Pumpe *f*, Zwischenpumpe *f*	pompe *f* intermédiaire, pompe de Booster	бустерный насос, вспо-могательный насос
I 111	**intermediate sampling**	Zwischenprobenent-nahme *f*	échantillonnage *m* intermédiaire	отбор промежуточных проб
I 112	**intermolecular**	zwischenmolekular	intermoléculaire	межмолекулярный
I 113	**internal friction,** internal viscosity	innere Reibung *f*	friction *f* interne	внутреннее трение
I 114	**internal heater**	Innenheizer *m*	chauffeur *m* intérieur	внутренний подогрева-тель, внутренний на-греватель
	internal thread, female thread	Innengewinde *n*	taraudage *m* [de vis intérieur]	внутренняя резьба
	internal viscosity	*s.* internal friction		
I 115	**International Union for Vacuum Science, Technique and Applications, IUVSTA**	Internationale Union der Forschung, Technik und Anwendung des Vakuums, IUFTAV	Union Internationale pour la Science, la Technique et les Applications du Vide, UISTAV	Международный вакуумный научно-технический союз
I 116	**interplanar spacing**	Netzebenenabstand *m*	distance *f* de plaine de réseau	межплоскостное рас-стояние решетки
	interstage condenser	*s.* intermediate condenser		
I 117	**intrinsic fatigue strength**	Schwellfestigkeit *f*	résistance *f* de seuil	предел прочности при знакопостоянном по-ложительном цикле, предел прочности при знакопостоянной периодической нагрузке
I 118	**intrinsic speed**	tatsächliches (wahres) Saugvermögen *n*	débit *m* effectif	истинная скорость откачки
I 119	**intrinsic speed,** theoretical speed	Eigensaugvermögen *n*	débit *m* intrinsèque	теоретическая быстрота откачки
I 120	**intrinsic throughput**	theoretischer Drucksatz *m*	débit *m* massique intrin-sèque, débit massique théorique	входной поток, теорети-ческий поток
I 121	**intrinsic viscosity**	Viskositätszahl *f*, Grundviskosität *f*	nombre *m* de viscosité	коэффициент внутрен-него трения
I 122	**inverse direction,** non-conducting sense	Sperrichtung *f*	sens *m* de blocage (non-conduction)	непроводящее (обратное) направление
	inverted evaporation, down evaporation	Abwärtsverdampfung *f*	vaporisation *f* descendante	испарение вниз
I 123	**inverted jet**	Umkehrstrahl *m*	rayon *m* d'inversion	обратная струя
	inverted jet nozzle, annular jet nozzle, umbrella jet	ringförmige Strahl-umlenkdüse *f*	diffuseur *m*, diffuseur-déflecteur *m*, tuyère *f* à jet inversé	зонтичная ступень, зон-тичное сопло, обра-щенное сопло Лаваля
I 124	**inverted magnetron gauge**	Vakuummeter *n* vom Typ des umgekehrten Magnetrons	vacuomètre *m* de type d'un magnétron inverse	инверсно-магнетронный манометр
I 125	**inverted magnetron sputter-ion pump**	Ionenzerstäuberpumpe *f* vom Typ des umge-kehrten Magnetrons	pompe *f* ionique à getter par pulvérisation catho-dique du type de mano-mètre magnétron ren-versé	инверсно-магнетронный электроразрядный насос
I 126	**ion baffle**	Ionenbaffle *n*	baffle *m* ionique, piège *m* ionique	отражатель ионов, ионная ловушка

I 127	ion beam deposition	Ionenstrahlbedampfung f	déposition f par faisceau ionique	ионнолучевое напыление
I 128	ion beam technique	Ionenstrahltechnik f	technique f de faisceau ionique	ионнолучевая техника
I 129	ion bombardment	Ionenbeschuß m	bombardement m ionique	ионная бомбардировка
I 130	ion burial	Ionenaufzehrung f <durch Einschießen in die Katode>	captage m d'ions, absorption f d'ions	замуровывание ионов
I 131	ion capture	Ioneneinfang m	captage m d'ions	ионный захват
I 132	ion-capture probability	Ioneneinfangwahrscheinlichkeit f	probabilité f de captage d'ions	вероятность ионного захвата
I 133	ion collector, plate, target	Ionenkollektor m, Ionenauffänger m	collecteur m, collecteur ionique	ионный коллектор
I 134	ion current transmission	Ionenstromtransmission f	transmission f de courant, transmission d'ions, transmission de courant ionique, transmission de flux ionique	трансмиссия ионного тока
I 135	ion efficiency	Ionenwirkungsgrad m	rendement m ionique	ионная эффективность
	ion etching, etching by ion bombardment	Ionenätzen n	corrosion f ionique	ионное травление
I 136	ion exchanger	Ionenaustauscher m	échangeur m ionique	ионообменник, ионит
I 137	ion gauge, ionization gauge	Ionisationsmanometer n, Ionisationsvakuummeter n	manomètre m à ionisation, jauge f à ionisation	ионизационный манометр
I 138	ion-gauge control	Ionisationsmeßgerät n	ionomètre m	измерительная схема ионизационного манометра
I 139	ion gauge rate monitor, ionization-type rate monitor	Ionisationsratemonitor m	moniteur m de quotepart d'ionisation	измеритель тока ионизационного манометра
	ion-getter pump, getter-ion pump	Ionengetterpumpe f	pompe f getter-ionique, pompe ionique à getter	геттеро-ионный насос, сорбционно-ионный насос
I 140	ionic[al]	in Ionenform, ionisch	ionique	ионный, в ионной форме
	ionic [bombardment] cleaning, discharge cleaning, glow discharge cleaning	Abglimmen n, Glimmreinigung f	nettoyage m par bombardement ionique, décharge f par effluves	очистка разрядом, очистка тлеющим разрядом, ионная очистка
I 141	ionic conductivity	Ionenleitfähigkeit f	conductibilité f ionique	ионная проводимость
I 142	ion impact	Ionenstoß m	choc m d'ions, impact m d'ions	ионный удар
I 143	ionization by collision	Stoßionisierung f	ionisation f par choc	ионизация при соударении, ударная ионизация
I 144	ionization chamber	Ionisationskammer f	chambre f d'ionisation	ионизационная камера
I 145	ionization detector	Ionisationsdetektor m	détecteur m d'ionisation	ионизационный детектор
I 146	ionization device	Ionisierungseinrichtung f	dispositif m (installation f) d'ionisation	ионизационная установка
I 147	ionization energy	Ionisierungsarbeit f	énergie f d'ionisation	энергия ионизации
	ionization gauge	s. ion gauge		
I 148	ionization potential, ionization voltage	Ionisierungsspannung f	potentiel m d'ionisation	потенциал ионизации
I 149	ionization probability	Ionisierungswahrscheinlichkeit f	probabilité f d'ionisation	вероятность ионизации
I 150	ionization pump, ion pump	Ionenpumpe f	pompe f ionique	ионный насос
I 151	ionization space	Ionisierungsraum m	espace m d'ionisation	ионизационное пространство
	ionization-type rate monitor	s. ion gauge rate monitor		
I 152	ionization vacuum gauge	Ionisationsvakuummeter n	jauge f à ionisation, vacuomètre m d'ionisation, manomètre m à ionisation	ионизационный вакуумметр
	ionization voltage	s. ionization potential		
I 153	ionization volume	Ionisierungsvolumen n	volume m d'ionisation	ионизационный объём
I 154	ionize	ionisieren	ioniser	ионизировать
I 155	ion-microprobe mass spectrometer	Ionenmikrosonden-Massenspektrometer n	spectromètre m de masse par micro-sondes ioniques	масс-спектрометр с ионным микрозондом
I 156	ionosorption tube	Ionensorptionsröhre f	tube m de sorption ionique	ионно-сорбционная трубка
I 157	ion plating-film	ionisch hergestellte Schicht f	filet m produit par des ions	плёнка, полученная ионным осаждением
	ion pump	s. ionization pump		
I 158	ion pumping rate	Ionenpumpgeschwindigkeit f	vitesse f de pompage ionique	быстрота действия ионного насоса
I 159	ion pump leak detector	Lecksuchgerät n mit Ionenpumpe	détecteur m de fuites par pompe ionique	течеискатель с ионным насосом
I 160	ion repeller	Gegenfeldelektrode f für Ionen	électrode f inverse pour ions	отражатель ионов
I 161	ion resonance	Ionenresonanz f	résonance f des ions	ионный резонанс
I 162	ion resonance spectrometer	Ionenresonanzspektrometer n	spectromètre m ionique à résonance	ионно-резонансный спектрометр
I 163	ion rocket	Ionenrakete f	fusée f ionique	ионная ракета, ионный ракетный двигатель
I 164	ion sorption	Ionensorption f	sorption f ionique	ионная сорбция

I 165	ion-sorption pump	Ionensorptionspumpe f	pompe f ionique à sorption	ионно-сорбционный насос
I 166	ion source	Ionenquelle f	source f d'ions	ионный источник
I 167	ion-yield	Ionenausbeute f	débit m d'ions	выход ионов
I 168	iridium filament	Iridium-Glühfaden m	filament m d'iridium	иридиевая нить накала
I 169	iris diaphragm	Irisblende f	diaphragme m iris	ирисовая диафрагма
I 170	iron slag	Eisenschlacke f	laitier m de hauts fourneaux	железный шлак
I 171	irradiation facilities	Bestrahlungstation f	installation f d'irradiation	установка для облучения
I 172	isentropic[al]	isentrop	isentropique	изентропический
I 173	island film	Inselschicht f	couche f insulaire	островная пленка
I 174	island structure	Inselstruktur f	structure f d'île	островная структура
I 175	isobaric, isopiestic	isobar	isobare	изобарический
I 176	isolation test, rate of rise test (method), pressure rise test	Druckanstiegsmethode f <zur Lecksuche>	méthode f d'accroissement de pression	метод измерения быстроты возрастания давления, метод изоляции <в течеискании>
	isolation valve, seal-off valve, cut-off valve	Verschlußventil n, Absperrventil n	soupape f de fermeture, vanne f obturatrice, robinet m obturateur, soupape d'arrêt, vanne d'arrêt	запорный клапан, запорный вентиль, изоляционный вентиль
	isopiestic	s. isobaric		
I 177	isopleth	Isoplethe f	isoplèthe f	изоплета
I 178	isothermal, isothermic	isotherm	isothermique	изотермический
I 179	isotope abundance ratio	Isotopenhäufigkeitsverhältnis n	relation f de fréquence des isotopes	относительная распространенность изотопа
I 180	isotope ratio	Isotopenverhältnis n	proportion f d'isotopes	изотопное отношение, изотопный состав
I 181	isotope separation	Isotopentrennung f	séparation f des isotopes	разделение изотопов
I 182	isotopic gas analysis	Isotopengasanalyse f, Isotopenhäufigkeitsbestimmung f	détermination f de la fréquence des isotopes, analyse f des gaz d'isotopes	радиоизотопный газовый анализ
	IUVSTA	s. International Union for Vacuum Science, Technique and Applications		

J

J 1	jacketed lift pipe, jacketed siphon	Mantelheber m	tube m d'ascension à double paroi, siphon m enveloppé, siphon avec enveloppe	сифон, заключенный в рубашку
J 2	jacketed shelf	Heizplatte f, Heizblech n	plaque f chauffante (de chauffage)	плита с обогревом
J 3	jacketed shelf dryer	Heizplattentrockner m	sécheur m par plaque de chauffage	сушилка с подогревными полками
	jacketed siphon	s. jacketed lift pipe		
J 4	jet assembly, nozzle assembly	Düsenstock m, Düsensatz m	ensemble m diffuseur	система сопел
J 5	jet cap, top cap, top jet cap	Düsenhut m	baffle m à chapeau, baffle-chapeau m	колпачок сопла
J 6	jet cap heater	Düsenhutheizer m	chauffeur m du baffle à chapeau	подогреватель колпачкового отражателя
J 7	jet clearance, nozzle clearance	Diffusionsspaltbreite f	largeur f de clivage de diffusion	ширина зазора сопла, ширина диффузионной поверхности
	jet clearance area, nozzle clearance area, pump throat area, aperture gap, annular gap, admittance area	Diffusionsspaltfläche f	plan m de clivage de diffusion	площадь кольцевого зазора, площадь поверхности диффузии
	jet condenser, injection condenser	Einspritzkondensator m	condenseur m par injection, condenseur à jet, injecteur m condenseur	конденсатор смешения с охлаждением впрыском
J 8	jet efficiency	Strahlwirkungsgrad m	rendement m du rayon	эффективность струи
J 9	jet exhauster, steam ejector [pump], steam ejector vacuum pump	Dampfstrahler m, Wasserdampfstrahlsauger m, Dampfstrahlsauger m, Dampfstrahlvakuumpumpe f	éjecteur m à vapeur d'eau, aspirateur m de vapeur, pompe f à vide à vapeur d'eau	паровой эжекторный насос, парожэжекторный насос, паровой (пароструйный) эжектор, пароструйный эжекторный вакуумный насос
J 10	jet formation in valves	Düsenvorgang m	formation f dans les tuyères	образование струи в сопле
J 11	jet ignition	Spritzzündung f	allumage m par injection	струйное зажигание
	jet nozzle, nozzle, ejector nozzle	Strahldüse f	tuyère f, buse f, diffuseur m, éjecteur m	эжектор, струйное сопло
J 12	jet propulsion	Düsenantrieb m, Strahlantrieb m	propulsion f à réaction	реактивное движение
J 13	jet stack, vapour pipe	Dampfsteigrohr n	tuyau m de vapeur, colonne f de vapeur, tube m d'amenée de vapeur	паропровод

J 14	jet system	Strahlsystem *n*	système *m* de rayon	струйная система
J 15	jet turbine engine	Strahltriebwerk *n*	mécanisme *m* moteur à injection	турбореактивный двигатель
J 16	jig	Aufspannvorrichtung *f*, Einspannvorrichtung *f*	dispositif *m* de serrage, dispositif de fixation	фиксирующее приспособление, фиксирующий зажим
J 17	joining	Aneinanderfügung *f*, Anschluß *m*	raccordement *m*, connexion *f*, jonction *f*	соединение, присоединение
J 18	joining by welding, joint welding, junction welding	Verbindungsschweißen *n*	soudure *f* de jonction	соединительная сварка, соединительная пайка
J 19	joint face, sealing joint, sealing surface, seal area	Dichtfläche *f*, Dichtungsfläche *f*	surface *f* de joint (contact du joint), contact *m* (surface) d'étanchéité, surface d'étanchage	поверхность соединения, уплотняющая поверхность, плоскость уплотнения
	joint welding	s. joining by welding		
	journal bearing, ball bearing	Kugellager *n*	roulement *m* à billes, palier *m* à billes	шариковый подшипник
	junction welding	s. joining by welding		

K

K 1	kerosene leak detection	Lecksuche *f* mit Petroleum	recherche *f* de fuites avec du pétrole	керосиновый течеискатель
K 2	Kerr effect, Kerr magneto-optic effect	Kerr-Effekt *m*	effet *m* Kerr	магнитооптический эффект Керра
K 3	ketone resins	Ketonharze *npl*	résines *fpl* de cétone	кетоновые смолы
K 4	key, plug	Hahnküken *n*, Stopfen *m*	clé *f*, vanne *f* à opercule, tournant *m*	пробка крана
K 5	key holder	Hahnfassung *f*	douille *f* du robinet	держатель ключа
K 6	killed steel	beruhigter Stahl *m*	acier *m* calciné à mort	спокойная сталь, успокоенная сталь, раскисленная сталь
K 7	kinematic viscosity	kinematische Zähigkeit *f*	viscosité *f* cinématique	кинематическая вязкость
K 8	kinetic theory of gases	kinetische Gastheorie *f*	théorie *f* cinétique des gaz	кинетическая теория газов
	kinetic vacuum system, dynamic vacuum system, pumped vacuum system	dynamisches Vakuumsystem *n*, dynamische Vakuumanlage *f*	système *m* dynamique à (du) vide	динамическая вакуумная система
K 9	Kirchhoff's law	Kirchhoffsches Gesetz *n*	loi *f* de Kirchhoff	закон Кирхгофа
K 10	kneading machine, masticator	Knetmaschine *f*	malaxeur *m*, pétrisseuse *f*	смесительная машина
K 11	knife-edge seal	Schneidendichtung *f*	joint *m* à couteau, jonction *f* à pénétration	уплотнение с ножевидными выступами, уплотнение острой кромкой
K 12	Knudsen flow, transition flow	Knudsen-Strömung *f* ‹Gasströmung *f* im Bereich zwischen laminarer und Molekularströmung›, Übergangsströmung *f*	écoulement *m* en régime intermédiaire, écoulement d'après Knudsen	Кнудсеновское течение ‹молекулярно-вязкостный режим течения газа›
K 13	Knudsen gauge, Knudsen radiometervacuummeter, radiometer gauge	Radiometer-Vakuummeter *n*, Knudsen-Vakuummeter *n*, Knudsensches Radiometer-Vakuummeter *n*	vacuomètre *m* radiométrique [de Knudsen]	манометр Кнудсена, радиометрический манометр [Кнудсена]
K 14	Knudsen number Knudsen radiometervacuummeter	Knudsen-Zahl *f* s. Knudsen gauge	nombre *m* de Knudsen	число Кнудсена
K 15	Knudsen rate of evaporation, maximum evaporation rate, Langmuir rate of evaporation	maximale Verdampfungsrate *f*, absolute Verdampfungsrate	taux *m* maximum d'évaporation	абсолютная скорость испарения
K 16	Knudsen's law, molecular law	Knudsensches Gesetz *n*	loi *f* de Knudsen	закон Кнудсена
K 17	Kovar seal	Glas-Kovar-Verschmelzung *f*	scellement *m* verre-kovar, joint *m* verre-kovar	спай стекла с коваром

L

L 1	labyrinth gland, labyrinth seal	Labyrinthdichtung *f*	joint *m* à labyrinthe, garniture *f* en cannelures	лабиринтное уплотнение
L 2	lacquer, varnish	lackieren	vernir, laquer	лакировать
L 3	lacquer coating	Lackschicht *f*, Lacküberzug *m*	couche *f* de laque	лаковое покрытие
L 4	lacquer mask	Fotolackmaske *f*	photo-masque *m* de vernis	маска из фотолака, маска из светочувствительного лака
L 5	ladle degassing	Pfannenentgasung *f*	dégagement *m* de poche	ковшовое обезгаживание
L 6	ladle drying plant	Pfannentrocknungsanlage *f*	installation *f* de séchage en poche de coulée	ковшовая сушилка

L 7	ladle-to-ladle vacuum stream degassing	Durchlaufentgasung f von Pfanne zu Pfanne	dégazage m continu de poêle à poêle	вакуумное обезгаживание с непрерывным переходом от одной позиции к другой
	Lambert's law, cosine law of emission	Lambertsches Kosinusgesetz n, Lambertsches Gesetz n	loi f cosinusoïdale de Lambert, loi de Lambert	закон Ламберта
L 8	laminar flow, laminar-viscous flow	laminare Strömung f, Laminarströmung f	écoulement m [en régime] laminaire, courant m laminaire	ламинарное течение
L 9	laminated plastic	Schichtkunststoff m	matière f plastique empilée	слоистая пластмасса
L 10	laminating	Schichtung f	stratification f, délit m	расслоение, расщепление
L 11	Langmuir-Dushman molecular gauge	Langmuir-Dushmansches Molekularvakuummeter n	vacuomètre m moléculaire de Langmuir-Dushman	мольвакуумметр Ленгмюра-Дэшмана
	Langmuir rate of evaporation, maximum evaporation rate, Knudsen rate of evaporation	maximale Verdampfungsrate f, absolute Verdampfungsrate	taux m maximum d'évaporation	абсолютная скорость испарения
L 12	lap	läppen	roder	пришлифовывать
L 13	lap joint, lap weld	Überlappstoß m	choc m de recouvrement	сварка (соединение) внахлестку
L 14	lapped joint lap weld	Falzverbindung f s. lap joint	raccord m plié	соединение внахлестку
L 15	large-aperture straight-through bakeable vacuum valve	ausheizbares Vakuumdurchgangsventil n großer Nennweite	vanne f étuvable à passage direct de grand diamètre nominal	прогреваемый проходной вентиль с большим диаметром условного прохода
L 16	laser mirror	Laser-Spiegel m	laser-miroir m	лазерное зеркало
L 17	lattice constant	Gitterkonstante f	constante f de grille (réseau moléculaire)	постоянная решетки
L 18	lattice defect	Gitterfehler m	défaut m de grille	дефект кристаллической решетки
L 19	lattice defect density	Gitterfehlstellendichte f	densité f des endroits défectueux d'un réseau moléculaire	плотность дефектов кристаллической решетки
L 20	lattice electron	Gitterelektron n	électron m de grille	электрон решетки
L 21	lattice imperfection	Gitterfehlstelle f	imperfection f de réseau moléculaire	несовершенство (нарушение) кристаллической решетки
L 22	lattice plane	Gitterebene f, Netzebene f	plaine f de réseau (grille)	плоскость кристаллической решетки
L 23	lattice vacancy, lattice vacant site	Gitterleerstelle f	point m non occupé du réseau moléculaire, place f non occupée du réseau moléculaire	вакантное место в решетке, вакансия в кристаллической решетке
L 24	lattice vibration	Gitterschwingung f	vibration f du treillis (réseau moléculaire)	колебание решетки
L 25	Laue back-reflection method	Rückstrahlverfahren n nach Laue	procédé m de réflexion de Laue	метод обратного отражения Лауэ
L 26	Laval valve	Laval-Düse f	Laval-injecteur m, tuyère f de Laval	сопло Лаваля
L 27	law of mass action	Massenwirkungsgesetz n	loi f d'action des masses	закон действующих масс
L 28	layer of tin foil, tin foil coating	Stanniolbelag m	couche f de feuilles d'étain	станиолевое покрытие
L 29	L-cathode	L-Katode f	L-cathode f	L-катод
L 30	lead-in, lead-through	Durchführung f	passage m, traversée f	ввод
L 31	lead-in cable	Einführungskabel n	câble m d'entrée	вводный кабель
L 32	leading-in wire	Einführungsdraht m	fil m d'entrée	вводный провод
L 33	lead monoxide and glycerine cement, litharge and glycerine cement, PbO and glycerine cement	Glyzerin-Bleiglätte-Kitt m	mastic m de glycérine et litharge	цемент (замазка) на основе глицерина и свинцового глета
L 34	lead seal	Bleidichtung f	joint m de plomb	свинцовое уплотнение (соединение)
	lead-through	s. lead-in		
L 35	leaf valve	Blattventil n	valve (soupape) f à lames	створчатый вентиль
L 36	leak, leakage	Leck n, Riß m, Verlust m durch Auslaufen, Undichtigkeit f	fuite f	течь
L 37	leakage current	Kriechstrom m	courant m de perte (fuite)	ток утечки
L 38	leakage detection, leak detection, leak hunting	Lecksuche f	détection f de fuites, recherche f de fuites	течеискание, обнаружение течей
L 39	leakage flow	Leckströmung f, Leckstrom m	courant m d'une fuite	поток утечки
L 40	leakage proving, leak checking	Leckprüfung f	essai (test) m de fuite	проверка на течи
	leakage rate, inleakage rate, leak rate	Leckrate f, Undichtigkeit f	défaut m d'étanchéité, débit m d'une fuite, taux m de fuite	натекание, скорость натекания
	leak checking	s. leakage proving		
L 41	leak detecting tube, leak detector head	Lecksuchröhre f	tube m à détection de fuites, tube détecteur de fuites	датчик течеискателя

	English	German	French	Russian
	leak detection	s. leakage detection		
L 42	leak detection device	Leckanzeigevorrichtung f	sonde f d'un détecteur de fuites, indicateur m de fuites	прибор для обнаружения течей
L 43	leak detector, leak-sensing device	Lecksucher m	détecteur m de fuites, appareil m pour la détection de fuites	течеискатель
	leak detector head	s. leak detecting tube		
	leak hunting	s. leakage detection		
L 44	leakproof, leak-tight	leckfrei, vakuumdicht	étanche, sans fuites	вакуумный
L 45	leak proving, leak testing, tightness control	Dichtigkeitsprüfung f	test m (épreuve f, essai m) d'étanchéité	испытание (проверка) на герметичность
	leak rate	s. leakage rate		
L 46	leak sealing material	Leckdichtungsmaterial n	matériel m d'étanchéité pour fuites	материал для замазки течей
	leak-sensing device	s. leak detector		
	leak testing	s. leak proving		
L 47	leak test mass spectrometer	Lecksuchmassenspektrometer n	spectromètre m de masse à détection de fuites	масс-спектрометрический течеискатель
	leak-tight	s. leakproof		
	leak valve, dosing valve, variable leak valve	Gaseinlaßventil n, Dosierventil n	soupape f réglable, vanne f de dosage, fuite f réglable	дозировочный вентиль, вентиль впуска газа
L 48	leather packing	Lederdichtung f	joint m (rondelle f) en cuir	кожаное уплотнение
	LEED	s. low-energy electron diffraction		
L 49	lens blooming, optics filming	Vergütung f optischer Gläser	revêtement m de surfaces optiques, application f de couches antiréfléchissantes	просветление оптических стекол (линз)
L 50	lens coating, optics filming, optics coating	Oberflächenvergütung f <optischer Elemente>	trempe f superficielle <des éléments optiques>	осветляющее покрытие <линзы>
L 51	level controller, liquid level controller	Niveaukonstanthalter m	niveau-stabilisateur m, stabilisateur m du niveau	регулятор уровня, устройство для поддержания уровня жидкости
L 52	level indicator	Füllstandanzeiger m	indicateur m du niveau	индикатор уровня
L 53	level monitor	Füllstandwächter m	régulateur m à niveau, contrôleur m automatique du niveau	автоматический прибор для контроля уровня
L 54	lever valve	Hebelventil n	vanne f à levier	рычажный вентиль
L 55	levitation evaporation	Schwebeverdampfen n	évaporation f en suspension, évaporation par lévitation	испарение во взвешенном состоянии
	levitation melting, free levitation method	Schwebeschmelzen n	fusion f par lévitation	плавка во взвешенном состоянии
	lift and force pump, sucking and forcing pump, double acting pump, suction and pressure pump	Saug- und Druckpumpe f	pompe f aspirante et refoulante	всасывающий и нагнетательный насос, насос двойного действия
L 56	lift force	Auftriebskraft f	poussée f, poussée verticale, poussée d'Archimède	подъемная сила
	lift tube, feed tube, lift pipe, siphon	Heber m	siphon m	сифон
L 57	limb <of the U-tube>	Schenkel m <des U-Rohres>	branche f <d'un tube en U>	колено <U-образной трубки>
L 57a	lime glass, soda lime glass	Natron-Kalk-Glas n	verre m caicaire-natron	известковое стекло
	limiting forepressure, critical backing pressure, critical forepressure, forepressure tolerance, tolerable forepressure	Vorvakuumbeständigkeit f, Vorvakuumgrenzdruck m, Grenzdruck m der Vorvakuumbeständigkeit	pression f limite d'amorçage, stabilité f au vide primaire	наибольшее выпускное (форвакуумное) давление, наибольшее противодавление
	limiting pressure, blank-off (ultimate, final) pressure, ultimate vacuum	Enddruck m	pression f finale	предельный вакуум, предельное (остаточное) давление
L 58	limiting resistance	Begrenzungswiderstand m	résistance f limite	ограничивающее сопротивление
L 59	limit switch	Endausschalter m	interrupteur m limiteur (de fin de course)	концевой выключатель
L 60	linac, linear accelerator	Linearbeschleuniger m	accélérateur m linéaire	линейный ускоритель
L 61	Linde cycle	Lindescher Kreisprozeß m	cycle m de Linde	цикл Линде
	line, duct, pipe, tubing	Rohrleitung f	canalisation f, tuyau m, conduite f	трубопровод
	linear accelerator	s. linac		
L 62	linear motion drive, linear motion feedthrough, thrust leadthrough	Schiebedurchführung f	traversée (entrée, douille) f glissante	ввод для передачи линейного перемещения
L 63	linear through-put plant	Anlage f mit linearem Durchlauf	installation f de passage linéaire	установка для линейной протяжки
L 64	line contact	Linienberührung f	contact m de ligne	линейный контакт
L 65	line-focus tube	Strichfokusröhre f	tube m à rayons X à foyer linéaire	рентгеновская трубка с штриховым фокусом
L 66	line spectrum	Linienspektrum n	spectre m de lignes	линейчатый спектр

L 67	line voltage fluctua-tion, mains voltage fluctuation	Netzspannungs-schwankung *f*	variation *f* de la tension de secteur, fluctuation *f* de la tension de secteur	колебания напряжения сети
L 68	line welding, seam welding	Nahtschweißung *f*	soudure *f* en ligne continue	роликовая сварка, сварка прямолиней-ным швом
L 69	lining	Auskleidung *f*, Futter *n*	revêtement *m*	обкладка, футеровка, облицовка
L 70	linkage	Verkettung *f*	interconnexion *f*, accou-plement *m*, enchaîne-ment *m*, liaison *f*	соединение, связь, сцепление
L 71	lip gasket, lip seal, lip washer	Lippendichtung *f*	joint *m* à lèvre, anneau *m* à lèvres	манжетное уплотнение
L 72	lip pour, pouring lip	Gießschnauze *f*	bec *m* de coulée	сливной носок
L 73	lip-pour ladle	Gießpfanne *f* mit Gieß-schnauze	poche *f* de coulée avec bec de coulée	разливочный ковш со сливным носком
	lip seal	*s.* lip gasket		
	lip washer	*s.* lip gasket		
	liquation, eliquation, segregation	Seigerung *f*	liquation *f*, ségrégation *f*	ликвация, сегрегация, зейгерование
L 74	liquefaction	Verflüssigung *f*	liquéfaction *f*, conden-sation *f*	сжижение, ожижение
L 75	liquefaction of gases	Gasverflüssigung *f*	liquéfaction *f* de gaz	сжижение газов
L 76	liquefier	Verflüssiger *m*	condenseur *m*, liqué-facteur *m*	сжижитель
	liquefier, condenser	Kondensor *m*, Kondensat-abscheider *m*, Konden-sator *m*	séparateur *m* de condensat, condensateur *m*	конденсатор
L 77	liquid air switch	Schalter *m* für flüssige Luft	commutateur *m* pour air liquide	реле жидкого воздуха
L 78	liquid-cooled	flüssigkeitsgekühlt	réfrigéré par liquide	с жидкостным охлаж-дением, охлаждаемый жидкостью
L 79	liquid level	Flüssigkeitsstand *m*	niveau *m* liquide	уровень жидкости
	liquid level controller	*s.* level controller		
L 80	liquid level indicator	Flüssigkeitsstandanzeiger *m*	indicateur *m* du niveau liquide	индикатор уровня жидкости
L 81	liquid level manometer	Flüssigkeitsmanometer *n*	manomètre *m* à niveau liquide	жидкостный манометр
L 82	liquid level sensing element	Füllstandsfühler *m*	indicateur *m* du niveau liquide	датчик (индикатор) уровня жидкости
L 83	liquid metal ultra-high vacuum valve	Ultrahochvakuumventil *n* mit Schmelzmetall-dichtung	soupape *f* à vide ultra-poussé avec une garni-ture de joint de métal liquide	сверхвысоковакуумный вентиль с уплотнением жидким металлом
L 84	liquid-metal vacuum valve	Schmelzmetall-Vakuum-ventil *n*	soupape *f* à vide de métal liquide	вентиль с уплотнением жидким металлом
L 85	liquid nitrogen	flüssiger Stickstoff *m*	azote *m* liquide	жидкий азот
L 86	liquid nitrogen consumption	Verbrauch *m* an flüssigem Stickstoff	consommation *f* d'azote liquide	расход жидкого азота
L 87	liquid nitrogen dispensing unit	Nachfüllvorrichtung *f* für flüssigen Stickstoff	appareil *m* de remplissage d'azote liquide	установка для разливки жидкого азота
	liquid ring pump, fluid-ring pump	Flüssigkeitsringpumpe *f*	pompe *f* à anneau de liquide	жидкостнокольцевой насос
L 88	liquid seal	Flüssigkeitsdichtung *f*	joint *m* liquide	жидкостное уплотнение
L 89	liquid-sealed mechani-cal pump	flüssigkeitsgedichtete mechanische Pumpe *f*	pompe *f* mécanique étanchéifiée par liquide	механический насос с жидкостным уплот-нением
L 90	liquidus [curve]	Liquiduslinie *f*	liquidus *m* dans le plan	ликвидус
L 91	liquidus temperature	Liquiduspunkt *m* <oberer Schmelzpunkt>	point *m* de fusion	температура ликвидуса
	litharge and glycerine cement	*s.* lead monoxide and glycerine cement		
L 92	litharge-glycerine <reaction cement>	Glyzerin-Bleiglätte *f* <Reaktionskitt>	litharge-glycérine *f* <mastic de réaction>	глицерино-свинцовый глет <реакционный цемент>
	load, capacity, through-put	Förderleistung *f*, Saug-leistung *f*, Fördermenge *f*, Durchsatz *m*, Durch-satzmenge *f*	flux *m* massique, débit *m* massique	скорость откачки
L 93	load evaporator	Leistungsverdampfer *m*	vaporisateur *m* de puissance	испаритель мощности
	loading, feeding, charging	Beschickung *f*	chargement *m*	питание, загрузка
L 94	loading bracket	Chargiervorrichtung *f*	appareil *m* de chargement	загрузочное приспосо-бление
	loading tray, charging tray	Fülltablett *n*	plateau *m* de remplissage	загрузочный лоток
L 95	lobster back, segment bend	Segmentbogen *m*	arc *m* à segment	составное колено
L 96	lock	Schleuse *f*, Formen-schleuse *f*	sas *m*, écluse *f* à gaz, dispositif *m* d'éclusage	шлюз
L 97	lock chamber, sluice chamber	Schleusenkammer *f*	chambre *f* d'écluse	шлюзовая камера
	locking, interlock, interlocking	Verriegelung *f*, Ver-blocken *n*, Verriegeln *n*	verrouillage *m*, blocage *m*	блокировка, запирание
	locking ring, clamping ring	Klemmring *m*, Spannring *m*	anneau *m* de serrage, collier *m* de serrage	кольцевой зажим, хомут

L 98	**lock valve**	Schleusenventil *n*	robinet *m* à sas, vanne-écluse *f*, robinet-vanne *m* à écluse	перепускной вентиль, затвор
L 99	**logarithmic pressure scale**	logarithmische Druck-skale *f*	graduation *f* logarith-mique de pression	логарифмическая шкала давления
L 100	**long glass**	langes Glas *n*	verre *m* long	«длинное» стекло
L 101	**long life tube**	Langlebensdauerröhre *f*	tube *m* de longévité	долговечная лампа
L 102	**long-time stability**	Langzeitstabilität *f*	stabilité *f* de longue durée	долговременная стабильность
L 103	**long-tube vertical evaporator**	Vertikalrohrverdampfer *m*	évaporateur *m* par tuyau-teries verticales	испаритель с длинной вертикальной трубкой
	loop dryer, festoon dryer	Schleifentrockner *m*, Girlandentrockner *m*, Hängetrockner *m*	sécheur *m* suspendu (à boucles)	фестонная сушилка, подвесная сушилка
L 104	**loose flange,** rotatable flange, rotating (rotary) flange	Drehflansch *m*	bride *f* rotative (tournante)	вращающийся фланец, свободный фланец
L 105	**Loschmidt number**	Avogadrosche Zahl *f*	nombre *m* d'Avogadro	число Лошмидта
L 106	**loss factor**	dielektrischer Verlust-faktor *m*	facteur *m* de pertes diélectriques	коэффициент потерь
L 107	**loss of head**	statischer Druckverlust *m*	perte *f* de pression statique	потеря напора
	louvre dryer, cascade dryer	Rieseltrockner *m*	installation *f* de séchage par ruissellement	каскадная сушилка
L 108	**low alloy,** mild steel, mild alloy <steels having a low carbon content and small pro-portions of other alloying elements>	Flußstahl *m*, Schmiede-eisen *n* <Stähle gerin-geren Kohlenstoffge-halts und geringer Anteile von Legierungs-bestandteilen>	acier *m* fondu, acier doux, fer *m* forgé	низкоуглеродистая сталь, мягкая сталь, ковкая сталь
L 109	**low-boiling**	tiefsiedend, leichtsiedend	à bas point d'ébullition	кипящий при низкой температуре
	low-energy electron diffraction, diffraction of low-energy electrons, low-voltage electron diffraction, LEED	Beugung *f* langsamer Elektronen	diffraction *f* des électrons lents (à basse énergie)	дифракция медленных электронов
L 110	**lowerable bottom**	absenkbarer Boden *m*	fond *m* à déplacement vers le bas	опускаемое дно
L 111	**low-loss cold trap**	Kühlfalle *f* mit geringem Kühlmittelverlust	piège *m* avec perte petite de réfrigérant	охлаждаемая ловушка с малым расходом хладоагента
L 112	**low pressure,** partial vacuum	Unterdruck *m*	dépression *f*, souspression *f*	разрежение, давление ниже атмосферного
	low-pressure chamber, altitude chamber	Höhenkammer *f*, Unter-druckkammer *f*	chambre *f* de dépression	барокамера
L 113	**low-pressure limit**	untere Druckmeßgrenze *f*	limite *f* inférieure de mesure de la pression	нижний предел измере-ния давления
L 114	**low-pressure stage**	Niederdruckstufe *f*	étage *m* à basse pression	ступень низкого давле-ния
L 115	**low-temperature installation,** refrigerating plant	Kälteanlage *f*	installation *f* frigorifique (de refroidissement)	низкотемпературная (холодильная) установка
L 116	**low vacuum** <760 ··· 25 Torr>	geringes Vakuum *n* <760 ··· 25 Torr>	vide *m* bas <760 ··· 25 Torr>	низкий вакуум <760—25 мм рт. ст.>
L 117	**low-vacuum mercury vapour lamp**	Niederdruckquecksilber-lampe *f*	lampe *f* à vapeur de mer-cure à basse pression	ртутная лампа низкого давления
L 118	**low vapour liquid**	Flüssigkeit *f* geringen Dampfdruckes	liquide *m* avec une tension de vapeur peu considé-rable	жидкость с малой упругостью паров
	low-voltage electron diffraction, diffraction of low-energy electrons, low-energy electron diffraction, LEED	Beugung *f* langsamer Elektronen	diffraction *f* des électrons lents (à basse énergie)	дифракция медленных электронов
L 119	**low-voltage welding**	Niederspannungs-schweißen *n*	soudage *m* à basse tension	низковольтная сварка
L 120	**low work function coating**	Schicht *f* niedriger Austrittsarbeit	couche *f* de travail bas de sortie	покрытие с низкой рабо-той выхода
	L-ring shaft seal, cap-collar gasket	Hutmanschette *f*	joint *m* à chapeau, em-bouti *m* à capuchon, joint d'arbre à section angulaire	колпачковая манжета, уплотняющее кольцо с L-образным сечением
L 121	**lubricant**	Schmiermittel *n*	lubrifiant *m*	смазка
L 122	**lubricant for use in vacuum**	Vakuumschmiermittel *n*	lubrifiant *m* à vide	смазка для применения в вакууме
L 123	**lubrication**	Schmierung *f*	graissage *m*, lubrification *f*	нанесение смазки, сма-зывание
L 124	**lubricity**	Schmierfähigkeit *f*	pouvoir *m* lubrifiant	смазывающая способ-ность
	lyophile, lyophilize, freeze-dry	gefriertrocknen	sécher par congélation, lyophiliser	сушить сублимационным способом
	lyophilization, cryo-drying, freeze-drying, sublimation from the frozen state	Gefriertrocknung *f*, Lyophilisation *f*	lyophilisation *f*, dessica-tion *f* par congélation, cryodessiccation *f*	сублимационная сушка из замороженного со-стояния, лиофилиза-ция

M

M 1	**Mach number**	Machzahl *f*	nombre *m* de Mach	число Маха
M 2	**Madras mica**	indischer Glimmer *m*	mica *m* des Indes	индийская слюда
M 3	**magnetically operated valve,** magnetic valve, solenoid valve	Mangnetventil *n*	vanne *f* magnétique (à commande magnétique), robinet *m* solénoïde	электромагнитный клапан
M 4	**magnetic ball-joint valve**	Kugelschliffmagnetventil *n*	soupape *f* magnétique à rodage sphérique	вентиль со сферическим шлифом и магнитным приводом
M 5	**magnetic force**	magnetische Feldstärke *f*	intensité *f* du champ magnétique	напряженность магнитного поля
M 6	**magnetic induction**	magnetische Induktion *f*	induction *f* magnétique	магнитная индукция
M 7	**magnetic storage film**	Magnetspeicherschicht *f*	couche *f* magnétique d'emmagasinage, couche magnétique à mémoire	магнитная пленка для запоминания
	magnetic valve	s. magnetically operated valve		
M 8	**magnetic yoke**	Magnetjoch *n*	culasse *f*	ярмо магнитопровода
M 9	**magnetron**	Magnetron *n*	magnétron *m*	магнетрон
M 10	**magnetron sputter-ion pump**	Ionenzerstäuberpumpe *f* vom Magnetrontyp	magnétron pompe *f* ionique à getter par pulvérisation	магнетронный электроразрядный насос
	magnetron vacuum gauge, cold cathode magnetron gauge	Magnetronvakuum-meter *n*	manomètre *m* magnétron	магнетронный вакуумметр
M 11	**main condenser**	Hauptkondensator *m*	condensateur *m* principal	основной конденсатор
M 12	**main drying,** primary drying	Haupttrocknung *f*	dessiccation *f* principale, séchage *m* principal	основная сушка, первичная сушка
	mains voltage fluctuation, line voltage fluctuation	Netzspannungs-schwankung *f*	variation *f* de la tension du secteur, fluctuation *f* de la tension de secteur	колебания напряжения сети
M 13	**maintenance-free**	wartungsfrei	sans surveillance, sans entretien	не требующий обслуживания
M 14	**male cone**	Kernschliff *m*	rodage (cône) *m* mâle	сердечник шлифа, пробка
	male thread, external thread	Außengewinde *n*	filet *m*, filetage *m*	наружная резьба
M 15	**malleability**	Dehnbarkeit *f*, Schmiedbarkeit *f*, Streckbarkeit *f*	extensibilité *f*, expansibilitè *f*, malléabilité *f*, dilatabilité *f*, forgeabilité *f*, ductilité *f*	ковкость
M 16	**malleable cast iron**	Temperguß *m*	fonte *f* malléable	ковкий чугун
M 17	**mandrel**	Dorn *m*	mandrin *m*	оправка, дорн
	mandrel drawing, bar drawing	Stopfenzug *m*	traction *f* de bouchage	волочение на оправке
	manifold, distributor	Verteilerrohr *n*, Verteiler-stück *n*, Verteiler *m*	distributeur *m*, répartiteur *m*, tuyau *m* collecteur	распределитель, загрузочное распределительное устройство, распределительная магистраль
M 18	**manograph,** pressure recorder	Druckschreiber *m*	manomètre *m* enregistreur	самопишущий манометр
M 19	**manometer,** pressure gauge	Druckmeßgerät *n*, Manometer *n*	manomètre *m*	манометр
M 20	**manometer leg**	Manometerschenkel *m*	branche *f* du manomètre	колено манометра
M 21	**manometric equivalent**	Manometeräquivalent *n*	équivalent *m* manométrique	манометрический эквивалент
M 22	**manually operated**	handbetätigt	opéré à la main, à commande manuelle	управляемый вручную
	manually operated valve, hand-operated valve	handbetätigtes Ventil *n*	valve *f* à commande manuelle, valve actionnée à la main	вентиль с ручным приводом
M 23	**marble cement**	Marmorkitt *m*	mastic *m* marbre	мраморная замазка, мраморный цемент
M 24	**mask**	abdecken	découvrir	загораживать, маскировать
M 25	**mask changer**	Maskenwechsler *m*	changeur *m* de masque	устройство для смены масок
M 26	**mask frame**	Maskenrahmen *m*	bâti *m* de masque	рамка маски
M 27	**masking**	Abdecken *n* <eines Lecks>	procédé *m* à masque	маскирование
M 28	**masking lacquer,** stop-off lacquer	Abdecklack *m*	laque *f* de couverture	лак для защиты при обработке
M 29	**mask storage**	Maskenspeicherung *f*	accumulation *f* de masque	масковое накопление
M 30	**mask turret**	Maskenkarussell *n*	carrousel *m* de masque	масковая карусель
	mass filter, electrical mass filter	Massenfilter *n*, elektrisches Massen-filter	filtre *m* [électrique] de masse	электрический фильтр масс
M 31	**mass flow**	Durchflußmenge *f*	débit *m* de passage	массовый расход
M 32	**mass flow,** mass throughput	Massendurchsatz *m*	débit *m* de masses	массовая производительность
M 33	**mass flowmeter**	Massendurchsatz-meßgerät *n*	appareil *m* de mesure du débit de masses	прибор для измерения массового расхода
M 34	**mass peak**	Massenlinie *f*	ligne *f* de masse	массовый пик
M 35	**mass range**	Massenbereich *m*	intervalle *m* de masses	диапазон масс

M 36	**mass scanning**	Massendurchlauf m	marche f continue de masse	развертка спектра масс
M 37	**mass separation**	Massentrennung f	séparation f des masses	массовая сепарация
M 38	**mass spectrometer**	Massenspektrometer n	spectromètre m de masse	масс-спектрометр
M 39	**mass spectrometer leak detector,** MSLD	Massenspektrometer-lecksuchgerät n	détecteur m de fuites par spectrométrie de masse	масс-спектрометрический течеискатель
	mass throughput, mass flow	Massendurchsatz m	débit m de masses	массовая производительность
	masticator, kneading machine	Knetmaschine f	malaxeur m, pétrisseuse f	смесительная машина
M 40	**matched seal**	angepaßte Verbindung f <Bestandteile mit gleichem Ausdehnungskoeffizienten>	jonction f adaptée	согласованный спай
	mating flange, counter flange	Gegenflansch m	contre-bride f	контрфланец
M 41	**matrix cathode**	Matrixkatode f	cathode f matrice	матричный катод
M 42	**maximum admissible water vapour pressure,** maximum safe inlet pressure of water vapour, water vapour tolerance pressure	Wasserdampfverträglichkeit f	résistance (insensibilité, compatibilité) f à la vapeur d'eau	максимальное допустимое впускное давление паров воды
	maximum backing pressure, tolerable forepressure, maximum forepressure, forepressure tolerance	„zulässiger" Vorvakuumdruck m <der einem um 10% größeren Ansaugdruck als bei normalem Vorvakuum entspricht>	pression f tolérée du vide préliminaire	максимально-допустимое выпускное давление <при превышении которого впускное давление возрастает более, чем на 10%>
M 43	**maximum capacity of treated water,** maximum rate of handling water vapour, water vapour handling capacity	Wasserdampfkapazität f	capacité f de vapeur d'eau, capacité de pompage de vapeur d'eau	наибольшая производительность по воде <газобалластного насоса>
	maximum compression pressure, pressure limit, compression limit	Verdichtungsenddruck m	compression f finie de sortie	предельное давление сжатия
	maximum evaporation rate, Knudsen rate of evaporation, Langmuir rate of evaporation	maximale Verdampfungsrate f, absolute Verdampfungsrate	taux m maximum d'évaporation	абсолютная скорость испарения
	maximum forepressure	s. maximum backing pressure		
	maximum rate of handling water vapour, maximum capacity of treated water, water vapour handling capacity	Wasserdampfkapazität f	capacité f de vapeur d'eau, capacité de pompage de vapeur d'eau	наибольшая производительность по воде <газобалластного насоса>
	maximum safe inlet pressure of water vapour	s. maximum admissible water vapour pressure		
	maximum surface stress in bend, flexural strength, transverse strength, modulus of rupture	Biegefestigkeit f	résistance f à la flexion	предел прочности при изгибе, прочность при изгибе
M 44	**maximum tolerable tensile strength,** yield point	Streckgrenze f	limite f d'allongement	предел текучести
M 45	**maximum working temperature**	Grenztemperatur f	température f limite	критическая температура
M 46	**Maxwell-Boltzmann distribution function**	Maxwell-Boltzmannsche Verteilungsfunktion f	fonction f de distribution de Maxwell-Boltzmann	функция распределения Максвелла-Больцмана
M 47	**Mc Leod gauge**	Mc Leodsches Kompressionsvakuummeter n, Mc Leod n	vacuomètre m de Mc Leod	компрессионный вакуумметр Мак-Леода
M 48	**mean free path,** mfp	mittlere freie Weglänge f	libre parcours m moyen	средний свободный пробег
	mean temperature, average temperature	Durchschnittstemperatur f, mittlere Temperatur f	température f moyenne	средняя температура
M 49	**mean velocity**	mittlere Geschwindigkeit f	vitesse f moyenne	средняя скорость
M 50	**measured backstreaming rate**	gemessene Rückströmungsrate f	débit m massique inverse mesuré, quantité mesurée de reflux	измеренная скорость обратного течения
M 51	**measured pumping speed**	gemessenes Saugvermögen n	vitesse f de pompage mesurée, capacité f d'absorption mesurée, susceptibilité f mesurée d'aspiration, succion (aspiration) f mesurée	измеренная скорость откачки
M 52	**measuring accuracy**	Meßgenauigkeit f	précision f de mesure	точность измерения

M 53	measuring buret	Meßbürette *f*	burette *f* de mesure	измерительная бюретка
M 54	measuring dome, test dome	Meßdom *m*	dôme *m* de mesure	испытательный колпак
M 55	measuring transducer	Meßumformer *m*	convertisseur *m* de mesure	измерительный преобразователь
M 56	measuring transformer	Meßwandler *m*	transformateur *m* de mesure	измерительный трансформатор
M 57	mechanical linkage	Hebelsystem *n*	système *m* de levier	рычажная система
M 58	mechanical pump	mechanische Pumpe *f*	pompe *f* mécanique	механический насос
M 59	mechanical refrigeration	Maschinenkühlung *f*	refroidissement *m* mécanique	машинное охлаждение
M 60	medium ladle	Zwischenpfanne *f*	poche *f* intermédiaire	промежуточный ковш
	medium vacuum, fine vacuum	Feinvakuum *n* (1 ··· 10⁻³Torr)	vide *m* moyen (intermédiaire)	средний вакуум
M 61	Meissner type cold trap	Meißner-Falle *f*	trappe *f* de Meißner	ловушка Майснера
	melt flux, complete fusion	Schmelzfluß *m*	matière *f* fusée, flux *m* de matière en fusion	расплав, плавка, флюс
M 62	melt head	Abschmelzkopf *m*	tête *f* fusible	прибыль
	melting, fusion	Schmelzen *n*, Schmelzprozeß *m*	fusion *f*	плавка, плавление, расплавление
M 63	melting bath, melting pool, molten pool	Schmelzbad *n*, Schmelzsumpf *m*	bain *m* de fusion	расплавленный металл, ванна с расплавом
M 64	melting cement, solvent-free melting adhesive	Schmelzkitt *m*	mastic *m* à fusion	плавкая замазка (мастика), расплавляемая замазка
M 65	melting efficiency	Schmelzleistung *f*	efficacité *f* de fusion	производительность плавильной печи, плавильная мощность
	melting electrode, consumable electrode, consutrode	Abschmelzelektrode *f*, selbstverzehrende Elektrode *f*	électrode *f* consommable, électrode fusible [pour soudage à l'arc]	расходуемый электрод [плавильной печи]
M 66	melting furnace	Schmelzofen *m*	four *m* de fusion	плавильная печь
	melting point, fusion point	Schmelzpunkt *m*	point *m* de fusion	точка плавления
	melting pool, melting bath, molten pool	Schmelzbad *n*, Schmelzsumpf *m*	bain *m* de fusion	расплавленный металл, ванна с расплавом
M 67	melting range	Gießbereich *m*	domaine *m* de coulée	температурный диапазон плавления
M 68	melting rate (speed)	Abschmelzgeschwindigkeit *f*, Stundenschmelzleistung *f*	vitesse *f* de fusion	скорость расплавления
M 69	melting stock, melt stock	Abschmelzstab *m*, Schmelzgut *n*	baguette *f* fusible	плавильный пруток (штаб)
M 70	melting tank	Schmelzkessel *m*	chaudière *f* à fusion	плавильный котел
M 71	melt lubrication	Schmelzflußschmierung *f*	graissage *m* par flux de masse fondue, lubrification *f* par matière fusée	смазка расплавлением
	melt stock, melting stock	Abschmelzstab *m*, Schmelzgut *n*	baguette *f* fusible	плавильный пруток (штаб)
M 72	melt stock cathode	Abschmelzkatode *f*	cathode *f* de fusion	расплавляемый катод
M 73	membrane leak	Membranleck *n*	fuite *f* de membrane	мембранная течь
	membrane vacuum manometer, diaphragm vacuum gauge	Membranvakuummeßgerät *n*, Membranvakuummeter *n*	vacuomètre *m* à membrane, manomètre *m* à membrane	мембранный вакуумметр
M 74	memory effect	Erinnerungseffekt *m*	effet *m* d'avertissement	эффект запоминания («памяти»)
M 75	mercury	Quecksilber *n*	mercure *m*	ртуть
M 76	mercury column	Quecksilbersäule *f*	colonne *f* de mercure	ртутный столб
M 77	mercury cut-off	Quecksilberverschlußventil *n*	robinet *m* à colonne de mercure	ртутный затвор
M 78	mercury-diffusion pump	Quecksilberdiffusionspumpe *f*	pompe *f* à diffusion de (à) mercure	паортутный диффузионный насос
M 79	mercury ejector pump	Quecksilberdampfstrahlpumpe *f*, Quecksilber-Ejektorpumpe *f*	pompe *f* à éjecteur de mercure	ртутный пароструйный насос
M 80	mercury level manometer, U-shaped manometer, U-tube manometer	U-Rohr-Manometer *n*	manomètre *m* en U tronqué	U-образный манометр
M 81	mercury sealed stopcock	Absperrhahn *m* mit Quecksilberabdichtung, Quecksilberverschlußhahn *m*	robinet *m* d'arrêt à mercure	кран с ртутным уплотнителем
M 82	metal bellows	Metallbalg *m*	peau *f* métallique	металлический сильфон
M 83	metal bellows seal	Metallbalgdichtung *f*	joint *m* métallique à soufflet	металлическое сильфонное уплотнение
M 84	metal ceramics	Metallkeramik *f*	céramique *f* métallique	металлокерамика
	metal-ceramic seal, ceramic-to-metal seal	Metall-Keramik-Verbindung *f*	scellement *m* céramique-métal	металлокерамическое соединение, спай металла с керамикой
M 85	metal-coating, metallizing	Metallisierung *f*, Metallisieren *n*	métallisation *f*	металлизация
M 86	metal degassing plant	Metallentgasungsanlage *f*	installation *f* de dégagement métallique, installation de dégazage de métaux	установка для обезгаживания металла

M 87	metal diaphragm filter	Metallmembranfilter *n*	filtre *m* à membrane métallique	металлический мембранный фильтр
M 88	metal diaphragm seal	Metallmembrandichtung *f*	joint *m* à membrane métallique	металлическое мембранное уплотнение
M 89	metal extraction	Metallgewinnung *f*	extraction *f* de métal	извлечение металла
M 90	metal gasket, metal seal	Metalldichtung *f*	joint *m* métallique	металлическая прокладка, металлический уплотнитель
	metal hose, flexible metallic tube	Metallschlauch *m*	tuyau *m* métallique flexible, tuyau flexible en métal	металлический шланг
	metallizing, metal-coating	Metallisieren *n*, Metallisierung *f*	métallisation *f*	металлизация
M 91	metal physics	Metallphysik *f*	physique *f* de métaux	физика металлов
M 92	metal screening	Metallgewebe *n*	tissu *m* métallique	металлическая сетка для экранировки
	metal seal	*s.* metal gasket		
M 93	metal sealed	metallgedichtet	étanché par métal	уплотненный металлом
M 94	metal shadowing	Metallbeschattung *f*	ombrage *m* métallique	затенение металлом
M 95	metal spraying	Metallspritzen *n*	métallisation *f* au pistolet, métallisation par pulvérisation, extrusion *f* métallique	металлизация распылением
M 96	metal termination	Metallanschluß *m*	connexion *f* métallique	металлический наконечник
M 97	metal-to-glass seal	Glas-Metall-Verschmelzung *f*, Glas-Metall-Einschmelzung *f*	fusion *f* verre-métal, scellement *m* verre-métal	спай металла со стеклом
M 98	metal working	Metallbearbeitung *f*	traitement *m* métallique, usinage (travail) *m* des métaux	обработка металла
M 99	metastable	metastabil	métastable	метастабильный
M 100	metastable state	metastabiler Zustand *m*	état *m* métastable	метастабильный уровень, метастабильное состояние
	mfp	*s.* mean free path		
M 101	mica	Glimmer *m*	mica *m*	слюда
M 102	mica flaking	Glimmerabblätterung *f*	écaillement *m* de mica	отслаивание слюды
M 103	mica insulation	Glimmerisolation *f*	isolation *f* de mica	слюдяная изоляция
M 104	micanite, pressed mica	Preßglimmer *m*, Mikanit *n*	mica *m* pressé	прессованная слюда, миканит
M 105	mica plate	Glimmerplatte *f*	plaque *f* en mica	слюдяная пластина
M 106	mica spacer	Glimmerscheibe *f*	lame *f* de mica	слюдяная прокладка
M 107	mica-to-metal seal	Metall-Glimmer-Verbindung *f*	joint (scellement) *m* mica-métal	соединение металла со слюдой
M 108	mica window	Glimmerfenster *n*	fenêtre *f* de mica	слюдяное окно
M 109	microbalance	Mikrowaage *f*	microbalance *f*	микровесы
M 110	microcapillary	Mikrokapillare *f*	tube *m* microcapillaire	микрокапилляр
M 111	microcircuit	Mikroschaltung *f*	microcircuit *m*	микросхема
M 112	microcircuit mask	Abdeckmaske *f* für Mikroschaltung	masque *m* à microcircuit	маска для нанесения микросхемы
M 113	microcircuit stencil	Mikroschaltkreis-schablone *f*	gabarit *m* du microcircuit	шаблон для микросхемы
M 114	microestimation	Mikrobestimmung *f*	microdosage *m*	микроопределение
M 115	microminiature circuit	Mikrominiaturschaltung *f*	circuit *m* microminiature	микроминиатюрная схема
M 116	migration	Kriechen *n*	migration *f*, fluage *m*	миграция
M 117	migration protector	Kriechschutz *m*	protection *f* de migration	устройство для защиты от миграции
M 118	migration velocity, velocity of migration	Wanderungsgeschwindigkeit *f*	vitesse *f* de migration	скорость миграции
	mild alloy, mild steel, low alloy <steels having a low carbon content and small proportions of other alloying elements>	Flußstahl *m*, Schmiedeeisen *n* <Stähle geringeren Kohlenstoffgehalts und geringer Anteile von Legierungsbestandteilen>	acier *m* fondu, acier doux, fer *m* forgé	низкоуглеродистая сталь, мягкая сталь, ковкая сталь
M 119	mimic, mimic diagram	Leuchtschaltbild *n*	plan *m* lumineux de montage, schéma *m* lumineux des connexions	мчемоническая схема
M 120	miniature gauge	Miniaturmeßröhre *f*	tube *m* gradué en miniature	миниатюрный датчик манометра
M 121	minimum detectable leak	kleinstes nachweisbares Leck *n*	plus petite fuite *f* détectable (décelable)	минимально обнаруживаемое натекание
M 122	minimum detectable pressure range	kleinste nachweisbare Druckänderung *f*	plus petit changement *m* décelable (détectable) de pression	минимально обнаруживаемое изменение давления
M 123	mirror coating, mirror lining, mirroring	Verspiegelung *f*	dépôt *m* de couches réfléchissantes, réalisation *f* de surfaces miroitantes	покрытие отражательным слоем
M 124	mirror coating, reflecting layer, specular layer	Spiegelbelag *m*, Reflexbelag *m*	couche *f* spéculaire, revêtement *m* réfléchissant	зеркальное покрытие
	mirroring, mirror lining	*s.* mirror coating		
M 125	miscibility gap	Mischungslücke *f*	lacune *f* de miscibilité	область несмесимости

	mixed gas, forming gas ‹mixture of hydrogen and nitrogen›	Formiergas n ‹Wasser-stoff-Stickstoff-Gemisch›	gaz m mixte ‹mélange d'hydrogène et d'azote›	формиргаз ‹смесь водо-рода с азотом›
M 126	mixer with agitator	Rührwerksmischer m	mélangeur m avec un agitateur	смеситель с мешалкой
M 127	mixing nozzle	Mischdüse f	tuyère f mélangeuse, mélangeur m	смесительное сопло
M 128	mixing steam, motive steam	Treibdampf m	vapeur f du fluide moteur, vapeur motrice	рабочий пар
M 129	mixing tube	Mischrohr n	tuyau m de mélange, buse f, diffuseur m	смесительная трубка
M 130	mobile desk	fahrbares Pult n	pupitre m roulant (mobile)	передвижной пульт
M 131	mobile film distillation	Mischfilmdestillation f	distillation f pelliculaire de surfaces mixtes	дистилляция с подвиж-ной пленкой
M 132	mobile tank ‹for lique-fied gases›	Transporttank m ‹für verflüssigte Gase›	réservoir m de transport ‹à gaz liquéfiés›	передвижной резервуар ‹для сжижаемых газов›
M 133	modulated beam photometer	Fotometer n mit modu-liertem Lichtstrahl	photomètre m avec rayon modulé	фотометр с модулиро-ванным световым пучком
M 134	modulator	Modulator m	modulateur m	модулятор
M 135	modulus of elasticity, Young's modulus	Elastizitätsmodul m	module m d'élasticité	модуль упругости при растяжении, модуль Юнга
M 136	modulus of rigidity, shear modulus	Schubmodul m	module m de rigidité	модуль сдвига
	modulus of rupture, flexural strength, maxi-mum surface stress in bend, transverse strength	Biegefestigkeit f	résistance f à la flexion	предел прочности при изгибе, прочность при изгибе
M 137	Mohs' hardness	Mohs-Härte f	Mohs-dureté f	твердость по Моосу
	moisture, humidity	Feuchtigkeit f	humidité f	влажность
M 138	moisture balance	Feuchtigkeitswaage f	balance f d'humidité	весовой гигрометр
M 139	moisture content	Feuchtigkeitsgehalt m	teneur m en humidité, degré m d'humidité	степень влажности, содержание влаги
M 140	molar fraction, mole fraction	Molenbruch m	fraction f de moles, concentration f molaire	молярная доля, моляр-ная долевая концен-трация
M 141	molar volume, molecular volume	Molvolumen n	volume m molaire	молекулярный объем, молярный объем
	mole, gramme molecule	Grammolekül n, Mol n	mole f	грамм-молекула, моль
M 142	molecular adhesion	Moleküladhäsion f	adhésion f moléculaire	молекулярная адгезия
M 143	molecular air pump	Molekularluftpumpe f	pompe f moléculaire	молекулярный воздуш-ный насос
M 144	molecular attraction	Molekularanziehung f	attraction f moléculaire	молекулярное притя-жение
M 145	molecular beam	Molekularstrahl m	rayon m moléculaire	молекулярный пучок
M 146	molecular beam source	Molekularstrahlquelle f	source f du rayon moléculaire	источник для создания молекулярного пучка
M 147	molecular collision	Molekülstoß m	percussion f (choc m) moléculaire	молекулярное соуда-рение
M 148	molecular conductance	Leitwert m bei Molekular-strömung	conductance f d'écoule-ment moléculaire	пропускная способность при молекулярном течении
M 149	molecular conductivity	molekulare Leitfähigkeit f	conductivité f moléculaire	молекулярная (мольная) проводимость
M 150	molecular distillation	Molekulardestillation f	distillation f moléculaire	молекулярная дистилля-ция
M 151	molecular drag gauge, molecular gauge, rotating disk gauge, rotating cylinder gauge	Molekularvakuummeter n	vacuomètre m moléculaire	молекулярный мано-метр, мольманометр
M 152	molecular drag pump, molecular pump	Molekularpumpe f	pompe f moléculaire	молекулярный насос
M 153	molecular effusion	Molekulareffusion f	effusion f moléculaire	молекулярная эффузия
M 154	molecular electronics	Molekularelektronik f	électronique f molécu-laire	молекулярная электро-ника, молектроника
M 155	molecular flow	Molekularströmung f	courant (écoulement) m moléculaire, flux m moléculaire	молекулярное течение
M 156	molecular force	Molekularkraft f	force f moléculaire	молекулярная сила
	molecular gauge	s. molecular drag gauge		
	molecular law, Knudsen's law	Knudsensches Gesetz n	loi f de Knudsen	закон Кнудсена
M 157	molecular leak	Molekularleck n	fuite f moléculaire	молекулярная течь
M 158	molecularly rough	molekular rauh	moléculaire brut	шероховатый на моле-кулярном уровне
	molecular pump, molecular drag pump	Molekularpumpe f	pompe f moléculaire	молекулярный насос
M 159	molecular sieve, sieve sorbent	Molekularsieb n	tamis m moléculaire, crible m moléculaire	молекулярное сито
	molecular sieve baffle, adsorption trap	Adsorptionsfalle f	piège m à adsorption, trappe f d'adsorption	адсорбционная ловушка
M 160	molecular sink	Molekülsenke f	point m d'annulation moléculaire	молекулярный сборник
M 161	molecular speed, molecular velocity	Molekulargeschwindig-keit f	vitesse f moléculaire	скорость молекул

M 162	**molecular still**	Molekulardestillations-anlage f	distillerie f moléculaire	аппарат для молекулярной перегонки
	molecular velocity	s. molecular speed		
	molecular volume, molar volume	Molvolumen n	volume m molaire	молекулярный объем, молярный объем
	mole fraction	s. molar fraction		
	molten pool, melting bath, melting pool	Schmelzsumpf m, Schmelzbad n	bain m de fusion	расплавленный металл, ванна с расплавом
M 163	**molybdenum boat**	Molybdänschiffchen n	nacelle f de molybdène	молибденовая лодочка
M 164	**moly-manganese ceramic-to-metal seal**	Metall-Keramik-Verbindung f nach dem Molybdän-Mangan-Verfahren	scellement m métal-céramique d'après la méthode molybdène-manganèse	спай металла с керамикой с металлизацией молибдено-марганцевым составом
M 165	**moly-manganese process, moly-manganese sealing method**	Molybdän-Mangan-Verfahren n	procédé m molybdène-manganèse, méthode f molybdène-manganèse	молибдено-марганцевый процесс, пайка с металлизацией молибдено-марганцевым порошком
M 166	**monitor crystal**	Kontrollkristall m	cristal m de contrôle	измерительный кристалл
M 167	**monitor resistance pick-up**	Kontrollwiderstandsabgriff m	prise f de résistance de contrôle	отвод от измерительного контрольного сопротивления
M 168	**monoatomic layer**	einatomare Schicht f	couche f monoatomique	моноатомная пленка, моноатомный слой
M 169	**monoblock pump,** single block pump	Einblockpumpe f	pompe f monobloque	моноблочный насос
M 170	**monocell,** single cell	Monozelle f	monocellule f, élément m mono	моноячейка
M 171	**monocrystalline wire**	Einkristalldraht m	fil m monocristallin	монокристаллическая проволока
M 172	**monolayer**	Monoschicht f	monocouche f	монослой
M 173	**monolayer coverage**	Monoschichtbedeckung f	couverture f mono-atomique	монослойное (мономолекулярное) покрытие
M 174	**monolayer evaporation**	Monoschichtverdampfung f	évaporation f des couches monomoléculaires	испарение мономолекулярного слоя
M 175	**monolayer time**	Wiederbedeckungszeit f für monomolekulare Bedeckung	temps m de recouvrement (réadsorption), temps de formation d'une couche monomoléculaire	время образования мономолекулярного слоя
M 176	**monomolecular**	monomolekular	monomoléculaire	мономолекулярный
M 177	**monomolecular adsorption**	monomolekulare Adsorption f	adsorption f monomoléculaire	мономолекулярная адсорбция
M 178	**monopole partial pressure analyzer**	Monopol-Partialdruckanalysator m	analyseur m monopole de pressions partielles	монополярный измеритель парциальных давлений
M 179	**monopole spectrometer**	Monopolspektrometer n	spectromètre m monopole	монополярный спектрометр
M 180	**most probable velocity**	wahrscheinlichste Geschwindigkeit f	vitesse f la plus probable	наиболее вероятная скорость
	motive steam	s. mixing steam		
M 181	**mould**	Gießform f	moule m	литейная форма
M 182	**moulded cathode,** pressed cathode	gepreßte Katode f	cathode f moulée (pressée)	прессованный катод
M 183	**mould rim**	Kokillenrand m	bord m de la coquille	буртик кокиля
	mounting plate, base plate	Grundplatte f, Montageplatte f	plaque f de base (montage), plaque-support f	опорная плита
M 184	**moving-coil galvanometer**	Drehspulgalvanometer n	galvanomètre m à cadre mobile	гальванометр с подвижной катушкой
	MSLD	s. mass spectrometer leak detector		
M 185	**muff coupling**	Muffenkupplung f	accouplement m par manchon	муфтовое соединение
M 186	**muffle furnace**	Muffelofen m	four m à moufle	муфельная печь
M 187	**multi-blade impeller**	Laufrad n <bei Wasserringpumpen>	rotor m, roue f mobile, roue de roulement	многолопастная крыльчатка
M 188	**multi-cellular anode**	wabenförmige Anode f	anode f multicellulaire	ячеистый анод
M 189	**multi-jet burner**	Vielstrahlbrenner m	brûleur m à plusieurs rayons	многоструйная горелка
M 190	**multi-layer adsorption**	Mehrschichtadsorption f	adsorption f de plusieurs couches	полимолекулярная адсорбция
M 191	**multi-layer capacitor,** multi-layered thin-film capacitor	Mehrfach-Dünnschicht-Kondensator m, Mehrschichtkondensator m	condensateur m à plusieurs couches [minces]	многослойный тонкопленочный конденсатор
M 192	**multi-layer coating**	Mehrfachschicht f, Vielfachschicht f	couche f multiple	многослойное покрытие
	multi-layered thin-film capacitor, multi-layer capacitor	Mehrfach-Dünnschicht-Kondensator m, Mehrschichtkondensator m	condensateur m à plusieurs couches [minces]	многослойный тонкопленочный конденсатор
	multi-layer glass, compound glass	Verbundglas n	verre m compound (à vitre collé, à couches multiples)	составное стекло
M 193	**multi-layer mirror**	Mehrschichtenspiegel m	miroir m à plusieurs couches	многослойное зеркало
M 194	**multi-meter**	Universalmeßinstrument n	multimètre m, instrument m de mesure universel	универсальный измерительный прибор

M 195	multi-nozzle	Mehrfachdüse f	tuyère f multiple, buse f multiple	сопло с многими отверстиями
M 196	multiple beam interferometry	Vielstrahlinterferometrie f	interférométrie f à plusieurs rayons	многолучевая интерферометрия
M 197	multiple belt conveyor vacuum drier	Mehrfachumlaufband-vakuumtrockner m	sécheur m à vide par bandes rotatives	многоленточная конвейерная сушилка
	multiple-graded glass seals, graded seal, graded glass seal tubulation	Schachtelhalm m, Übergangsglasrohr n	tube m en verre de transition	стеклянный переход, соединение, состоящее из нескольких переходных стекол
M 198	multiple ring baffle	Dampfsperre f mit Ringblechen	piège m à vapeur avec tôle annulaire	ловушка с множественными кольцами
M 199	multiple scattering	Vielfachstreuung f	dispersion f multiple	многократное рассеяние
M 200	multiple sliding vane rotary pump, rotary mechanical pump with multiple vanes, multiple vane pump	Vielschieberpumpe f	pompe f multiple (à palettes multiples)	многопластинчатый насос
M 201	multiple thread	mehrgängiges Gewinde n	filetage m à pas multiple	многониточная (многозаходная, многоходовая) резьба
M 202	multiple-type valve tube	me hrteilige Ventilröhre f	tube m de vanne à plusieurs parties	комбинированная выпрямительная лампа
	multiple vane pump	s. multiple sliding vane rotary pump		
M 203	multiple-wire glassed header	Durchführung f mit mehreren eingeglasten Drähten	ensemble m base	ножка-цоколь с остеклованными выводами
M 204	multi-point pen recorder, multi-point recorder	Mehrfach-Punkt-Schreiber m	enregistreur m multi-courbe par points	регистрирующий прибор для записи нескольких величин
M 205	multi-purpose furnace multi-stage, series single stage, compound	Mehrzweckofen m mehrstufig	four m à usage multiple à plusieurs étages	многоцелевая печь многоступенчатый
M 206	multi-stage degassing	Vielstufenentgasung f	dégazage m à plusieurs étages, dégagement m de gaz à plusieurs étages	многоступенчатое обезгаживание
M 207	multi-stage pump	mehrstufige Pumpe f	pompe f à plusieurs étages, pompe multiétagée	многоступенчатый насос, многоступенчатый насос
	multi-vibrator, flip-flop circuit	Flip-Flop-Schaltung f	montage m flip-flop, circuit m flip-flop, multivibrateur m	мультивибратор

N

N 1	natural convection	freie Konvektion f	convection f libre	свободная (есте ственная конвекция
	natural steatite, block talc	Naturspeckstein m, Talk m	stéatite f naturelle, talc m	природный стеатит, тальк
N 2	needle valve	Nadelventil n	vanne f à pointeau, robinet m à aiguille	игольчатый вентиль
N 3	negative glow	negatives Glimmlicht n	décharge f luminescente négative	катодное (отрицательное) тлеющее свечение, второе катодное свечение
N 4	negative temperature coefficient resistor, NTC resistor	NTC-Widerstand m	NTC-résistance f	резистор с отрицательным температурным коэффициентом
N 5	net density	Reindichte f	densité f nette	нетто-плотность
	net evaporation rate, effective evaporation rate	effektive Verdampfungsrate f	quote-part f effective d'évaporation	эффективная скорость испарения
	net plane, crystal plane, grate plane	Netzebene f	plaine f de réseau	плоскость [кристаллической] решетки
N 6	net pumping speed, overall pumping speed	effektives Saugvermögen n, wirksames Saugvermögen	puissance f effective d'une pompe, capacité f effective d'absorption	быстрота разрежения объекта, эффективная скорость откачки
N 7	net-shaped electrode, wire gauze electrode	Netzelektrode f	électrode f réticulaire (de réseau), électrode en toile métallique	электрод из проволочной сетки
	net speed, operational speed, true speed, effective speed	wirksames Saugvermögen n	débit m utile, débit effectif	эффективная скорость откачки, быстрота разрежения объекта
N 8	nipple	Nippel m	raccord m fileté, nipple m	соединительная трубка с резьбой, ниппель
N 9	nitrogen equivalent noble gas, rare (inert) gas	Stickstoffäquivalent n Edelgas n	équivalent m d'azote gaz m inerte (rare)	азотный эквивалент инертный (редкий, благородный) газ
N 10	no-creep type baffle	Dampfsperre f mit Kriechschutz	chicane f non rampante	механическая ловушка, в которой исключена миграция
N 11	noise damping	Entdröhnen n, Schalldämpfung f	amortissement m de son	ослабление звука, звукоизоляция
N 12	noise level	Rauschpegel m, Störpegel m	niveau m de bruit	уровень шума

N 13	**noise muffler,** silencer	Schalldämpfer *m*	sourdine *f*, silencieux *m*, amortisseur *m* de son	глушитель шума, глушитель [звука]
N 14	**noise test**	Rauschtest *m*	essai *m* de bruit	испытание на шумы
N 15	**nominal bore,** nominal width	Nennweite *f*, NW	diamètre *m* nominal	диаметр условного прохода
N 16	**nominal pressure,** rated pressure	Nenndruck *m*	pression *f* nominale	номинальное давление
N 17	**nominal throughput,** rated throughput	Nennsaugleistung *f*	puissance *f* d'aspiration (d'admission) nominale	номинальная производительность
	nominal width, nominal bore	Nennweite *f*, NW	diamètre *m* nominal	диаметр условного прохода
N 18	**non-condensable gas,** permanent gas	nichtkondensierbares Gas *n*, permanentes Gas	gaz *m* incondensable (permanent)	неконденсируемый газ
	non-conducting sense, inverse direction	Sperrichtung *f*	sens *m* de blocage (non-conduction)	непроводящее (обратное) направление
N 19	**non-consumable electrode**	Permanentelektrode *f*, nicht abschmelzbare Elektrode *f*, Dauerelektrode *f*	électrode *f* permanente	нерасходуемый электрод
N 20	**non-destructive**	zerstörungsfrei	non destructif	неразрушающий
N 21	**non-magnetic ionization gauge**	Ionisationsvakuummeter *n* ohne Magnet	manomètre *m* à ionisation sans aimant	немагнитный ионизационный манометр
N 22	**non-positive displacement pump,** turbo-pump	Turbopumpe *f*	turbopompe *f*, pompe *f* mécanique non volumétrique	турбонасос
N 23	**non-refrigerated isolation trap**	ungekühlte Dampfsperre *f* ‹Zeolithfalle›	chicane *f* non réfrigérée	неохлаждаемая ловушка
	non-return valve, back-pressure valve	Rückschlagventil *n*	valve *f* de rebondissement, soupape *f* de rebondissement (retenue)	обратный клапан, возвратный клапан
N 24	**non-self-maintained discharge**	unselbständige Entladung *f*	décharge *f* non indépendante	несамостоятельный разряд
N 25	**non-volatile**	nichtflüchtig	non volatil	нелетучий
N 26	**non-volatilized getter**	Sorptionsgetter *m*	getter (reducteur) *m* de sorption	неиспаряемый (нераспыляемый) газопоглотитель
	normal atmosphere, standard atmosphere, atmospheric pressure ‹sea level, °C›, atmosphere, A.P., atm	normaler Atmosphärendruck *m* ‹760 Torr›, physikalische Atmosphäre *f*, Normdruck *m*, Atm	pression *f* atmosphérique normale, atmosphère *f* physique	физическая атмосфера, атм
	normal breakdown forepressure, dynamic breakdown forepressure	„normaler Durchbruchs"-Vorvakuumdruck *m*	pression *f* normale du débit du vide préliminaire	нормальное выпускное давление срыва
	normal forepressure, dynamic forepressure, operating forepressure	Vorvakuumdruck *m* ‹den eine Vorpumpe am Vorvakuumstutzen einer Treibmittelpumpe erzeugt›	pression *f* du vide préliminaire	выпускное (форвакуумное) давление
N 27	**normal glow discharge**	normale Glimmentladung *f*	décharge *f* luminescente normale	нормальный тлеющий разряд
N 28	**notched bar impact test**	Kerbschlagversuch *m*	essai *m* de résilience	испытание на удар
N 29	**notched impact strength**	Kerbschlagzähigkeit *f*	résilience *f*, résistance *f* à l'entaille	ударная вязкость образца с надрезом
N 30	**notch sensitivity**	Kerbempfindlichkeit *f*	sensibilité *f* à l'entaille, sensibilité à l'encroche	чувствительность к насечке.
N 31	**nozzle**	Düse *f*	tuyère *f*, diffuseur *m*, buse *f* ‹d'une pompe à fluide moteur›, tube *m* de Venturi, éjecteur *m*	сопло
	nozzle, ejector nozzle, jet nozzle	Strahldüse *f*	tuyère *f*, buse *f*, diffuseur *m*, éjecteur *m*	эжектор, струйное сопло
	nozzle assembly, jet assembly	Düsenstock *m*, Düsensatz *m*	ensemble *m* diffuseur	система сопел
	nozzle clearance, jet clearance	Diffusionsspaltbreite *f*	largeur *f* de clivage de diffusion	ширина зазора сопла, ширина диффузионной поверхности
	nozzle clearance area, pump throat area, jet clearance area, aperture gap, annular gap, admittance area	Diffusionsspaltfläche *f*	plan *m* de clivage de diffusion	площадь кольцевого зазора, площадь поверхности диффузии
N 32	**nozzle head**	Düsenkopf *m*	tête *f* de tuyère	головка сопла
N 33	**nozzle mixing burner**	Kreuzstrombrenner *m*	brûleur *m* à courant en croix	горелка со смесительным соплом
N 34	**nozzle mouth, nozzle opening**	Düsenmund *m*, Düsenmündung *f*	embouchure *f* de tuyère (diffuseur)	устье сопла
N 35	**nozzle throat,** throat of nozzle	Düsenhals *m*, kleinste Düsenspaltfläche *f*	col *m* de tuyère (diffuseur), buse *f* de diffuseur	горловина сопла
	NTC resistor	*s.* negative temperature coefficient resistor		
N 36	**nuclear fission**	Kernspaltung *f*	fission *f* nucléaire (du noyau)	деление (расщепление) ядра
N 37	**nuclear fusion**	Kernfusion *f*	fusion *f* nucléaire, réaction *f* thermonucléaire	синтез (слияние) ядер
N 38	**nucleation**	Kernbildung *f*, Keimbildung *f*	formation *f* d'un noyau (nucléus)	образование центров (ядер)

N 39	**nucleon**	Kernteilchen *n*, Nukleon *n*	particule *f* nucléaire, nucléon *m*	нуклон
N 40	**nucleus**	Kern *m*, Keim *m*, Kristallisationskern *m*	nucléus *m*	ядро, центр
	nude gauge, nude ion gauge, bare gauge tube, high speed gauge	Einbausystem *n*, Eintauchsystem *n* <eines Ionisationsvakuummeters>	cellule *f* de mesure insérée, système *m* d'immersion	ионизационный манометр без оболочки, открытый ионизационный манометр
N 41	**nude hot filament ionization gauge tube feedthrough**	Durchführung *f* für Eintauch-Ionisationsvakuummeter-System	passage *m* pour un système de vacuomètre d'ionisation d'immersion	ввод для открытого ионизационного манометра с накаленным катодом
	nude ion gauge, nude gauge, bare gauge tube, high speed gauge	Einbausystem *n*, Eintauchsystem *n* <eines Ionisationsvakuummeters>	cellule *f* de mesure insérée, système *m* d'immersion	ионизационный манометр без оболочки, открытый ионизационный манометр
	nude ion source, insertion ion gauge	Eintauchionenquelle *f*	source *f* ionique d'immersion	открытый ионный источник
N 42	**nude omegatron mass spectrometer**	Einbau-Omegatron-Massenspektrometer *n*	omégatron-spectromètre *m* de masse incorporé	открытый омегатронный масс-спектрометр
N 43	**nude system**	Eintauchsystem *n*	système *m* à immersion	открытая система, система без оболочки
N 44	**null-setting device**	Nullabgleichgerät *n*	appareil *m* de réglage du zéro	нуль-индикатор
N 45	**number density of molecules**	Moleküldichte *f*	densité *f* moléculaire	молекулярная плотность
	number of collisions on a wall, impingement rate, rate of incidence	Flächenstoßhäufigkeit *f*, Stoßzahlverhältnis *n*, mittlere Wandstoßzahl *f* pro Zeit- und Flächeneinheit, mittlere spezifische Wandstoßrate *f*	fréquence *f* du choc de surface, nombre *m* de collisions avec une paroi, fréquence de collision, taux *m* d'incidence	частота соударений с поверхностью
N 46	**number of stages**	Stufenzahl *f*	nombre *m* des plots	число ступеней (каскадов)
N 47	**Nusselt's number**	Nusselt-Zahl *f*	nombre *m* de Nusselt	число Нуссельта

O

O 1	**object carrier,** object holder, riser rod	Objekthalter *m*	barre-support *f*, porte-objets *m*	держатель объекта
O 2	**object carrier drum**	Objektträgertrommel *f*	tambour *m* du support d'objet	барабан для закладки объекта
	object holder	*s.* object carrier		
O 3	**oblique-incidence coating,** shadow casting, shadowing	Schrägbedampfung *f*, Beschatten *n*, Beschattung *f*	évaporation *f* oblique, ombrage *m*, protection *f*	наклонное напыление, осаждение под углом
O 4	**oblique shock**	schräger Verdichtungsstoß *m*, Aufspreizwinkel *m*	choc *m* oblique de compression	наклонный удар
	observation window, inspection glass, sight glass, sight port, sight window, viewing glass, viewing port, viewing window	Beobachtungsfenster *n*, Schauglas *n*, Einblickfenster *n*	fenêtre *f* de contrôle, regard *m*, regard de contrôle, fenêtre d'observation, fenêtre de coup d'œil, hublot *m*	контрольное (смотровое) стекло, глазок, смотровое отверстие (окно), наблюдательное окно
O 5	**occlusion**	Okklusion *f*	occlusion *f*	окклюзия
	OD	*s.* outer diameter		
O 6	**off-cycle defrosting**	natürliches Abtauen *n*	dégel *m* naturel, décongélation *f* naturelle	оттаивание во время нерабочей части цикла, естественное оттаивание
O 7	**offset yield strength**	Dehngrenze *f* <0,1 %>	limite *f* de dilatation	условный предел текучести <0,1 %>
	OFHC copper	*s.* oxygen-free high-conductivity copper		
O 8	**oil backstreaming**	Ölrückströmung *f*	reflux *m* d'huile	обратное проникновение масла
O 9	**oil baffle sheet,** oil spray arrester	Ölfangblech *n*	chicane *f* pour retenue d'huile	маслоуловительный экран
O 10	**oil change**	Ölwechsel *m*	changement (renouvellement) *m* d'huile	смена масла
O 11	**oil charge,** oil filling	Ölfüllung *f*	remplissage *m* d'huile, charge *f* d'huile	порция залитого масла
O 12	**oil charge valve**	Öleinfüllventil *n*	soupape *f* de remplissage d'huile	маслоналивной вентиль, маслоналивной клапан
O 13	**oil circulation heating**	Ölumlaufheizung *f*	chauffage *m* à circulation d'huile	нагрев циркуляцией масла
O 14	**oil clarifier,** oil purifier, oil stripper, oil evaporizer	Ölreiniger *m*	épurateur *m* d'huile	маслоочиститель
O 15	**oil-conditioning plant,** oil treating plant	Ölaufbereitungsanlage *f*	installation *f* de préparation d'huile	установка для приготовления масла
O 16	**oil creepage**	Ölkriechen *n*	fluage *m* d'huile	поверхностная утечка масла
	oil creep barrier, anticreep barrier, antimigration	Kriechsperre *f*, Ölkriechsperre *f*	barrière *f* de fuites d'huile	барьер для защиты от проникновения масла по поверхности, антимиграционный барьер

	oil evaporizer	s. oil clarifier		
O 17	oil diffusion pump	Öldiffusionspumpe f	pompe f à diffusion de vapeur d'huile, pompe à diffusion à huile	масляный диффузионный насос
O 18	oil dip rod, oil dip stick	Ölmeßstab m	jauge f à huile	маслощуп, маслоизмерительный стержень
O 19	oil draining, oil purging	Ölablaß m	vidange f d'huile	слив (спуск) масла
O 20	oil drain plug	Ölablaßschraube f	vis f (bouchon m) de vidange d'huile, vis de purge d'huile	маслосливная пробка, маслосливной винт
O 21	oil drain valve	Ölablaßventil n	soupape f de vidange d'huile	маслоспускной вентиль (клапан)
O 22	oil ejector pump	Öldampfstrahlpumpe f, Ölejektorpumpe f	pompe f à éjecteur d'huile, éjecteur m d'huile	масляный эжекторный насос
O 23	oil feed	Ölumlauf m	circulation f d'huile	циркуляция масла, подача масла
	oil filling, oil charge	Ölfüllung f	remplissage m d'huile, charge f d'huile	порция залитого масла
O 24	oil filling plug	Öleinfüllschraube f	vis f de remplissage d'huile	винт (винтовая пробка) для заливки масла
O 25	oil filter, oil strainer	Ölfiltereinrichtung f, Ölfilter n	filtre m pour (à) huile	масляный фильтр
O 26	oil-immersed, oil-submerged	ölüberlagert	baignant dans l'huile, submergé d'huile, recouvert d'huile	погруженный в масло
O 27	oil level	Ölstand m	niveau m d'huile	уровень масла
O 28	oil-level plug	Ölstandschraube f	vis f du niveau d'huile	винтовая пробка для регулировки уровня масла
O 29	oil mist filter	Ölnebelfilter n	filtre m antibrouillard (de nuage) d'huile	фильтр для улавливания паров масла
	oil purging	s. oil draining		
	oil purifier	s. oil clarifier		
O 30	oil recirculating filter	Ölfilter n mit Ölrücklauf, Ölrücklauffilter n	filtre m pour l'épuration de l'huile	масляный рециркуляционный фильтр
O 31	oil regenerating plant	Ölregenerieranlage f	dispositif m de régénération d'huile, purificateur (épurateur)m d'huile	установка для регенерации масла
	oil return, back-flow of oil, oil suck-back	Ölrückfluß m	reflux (retour) m d'huile	обратный поток масла, обратное течение масла
O 32	oil return tube	Ölrücklaufleitung f	conduite f à reflux d'huile	трубка для возврата масла
O 33	oil-seal	Öldichtung f	joint m d'huile	масляное уплотнение
O 34	oil separator	Ölabscheider m	séparateur m d'huile	маслоотделитель, маслоуловитель
O 35	oil sight glass	Ölstandsschauglas n, Ölstandsauge n	regard m de niveau d'huile, indication f de niveau d'huile	стекло для наблюдения за уровнем масла
O 36	oil splash baffle	auspuffseitiger Ölfänger m	séparateur m d'huile d'échappement	ловушка для защиты от выплесков масла
	oil spray arrester	s. oil baffle sheet		
	oil strainer	s. oil filter		
	oil stripper	s. oil clarifier		
	oil-submerged, oil-immersed	ölüberlagert	baignant dans l'huile, submergé d'huile, recouvert d'huile	погруженный в масло
	oil suck-back	s. oil return		
O 37	oil trap	Ölfänger m	collecteur m d'huile, piège m à huile	масляная ловушка
	oil treating plant	s. oil-conditioning plant		
O 38	oil vapour pump	Öltreibdampfpumpe f	pompe f à flux de vapeur d'huile	паромасляный насос
O 39	omegatron, omegatron-type mass spectrometer	Omegatron n	omégatron m	омегатрон
O 40	omegatron gauge	Vakuummeter n nach dem Omegatronprinzip	omégatron-vacuomètre m	омегатронный манометр
	omegatron-type mass spectrometer, omegatron	Omegatron n	omégatron m	омегатрон
O 41	one-lever valve block	Einhebelventilblock m	bloc m de soupape à un levier	однорычажный блок клапанов
O 42	opacity	Undurchsichtigkeit f, Opazität f	opacité f	непрозрачность
O 43	opaque, optically dense	optisch dicht	opaque, optiquement étanche	непрозрачный
O 44	opaque quartz, quartz ware, vitreous silica	Quarzgut n	silice f vitreuse (fondue)	кварцевая порода, непрозрачный кварц
O 45	open conductance	Leitwert m des geöffneten Ventils	conductance f du clapet ouvert	пропускная способность открытого вентиля
O 46	open hearth furnace	Siemens-Martin-Ofen m, Herdofen m	four m Martin	мартеновская печь
O 47	open-path distillation, short-path distillation	Kurzwegdestillation f	distillation f à court trajet	молекулярная перегонка
O 48	open path still, short path still, unobstructed path still	Kurzwegdestillationsanlage f	distillerie f à court trajet, installation f de distillation à court trajet	установка для молекулярной перегонки
O 49	open surface degassing	Rieselentgasung f	dégagement m gazeux par ruissellement, dégazage m par ruissellement	пленочная дегазация

O 50	**operating fluid,** pump fluid, pump oil, working medium, working fluid	Pumpentreibmittel *n*, Pumpenöl *n*	agent *m* moteur, fluide *m* moteur d'une pompe, huile *f* à pompe	вакуумное масло, масло для высоковакуумного насоса
	operating forepressure, normal forepressure, dynamic forepressure	Vorvakuumdruck *m* ‹den eine Vorpumpe am Vorvakuumstutzen einer Treibmittelpumpe erzeugt›	pression *f* du vide préliminaire	выпускное (форвакуумное) давление
O 51	**operating pressure,** working pressure	Betriebsdruck *m*, Arbeitsdruck *m*	pression *f* de service (travail)	эксплуатационное (рабочее) давление
	operational reliability, foolproofness	Betriebssicherheit *f*	sécurité *f* de marche (service), sécurité d'exploitation	эксплуатационная надежность, безопасность в работе
	operational speed, true speed, effective speed, net speed	wirksames Saugvermögen *n*	débit *m* utile, débit effectif	эффективная скорость откачки, быстрота разрежения объекта
	operation cycle, action cycle	Arbeitstakt *m*	phase *f* de travail, cycle *m* de travail, période *f* de travail, rythme *m* de travail	рабочий такт
O 52	**optical baffle,** optical opaque baffle, optical tight baffle	Dampfsperre *f* ohne axialen Durchblick, optisch dichte Dampfsperre	piège *m* à vapeur opaque (optiquement étanche)	оптически непрозрачная ловушка, оптически непросматриваемая ловушка
O 53	**optically active body**	optisch aktiver Körper *m*	corps *m* d'activité optique	оптически активное тело
	optically dense, opaque	optisch dicht	opaque, optiquement étanche	непрозрачный
O 54	**optical opaque baffle**	s. optical baffle		
	optical pyrometer	Strahlungshitzemesser *m*	pyromètre *m* optique	оптический пирометр
O 55	**optical thin-film monitor**	optisches Schichtdickenmeßgerät *n*	appareil *m* optique de mesure de l'épaisseur de couches	оптический измеритель толщины пленки
	optical tight baffle	s. optical baffle		
	optics coating (filming), lens coating	Oberflächenvergütung *f* ‹optischer Elemente›	trempe *f* superficielle ‹des éléments optiques›	осветляющее покрытие ‹линзы›
	optics filming, lens blooming	Vergütung *f* optischer Gläser	revêtement *m* de surfaces optiques, application *f* de couches antiréfléchissantes	просветление оптических стекол (линз)
O 56	**orb-ion pump,** orbitron vacuum pump	Orbitronvakuumpumpe *f*, Orbitronionenpumpe *f*	orbitron-pompe *f* ionique (à vide)	орбитронный ионный насос
O 57	**orbit**	Umlaufbahn *f*	orbite *m*	орбита
O 58	**orbitron ionization gauge**	Orbitron-Ionisationsvakuummeterröhre *f*	tube *m* d'orbitron-vacuomètre *m* à ionisation	орбитронный ионизационный манометр
	orbitron vacuum pump, orb-ion pump	Orbitronvakuumpumpe *f*, Orbitronionenpumpe *f*	orbitron-pompe *f* ionique (à vide)	орбитронный ионный насос
O 59	**ore dressing**	Erzaufbereitung *f*	installation *f* de préparation des minerais, traitement *m* de minerai	обогащение руды
O 60	**orientation effect**	Richtungseffekt *m*	effet *m* d'orientation, effet de sens	эффект ориентации
O 61	**oriented overgrowth**	orientiertes Aufwachsen *n*	accroissance *f* orientée	ориентированное (направленное) наращивание
O 62	**orifice**	Öffnung *f*, Blende *f*	orifice *m*, écran *m*	отверстие, дыра, дырка
O 63	**orifice method** ‹of measuring speed›	Blendenmethode *f* ‹zur Messung des Saugvermögens›	méthode *f* de diaphragme	метод диафрагмы ‹измерения скорости откачки›
O 64	**O-ring, O-ring gasket, O-ring washer**	Rundring *m*, Rundschnurring *m*, O-Ring *m*	anneau *m* à section ronde, joint *m* torique, rondelle *f* à section circulaire	кольцо круглого сечения
	oscillating quartz crystal, high-frequency quartz crystal, crystal transducer, quartz crystal oscillator, vibrating quartz crystal	Schwingquarz *m*	quartz *m* oscillatoire (de résonance)	кристалл кварцевого генератора, генерирующий кварцевый кристалл, генерирующий кварц
O 65	**oscillating-vane vacuum gauge**	Reibungsvakuummeter *n*	vacuomètre *m* à frottement (amortissement)	динамический манометр с колеблющейся пластиной
	outer diameter, OD, outside diameter, external diameter	Außendurchmesser *m*	diamètre *m* extérieur	наружный диаметр
O 66	**outer metal coating**	Außenmetallisierung *f*	métallisation *f* extérieure	наружная металлизация
O 67	**outer space simulator,** space simulation chamber	Weltraumsimulierkammer *f*	chambre *f* de simulateur d'espace	камера для имитации условий космического пространства
	outgassing, degassing, degasification	Entgasung *f*	dégazage *m*, dégazation *f*	обезгаживание
O 68	**outgassing**	Gasabgabe *f*	dégazage *m*, dégagement *m* gazeux, flux *m* de dégazage, désorption *f* de gaz	газоотдача
	outlet, discharge port, exhaust port, exhaust	Auspuff *m*, Auspufföffnung *f*	échappement *m*, orifice *m* de refoulement, orifice d'échappement, orifice d'expulsion	выхлопное окно, выпускной патрубок

	outlet flange, fore-vacuum connection, fore-arm connection, backing line connection, fore-pump connection, exhaust fitting	Vorvakuumanschluß *m*	raccord *m* de vide préliminaire	соединение [насоса] с форвакуумом
	outlet pipe, exhaust pipe, discharge pipe	Auspuffleitung *f* <rotierende Pumpe>	tube *m* d'échappement, conduite *f* d'échappement	выхлопная труба, выпускной патрубок, выпускная труба
	outlet pressure, discharge pressure, exhaust pressure, head pressure	Verdichtungsdruck *m*, Auspuffdruck *m*, Ausstoßdruck *m*	pression *f* de sortie (refoulement), pression du vide préliminaire, pression d'échappement, pression finale	давление сжатия, выпускное давление, давление на выхлопе
O 69	**output signal-to noise ratio**	Ausgangssignal-Rauschverhältnis *n*	rapport *m* signal sur bruit du signal de sortie	отношение выходного сигнала к шуму, отношение сигнал — шум
	outside diameter, outer diameter, OD, external diameter	Außendurchmesser *m*	diamètre *m* extérieur	наружный диаметр
O 70	**outside lap seal**	Außenzylinderlötung *f*	brasage *m* (soudage *m*, soudure *f*) de cylindre extérieur	наружное паянное цилиндрическое соединение
O 71	**outside tubular seal**	Außenanglasung *f*	scellement *m* de verre extérieur	наружный трубчатый спай, спай с наружной трубкой
O 72	**overall coefficient of heat transfer,** thermal transmittance	Wärmedurchgangszahl *f*	coefficient *m* du passage de chaleur	коэффициент теплопередачи
	overall pumping speed, net pumping speed	effektives Saugvermögen *n*, wirksames Saugvermögen	puissance *f* effective d'une pompe, capacité *f* effective d'absorption	быстрота разрежения объекта, эффективная скорость откачки
	overall test, chamber test, envelope test, hood pressure test	Hüllentest *m*, Haubenlecksuchverfahren *n*, Haubenleckprüfung *f*	test *m* d'enveloppe, épreuve *f* d'étanchéité	проверка помещением объекта в оболочку
O 73	**overcoat,** top coat, top layer	Deckschicht *f*	couche *f* de recouvrement	наружный слой
O 74	**overexpansion**	Überexpansion *f*	surexpansion *f*	перерасширение, сверхрасширение, форсированное расширение
O 75	**overload**	Überlast *f*	surcharge *f*	перегрузка
O 76	**overload protection**	Überlastungsschutz *m*	protection *f* contre les surcharges	защита от перегрузок
	overload relay, cut-off relay	Überstromrelais *n*	relais *m* à excès de courant, relais de surintensité	реле максимального тока, максимальное реле, реле перегрузки
O 77	**overpressure,** positive pressure	Überdruck *m*	pression *f* effective, surpression *f*	избыточное давление
O 78	**overpressure method**	Überdruckverfahren *n*	méthode *f* de surpression	метод проверки под избыточным давлением
O 79	**overpressure testing method**	Überdruckdichtigkeitsprüfung *f*	épreuve *f* d'étanchéité à surpression	метод поиска течей опрессовкой
O 80	**oversaturated vapour,** supersaturated vapour	übersättigter Dampf *m*	vapeur *f* sursaturée	пересыщенный пар
O 81	**oxide cathode,** oxide coated cathode	Oxidkatode *f*	cathode *f* d'oxyde	оксидный катод
O 82	**oxide-ceramic stamping mass**	oxidkeramische Stampfmasse *f*	pisé *m* réfractaire damé de céramique oxydée	масса для прессования оксидной керамики
	oxide coated cathode, oxide cathode	Oxidkatode *f*	cathode *f* d'oxyde	оксидный катод
O 83	**oxygen-free high-conductivity copper,** OFHC copper	OFHC-Kupfer *n*, Reinstkupfer *n*	OFHC-cuivre *m*, cuivre *m* très pur	бескислородная медь, медь OFHC
O 84	**oxygen leak detector**	Sauerstofflecksuchgerät *n*	renifleur (reniflar) *m* par oxygène, détecteur *m* de fuites par oxygène	кислородный течеискатель
O 85	**ozalid paper**	Ozalidpapier *n*	papier *m* héliographique « Ozalid »	диазотипная (светочувствительная) бумага

P

	packaged pumping system, compact pumping set, pumping unit	betriebsfertiger Pumpstand *m*	poste (bâti, banc, stand) *m* de pompage	откачной стенд
P 1	**packed column,** packing tower	Füllkörperkolonne *f*	colonne *f* à corps de remplissage	набивная колонка, насадка
P 2	**packed valve**	Stopfbuchsenventil *n*, Ventil *n* mit Stopfbuchse	vanne *f* à presse-étoupe	сальниковый вентиль
	packing, seal, sealing, sealing-in, gasket, sealing element	Dichtung *f*, Dichtungselement *n*, Einschmelzstelle *f*, Abdichtung *f*, Einschmelzung *f*, Verschmelzung *f*	joint *m*, joint étanche (scellé, d'étanchéité), scellement *m*, étanchement *m*, jonction *f* étanche, point *m* de scellement, élément *m* d'étanchéité	уплотнение, соединение, уплотняющий элемент, спай, место спая, запайка, впайка

P 3	**packing density**	Packungsdichte f	densité f de garniture	плотность набивки (упаковки)
	packing fluid, confining liquid	Sperrflüssigkeit f	liquide m obturateur	жидкость для [перекрытия] затвора, уплотняющая жидкость
P 4	**packing gland,** stuffing box	Stopfbuchse f	presse-étoupe m, boîte f à bourrage	сальник, сальниковая коробка
P 5	**packing grease,** sealing grease	Dichtungsfett n	graisse f d'étanchéité, graisse à vide	уплотняющая смазка
P 6	**packingless vacuum valve**	stopfbuchsloses Vakuumventil n	robinet m sans presse-étoupe	вакуумный вентиль без сальника
	packing ring, gasket, gasket ring, sealing ring	Dichtungsring m	anneau m d'étanchéité, rondelle f de joint	уплотнительное (уплотняющее, прокладочное) кольцо, прокладка
	packing tower, packed column	Füllkörperkolonne f	colonne f à corps de remplissage	набивная колонна, насадка
P 7	**packless valve**	stopfbuchsloses Ventil n	robinet m sans presse-étoupe	бессальниковый вентиль
P 8	**paddle dryer**	Schaufeltrockner m	sécheur m à palettes (ailettes)	сушилка с перелопачивающей мешалкой
P 9	**palladium barrier ionization gauge**	Palladium-Wasserstoff-Ionisationsvakuummeter n	vacuomètre m d'ionisation palladium-hydrogène	ионизационный манометр с палладиевой перегородкой
P 10	**palladium barrier leak detector,** palladium leak detector	Palladium-Wasserstoff-Lecksucher m	détecteur m de fuites d'un récipient par palladium et hydrogène	течеискатель с палладиевым барьером, палладиевый течеискатель
P 11	**palladium black**	Palladiumschwarz n	palladium m noir	палладиевая чернь
P 12	**palladium hydrogen leak**	Palladium-Wasserstoff-Leck n	fuite f palladium-hydrogène	палладиевый водородный натекатель
	palladium leak detector	s. palladium barrier leak detector		
P 13	**palladium leak detector**	Palladiumlecksuchgerät n	détecteur m de fuites à [filtres de] palladium	палладиевый течеискатель ⟨детектор⟩
P 14	**partial measure gauge,** partial pressure measuring device	Partialdruckmeßgerät n	appareil m de mesure de pression partielle, manomètre m de pression partielle	измеритель парциального давления
P 15	**partial pressure**	Partialdruck m	pression f partielle	парциальное давление
P 16	**partial pressure analyzer**	Partialdruckanalysator m	analyseur m de pressions partielles	анализатор парциального давления
	partial pressure measuring device, partial measure gauge	Partialdruckmeßgerät n	appareil m de mesure de pression partielle, manomètre m de pression partielle	измеритель парциального давления
P 17	**partial pressure sensitivity**	Partialdruckempfindlichkeit f	sensibilité f à la pression partielle	чувствительность по парциальному давлению
P 18	**partial pressure vacuum gauge**	Partialdruckvakuummeter n	vacuomètre m de pression partielle	вакуумметр парциального давления
	partial vacuum, low pressure	Unterdruck m	dépression f, souspression f	разрежение, давление ниже атмосферного
	particle, corpuscle	Masseteilchen n	corpuscule m, particule f, élément m de masse	корпускула
P 19	**particle density**	Teilchendichte f	densité f de particules	плотность частиц
P 20	**Paschen's law**	Paschensches Gesetz n	loi f de Paschen	закон Пашена
P 21	**Paul quadrupole spectrometer**	Quadrupol-Massenspektrometer n nach Paul	spectromètre m quadrupôle de Paul, spectromètre de masse à quadrupôle de Paul	квадрупольный масс-спектрометр Пауля
	PbO and glycerine cement, litharge and glycerine cement, lead monoxide and glycerine cement	Glyzerin-Bleiglätte-Kitt m	mastic m de glycérine et litharge	цемент (замазка) на основе глицерина и свинцового глета
P 22	**peak-to-valley value**	Rauhwert m	valeur f de la rugosité	величина шероховатости (неровности)
P 23	**pechblende**	Pechblende f	pechblende f, uranine f	урановая смолка, уранинит
P 24	**pellicular electronics**	Folienelektronik f	électronique f de feuilles minces	тонкопленочная электроника
P 25	**Peltier trap,** thermoelectric baffle	Peltier-Falle f, Peltier-Baffle n	piège m à effet Peltier, piège (baffle) m thermoélectrique	термоэлектрическая ловушка
P 26	**pendulum type movement gate valve**	Pendelschieberventil n	soupape f (vanne f, robinet m) à coulisse oscillante	вентиль с качающейся заслонкой
	penetrating depth, depth of penetration	Eindringtiefe f	profondeur f de pénétration	глубина проникновения
P 27	**penetration factor of anode**	Anodendurchgriff m	coefficient m d'entrée anodique	проницаемость анода, коэффициент проницаемости анода
P 28	**penetrometer**	Penetrationsmesser m	pénétromètre m	пенетрометр, измеритель проникающей способности излучения

	Penning gauge, cold-cathode ionization gauge, Philips ionization gauge	Kaltkatodenvakuum-meter *n*, Philips-Vakuummeter *n*, Penning-Vakuum-meter *n*	manomètre *m* à ionisation à cathode froide	[магнитный] электрораз-рядный манометр с хо-лодным катодом, ма-нометр Пеннинга
	penning-type pump, cold-cathode ion pump, sputter-ion pump	Ionenzerstäuberpumpe *f*	pompe *f* ionique à cathode froide, penning-pompe *f*, pompe ionique de pulvérisation	магнитный электрораз-рядный насос, ионный насос с холодным ка-тодом, электрораз-рядный насос, насос Пеннинга, ионно-рас-пылительный насос
P 29	**pen recorder,** recording instrument	Registriergerät *n*, Schreiber *m*	enregistreur *m*, appareil *m* enregistreur, dispositif *m* d'enregistrement	самописец, регистрирую-щий прибор
P 30	**penumbral region**	Halbschattenbereich *m*	zone *f* de pénombre	область полутени, область полусвета
	percussion welding, electropercussive welding	Stoßschweißung *f*	soudure *f* à électro-percussion	ударная сварка
	perfect crystal, ideal crystal	Idealkristall *m*	cristal *m* idéal	идеальный кристалл
	perfect gas, ideal gas	ideales Gas *n*	gaz *m* idéal (parfait)	идеальный газ
	perfect pump, ideal pump	ideale Pumpe *f* <Pumpe mit konstantem Saug-vermögen über den ge-samten Druckbereich>	pompe *f* idéale	идеальный насос <с не-зависящей от давле-ния быстротой действия>
	perfect vacuum, absolute vacuum	absolutes Vakuum *n*	vide *m* absolu	абсолютный (идеальный) вакуум
	peripheral speed (velocity), circumferential speed	Umfangsgeschwindig-keit *f*	vitesse *f* périphérique	окружная скорость
	permanent gas, non-condensable gas	permanentes Gas *n*, nicht-kondensierbares Gas	gaz *m* permanent (incondensable)	неконденсируемый газ
P 31	**permanent mould**	Dauergußform *f*	moulage *m* permanent	постоянная литейная форма
P 32	**permanent stability**	Dauerstandfestigkeit *f*	stabilité *f* permanente	усталостная прочность
P 33	**permeability [coefficient]**	Durchlässigkeit *f*	perméabilité *f*	коэффициент проницае-мости, проницаемость
P 34	**permeation**	Permeation *f*	pénétration *f*	проникновение
P 35	**permissible limit of impurities**	Zulässigkeitsgrenze *f* der Verunreinigungen	limite *f* de tolérance des impuretés	допустимый предел при-месей
P 36	**persorption**	Adsorption *f* in Poren, Persorption *f*	persorption *f*	адсорбция порами, персорбция
	PG-wire, core wire	Seelendraht *m*, PG-Draht *m*	fil *m* d'âme	провод сердечника
P 37	**phase diagram**	Phasendiagramm *n*	diagramme *m* de phases	фазовая диаграмма, диаграмма состояния
P 38	**phase displacement**	Phasenverschiebung *f*	déphasage *m*, décalage *m* de phases	сдвиг фаз, смещение фаз
P 39	**phase rule**	Phasenregel *f*	règle *f* de phase	правило фаз
	Philips ionization gauge, cold-cathode ionization gauge, Penning gauge	Kaltkatodenvakuum-meter *n*, Philips-Vakuummeter *n*, Penning-Vakuummeter *n*	manomètre *m* à ionisation à cathode froide	[магнитный] электрораз-рядный манометр с хо-лодным катодом, манометр Пеннинга
P 40	**photocell,** photo-electric cell	Fotozelle *f*	cellule *f* (tube *m*) photo-électrique	фотоэлемент
P 41	**photoconductive effect**	innerer Fotoeffekt *m*	effet *m* photoélectrique intérieur	внутренний фотоэффект
P 42	**photoelectric cathode**	Fotokatode *f*	cathode *f* photoélectrique	фотоэлектронный катод, фотокатод
	photoelectric cell, photocell	Fotozelle *f*	cellule *f* (tube *m*) photo-électrique	фотоэлемент
P 43	**photoelectric counter**	Lichtzählrohr *n*	tube *m* compteur de lumière, compteur *m* lumineux	счетчик фотонов, фото-электрический счет-чик
P 44	**photoelectric emission**	äußerer lichtelektrischer Effekt *m*, äußerer Foto-effekt *m*	effet *m* photoélectrique ex-terne, émission *f* photo-électrique	фотоэлектронная эмис-сия, внешний фото-эффект
P 45	**photoelectron constant**	Fotoelektronenkonstante *f*	constante *f* des photo-électrons	фотоэлектронная постоянная
P 46	**photoetching**	Fotoätzen *n*	photocorrosion *f*	фототравление
P 47	**photoionization**	Fotoionisation *f*	ionisation *f* photo-électrique	фотоионизация
P 48	**photoluminescence**	Fotolumineszenz *f*	photoluminescence *f*	фотолюминесценция
P 49	**photon**	Photon *n*	photon *m*	фотон
P 50	**photovoltaic cell,** rectifier photocell	Sperrschichtfotozelle *f*	cellule *f* photoélectrique de couche d'arrêt	вентильный фото-элемент, фотоэлемент с запорным слоем
P 51	**physisorption**	Physisorption *f*	physisorption *f*	физическая сорбция
P 52	**pickling**	Mattbeizen *n*	décapage *m*	травление поверхности
P 53	**piezoelectric**	druckelektrisch, piezoelektrisch	piézoélectrique	пьезоэлектрический
P 54	**pig iron**	Roheisen *n*	fonte *f* brute	доменный чугун
P 55	**pile in layers**	anschichten	déposer en couches suc-cessives	последовательно наносить слои
P 56	**pilot plant**	Versuchsanlage *f*	installation *f* d'essai	опытная установка

	English	German	French	Russian
P 57	pilot valve	Schaltventil n	soupape f de commande	клапан управления
P 58	pinch, press, stem, pinched base	Quetschfuß m	socle m pressant, tige f	штампованная ножка
	pinchcock, clip, hose press clamp, squeezing cock	Quetschhahn m	pince f pressante	зажим для шланга
	pinched base	s. pinch		
P 59	pinched joint	Klemmverbindung f, Einklemmen n	jonction f de serrage, assemblage m par pinces	зажимное соединение
P 60	pinch effect	Pincheffekt m	pinch-effet m	пинч-эффект
P 61	pinch off, squeeze off	abquetschen, abpressen, abklemmen	extraire (faire sortir) par pression, écraser, éprouver à l'air comprimé	отделять сжатием, отжимать, отдавливать, отсоединять пережатием
	pinch-off seal, copper tubing seal	Abquetschdichtung f <durch Kaltpreß-schweißen>	joint m à compression	отсоединение <методом холодной сварки>
P 62	pin hole	feines Loch n, Pore f	soufflure f	пора
P 63	pin seal	Stifteinschmelzung f	fondage m de cheville	штифтовой спай
	pipe, duct, line, tubing	Rohrleitung f	canalisation f, tuyau m, conduite f	трубопровод
P 64	pipette degassing by lifting, vacuum-siphon degassing	Vakuumheberentgasung f	dégagement m par siphon sous vide	сифонное обезгаживание в вакууме
P 65	pipette method	Pipettenmethode f	méthode f de pipette	метод «пипетки»
P 66	Pirani gauge	Pirani-Vakuummeter n	jauge f (manomètre m) de Pirani	манометр сопротивления, манометр Пирани
P 67	piston pump, reciprocating pump	Kolbenpumpe f	pompe f à piston	поршневой насос
P 68	piston ring	Kolbenring m	anneau (segment) m de piston	поршневое кольцо
P 69	pitch circle	Lochkreis m	cercle m des trous	окружность центров отверстий
P 70	pitch circle diameter	Lochkreisdurchmesser m	diamètre m du cercle des trous	диаметр окружности центров отверстий
P 71	pitching moment	Kippmoment n	couple m de décrochage	опрокидывающий момент
P 72	pit furnace	Schachtofen m	four m à cuve	шахтная печь
P 73	pitting	Grübchenbildung f	cavitation f	образование раковин, точечная коррозия
P 74	plain butt-weld	I-Naht f	joint m en I, soudure f en I	плоский стыковой сварной шов
P 75	plain connector	Flachflansch m	bride f plane	плоский соединитель
P 76	Planck's constant	Plancksches Wirkungsquantum n	constante f de Planck	постоянная Планка
P 77	Planck's formula	Plancksches Strahlungsgesetz n	loi f du rayonnement de Planck	формула Планка, закон излучения Планка
P 78	planetary cage	Planetenkäfig m, Drehtrommel f	cage f planétaire, tambour m à double rotation	коробка планетарной передачи
	plant [equipment], plant, installation	Anlage f	installation f, appareillage m	установка, устройство
P 79	plant volume	Volumen n der Anlage	volume m de l'installation (l'appareillage)	объем установки
P 80	plasma gas	Plasmagas n	gaz m de plasma	плазменный газ, газ в состоянии плазмы
P 81	plasma ion source	Plasmaionenquelle f	source f d'ions du plasma	плазменный ионный источник
P 82	plasma [jet] spraying	Plasmaspritzen n	injection f de plasma	плазменное (струйное) напыление
P 83	plasma torch	Plasmabrenner m	brûleur m de plasma	плазменная горелка
P 84	plasticizer	Weichmacher m	plastifiant m	пластификатор
P 85	plastics	Plaste mpl	matière f plastique, produit m synthétique	пластмассы
P 86	plastics coating	Kunststoffmetallisierung f	métallisation f d'une matière plastique	покрытие пластмассой
	plate, ion collector, target	Ionenkollektor m, Ionenauffänger m	collecteur m, collecteur ionique	ионный коллектор
P 87	plate baffle	Plattendampfsperre f	baffle m à disque (plateau, écran plan)	пластинчатый отражатель
P 88	plate dryer	Tellertrockner m	sécheur m à disque	тарельчатая сушилка
P 89	plate freezer	Plattengefrieranlage f	installation f de congélation en forme de plaques	плиточная морозильная установка
	plate spring, cup spring	Tellerfeder f	ressort m en disque, rondelle f élastique	дисковая (тарельчатая) пружина
	plate valve	s. disk valve		
	plug, key	Hahnküken n, Stopfen m	clé f, vanne f à opercule, tournant m	пробка крана
P 90	plug valve, stopcock	Absperrhahn m	robinet m d'arrêt, robinet à boisseau	стопорный кран, кран
	ply metal, composite metal	Verbundmetall n	métal m mixte	металлокерамическая композиция, плакированный металл
P 91	ply separation	Schichtspaltung f	clivage m (séparation f) des couches	разделение слоев
	pointer manometer, dial gauge	Zeigervakuummeter n	manomètre m à auguille (cadran)	стрелочный манометр (вакуумметр)
P 92	point source	Punktquelle f	source f ponctuelle	точечный источник

P 93	**Poiseuille flow**	Poiseuille-Strömung f	écoulement m de Poiseuille	течение Пуазейля
P 94	**Poiseuille law**	Poiseuillesches Gesetz n	loi f de Poiseuille	закон Пуазейля
P 95	**poison**	vergiften	empoisonner	отравлять
P 96	**poisoning**	Vergiften n, Vergiftung f	empoisonnement m	отравление
P 97	**Poisson's ratio**	Poissonsche Konstante f, Querdehnungsziffer f	constante f de Poisson	коэффициент Пуассона
P 98	**polar diagram**	Polardiagramm n	diagramme m polaire	полярная диаграмма, диаграмма в полярных координатах
P 99	**polarizing film,** polaroid polarizer	Polaroidfolie f als Polarisator	polariseur m de feuille polaroïdale	поляроидный поляризатор
	polaroid analyzer, analyzing film	Polaroidfolie f als Analysator	analyseur m de feuille polaroïdale	поляроидный анализатор
P 100	**polaroid foil**	Polaroidfolie f	feuille f polaroïdale	поляризационная фольга
	polaroid polarizer, polarizing film	Polaroidfolie f als Polarisator	polariseur m de feuille polaroïdale	поляроидный поляризатор
P 101	**pole-piece**	Polschuh m	corne (pièce) f polaire	полюсный наконечник
P 102	**polyaddition**	Additionspolymerisation f	polymérisation f d'addition	собственно полимеризация
P 103	**polymerization**	Polymerisation f	polymérisation f	полимеризация
P 104	**polymorphism**	Polymorphie f	polymorphisation f	полиморфизм
P 105	**polytropic**	polytrop	polytropique	политропный
	pool melt, melting bath	Schmelzbad n	bain m de fusion	расплав
	pool melting, consutrode melting	Badschmelzen n, Schmelzen n mit flüssiger Katode	fusion f à cathode liquide	плавка с переплавляемым электродом
	poppet valve, disk valve	Tellerventil n	soupape f à disque	тарельчатый вентиль
P 106	**porosity**	Porigkeit f, Porosität f	porosité f	пористость
P 107	**porous getter,** porous-type getter	Porengetter m	getter m de pores	пористый (губчатый) газопоглотитель
P 108	**portable**	transportabel, fahrbar	portatif, roulant	переносный, транспортабельный
P 109	**positioner**	Positioniervorrichtung f	indicateur m de position	позиционер
P 110	**positive column**	positive Säule f	colonne f positive	положительный столб
P 111	**positive displacement compressor**	Verdrängungsverdichter m	compresseur m de déplacement	объемный компрессор
	positive pressure, overpressure	Überdruck m	pression f effective, surpression f	избыточное давление
P 112	**pot annealing furnace**	Topfglühofen m	four m à recuire en forme de pot	ящичная печь для отжига
P 113	**potential barrier**	Potentialberg m, Potentialschwelle f	barrière f de potentiel	потенциальный барьер
P 114	**potential distribution**	Potentialverteilung f	distribution f de potentiel	распределение потенциала
P 115	**potential trough**	Potentialtopf m	puits m de potentiel	потенциальная яма
P 116	**pot furnace**	Hafenofen m <Glas>	four m à pots	горшковая печь
P 117	**pot still**	Destillierblase f	bulle f de distillation	вертикальный перегонный куб
	potting, forming	Verformung f	déformation f, formage m, façonnage m	деформация
P 118	**pourable sealing**	Vergußmasse f	masse f plastique (isolante), compound m	заливочная масса
	pouring ladle, casting ladle, chill, foundry ladle	Gießpfanne f	cuvette (coupe, écuelle, poche) f de coulée	литейный (разливочный) ковш, кокиль, форма
	pouring lip, lip pour	Gießschnauze f	bec m de coulée	сливной носок
P 119	**pouring spout**	Gießrinne f	rigole f de coulée	разливочный желоб
	pour-off, decant	dekantieren	décanter	декантировать
P 120	**pour point,** setting (solidification, solidifying) point	Stockpunkt m	point m de décongélation (solidification)	температура застывания
P 121	**powder-fed flash evaporation source**	pulvergespeiste Flash-Verdampfungsquelle f	source f de flash évaporation	взрывной испаритель с порошковым питанием
P 122	**powder metallurgy**	Pulvermetallurgie f, Sintermetallurgie f	métallurgie f des poudres	порошковая металлургия
P 123	**power density**	Energiedichte f	densité f d'énergie	плотность энергии
P 124	**power density**	Leistungsdichte f	densité f de puissance	удельная мощность
P 125	**power factor**	Verlustwinkel m	angle m de pertes	угол диэлектрических потерь
	power feedthrough, heavy current lead-in, high-current feedthrough	Hochstromdurchführung f	traversée f à courant de haute intensité, passage m à courant fort	сильноточный ввод
P 126	**power gas**	Treibgas n	gaz m pauvre, carburant m gazeux	генераторный газ
P 127	**power input**	Leistungsaufnahme f, Anschlußleistung f	puissance f de raccordement	подводимая мощность, входная мощность
P 128	**power supply unit**	Netzteil n, Netzgerät n	coffret m d'alimentation	блок питания
P 129	**Prandtl's number**	Prandtl-Zahl f	nombre m de Prandtl	число Прандтля
P 130	**precooler**	Vorkühler m	préréfroidisseur m, réfrigérant m préliminaire	предварительный охладитель
P 131	**precooling**	Vorkühlen n	préréfroidissement m	предварительное охлаждение
P 132	**pre-dry**	vortrocknen	présécher	предварительно сушить
P 133	**pre-evacuate**	vorevakuieren, grobpumpen	prévider	предварительно удалять

P 134	**preheat**	vorheizen, vorwärmen	préchauffer	подогревать
P 135	**preheat**	Vorglühen n	calcination f préliminaire	предварительный нагрев
P 136	**preheater**	Vorwärmer m	préchauffeur m	подогреватель
P 137	**preheating**	Vorwärmung f	préchauffage m	подогрев, предварительный нагрев
P 138	**preliminary drying**	Vortrocknung f	préséchage m, séchage m primaire	подсушивание
P 139	**preliminary mechanical treatment**	mechanische Vorbehandlung f	traitement m mécanique préalable, préparation f mécanique	предварительная механическая обработка
P 140	**preliminary sintering**	Vorsinterung f	concrétion f préliminaire	предварительное спекание
	pre-pump, fore-pump	vorpumpen	prévider, faire le vide primaire	предварительно откачивать
P 141	**preservation method**	Konservierungsmethode f	méthode f de conservation	способ хранения
	press, pinch, stem, pinched base	Quetschfuß m	socle m pressant, tige f	штампованная ножка
P 142	**pressed amber**	Preßbernstein m	ambre m pressé	прессованный янтарь
	pressed cathode, moulded cathode	gepreßte Katode f	cathode f pressée, cathode moulée	прессованный катод
P 143	**pressed glass plate**, sintered glass plate	Preßglasteller m	platine f en fritté	штампованная стеклянная пластина
	pressed mica, micanite	Preßglimmer m, Mikanit n	mica m pressé	прессованная слюда, миканит
P 144	**pressing jaw**	Quetschfußzange f	pince f à tige (socle pressant)	щипцы для прессовки ножек
P 145	**press sintering furnace**	Drucksinterofen m	four m de frittage sous pression	печь для спекания под давлением
P 146	**pressure**	Druck m	pression f	давление
P 147	**pressure burst**	Druckstoß m	choc m de pression	бросок (скачок) давления
P 148	**pressure butt welding**	Preßstumpfschweißen n	soudage m à rapprochement sous pression	стыковая сварка давлением
P 149	**pressure casting**	Druckguß m, Preßguß m	coulage (moulage) m par pression	литье под давлением
P 150	**pressure change**	Druckänderung f	changement m de pression	изменение давления
P 151	**pressure compensation, pressure equalizing**	Druckausgleich m	compensation f de pression	выравнивание давления
P 152	**pressure controller**	Druckregler m	régulateur m de pression	регулятор давления
P 153	**pressure conversion constant**	Druckumrechnungskonstante f	constante f de conversion de la pression	постоянная пересчета давления
P 154	**pressure-dependent**	druckabhängig	dépendant de la pression	зависимый от давления
P 155	**pressure difference (differential)**	Druckdifferenz f	différence f de pression	перепад давлений, дифференциальное давление
P 156	**pressure diffusion**	Druckdiffusion f	diffusion f sous pression	диффузия под давлением
P 157	**pressure division**	Druckteilung f	partage m de pression	разделение давления
P 158	**pressure drop**	Druckabfall m	chute f de pression	понижение (падение) давления
P 159	**pressure drop test**	Druckabfallmethode f	méthode f par décroissance de pression	метод падения давления
	pressure equalizing, pressure compensation	Druckausgleich m	compensation f de pression	выравнивание давления
P 160	**pressure equalizing valve**, pressure relief valve	Druckausgleichsventil n	soupape f d'équilibrage de pression	уравнительный вентиль
	pressure gauge, manometer	Druckmeßgerät n, Manometer n	manomètre m	манометр
P 161	**pressure head**	Druckhöhe f, Förderhöhe f	hauteur f de pression	гидростатический напор, высота давления
	pressure limit, compression limit, maximum compression pressure	Verdichtungsenddruck m	compression f finie de sortie	предельное давление сжатия
P 162	**pressure loss**	Druckverlust m	perte f de pression	потеря (снижение) давления
	pressure lubrication, flood lubrication	Druckumlaufschmierung f	graissage m central sous pression, graissage par circulation sous pression	централизованная (циркуляционная) смазка под давлением
	pressure lubrication, forced lubrication	Druckschmierung f	graissage m sous pression, graissage forcé	смазка под давлением, принудительная смазка
P 163	**pressure monitor**	Druckwächter m	garde m de pression, dispositif m de surveillance de pression	реле давления
P 164	**pressure probe**, sampling probe, sniffer [probe]	Schnüffelsonde f, Schnüffler m, Leckschnüffler m	sonde f reniflante, renifleur m	всасывающий щуп
P 165	**pressure range**	Druckbereich m	domaine m de pression	область (диапазон) давлений
	pressure ratio, compression ratio	Druckverhältnis n, Kompressionsverhältnis n	rapport m de compression, taux m de pression	степень сжатия, степень компрессии
P 166	**pressure reading**	Druckablesung f	lecture f de la pression	отсчет давления

	pressure recorder, manograph	Druckschreiber *m*	manomètre *m* enregistreur	самопишущий манометр
P 167	**pressure reducing valve, pressure regulator**	Druckminderventil *n*, Druckreduzierventil *n*	soupape *f* réducteur de pression, détendeur *m* de pression	редукционный клапан, манодетандер
P 168	**pressure relief line**	Druckausgleichsleitung *f*	conduite *f* de dérivation	коммуникация для выравнивания давления
	pressure relief valve	*s.* pressure equalizing valve		
P 169	**pressure rise**	Druckanstieg *m*	accroissement *m* (augmentation *f*) de pression	возрастание давления
	pressure rise test, isolation test, rate of rise test (method)	Druckanstiegsmethode *f* <zur Lecksuche>	méthode *f* d'accroissement de pression	метод измерения быстроты возрастания давления, метод изоляции <в течеискании>
P 170	**pressure sensitivity**	Druckempfindlichkeit *f*	sensibilité *f* de pression	чувствительность по давлению
P 171	**pressure setting**	Druckeinstellung *f*	réglage *m* (régulation *f*) de pression	установка давления
P 172	**pressure stage**	Druckstufe *f*	étage *m* de pression	ступень давления
P 173	**pressure-stage packing**	Druckstufendichtung *f*	garniture *f* étanche à étage de pression	уплотнение ступени давления
P 174	**pressure switch**	Druckschalter *m*	interrupteur *m* à pression	выключатель, срабатывающий от давления
P 175	**pressure test**	Drucktest *m* Druckprüfung *f* <Lecksuche>	test *m* de pression	испытание под давлением < в течеискании>
P 176	**pressure-time curve**	Druck-Zeit-Kurve *f*	courbe *f* de pression en fonction du temps	кривая изменения давления во времени
P 177	**pressure transducer**	Druckumwerter *m*	changeur (transformateur) *m* de pression	преобразователь давления
P 178	**pressure unit**	Druckeinheit *f*	unité *f* de pression	единица давления
P 179	**pressure values in pumps**	Druckverhältnisse *npl* in Pumpen	taux *mpl* de pression en pompes	отношения давлений в насосах
P 180	**pressure window**	Druckfenster *n*	fenêtre *f* à la pression	окно, находящееся под давлением
P 181	**pressuring**	Abpressen *n*, Abdrücken *n*	épreuve *f* à l'air comprimé	отжатие
P 182	**pressurize**	abdrücken	mettre à l'épreuve de pression, dégager, repousser	отжать
P 183	**pressurized air**	Druckluft *f*	air *m* comprimé	сжатый воздух
P 184	**pressurized plant**	Druckanlage *f*	installation *f* de pression	установка под давлением
P 185	**pre-treatment,** prior treatment, prior processing	Vorbehandlung *f*	traitement *m* préparatoire (préalable)	предварительная обработка
	primary drying, main drying	Haupttrocknung *f*	dessiccation *f* principale, séchage *m* principal	основная сушка, первичная сушка
	primary vacuum, fore vacuum	Vorvakuum *n*	vide *m* préliminaire	форвакуум, предварительный вакуум
P 186	**prime coat,** priming coat, primer	Grundschicht *f*, Grundanstrich *m*	couche *f* de fond, première couche	основной слой, грунтовка
	prime lacquer, base lacquer	Grundlack *m*	laque *f* d'accrochage, couche *f* support de laque	грунтовочный лак
	primer	*s.* prime coat		
	priming coat	*s.* prime coat		
	prior processing	*s.* prior treatment		
P 187	**prior treatment,** pre-treatment, prior processing	Vorbehandlung *f*	traitement *m* préparatoire (préalable)	предварительная обработка
P 188	**probability of association**	Assoziationswahrscheinlichkeit *f*	probabilité *f* d'association	вероятность ассоциации
	probability of collision, impact probability	Stoßwahrscheinlichkeit *f*	taux *m* de collision, probabilité *f* du choc	вероятность соударения (столкновения)
P 189	**probe**	Sonde *f*, Absprühsonde *f*	sonde *f*	зонд
P 190	**probe gas,** search gas, test gas, tracer gas	Testgas *n*, Prüfgas *n*	gaz *m* témoin (traceur, d'épreuve), gaz-sonde *m*	пробный газ
P 191	**probe technique,** probe testing	Lecksuchtechnik *f* mit Absprühsonde	technique *f* de détection de fuites par un jet de gaz-témoin	техника течеискания путем обдувания пробным газом
P 192	**process annealing**	Zwischenglühen *n*	incandescence *f* intermédiaire	промежуточный отжиг
P 193	**process engineering**	Verfahrenstechnik *f*	technologie *f* des procédés industriels	технологический процесс
P 194	**process pump**	Haltepumpe *f* <für den Betriebsdruck>	pompe *f* de retenue, pompe d'entretien	поддерживающий насос <для поддержания форвакуумного давления на выходе пароструйного насоса>
P 195	**profile gasket**	Profildichtung *f*	joint *m* profilé, anneau *m* profilé	профильный уплотнитель, профильное кольцо

P 196	**programming valve**	Programmventil *n*	vanne *f* (soupape *f*, robinet *m*) à programme	программный вентиль
P 197	**propellant**	Antrieb *m*, Treibmittel *n*	commande *f*, entraînement *m*, fluide *m* moteur, fluide actif	топливо, реактивное топливо
P 198	**protecting sphere** \<of a molecule>	Deckungssphäre *f* \<eines Moleküls>	sphère *f* à pouvoir couvrant \<d'une molécule>	защитная сфера \<молекулы>
	protection against implosion, implosion guard	Implosionsschutz *m*	protection *f* contre l'implosion	защита от имплозии
P 199	**protection circuit,** safety circuit	Schutzschaltung *f*	protection *f* électrique, connexion *f* de protection	схема защиты
P 200	**protective coat[ing],** protective film	Schutzschicht *f*	couche *f* protectrice (de protection)	защитная облицовка (пленка), защитный слой
P 201	**protective device**	Schutzvorrichtung *f*	dispositif *m* protecteur (de protection, de sécurité)	защитное устройство (приспособление)
	protective film	*s.* protective coat[ing]		
P 202	**protective gas atmosphere**	Schutzgasatmosphäre *f*	atmosphère *f* gazeuse protectrice, atmosphère de gaz protecteur	защитная атмосфера
P 203	**protective gas contact**	Schutzgaskontakt *m*	contact *m* de gaz protecteur (protectif)	контакт в защитном газе
P 204	**protective lacquer**	Schutzlack *m*	laque *f* de protection	защитный лак
P 205	**protective metal**	metallischer Schutzüberzug *m*	recouvrement *m* métallique de protection	защитное металлическое покрытие
P 206	**protective water flow relay**	Wasserströmungswächter *m*	garde *m* du courant d'eau	водяное защитное реле, защитное реле водяного потока
P 207	**puffing**	Blasenentwicklung *f*	formation *f* de bulles [d'air]	образование пузырей
	pull, hot tear, heat crack	Wärmeriß *m*	fente *f* thermique	термическая трещина, тепловой разрыв
P 208	**pump,** suck	saugen, ansaugen, pumpen	aspirer, sucer, pomper	всасывать, засасывать, качать
P 209	**pump**	Pumpe *f*	pompe *f*	насос
P 210	**pump casing,** pump housing	Pumpengehäuse *n*	carter *m* (carcasse *f*, corps *m*) de pompe	корпус насоса
	pump down, evacuate, exhaust, pump out	evakuieren, auspumpen, abpumpen	évacuer, vider, faire le vide	откачивать, разрежать
P 211	**pump-down curve**	Pumpcharakteristik *f*, Auspumpkurve *f*	caractéristique (courbe) *f* de pompage	кривая откачки
P 212	**pump-down time,** time of evacuation, time of exhaust	Pumpzeit *f*, Evakuierungszeit *f*, Auspumpzeit *f*	durée *f* d'évacuation, temps *m* de pompage	время откачки \<до заданного давления>
	pumped vacuum system, dynamic vacuum system, kinetic vacuum system	dynamisches Vakuumsystem *n*, dynamische Vakuumanlage *f*	système *m* dynamique à (du) vide	динамическая вакуумная система
	pump fluid, operating fluid, pump oil, working medium, working fluid	Pumpentreibmittel *n*, Pumpenöl *n*	agent *m* moteur, fluide *m* moteur d'une pompe, huile *f* à pompe	вакуумное масло, масло для высоковакуумного насоса
	pump housing	*s.* pump casing		
P 213	**pumping action**	Pumpwirkung *f*	effet *m* de pompage	откачивающее действие
P 214	**pumping action**	Saugwirkung *f*	aspiration *f*, succion *f*	всасывающее действие
P 215	**pumping capacity,** pumping speed, pump speed, speed of a pump, suction speed	Saugvermögen *n*, Pumpgeschwindigkeit *f*, Förderleistung *f*	débit *m* de pompage, vitesse *f* d'aspiration, vitesse de pompage, débit d'une pompe, débit effectif de pompage	скорость откачки, быстрота действия насоса, эффективная способность всасывания
P 216	**pumping combination**	Pumpkombination *f*	combinaison *f* de pompage	соединение насосов, насосное соединение
	pumping hole, air-relieving hole	Entlüftungsloch *n*	foration *f* à évacuation de l'air, foration de désaération	отверстие для откачки воздуха
P 217	**pumping method**	Pumpverfahren *n*	procédé *m* de pompage	процесс откачки
	pumping speed	*s.* pumping capacity		
	pumping stem, exhaust stem, exhaust tubulation, pump-out tubulation	Pumpstengel *m*, Vakuumstutzen *m*	queusot *m*, tubulure *f* de pompage	откачной патрубок, штенгель
P 218	**pumping system**	Pumpsystem *n*	système *m* de pompage	насосная система
P 219	**pumping time constant** \<ratio of volume to pumping speed>	Pumpzeitkonstante *f* \<Verhältnis von Volumen und Saugvermögen>	constante *f* de temps de pompage	постоянная времени откачки \<отношение объема к скорости откачки>
	pumping unit, compact pumping set, packaged pumping system	betriebsfertiger Pumpstand *m*	poste (bâti, banc, stand) *m* de pompage	откачной стенд
P 220	**pumping unit control**	Pumpstandsteuerung *f*	réglage *m* du bâti de pompage, commande *f* du bâti de pompage	управление откачным стендом
	pump inlet, suction port, intake port	Ansaugöffnung *f*	fente *f* d'aspiration, ouverture *f* d'aspiration	впускной патрубок
	pump oil	*s.* pump fluid		

	pump out, evacuate, exhaust, pump down	evakuieren, auspumpen, abpumpen	évacuer, vider, faire le vide	откачивать, разрежать
	pump-out tubulation	s. pumping stem		
	pump over, circulate by pumping	umpumpen	transvaser par pompage, faire circuler par pompage	перекачивать
	pump set, group of pumps	Pumpsatz m	groupe (agrégat) m de pompage	насосный (откачной) агрегат
	pump speed	s. pumping capacity		
	pump throat area, nozzle clearance area, jet clearance area, aperture gap, annular gap, admittance area	Diffusionsspaltfläche f	plan m de clivage de diffusion	площадь кольцевого зазора, площадь поверхности диффузии
P 221	**pump vent valve**	Belüftungsventil n der Vorpumpe, Pumpenbelüftungsventil n	valve f de ventilation de la pompe préliminaire, robinet m d'entrée d'air de la pompe préliminaire	вентиль для пуска атмосферы в форвакуумный насос
P 222	**pure metal cathode**	Reinmetallkatode f	cathode f de métal pur	чистометаллический катод
	purge, draining	Entleerung f	vidange f, évacuation f	продувка
	purge gas, bleed gas, flush gas	Spülgas n	gaz m de balayage (rinçage)	газ для продувки, промывочный газ
	purge gas flange, gas bleed flange	Spülgasflansch m	bride f de gaz de balayage	фланец для подсоединения промывочного газа
	purger, drain valve	Entleerungsventil n	soupape f de vidange	выпускной вентиль, спускной вентиль
P 223	**purge valve**	Ablaßventil n	soupape f de vidange	спускной вентиль
P 224	**purging cock**	Reinigungshahn m	purgeur m	кран для продувки
	purification, cleaning	Reinigung f	purification f, épuration f, rectification f, nettoyage m	очистка
P 225	**purification plant**	Reinigungsanlage f	installation f d'épuration	очистительная установка
	purifying pump, fractionating pump, self-fractionating pump	fraktionierende Pumpe f	pompe f fractionnante (à rectification)	разгоночный насос, фракционирующий насос
	push-in seal, edge seal	Schneidenanglasung f, Innen- und Außenanglasung f	scellement m de verre à pénétration	лезвенный спай
P 226	**push-rod**	Stößel m	poussoir m, pilon m	толкатель, стержень (штанга) толкателя
P 227	**pyrolysis**	Pyrolyse f	pyrolyse f	пиролиз
	pyrolytic plating, coating by vapour decomposition	pyrolytische Beschichtung f	revêtement m pyrolytique	пиролитическое нанесение

Q

Q 1	**quadrupole field**	Quadrupolfeld n	champ m quadrupôle	квадрупольное поле
Q 2	**quadrupole high-frequency mass spectrometer**	Quadrupol-Hochfrequenzmassenspektrometer n, Massenfilter n	spectromètre m de masse quadripolaire à haute fréquence	масс-спектрометр с квадрупольным анализатором, квадрупольный масс-спектрометр, электрический фильтр масс
Q 3	**quadrupole ionization gauge**	Quadrupolionisationsvakuummeter n	quadrupôle-vacuomètre m d'ionisation	квадрупольный ионизационный манометр
	quality factor, coefficient of performance, figure of merit	Leistungsziffer f, Gütegrad m, Gütezahl f	coefficient (facteur) m de puissance	коэффициент мощности
Q 4	**quantity of gas**	Gasmenge f	quantité f de gaz	количество газа
Q 5	**quantum mechanical**	quantenmechanisch	quantummécanique	квантомеханический
Q 6	**quantum yield**	Quantenausbeute f	rendement m quantique	квантовый выход
	quarter swing valve, butterfly valve, throttle valve	Drosselventil n, Drosselklappe f	papillon m d'obturation, robinet m d'étranglement	дроссельный вентиль
	quartz crystal oscillator, high-frequency quartz crystal, crystal transducer, oscillating quartz crystal, vibrating quartz crystal	Schwingquarz m	quartz m oscillatoire (de résonance)	кристалл кварцевого гэнератора, генерирующий кварцевый кристалл, генерирующий кварц
Q 7	**quartz crystal thin film monitor,** vibrating crystal thickness monitor	Schwingquarzschichtdickenmeßgerät n	appareil m de mesure des couches minces à quartz oscillatoire	прибор с кварцевым датчиком для измерения толщин пленок
Q 8	**quartz-fibre gauge**	Reibungsvakuummeter n mit Quarzfadenpendel	vacuomètre m adhérent (à amortissement) à fil de quartz	динамический (вязкостный) манометр с кварцевой нитью, манометр с кварцевой нитью
	quartz glass, fused silica	Quarzglas n	verre m quartzeux	кварцевое стекло
Q 9	**quartz tube**	Quarzröhre f	tube m de quartz	кварцевая лампа

Q 10	**quartz vacuum microbalance**	Quarz-Vakuum-Mikrowaage *f*	microbalance *f* en quartz-vide	вакуумные микровесы с кварцевой нитью
	quartz ware, opaque quartz, vitreous silica	Quarzgut *n*	silice *f* vitreuse (fondue)	кварцевая порода, непрозрачный кварц
Q 11	**quenching**	Abschrecken *n*, Löschen *n*	trempe *f*	гашение, закалка
Q 12	**quick-acting hand valve**	Schnellschlußhandventil *n*	soupape *f* à main à fermeture instantanée (rapide)	быстродействующий вентиль с ручным приводом
	quick-acting valve, fast-acting valve, rapid-action valve	Schnellschlußventil *n*	soupape *f* à fermeture instantanée, robinet *m* à fermeture rapide	быстродействующий вентиль (клапан)
Q 13	**quick cooling,** rapid cooling	Schnellkühlung *f*	refroidissement *m* rapide	быстрое охлаждение
Q 14	**quick discharging**	Schnellentleerung *f*	vidange *f* rapide	быстрое опорожнение (разрежение)
Q 15	**quick flange**	Kleinflansch *m*	petite bride *f*, bride miniature	малый фланец
Q 16	**quick freezing**	Schnellgefrieren *n*	congélation *f* à grande vitesse	быстрое замораживание
Q 17	**quick vacuum coupling**	Schnellverschluß *m*	fermeture (obturation) *f* instantanée (rapide)	быстроразборное соединение

R

R 1	**rack gear**	Zahnstangentrieb *m*	engrenage *m* à crémaillère	зубчато-реечная передача
	radial flow compressor, centrifugal compressor	Kreiselverdichter *m*, Radialverdichter *m*	compresseur *m* centrifuge	центробежный компрессор
R 2	**radial flow pump**	Radialpumpe *f*	pompe *f* à flux radial	центробежный насос
R 3	**radial interference**	Spaltbreite *f* <beim Spaltlöten>	largeur *f* de fente	ширина зазора <при пайке по зазору>
R 4	**radial oil seal with gasket for shafts**	Radialwellendichtring *m* mit Manschette	anneau-joint *m* radial avec embouti	радиальное сальниковое кольцо с манжетой для уплотнения вала
R 5	**radial seal [for rotating shaft],** shaft seal	Radialdichtring *m* für Welle, Wellendichtung *f*, Wellendichtring *m*	anneau-joint *m* radial, joint *m* d'arbre, garniture *f* étanche d'arbre	радиальное уплотнение вращающегося вала, уплотнение вала
R 6	**radiant gas burner**	Strahlungsgasbrenner *m*	brûleur *m* à gaz à rayonnement	излучающая газовая горелка
R 7	**radiant heater,** radiation heater	Strahlungsheizer *m*	appareil *m* de chauffage par rayonnement, chauffeur *m* à rayonnement	лучеиспускательный нагреватель, излучательный нагреватель, лучистый радиатор
R 8	**radiant heating,** radiation heating	Strahlungsheizung *f*	chauffage *m* par rayonnement, chauffage à rayonnement	радиационный нагрев, лучистый нагрев
R 9	**radiate**	strahlen	rayonner	излучать
R 10	**radiation**	Strahlung *f*	rayonnement *m*, radiation *f*	излучение, радиация, лучеиспускание
R 11	**radiation emittance,** thermal emissivity	thermisches Emissionsvermögen *n*	pouvoir *m* d'émission thermique	термическая эмиссионная способность
R 12	**radiation furnace**	Strahlungsofen *m*	four *m* de rayonnement	отражательная печь
	radiation heater	s. radiant heater		
	radiation heating	s. radiant heating		
R 13	**radiation permeability**	Strahlendurchlässigkeit *f*	perméabilité *f* à rayons	проницаемость для лучей
R 14	**radiation pressure**	Strahlungsdruck *m*	pression *f* à rayonnement	давление излучения
	radiation shield, heat shield	Strahlungsschirm *m*, Strahlungsschutzschirm *m*	écran *m* protecteur contre le rayonnement, écran contre la radiation	тепловой экран, экран для защиты от излучения
R 15	**radiative interchange**	Strahlungsaustausch *m*	échange *m* de rayonnement	обмен лучеиспусканием, излучательный обмен
R 16	**radioactive ionization gauge**	Kernstrahlungsionisationsvakuummeter *n*, Ionisationsvakuummeter *n* mit radioaktivem Präparat	manomètre *m* à ionisation par radiation nucléaire, vacuomètre *m* d'ionisation avec produit radio-actif	радиоактивный ионизационный манометр
R 17	**radioactive source**	radioaktive Quelle *f*	source *f* radioactive	радиоактивный источник
R 18	**radioactive tracer**	radioaktiver Indikator *m*	indicateur *m* radioactif	радиоактивный индикатор, меченый атом
R 19	**radioactive tracer leak detection**	Lecksuche *f* mit radioaktivem Indikator	détection *m* de fuites avec indicateur radio-actif	течеискатель с радиоактивным индикатором
	radio-frequency heating, induction heating, rf heating	Induktionsheizung *f*, induktive Wärmebehandlung *f*	chauffage *m* par induction	индукционный нагрев, высокочастотный нагрев
R 20	**radio-frequency sputtering**	Hochfrequenzzerstäubung *f*	pulvérisation *f* à haute fréquence	высокочастотное распыление
	radiometer gauge, Knudsen radiometervacuummeter, Knudsen gauge	Radiometer-Vakuummeter *n*, Knudsen-Vakuummeter *n*, Knudsensches Radiometer-Vakuummeter *n*	vacuomètre *m* radiométrique [de Knudsen]	манометр Кнудсена, радиометрический манометр [Кнудсена]
R 21	**random distribution**	statistische Verteilung *f*	distribution *f* statistique	беспорядочное (хаотическое, статистическое) распределение

R 22	range switching	Meßbereichsumschaltung f	commutation f des zones de mesure	переключение диапазона
R 23	Rankine efficiency	Rankinescher Wirkungs-grad m	rendement m de Rankine	коэффициент полезного действия цикла Ренкина
	Rankine temperature scale, absolute Fahren-heit scale	absolute Temperaturskale f, gemessen in Grad Rankine, °R ⟨T/°R = 1,8 T/ ° Kelvin⟩	échelle f de la température absolue	абсолютная температур-ная шкала Фарен-гейта, температурная шкала Ренкина
	rapid-action valve, fast-acting valve, quick-acting valve	Schnellschlußventil n	soupape f à fermeture instantanée, robinet m à fermeture rapide	быстродействующий вентиль (клапан)
	rapid cooling, quick cooling	Schnellkühlung f	refroidissement m rapide	быстрое охлаждение
R 24	rapid scanning	Schnelldurchlauf m, Schnellabtastung f	passage m rapide	быстрое сканирование
R 25	rarefied gas	verdünntes Gas n	gaz m raréfié	разреженный газ
R 26	rarefy	verdünnen	raréfier	разрежать
	rare gas, noble gas, inert gas	Edelgas n	gaz m inerte, gaz rare	инертный газ, редкий газ, благородный газ
R 27	rare gas tube	Edelgasröhre f	tube m à gaz inerte	лампа, наполненная инертным газом
R 28	Raschig ring	Raschig-Ring m	anneau m Raschig	кольцо Рашига
R 29	ratchet	Rastgesperre n, Sperr-vorrichtung f	dispositif (appareil) m d'arrêt	стопорное устройство
R 30	ratchet gear	Schaltwerk n, Klinkwerk n	station f (dispositif m) de commande, mécanisme m à mettre en circuit, encliquetage m	включающий механизм
R 31	rated fatigue limit	Gestaltfestigkeit f, Dauerhaltbarkeit f	résistance f à la fatigue	предел выносливости, усталостная прочность
R 32	rated heater input	Heizleistung f bei Netz-spannungsanschluß	puissance f calorifique (de chauffage)	номинальная мощность накала
	rated pressure, nominal pressure	Nenndruck m	pression f nominale	номинальное давление
R 33	rated speed	Nennsaugvermögen n	capacité f d'aspiration nominale, débit m nominal	номинальная скорость откачки
	rated throughput, nominal throughput	Nennsaugleistung f	puissance f d'aspiration nominale, puissance d'admission nominale	номинальная произво-дительность
R 34	rate of flow	Mengenstrom m, Ausflußrate f	vitesse f d'écoulement, débit m du courant, vitesse du courant	расход
	rate of flow, flow speed, flow rate	Strömungsgeschwindig-keit f	vitesse f d'écoulement	скорость течения (потока)
	rate of growth, growth rate	Wachstumsrate f, Wachs-tumsgeschwindigkeit f	vitesse f de croissance	скорость роста
	rate of incidence, impingement rate, number of collisions on a wall	Flächenstoßhäufigkeit f, Stoßzahlverhältnis n, mittlere Wandstoßzahl f pro Zeit- und Flächen-einheit, mittlere spezifi-sche Wandstoßrate f	fréquence f du choc de surface, nombre m de collisions avec une paroi, fréquence de col-lision, taux m d'incidence	частота соударений с по-верхностью
R 35	rate of rise	Druckanstiegsgeschwin-digkeit f	écoulement m d'augmen-tation de pression	временной ход повы-шения давления
	rate of rise method (test), isolation test, pressure rise test	Druckanstiegsmethode f ⟨zur Lecksuche⟩	méthode f d'accroisse-ment de pression	метод измерения быстро-ты возрастания давле-ния, метод изоляции ⟨в течеискании⟩
R 36	ratio of penetration-to-width	Schweißtiefe-Breite-Verhältnis n	relation f de la profondeur de soudure et de la largeur	отношение глубины провара к ширине
R 37	ray proofing	Strahlenschutz m	protection f contre le rayonnement, protec-tion anti-rayonnante	защита от излучения
	γ rays, gamma rays	Gammastrahlen mpl	rayons mpl gamma, rayons γ	гамма-лучи
	Re	s. Reynolds number		
R 38	reaction cement	Reaktionskitt m	mastic m de réaction	реакционный цемент
R 39	reactivation	Reaktivierung f	réactivation f	реактивация
R 40	reactive	reaktionsfähig	réactif	реагирующий
R 41	reactive evaporation	reaktives Verdampfen n	évaporation f réactive	реактивное испарение
R 42	reactive getter	reaktiver Getter m	getter m réactif	реактивный газопогло-титель
R 43	reactive sputtering	reaktive Zerstäubung f	atomisation m réactif	реактивное распыление
	real gas, imperfect gas	reales Gas n	gaz m réel	реальный (неидеаль-ный) газ
R 44	Réaumur scale	Réaumur-Skale f	échelle f Réaumur	шкала Реомюра
R 45	reboiler, re-evaporator	Nachverdampfer m	réévaporateur m	дополнительный испари-тель
R 46	reboiler	Rücksieder m	rebouilleur m	дополнительный кипятильник
R 47	rebound hardness, scleroscopic hardness	Rücksprunghärte f, Rückprallhärte f	dureté f de rebondissement	упругая твердость, твердость по методу отскока, твердость по Шору

R 48	recalibration	Nachkalibrierung f	recalibrage m, réétalonnage m, vérification f de l'étalonnage	проверочная калибровка
R 49	recarburization	Aufkohlung f	recarbonisation f, recarburation f	повторное науглероживание, рекарбонизация
R 50	reciprocating compressor	Kolbenverdichter m	compresseur m à piston	поршневой компрессор
R 51	reciprocating piston pump	Hub-Kolbenpumpe f	pompe f à piston élévatoire	поршневой насос, возвратно-поступательный насос
	reciprocating pump, piston pump	Kolbenpumpe f	pompe f à piston	поршневой насос
R 52	reciprocating rolling process	Pilgerschrittverfahren n	procédé m à pas de pèlerin	прокатка труб на пилигримовом (пильгерном) станке
R 53	reciprocating vacuum seal	Schiebedurchführung f	traversée f coulissante (glissante)	скользящий ввод, вакуумный ввод с осевым перемещением
R 54	reciprocating wet vacuum pump	Naßluftpumpe f	pompe f à air humide	мокровоздушный насос
	recirculating pump, circulation pump	Umwälzpumpe f	pompe f de circulation	рециркуляционный насос, циркуляционный насос
R 55	reclamation, recovery	Rückgewinnung f	récupération f	рекуперация, регенерация
R 56	recombination	Rekombination f, Wiedervereinigung f	recombinaison f	рекомбинация
R 57	recording facility	Registriervorrichtung f	appareil m enregistreur	регистрирующее устройство
	recording instrument, pen recorder	Registriergerät n, Schreiber m	enregistreur m, appareil m enregistreur, dispositif m d'enregistrement	самописец, регистрирующий прибор
R 58	recording time	Registrierzeit f	temps m d'enregistrement	время регистрации
	recovery	s. reclamation		
	recovery time, clean-up time, settling (rest) time	Erholungszeit f, Wiederansprechzeit f	temps m régénérateur, temps de régénérabilité	время [для] восстановления, период восстановления
R 59	rectangular groove	Rechtecknut f	gorge (rainure) f rectangulaire	прямоугольный паз
	rectifier photocell, photo-voltaic cell	Sperrschichtfotozelle f	cellule f photoélectrique de couche d'arrêt	вентильный фотоэлемент, фотоэлемент с запорным слоем
R 60	rectify	rektifizieren	rectifier	ректифицировать
R 61	recrystallization	Rekristallisation f	recristallisation f	рекристаллизация
R 62	recrystallization temperature	Rekristallisationstemperatur f	température f de recristallisation	температура рекристаллизации
R 63	recrystallized threshold	Kristallisationsschwelle f	seuil m de cristallisation	порог рекристаллизации (кристаллизации)
R 64	reducer, reducing piece	Reduzierstück n	réducteur m, pièce f de réduction, pièce d'adaptation	редуктор, редукторный переход
R 65	reducing coupling	Reduzierverbindungsstück n	pièce f de raccord de réduction	переходная соединительная деталь
R 66	reducing coupling	Reduzierkupplung f	accouplement m de réduction	переходная муфта, переходник
R 67	reducing cross	Reduzier-Kreuzstück n, Kreuzstück n mit Reduzierflanschen	pièce f en croix avec bride de réduction	крестовина с переходным фланцем
R 68	reducing furnace	Reduktionsofen m	four m à réduction	восстановительная печь
	reducing piece	s. reducer		
R 69	reed valve	Zungenventil n	valve (soupape) f à languette	языковый вентиль (клапан)
R 70	re-emission	Reemission f	réémission f	реэмиссия
R 71	re-emission rate	Wiedergewinnungsrate f	taux m de récupération	коэффициент реэмиссии
R 72	re-evaporate	wiederverdampfen	réévaporiser	повторно испаряться
	re-evaporator, reboiler	Nachverdampfer m	réévaporateur m	дополнительный испаритель
R 73	reference circuit	Bezugsstromkreis m	circuit m de référence	эталонная цепь, эталонный контур
	reference leak, calibrated leak, sensitivity calibrator, standard leak, test leak	Eichleck n, Testleck n, Vergleichsleck n, Leck n bekannter Größe, Bezugsleck n	fuite f calibrée	стандартная течь, калиброванная течь, эталонная течь
R 74	reference point	Eichmarke f, Bezugspunkt m	point m de référence, marque f de calibrage	начальная точка отсчета, точка отсчета, исходная точка
R 75	reference pressure	Bezugsdruck m, Vergleichsdruck m	pression f de référence (comparaison)	эталонное (отсчетное) давление
R 76	reference temperature	Bezugstemperatur f	température f de référence	эталонная температура
R 77	reference vacuum	Vergleichsvakuum n	vide m de comparaison	эталонный вакуум
R 78	reference voltage level	Bezugsspannung f, Bezugspotential n	tension f de référence	опорное напряжение, опорный потенциал
R 79	refilling device, refilling unit	Nachfüllvorrichtung f	appareil m de remplissage, dispositif m pour ajouter	аппарат (устройство) для пополнения
R 80	refined gold cathode	Feingoldkatode f	cathode f en or fin	катод из рафинированного золота

R 81	refining	Raffinieren n	raffinage m	очистка
R 82	reflectance, reflectivity	Reflexionsvermögen n	pouvoir m réflecteur (réfléchissant)	отражательная способность
	reflecting layer, mirror coating, specular layer	Spiegelbelag m, Reflexbelag m	couche f spéculaire, revêtement m réfléchissant	зеркальное покрытие
R 83	reflection coefficient	Reflexionskoeffizient m	coefficient m de réflexion	коэффициент отражения
R 84	reflection diffraction	Reflexionsbeugung f	diffraction f de réflexion	дифракция в отраженных лучах
R 85	reflection electron diffraction	Elektronenbeugung f in Reflexion	diffraction f des électrons par réflexion	дифракция электронов в отраженном пучке
R 86	reflection high energy electron diffraction, RHEED	Beugung f schneller Elektronen in Reflexion	diffraction f des électrons à grande vitesse par réflexion	дифракция быстрых электронов в отраженном пучке
	reflection reduction, anti-reflection	Reflexionsverminderung f	abaissement m réfléchissant	ослабление отражения
	reflectivity	s. reflectance		
	reflux, flow back	zurückströmen, zurückfließen	refluer	течь обратно
R 87	reflux	Rückfluß m	reflux m	обратное течение
R 88	reflux locking mechanism	Rücklaufsperre f	blocage m de retour	механизм блокировки обратного хода
R 89	reflux pump	Rücklaufpumpe f	pompe f à reflux	рециркуляционный насос
R 90	refractive index	Brechungsindex m	indice m de réfraction	показатель преломления
R 91	refractometric vacuum meter	Vakuummeter n mit optischer Refraktionsmessung	vacuomètre m réfractométrique	рефрактометрический вакуумметр
R 92	refractoriness	Wärmebeständigkeit f	thermostabilité f	термостойкость, теплостойкость
R 93	refractory metals <metals with melting points above 2,000 °C>	schwerschmelzende Metalle npl <Metalle, deren Schmelzpunkt über 2 000 °C liegt>	métaux mpl difficilement fusibles <point de fusion minimum 2 000 °C>	тугоплавкие металлы <с температурой плавления выше 2.000 °C>
R 94	refreezing	Wiedereinfrieren n	recongélation f	повторное замораживание
	refrigerant, coolant, cooling agent	Kühlmittel n, Kühlflüssigkeit f, Kältemittel n	réfrigérant m, agent m frigorifique, frigorigène m, moyen m réfrigérant	охладитель, охлаждающий агент, охлаждающая среда, холодильный агент, хладоагент
	refrigerated trap, cold trap	Kühlfalle f	piège m, piège refroidi (à froid, cryogénique, réfrigérant, de refroidissement)	охлаждаемая ловушка
R 95	refrigerating capacity, refrigerating effect	Kälteleistung f	pouvoir m (puissance f) frigorifique	холодопроизводительность
R 96	refrigerating engineering	Kältetechnik f	technique f du froid	холодильная техника
R 97	refrigerating machine, refrigerator	Kältemaschine f	machine f frigorifique	рефрижератор,холодильная машина
	refrigerating plant, low-temperature installation	Kälteanlage f	installation f de refroidissement, installation frigorifique	холодильная установка, низкотемпературная установка
R 98	refrigerating unit, refrigeration machine (system), refrigerator	Kälteaggregat n	groupe m frigorifique, machine f (système m) frigorifique	холодильная система (машина)
R 99	refrigeration	Kälteerzeugung f	réfrigération f	искусственное охлаждение
	refrigeration duty (load), heat load, cooling load	Kältebedarf m	besoin m frigorifique	расход холода
	refrigeration machine (system)	s. refrigerating unit		
	refrigerator, refrigerating machine	Kältemaschine f	machine f frigorifique	рефрижератор, холодильная машина
	refrigerator	s. refrigerating unit		
	regulating valve, control valve	Regelventil n	soupape (vanne) f de réglage	регулирующий вентиль
R 100	rehydration	Wiederbefeuchtung f	remouillage m	регидратация, повторная гидратация
R 101	relative humidity, RH	relative Feuchtigkeit f	humidité f relative	относительная влажность
R 102	relaxation time	Relaxationszeit f	temps m de relaxation	время релаксации
	release, blow off, bleed, valve	abblasen	souffler, laisser échapper, lâcher, évacuer	выпускать
R 103	release time	Abfallzeit f	temps m tombant	время спада
	reliable, fail-safe	betriebssicher	sûr, éprouvé, de fonctionnement sûr	надежный, безопасный
R 104	relief pressure valve	Überdruckventil n	soupape f de surpression	предохранительный клапан, редукционный клапан
R 105	remelt	umschmelzen	refondre	переплавлять
R 106	remelting hardness	Umschmelzhärte f	dureté f de refonte	твердость после переплавки
	remote cathode, distant cathode, distance cathode	Fernkatode f	cathode f à distance	удаленный катод

R 107	remote control	Fernsteuerung f	télécontrôle m, télé-commande f, commande f à distance, réglage m à distance	дистанционное управление, телеуправление
R 108	remote gun	Fernkanone f	canon m à distance	удаленная электронная пушка
R 109	remote gun electron beam melting furnace	Elektronenstrahlschmelzofen m mit Fernkanone	four m de fusion à bombardement électronique avec canon à distance	электроннолучевая плавильная печь с удаленной пушкой
R 110	remote indication	Fernanzeige f	indication f à distance	дистанционная индикация
R 111	renewal	Erneuerung f	renouvellement m	обновление, реконструирование
	repeller, baffle plate, deflector	Prallplatte f	déflecteur m, écran m, chicane f	отражательная пластина
	replaceable, interchangeable	austauschbar	échangeable	взаимозаменяемый
R 112	replica	Abdruck m, Kopie f	copie f, calque m	оттиск, реплика
R 113	reradiation	Rückstrahlung f	réverbération f	обратное излучение
R 114	reservoir-type freezing trap, thimble type cold trap	Behälterkühlfalle f, Kühlfinger m	réservoir-piège m à refroidissement, sonde f de refroidissement, doigt m refroidisseur	охлаждаемая ловушка с резервуаром для хладоагента
R 115	resetting of zero, zero resetting	Nullabgleich m, Rückstellung f auf den Nullpunkt	réglage m du point zéro	сброс на нуль
	residence time, adherence time, dwelling time, retention time, stay period, sticking time	Verweilzeit f	durée f, temps m de contact, temps d'exposition	время удерживания (выдержки, контакта, пребывания)
R 116	residual activity	Restaktivität f	activité f résiduelle	остаточная активность
R 117	residual conductance	Restleitwert m	conductance f résiduelle	остаточная проводимость
R 118	residual current	Reststrom m	courant m de fuite, courant m résiduel	остаточный ток
R 119	residual gas	Restgas n	gaz m résiduel	остаточный газ
R 120	residual gas analyzer	Restgasanalysator m	analyseur m de gaz résiduel	анализатор остаточного газа
R 121	residual gas composition	Restgaszusammensetzung f	composition f de gaz résiduel	состав остаточного газа
R 122	residual gas pressure	Restgasdruck m	pression f de gaz résiduel	давление остаточных газов
R 123	residual moisture	Restfeuchtigkeit f	humidité f résiduelle	остаточная влажность
R 124	residual valve conductance, sealed conductance	Leitwert m des geschlossenen Ventils	conductance f du robinet fermé, valeur f de conductivité du clapet fermé	остаточная пропускная способность вентиля, пропускная способность закрытого вентиля
R 125	residual vapour pressure	Restdampfdruck m	pression f de vapeur résiduelle	давление остаточных паров
R 126	residue	Rückstand m	résidu m	остаток, осадок
R 127	resilience	elastischer Wirkungsgrad m, Rückprallelastizität f	rendement m d'élasticité, rebondissement m élastique	эластичность, упругость
R 128	resin bonded	kunstharzgetränkt	imprégné de résine synthétique	пропитанный синтетической смолой
R 129	resin casting plant	Gießharzanlage f	installation f pour la coulée de résines synthétiques	установка для литья смолы
R 130	resin-hardener	Harzhärter m	trempeur m de résine	отвердитель смолы
R 131	resistance film thickness monitor	Widerstandsschichtdickenmeßgerät n	appareil m de mesure de la résistance des couches minces	прибор для измерения толщины тонкой пленки по омическому сопротивлению
R 132	resistance furnace	Widerstandsofen m	four m à résistance	печь сопротивления
R 133	resistance heating	Widerstandheizung f	chauffage m par résistance	резистивный нагрев
R 134	resistance heating	Widerstanderwärmung f	échauffement m par résistance	нагрев сопротивления
R 135	resistance monitor	Schichtwiderstandsmeßgerät n	appareil m de mesure de la résistance de couche	прибор для измерений пленочных сопротивлений
R 136	resistance ratio	Widerstandsverhältnis n	proportion f de résistances	отношение сопротивлений
	resistance strain gauge, electrical resistance strain gauge	Dehnungsmeßstreifen m	bande f de mesure de dilatation	тензодатчик омического сопротивления
R 137	resistance thermometer	Widerstandsthermometer n	thermomètre m à résistance	термометр сопротивления
	resistance to creepage, creeping strength, creep resistance	Kriechfestigkeit f, Dauerstandsfestigkeit f	résistance f au fluage	сопротивление ползучести, предел ползучести
	resistance to lateral bending, buckling resistance, cross breaking strength	Knickfestigkeit f	résistance f au flambage	прочность при продольном изгибе
	resistance to thermal shock, heat-shock resistance	Temperaturwechselbeständigkeit f, Abschreckfestigkeit f	résistance f à choc de température, résistance de trempe	устойчивость к термоудару

R 138	**resistance to wear**	Verschleißfestigkeit *f*	résistance *f* à l'usure	прочность на износ, износоустойчивость
R 139	**resistance welding**	Widerstandsschweißen *n*	soudage *m* par résistance	сварка сопротивлением
R 140	**resist coating**	Abdeckschicht *f*	couche *f* de fermeture (couverture)	непроводящее (кислото-упорное) покрытие
R 141	**resistive network**	Widerstandsnetz *n*	réseau *m* de résistance	резистивная схема, резистивная цепь
R 142	**resolution power, resolving power**	Auflösungsvermögen *n*	pouvoir *m* séparateur, pouvoir de résolution, pouvoir dissolvant	разрешающая способность
R 143	**resonance frequency**	Resonanzfrequenz *f*	fréquence *f* de résonance	резонансная частота
R 144	**resonating quartz**	Quarzoszillator *m*	oscillateur *m* à cristal, oscillateur de quartz, oscillateur commandé par quartz	кварцевый генератор
R 145	**respiratory mass spectrometer**	Respirationsmassen-spektrometer *n*	spectromètre *m* de masse de respiration	масс-спектрометр для анализа выхлопных газов
R 146	**respond**	ansprechen, reagieren	réagir	реагировать
R 147	**response lag**	Anzeigeverzögerung *f*	retard *m* d'indication	задержка показаний
R 148	**response time,** time of response	Ansprechzeit *f*	durée *f* de réponse, temps *m* de réaction, temps de réponse	время срабатывания, время запаздывания
R 149	**responsiveness**	Ansprechempfindlich-keit *f*	sensibilité *f* de mise en action	скорость реагирования, отзывчивость ‹прибора›
R 150	**restoring moment**	Rückstellmoment *n*	moment *m* de rappel	восстанавливающий момент
	rest time	*s.* recovery time		
R 151	**retaining gasket ring**	Dichtungstragring *m*	anneau *m* porteur d'étan-chéité	опорный кольцевой уплотнитель, ограничительное уплотнительное кольцо
R 152	**retaining ring**	Sprengring *m*	circlips *m*, anneau *m* de retenue	замковое кольцо
	retardation, deceleration, delay, time lag	Verzögerung *f*	retard *m*, retardation *f*, ralentissement *m*, temporisation *f*	задержка, запаздывание
R 153	**retarding field**	Bremsfeld *n*	champ *m* de freinage	задерживающее поле
	retention time	*s.* residence time		
R 154	**retort furnace,** tube furnace	Retortenofen *m*	four *m* tubulaire (à cornue), alambicfour *m*	реторная печь
	reverberating furnace, air furnace	Flammofen *m*	four *m* à réverbère	пламенная печь, отражательная печь
R 155	**Reynolds number,** Re	Reynoldssche Zahl *f*, Re	nombre *m* de Reynolds, Re	число Рейнольдса, Re
	rf heating	*s.* radio-frequency heating		
	RH	*s.* relative humidity		
	RHEED	*s.* reflexion high energy electron diffraction		
R 156	**rheology**	Fließkunde *f*	rhéologie *f*	реология
	right angle valve, angle valve, corner valve	Eckventil *n*	soupape *f* à passage angu-laire, vanne *f* d'équerre	угловой вентиль
R 157	**rimming steel**	unberuhigter Stahl *m*	acier *m* non calmé	неуспокоенная сталь, кипящая сталь
R 158	**ring beam system**	Ringstrahlsystem *n*	système *m* de rayon annu-laire	система с кольцевым лучом
	ring burner, annular burner	Ringbrenner *m*	brûleur *m* circulaire (annulaire)	кольцевая горелка
R 159	**ring cathode,** ring-shaped cathode	ringförmige Nahkatode *f*, Nahkatode, Ringkatode *f*, Ringstrahlnahkatode *f*	cathode *f* cyclique (annulaire)	кольцевой катод
R 160	**ring cathode electron bombardment source**	Ringkatoden-Elektronen-stoßquelle *f*	source *f* de choc d'élec-trons en forme de cathode annulaire	источник для электрон-ной бомбардировки с кольцевым катодом
R 161	**ring gasket**	Ringdichtung *f*	joint *m* annulaire	кольцевое уплотнение, кольцевая прокладка
R 162	**ring getter**	Ringgetter *m*	getter *m* annulaire	кольцевой газопо-глотитель
	ring-shaped cathode, ring cathode	ringförmige Nahkatode *f*, Nahkatode, Ringka-tode *f*, Ringstrahlnah-katode *f*	cathode *f* cyclique (annulaire)	кольцевой катод
R 163	**rinse**	ausspülen, spülen	rincer	промывать, споласкивать
R 164	**rinsing electrode**	Spülelektrode *f*	électrode *f* de balayage	промывочный электрод
	riser rod, object holder, object carrier	Objekthalter *m*	barre-support *f*, porte-objets *m*	держатель объекта
R 165	**Rockwell hardness**	Rockwell-Härte *f*	dureté *f* Rockwell	твердость по Роквеллу
	rod heater, cartridge heater	Heizpatrone *f*	cartouche *f* chauffante, élément *m* chauffant en forme de barre	патронный нагреватель
R 166	**rod melting**	Stabschmelzen *n*	fusion *f* de barre	плавка прутка
R 167	**rod-sintering device**	Stabsintereinrichtung *f*	installation *f* d'agglomé-ration à barres	установка для спекания штабика
	roll coating, band coating, strip coating	Bandbedampfung *f*	vaporisation *f* à bande, métallisation *f* au déroulé	нанесение покрытия на ленту
R 168	**roll down**	abwalzen	calandrer, laminer	прокатывать

	English	German	French	Russian
	roller dryer, rotary dryer, drum dryer, cylinder dryer	Walzentrockner *m*, Zylindertrockner *m*, Trommeltrockner *m*, Röhrentrockner *m*	sécheur *m* cylindrique (tournant, tubulaire)	вальцовая сушилка, [вращающаяся] барабанная сушилка, сушильный барабан
	rolling diaphragm, cylindrical diaphragm	Rollmembran *f*	soufflet *m*, membrane *f* cylindrique	цилиндрическая мембрана, сильфон
R 169	root mean square velocity	mittleres Geschwindigkeitsquadrat *n*	vitesse *f* moyenne quadratique	средняя квадратичная скорость
R 170	Roots blower pump, Roots pump, rotary blower pump	Roots-Pumpe *f*, Wälzkolbenpumpe *f*	dépresseur *m* Roots (radial), pompe *f* Roots	двухроторный насос, насос Рутса, ротационный компрессор
	rosin, colophony	Kolophonium *n*	colophane *f*	канифоль
	rotary blower, centrifugal airpump, centrifugal blower	Kreiselgebläse *n*, Fliehkraftgebläse *n*, Zentrifugalgebläse *n*	soufflante *f* centrifuge	центробежная воздуходувка
	rotary blower pump, Roots pump, Roots blower pump	Roots-Pumpe *f*, Wälzkolbenpumpe *f*	dépresseur *m* Roots (radial), pompe *f* Roots	двухроторный насос, насос Рутса, ротационный компрессор
R 171	rotary cage	Drehkorb *m*	panier *m* rotatif	вращающаяся клетка
R 172	rotary compressor	Drehkolbenverdichter *m*	compresseur *m* à piston rotatif	ротационный компрессор, компрессор с вращающимся ротором
R 173	rotary cross flow blower pump	Radialgebläse *n*	soufflerie *f* radiale	радиальный компрессор
R 174	rotary drum, rotary jig	drehbare Vorrichtung *f*, Drehtrommel *f*	appareil (dispositif) *m* rotatif	поворотный барабан
	rotary dryer	*s.* roller dryer		
R 175	rotary exhaust machine, rotary exhaust system	Pumpkarussell *n*, rotierendes Pumpsystem *n*	banc *m* de pompage rotatoire, carrousel *m* de pompage	карусельный откачной автомат
	rotary flange	*s.* rotatable flange		
R 176	rotary gate valve pump	Drehschieberpumpe *f*	pompe *f* à tiroirs rotatifs	пластинчато-роторный насос
	rotary jig	*s.* rotary drum		
	rotary mechanical pump with multiple vanes, multiple sliding vane rotary pump, multiple vane pump	Vielschieberpumpe *f*	pompe *f* multiple (à palettes multiples)	многопластинчатый насос
R 177	rotary mercury pump	Rotationsquecksilberluftpumpe *f*	pompe *f* rotative à mercure	вращательный ртутный насос Геде
R 178	rotary motion feedthrough, rotary seal, rotary shaft feedthrough	Drehdurchführung *f*	passage *m* tournant, traversée *f* rotative	ввод вращения, вращающийся ввод
R 179	rotary oil air pump, rotary oil-sealed mechanical pump	Rotationsölluftpumpe *f*	pompe *f* rotative à huile	вращательный масляный воздушный насос
R 180	rotary oil-sealed pump	ölgedichtete Rotationspumpe *f*	pompe *f* rotative (tournante) à joint d'huile	масляный вращательный насос
	rotary piston, impeller	Drehkolben *m*	rotor *m*, piston *m* rotatif	поворотный плунжер, ротор
R 181	rotary piston blower	Drehkolbengebläse *n*	soufflerie *f* à pistons rotatifs	ротационная воздуходувка
R 182	rotary piston pump, rotary plunger type pump, rotating plunger vacuum pump	Drehkolbenpumpe *f*, Sperrschieberpumpe *f*	pompe *f* à piston tournant (rotatif)	насос с вращающимся поршнем, пластинчато-статорный насос, плунжерный насос
R 183	rotary pump	Rotationspumpe *f*	pompe *f* rotative	вращательный насос
R 184	rotary pump oil	Vorpumpenöl *n*	huile *f* à pompe primaire	масло для вращательного насоса
	rotary seal, rotary shaft feedthrough	*s.* rotary motion feedthrough		
R 185	rotary shaft seal	Wellendurchführung *f*	traversée *f* d'arbre, passage *m* d'arbre	ввод вращающегося вала
R 186	rotary sliding vane type, rotary vane type	Drehschieberbauart *f*	type *m* de curseur rotatif, type de palettes rotatives	пластинчато-роторная конструкция
R 187	rotary vacuum filter	Vakuumdrehfilter *n*	filtre *m* tournant à vide	вращающийся вакуумный фильтр
R 188	rotary vacuum pump	Rotationsvakuumpumpe *f*	pompe *f* rotative à vide	вращательный насос
R 189	rotary vacuum seal	Vakuumdrehdichtung *f*	joint *m* rotatif à vide, joint d'arbre étanche au vide	вращающееся вакуумное уплотнение, вакуумно-плотный ввод вала
	rotary vane type	*s.* rotary sliding vane type		
	rotatable flange, rotating (rotary, loose) flange	Drehflansch *m*	bride *f* rotative (tournante)	вращающийся фланец, свободный фланец
R 190	rotating anode	Drehanode *f*	anode *f* rotative	вращающийся анод
	rotating cylinder gauge, rotating disk gauge, molecular drag gauge, molecular gauge	Molekularvakuummeter *n*	vacuomètre *m* moléculaire	молекулярный манометр, мольманометр
R 191	rotating field	Drehfeld *n*	champ *m* tournant (rotatif)	вращающееся поле

R 192	**rotating flange**	s. rotatable flange		
	rotating mix-condenser	rotierender Mischkondensator m	condenseur m rotatif à mélange	вращающийся смесительный конденсатор
	rotating plunger vacuum pump	s. rotary piston pump		
R 193	**rotating water blast pump**	Rotationswasserstrahlpumpe f	pompe f rotative à jet d'eau	вращательный водоструйный насос
R 194	**rotational freezing,** shell-freezing	Rollschichtgefrieren n, Rotationsgefrieren n	congélation f à rotation	ротационное замораживание
R 195	**rotor**	Rotor m	rotor m	ротор
R 196	**roughening** <of glass>	Runzelbildung f, Rauhwerden n, Mattieren n <bei Glas>	dépolissage m <du verre>	матирование <стекла>
	roughing, coarse pumping	Grobevakuieren n, Vorpumpen n <des Rezipienten>	pompage m primaire, évacuation f préliminaire, prévidage m, évacuation grossière	откачка до предварительного разрежения
R 197	**roughing line**	Grobvakuumleitung f, Leitung f zur Grobevakuierung <des Rezipienten>	conduite f d'évacuation préliminaire, canalisation f de prévidage, conduite de prévide	магистраль низкого вакуума
R 198	**roughing pump,** rough vacuum pump	Grobvakuumpumpe f	pompe f de prévidage, pompe à vide préliminaire	насос для получения грубого вакуума
	roughing time, forepumping time	Vorpumpzeit f, Grobpumpzeit f	temps m de pompage primaire, temps de prévidage	время предварительной откачки, время достижения форвакуума
	roughing valve, bypass valve	Grobpumpventil n, Ventil n in der Umgehungsleitung, Ventil in der Umwegleitung	soupape f dans la conduite de dérivation	байпасный вентиль
R 199	**roughing valve**	Grobvakuumventil n	robinet m pour l'évacuation préliminaire, robinet de prévidage, vanne f dans la canalisation de prévidage	форвакуумный вентиль, вентиль в коммуникации предварительной откачки
R 200	**rough vacuum**	Grobvakuum n	vide m grossier, vide primaire	низкий вакуум
	rough vacuum pump	s. roughing pump		
R 201	**rubber connector**	Gummiverbindungsstück n	raccordement m en caoutchouc, raccord m (jonction f) en caoutchouc	резиновый соединитель
R 202	**rubber finger-cot**	Gummifingerling m	doigtier (capuchon) m en caoutchouc	резиновый напальчник
R 203	**rubber gasket,** rubber seal	Gummidichtung f	joint m en caoutchouc	резиновая прокладка, резиновый уплотнитель
R 204	**rubber hose,** rubber tube	Gummischlauch m	tuyau m de caoutchouc	резиновый шланг, резиновая трубка
	rubber seal	s. rubber gasket		
R 205	**rubber-type elastic material**	gummielastischer Werkstoff m	matériel m gomme élastique	резиноподобный эластичный материал
	rubber tube	s. rubber hose		
	rupture disk, bursting disk	Berstscheibe f, Bruchplatte f	plaque f de rupture	предохранительная мембрана, предохранительная диафрагма

S

	safety circuit, protection circuit	Schutzschaltung f	connexion f de protection protection f électrique	схема защиты
S 1	**safety device**	Sicherheitsvorrichtung f	dispositif m de sécurité (protection)	предохранительное устройство, защитное устройство
S 2	**safety head**	gefederte Ventilplatte f	disque f élastique de vanne, obturateur m élastique d'un robinet, clapet m (platine f) élastique de vanne	упругий элемент вентиля
S 3	**safety hoist**	Sicherheitshebevorrichtung f	appareil m de levage de sécurité	предохранительное подъемное устройство
S 4	**safety interlock circuit**	Sicherheitsverriegelung f, Schutzverriegelung f	verrouillage m de sécurité, blocage m de sécurité (protection)	схема защитной блокировки
S 5	**safety interlock system**	Schutzverriegelungssystem n	système m de verrouillage protecteur	система защитной блокировки
S 6	**safety pressure trip**	Sicherheitsdruckauslösung f	déclenchement m de pression de sécurité	предохранительное выключение от давления
S 7	**safety valve**	Sicherheitsventil n	soupape f de sécurité	предохранительный вентиль (клапан)
S 8	**sample**	Probe f, Prüfling m, Muster n	essai m, échantillon m	проба, образец
S 9	**sample changer**	Probenwechsler m	changeur m d'échantillons	устройство для смены образцов
S 10	**sample taking,** sampling	Probenahme f, Probenehmen n, Musternehmen n	échantillonnage m, prise f d'essai	отбор проб

	sampling	s. sample taking		
	sampling probe, pressure probe, sniffer [probe]	Schnüffelsonde f, Schnüffler m, Leckschnüffler m	sonde f reniflante, renifleur m	всасывающий щуп
S 11	**sampling probe test**	Schnüfflermethode f, Schnüffeltest m, Schnüffelmethode f	méthode f reniflante	метод всасывающего щупа
S 12	**sand-blasting**	Sandstrahlen n, Sanden n	sablage m	пескоструйная очистка
S 13	**sand shell moulding**	Formmaskenverfahren n	procédé m de masque de moulage	литье в оболочковые формы
S 14	**sapphire-to-metal seal**	Metall-Saphir-Verbindung f	scellement m métal-saphir	соединение металл-сапфир
S 15	**saturate**	sättigen	saturer	насыщать
S 16	**saturated vapour**	gesättigter Dampf m	vapeur f saturée	насыщенный пар
S 17	**saturated vapour pressure,** saturation vapour pressure	Sattdampfdruck m, Sättigungsdampfdruck m	pression (tension) f de vapeur saturée, pression f de vapeur satuerant	давление насыщенных паров
S 18	**saturation level**	Sättigungsniveau n	niveau m de saturation	уровень насыщения
	saturation vapour pressure	s. saturated vapour pressure		
	saucer spring, disk-spring column	Tellerfedernsäule f	colonne f de rondelle élastique	колонка с дисковыми пружинами
S 19	**scale effect**	Ablagerungs-erscheinung f	effet m d'incrustation	эффект осадка, эффект отложения
S 20	**scale etching**	Zunderbeizen n	calamine-décapage m	травление накипи, травление окалины
	scale-off, chip-off, flake-off	abblättern <von Schichten>	effeuiller	отслаиваться, отпадать
	scale trap, dirt trap	Schmutzfänger m	piège m à impuretés, collecteur m d'impure-tés	грязеуловитель
S 21	**scan**	zerlegen, abtasten	balayer	развертывать, сканировать
S 22	**scanning electron diffraction**	Abtastelektronen-beugung f	diffraction f électronique d'exploration	сканированная элек-тронная дифракция
S 23	**scanning electron microscope**	Abtastelektronenmikro-skop n	microscope m électronique d'exploration	растровый электронный микроскоп
S 24	**scanning speed**	Abtastgeschwindigkeit f	vitesse f d'exploration, vitesse de balayage	скорость развертки
	scattering, dispersion	Streuung f, Zerstreuung f, Dispersion f	dispersion f, diffusion f	рассеяние, дисперсия
	scavenge, flush	ausschwemmen, aus-spülen <mit Gas>	balayer	промывать, вымывать, смывать <струей>
S 25	**scheduling**	Arbeitsplanung f	programme m, planning m	технологическая про-работка, составление графика технологиче-ского процесса
	scleroscopic hardness, rebound hardness	Rücksprunghärte f, Rückprallhärte f	dureté f de rebondisse-ment	упругая твердость, твердость по методу отскока, твердость по Шору
S 26	**scrap**	Schrott m	grenaille f, ferailles fpl, mitraille f	скрап
S 27	**scraper**	Schaber m, Spachtel f	grattoir m, spatule f	скребок
	scratch hardness, abrasive hardness	Ritzhärte f	dureté f sclérométrique, résistance f à la rayure	твердость по царапанию, склерометрическая твердость
S 28	**scratch tester**	Kratzprüfgerät n	dispositif m d'épreuve de friture	склерометр <прибор для испытания на склеро-метрическую твер-дость>
S 29	**screen belt dryer**	Siebbandtrockner m	sécheur m à bande de filtràge	конвейерная сушилка на ситах
S 30	**screen grid**	Schirmgitter n	grille f de protection, grille-écran f	экранирующая сетка, экранная сетка
S 31	**screw coupling,** screwed joint	Schraubverbindung f	assemblage m à filet, raccord m à vis	винтовое соединение, болтовое соединение
S 32	**screw dislocation**	Schraubenversetzung f	dislocation f de vis	винтовая дислокация
S 33	**screw dryer,** trough dryer	Muldentrockner m	sécheur m en auge	лотошная сушилка
	screwed joint, screw coupling	Schraubverbindung f	assemblage m à filet, raccord m à vis	винтовое соединение, болтовое соединение
S 34	**screwed pipe joint,** screwed sleeve	Schraubmuffen-verbindung f	assemblage m par man-chon à vis	трубчатое резьбовое соединение
S 35	**screw extruder,** screw-type extrusion machine	Schneckenpresse f	presse f de vis sans fin	шнековый пресс
S 36	**seal,** seal off, tighten, tip off	abdichten, dichten, abschmelzen	étancher, calfeutrer, bou-cher, rendre étanche, assurer l'étanchéité, séparer par fusion	уплотнять, гермети-зировать, заделывать, изолировать, отпаи-вать
	seal, sealing, sealing-in, gasket, packing, sealing element	Dichtung f, Dichtungs-element n, Einschmelz-stelle f, Abdichtung f, Einschmelzung f, Verschmelzung f	joint m, joint étanche (scellé, d'étanchéité), scellement m, étanche-ment m, jonction f étanche, point m de scellement, élément m d'étanchéité	уплотнение, соединение, уплотняющий эле-мент, спай, место спая, запайка, впайка

S 37	**sealant, sealant material**	Dichtungsmaterial *n*	matériel *m* d'étanchéité	уплотняющий (прокладочный, набивочный) материал
	seal area, joint face, sealing joint, sealing surface	Dichtfläche *f*, Dichtungsfläche *f*	surface *f* de joint (contact du joint), contact *m* (surface) d'étanchéité, surface d'étanchage	поверхность соединения, уплотняющая поверхность, плоскость уплотнения
S 38	**seal conductance**	Leitwert *m* einer Verbindung	conductance *f* d'une combinaison	пропускная способность соединения
	sealed conductance, residual valve conductance	Leitwert *m* des geschlossenen Ventils	valeur *f* de conductivité du clapet fermé, conductance *f* du robinet fermé	пропускная способность перекрытого вентиля
S 39	**sealed-off vacuum system,** vacuum device, vacuum sealed device	abgeschlossenes Vakuumsystem *n*	installation *f* scellée de vide, système *m* séparé par fusion	отпаянный вакуумный прибор, отпаянная вакуумная система
	sealing	*s.* seal		
S 40	**sealing cement**	Dichtungswachs *n*, Dichtungskitt *m*	cire *f* pour joints, mastic *m* d'étanchéité	уплотняющая замазка
S 41	**sealing collar,** sealing sleeve, sleeve gasket	Dichtungsmanschette *f*	garniture *f* cylindrique, embouti *m* de joint, manchon *m* d'étanchéité	уплотняющая манжета, набивочная манжета
S 42	**sealing disk**	Dichtscheibe *f*	rondelle *f* joint (d'étanchéité)	уплотняющая шайба
	sealing element	*s.* seal		
S 43	**sealing fluid,** sealing liquid	Dichtflüssigkeit *f*	liquide *m* de joint, liquide à rendre étanche	уплотняющая жидкость
S 44	**sealing force**	Dichtkraft *f*	puissance *f* à rendre étanche, puissance à assurer l'étanchéité	уплотняющая сила
S 45	**sealing glass**	Einschmelzglas *n*	verre *m* à refondre	стекло для запайки
	sealing grease, packing grease	Dichtungsfett *n*	graisse *f* d'étanchéité, graisse à vide	уплотняющая смазка
	sealing-in	*s.* seal		
	sealing joint, seal area, sealing surface, joint face	Dichtungsfläche *f*, Dichtfläche *f*	surface *f* de contact du joint, surface d'étanchéité, surface d'étanchage, surface de joint, contact *m* d'étanchéité	плоскость уплотнения, поверхность соединения, уплотняющая поверхность
	sealing liquid	*s.* sealing fluid		
S 46	**sealing paint,** vacuum paint	Dichtungslack *m*, Vakuumlack *m*	laques *fpl* à étanchéifier, laques d'étanchéité	вакуумный лак
S 47	**sealing point**	Anschmelzstelle *f*	joint *m* par fusion, joint scellé	место припайки, место спая
S 48	**sealing pressure**	Dichtungsdruck *m*	pression *f* de joint	уплотняющее давление
S 49	**sealing ridge**	Dichtungswulst *m*	bourrelet *m* joint	выступ (буртик, ребро) уплотнения
	sealing ring, packing ring, gasket ring	Dichtungsring *m*	anneau *m* d'étanchéité, rondelle *f* de joint	уплотнительное (уплотняющее, прокладочное) кольцо, прокладка
	sealing sleeve	*s.* sealing collar		
	sealing surface	*s.* seal area		
S 50	**sealing wax,** vacuum wax	Dichtungswachs *n*, Vakuumwachs *n*	cire *f* pour joint étanche, cire d'étanchéité, cire-laque *f*	уплотняющий воск, вакуумная замазка
S 51	**sealing wire,** seal-in wire	Einschmelzdraht *m*	fil *m* scellé	проволока для впая
	seal off, seal, tighten, tip off	abdichten, dichten, abschmelzen	étancher, calfeutrer, boucher, rendre étanche, assurer l'étanchéité, séparer par fusion	уплотнять, герметизировать, заделывать, изолировать, отпаивать
S 52	**seal-off capillary**	Abschmelzkapillare *f*	capillaire *m* séparé par fusion	отпаянный (запаянный) капилляр
	seal-off valve, isolation valve, cut-off valve	Verschlußventil *n*, Absperrventil *n*	soupape *f* de fermeture, vanne *f* obturatrice, robinet *m* obturateur, soupape d'arrêt, vanne d'arrêt	запорный клапан (вентиль), изоляционный вентиль
	seam welding, line welding	Nahtschweißung *f*	soudure *f* en ligne continue	роликовая сварка, сварка прямолинейным швом
S 53	**seam welding**	Nahtschweißen *n*	soudure *f* par couture, soudure en filet	роликовая сварка
	search gas, probe gas, test gas, tracer gas	Testgas *n*, Prüfgas *n*	gaz *m* témoin (traceur, d'épreuve), gaz-sonde *m*	пробный газ
S 54	**secondary air**	Beiluft *f*	air *m* secondaire	вторичный воздух
	secondary drying, afterdrying, final drying, subsequent drying	Nachtrocknung *f*, Endtrocknung *f*	dessiccation *f* secondaire (finale), séchage *m* final (secondaire)	досушивание, досушка, вторичная осушка, конечная (окончательная) сушка
S 55	**second melt**	Zweitschmelze *f*	fonte *f* secondaire	вторая плавка, вторичная плавка
S 56	**sector-field spectrometer**	Spektrometer *n* mit magnetischem Sektorfeld	spectromètre *m* de masse à déflexion magnétique, spectromètre de masse à secteur magnétique	секторный масс-спектрометр

S 57	sector instrument	Sektorgerät n	instrument m en forme de secteur	секторный прибор
	seed crystal, crystal nucleous	Kristallisationskeim m, Kristallisationskern m, Impfling m, Kristallkeim m	germe m cristallin, amorce f de cristallisation	затравочный кристалл, зародыш кристалла
	segment bend, lobster back	Segmentbogen m	arc m à segment	составное колено
	segregation, eliquation, liquation	Seigerung f	liquation f, ségrégation f	ликвация, сегрегация, зейгерование
S 58	seizure	Festfressen n	grippage m	заедание
S 59	selective getter	selektiver Getter m	getter m sélectif	селективный (избирательный) газопоглотитель
S 60	self-accelerated electron gun	Elektronenstrahler m mit Eigenbeschleunigung	canon m à électrons d'accélération propre	электронная пушка с самоускорением
S 61	self-cleaning, self-purifying	Selbstreinigung f	auto-épuration f	самоочистка
S 62	self-consuming electrode	selbstverzehrende Elektrode f	électrode f auto-consommateur	саморасходующийся электрод
S 63	self-contamination	Wiederbedeckung f	recouverture f, recouvercle m	самозагрязнение
S 64	self-diffusion	Eigendiffusion f	auto-diffusion f	самодиффузия
S 65	self-evacuation of the tube	Selbstevakuierung f der Röntgenröhre	auto-durcissement m du tube pour rayons X	саможестчение [рентгеновской] трубки
S 66	self-evaporation	Eigenverdampfung f	vaporisation f propre	самоиспарение
S 67	self-fractionating device	Selbstfraktionierungseinrichtung f	dispositif m de fractionnement automatique	устройство с автоматическим фракционированием
	self-fractionating pump, fractionating pump, purifying pump	fraktionierende Pumpe f	pompe f fractionnante (à rectification)	разгоночный насос, фракционирующий насос
S 68	self-healing	selbstheilend	à purification automatique, d'épuration automatique	самоочищающийся
S 69	self-locking	Selbstsperrung f	blocage m automatique	автоблокировка
S 70	self-maintained discharge, self-sustaining discharge	selbständige Entladung f	décharge f automatique, déchargement m spontané	самостоятельный разряд
S 71	self-priming	selbstansaugend	auto-aspirant	самозаполняющийся, самовсасывающий
	self-purifying, self-cleaning	Selbstreinigung f	auto-épuration f	самоочистка
S 72	self-purifying pump, semifractionating pump	teilfraktionierende Pumpe f	pompe f semi-fractionnante	насос с частичным фракционированием
S 73	self-scattering	Eigenstreuung f	self-dispersion f	саморассеяние
S 74	self-sticking coefficient	Selbsthaftkoeffizient m	coefficient m d'auto-capture, coefficient d'auto-collage, coefficient d'auto-fixation	коэффициент самоприлипания
S 75	self-structure	Eigenstruktur f	structure f spécifique	собственная структура
	self-sustaining discharge	s. self-maintained discharge		
S 76	semiconductor cathode	Halbleiterkatode f	cathode f à semi-conducteur (thermistance)	полупроводниковый катод
S 77	semiconductor device	Halbleiterbauelement n	composant m semiconducteur	полупроводниковый компонент
S 78	semi-finished goods	Halbzeug n	produit m demi-fini	полуфабрикат
	semifractionating pump	s. self-purifying pump		
S 79	sensing element, sensor	Fühler m, Meßfühler m	tâteur m, élément m sensible, sonde f de mesure	чувствительный (воспринимающий) элемент, детектор
	sensing head, gaugehead, head, gauge	Meßröhre f, Meßkopf m	burette f, tube m de mesure, cellule f de mesure, tête f de mesure, tête manométrique	датчик манометра
S 80	sensitivity	Empfindlichkeit f, Ansprechwahrscheinlichkeit f, Feinfühligkeit f	sensibilité f	чувствительность
	sensitivity calibrator, calibrated leak, standard leak, reference leak, test leak	Eichleck n, Testleck n, Vergleichsleck n, Leck n bekannter Größe, Bezugsleck n	fuite f calibrée	стандартная течь, калиброванная течь, эталонная течь
	sensitivity of a gauge, gauge sensitivity	Röhrenempfindlichkeit f	sensibilité f d'un tube électronique	чувствительность манометра
S 81	sensitivity setting	Empfindlichkeitseinstellung f, Empfindlichkeitsstufe f	réglage m de sensibilité, degré m de sensibilité	регулировка чувствительности
S 82	sensitizer	Sensibilisator m	sensibilisateur m	сенсибилизатор
	sensor, sensing element	Fühler m, Meßfühler m	tâteur m, élément m sensible, sonde f de mesure	чувствительный (воспринимающий) элемент, детектор
	separate field, external field	Fremdfeld n	champ m étrangeur (perturbateur)	постороннее поле, внешнее поле
S 83	separating factor	Trennfaktor m	facteur m séparateur (de séparation)	коэффициент сепарации

		English	German	French	Russian
		separation column, fractionating column	Trennsäule *f*	colonne *f* de séparation	разделительная колонна
S	84	separation effect	Entmischungseffekt *m*	effet *m* de démixtion	эффект разделения
S	85	separation efficiency	Trennwirkung *f*	effet *m* de séparation	эффективность сепарации
S	86	separator	Abscheider *m*	séparateur *m*	сепаратор
		series single stage, compound, multi-stage	mehrstufig	à plusieurs étages	многоступенчатый
S	87	service instrument	Betriebsmeßinstrument *n*	appareil *m* de mesure de service	рабочий измерительный прибор
S	88	service valve	Bedienungsventil *n*	soupape *f* de service (réglage, commande)	регулировочный вентиль
S	89	set screw	Stellschraube *f*	vis *f* de réglage, vis calante (régulatrice)	установочный винт
S	90	setting, timing	Einstellung *f*	réglage *m*, ajustage *m*	регулировка, установка
S	91	setting, solidification	Erstarrung *f*, Solidifikation *f*, Verfestigung *f*	solidification *f*	кристаллизация, затвердевание
S	92	setting cement	Abbindekitt *m*	mastic *m* à prendre	схватывающая замазка
S	93	setting knob	Justiervorrichtung *f*	appareil *m* d'ajustage	приспособление для регулировки
		setting point, pour point, solidification point, solidifying point	Stockpunkt *m*	point *m* de décongélation (solidification)	температура застывания
S	94	setting point <η glass ≈ 10^{14} poises>	Temperpunkt *m* <η Glas ≈ 10^{14} Poise>	point *m* à recuire <η verre ≈ 10^{14} poise>	точка натяжения <η стекла ≈ 10^{14} пз>
S	95	setting range	Einstellbereich *m*	gamme *f* de réglage, gamme d'ajustage, étendue *f* de réglage, étendue d'ajustage	диапазон регулировки, диапазон установки
S	96	setting time	Einstellzeit *f*	temps *m* de réglage, temps de réponse	время регулирования, время установления
S	97	settle	absetzen, sich ablagern, sedimentieren	se déposer	осаждаться
		settling time, clean-up time, recovery time, rest (relaxation) time	Erholungszeit *f*, Wiederansprechzeit *f*	temps *m* régénérateur, temps de régénérabilité	время [для] восстановления, период восстановления
		shadow casting, shadowing, oblique-incidence coating	Beschatten *n*, Beschattung *f*, Schrägbedampfung *f*	ombrage *m*, protection *f*, évaporation *f* oblique	наклонное напыление, осаждение под углом
S	98	shadow mask	Abdeckmaske *f*	masque *m* de protection	теневая маска
S	99	shaft oil drip collar	Wellen-Öltropfkragen *m*	collet *m* d'égouttage ondulatoire	буртик вала, смазываемый капельной масленкой
		shaft seal, radial seal [for rotating shaft]	Wellendichtung *f*, Radialdichtring *m* für Welle, Wellendichtring *m*	garniture *f* étanche d'arbre, joint *m* d'arbre, anneau-joint *m* radial	уплотнение вала, радиальное уплотнение вращающегося вала
S	100	shaft sealing gasket	Wellendichtung *f*	joint *m* d'arbre	прокладка для уплотнения вала
S	101	shape casting	Formguß *m*	moulage *m*	фасонное литье
S	102	shape factor	Formfaktor *m*, Gestaltfaktor *m*	facteur *m* de forme	формфактор
S	103	shaping	Formgebung *f*	façonnage *m*	формовка, обработка давлением
S	104	sharp freezer	Tiefgefrieranlage *f*	installation *f* de congélation à basse température	установка для быстрого замораживания
S	105	shear flow	Scherströmung *f*	flux *m* de cisaillement	поперечное течение, течение с поперечным градиентом скорости
S	106	shear gasket	Scherdichtung *f*	joint *m* à cisaillement	ножничное уплотнение
S	107	shearing strength, shear strength	Abscherfestigkeit *f*, Scherfestigkeit *f*	résistance *f* au cisaillement	предел прочности при сдвиге, прочность при сдвиге
		shear modulus, modulus of rigidity	Schubmodul *m*	module *m* de rigidité	модуль сдвига
		shear strength	*s.* shearing strength		
S	108	sheath electron	Hüllenelektron *n*	électron *m* périphérique	электрон оболочки
		sheath replica, envelope replica	Hüllabdruck *m*	impression *f* enveloppante, reproduction *f* enveloppante, épreuve *f* enveloppante	отпечаток оболочки
S	109	sheath rolling	Deckwalzen *n*, Mantelwalzen *n*	laminage *m* d'enveloppe	прокатка в контейнере
S	110	sheet blowing	Folienblasen *n*	soufflage *m* de lamelles (feuilles)	получение пленок выдуванием
S	111	sheet capacitance	Flächenkapazität *f*	capacité *f* de face	поверхностная емкость
S	112	sheeting dryer, web dryer	Bahnentrockner *m*	sécheur *m* en lés	листовая сушилка
S	113	sheet mica	Plattenglimmer *m*	mica *m* en feuilles	листовая слюда
S	114	shelf area	Stellfläche *f*	plaine *f* (plan *m*) de réglage	площадка регулировки
S	115	shelf-drum temperature	Plattentrommeltemperatur *f*	température *f* de tambour à disques	температура полочного барабана
S	116	shelf dryer	Hordentrockner *m*, Trockenschrank *m*	grille *f* de séchage	полочная (решетчатая) сушилка
S	117	shelf freezer	Hordengefrieranlage *f*	installation *f* de congélation à claies	морозильный аппарат с замораживанием на стеллажах

	shelf spacing, clearance between the plates	Plattenabstand *m*	écartement *m* des plaques	расстояние между пластинами
S 118	**shellac**	Schellack *m*	shellac *m*	шеллак
	shell-freezing, rotational freezing	Rollschichtgefrieren *n*	congélation *f* à rotation	ротационное замораживание
S 119	**shell moulding**	Formmaskenguß *m*	fonte *f* par moulage système Croning	изготовление корковых литейных форм
S 120	**shell thermocouple**	Mantelthermoelement *n*	bilame *m* enrobé, thermo-élément *m* enrobé	термопара в кожухе
S 121	**shock**	Verdichtungsstoß *m*	choc *m*	удар, скачок уплотнения
S 122	**shock front**	Stoßfront *f*	front *m* de choc	фронт ударной волны
S 123	**shock polaric diagram**	Stoßpolarendiagramm *n*	diagramme *m* polaire du choc	полярная диаграмма удара, диаграмма удара в полярных координатах
S 124	**shock sensitive**	stoßempfindlich	sensible au choc	восприимчивый к удару
S 125	**shock tube**	Stoßwellenrohr *n*	tuyau *m* à onde de choc	ударная труба
S 126	**shock wave**	Stoßwelle *f*	onde *f* de choc	ударная волна
S 127	**shock wave system**	Verdichtungsstoßsystem *n*	système *m* de choc de compression	ударноволновая система
S 128	**Shore hardness**	Shorehärte *f*	dureté *f* Shore	твердость по Шору
S 129	**short glass**	kurzes Glas *n*	verre *m* court	«короткое» стекло
	short-path distillation, open-path distillation	Kurzwegdestillation *f*	distillation *f* à court trajet	молекулярная перегонка
	short path still, open path still, unobstructed path still	Kurzwegdestillationsanlage *f*	distillerie *f* à court trajet, installation *f* de distillation à court trajet	установка для молекулярной перегонки
S 130	**short path wiped wall distillation unit**	Kurzwegdestillationsgerät *n* mit mechanischer Wandreinigung	installation *f* de distillation à court trajet avec purification murale mécanique	установка для дистилляции при минимально-коротком расстоянии с механической очисткой стенки
S 131	**short time measurement of residual gases**	Kurzzeit-Restgasmessung *f*	mesure *f* à court temps du gaz résiduel	экспресс-анализ остаточных газов
S 132	**shot blasting**	Freistrahlen *n* mit Stahlsand	sablage *m* à jet de sable d'acier	дробеструйная обработка
S 133	**shot noise**	Schrotrauschen *n*	bruit *m* de grenaille	дробовой шум
S 134	**shrinkage,** shrinking	Schrumpfen *n*	contraction *f*	стягивание, сжатие
	shrinkage crack, check crack	Schrumpfriß *m*	fente *f* par contraction, fissure *f* de contraction	усадочная трещина, трещина при усадке
	shrinking	*s.* shrinkage		
S 135	**shrunk-on sleeve**	Schrumpfmuffe *f*	manchon *m* de contraction	муфта горячей посадки
	shunt, bridge	überbrücken	shunter, dériver	шунт
S 136	**shut-off valve**	Absperrventil *n*	vanne *f* d'arrêt, robinet *m* de fermeture	запорный клапан, запорный вентиль
S 137	**shutter**	Abdeckblende *f*, Unterbrecherblende *f*	interrupteur *m* de diaphragme	заслонка ‹для диафрагмы›
	siccative, drier	Trockenstoff *m*, Sikkativ *n*	siccatif *m*	сиккатив, обезвоживатель
	side arm, fore-arm	Vorvakuumstutzen *m*	raccord *m* au vide primaire	форвакуумный патрубок
S 138	**side ejector stage**	Seitendampfstrahlstufe *f*	injecteur *m* latéral, tuyère *f* latérale, étage *m* du jet à vapeur, étage d'injecteur à vapeur	эжекторная ступень с боковым расположением
	sieve sorbent, molecular sieve	Molekularsieb *n*	tamis *m* moléculaire, crible *m* moléculaire	молекулярное сито
	sight glass, sight port, sight window, inspection glass, observation window, viewing glass, viewing port, viewing window	Beobachtungsfenster *n*, Schauglas *n*, Einblickfenster *n*	fenêtre *f* de contrôle, regard *m*, regard de contrôle, fenêtre d'observation, fenêtre de coup d'œil, hublot *m*	контрольное (смотровое) стекло, глазок, смотровое отверстие (окно), наблюдательное окно
	sigma welding, inert gas consumable-electrode arc welding	Metall-Inertgas-Schweißen *n*	soudage *m* gaz inerte-métal	дуговая сварка в среде инертного газа с расходуемым электродом
S 139	**signal-to-noise ratio**	Signal-Rauschverhältnis *n*, Störfaktor *m*	rapport *m* signal-bruit	отношение сигнал-шум
S 140	**silastic,** silicone rubber	Silikongummi *m*	caoutchouc *m* de silicone	силиконовый каучук, кремнекаучук
	silencer, noise muffler	Schalldämpfer *m*	sourdine *f*, silencieux *m*, amortisseur *m* de son	глушитель [звука], глушитель шума
S 141	**silica gel**	Silikagel *n*	gel *m* de silice, silicagel *m*	силикагель
S 142	**silica gel leak detector**	Silikagel-Lecksuchgerät *n*	détecteur *m* de fuites par gel silice	силикагелевый течеискатель
S 143	**silica liner**	Quarzauskleidung *f*	manchon *m* de quartz	кварцевая футеровка
S 144	**silicone grease**	Silikonfett *n*	graisse *f* silicone	силиконовая смазка
S 145	**silicone oil**	Silikonöl *n*	huile *f* de silicone	силиконовое (кремний органическое) масло
	silicone rubber	*s.* silastic		
S 146	**silicon steel**	Siliziumstahl *m*	acier *m* au silicium	кремнистая сталь
S 147	**silver,** silver-plate	versilbern	argenter	серебрить
S 148	**silver-oxygen leak**	Silber-Sauerstoff-Leck *n*	fuite *f* argent-oxygène	серебрянокислородный натекатель
	silver-plate	*s.* silver		

S 149	**silver-plating**	Versilberung *f*	argenture *f*	серебрение
S 150/1	**single acting**	einfachwirkend	à simple effet	однократно действующий, однократного действия
	single block pump, monoblock pump	Einblockpumpe *f*	pompe *f* monobloque	моноблочный насос
	single cell, monocell	Monozelle *f*	mono-cellule *f*, élément *m* mono	моноячейка
S 152	**single-cell anode**	Anodeneinzelzelle *f*	simple-cellule *f* anodique	одноячеистый анод
S 153	**single coil**	Einfachwendel *f*	filament *m* spiral	моноспираль
S 154	**single core**	einadrig	à unique conducteur	одножильный
S 155	**single crystal pulling apparatus**	Einkristallziehmaschine *f*	machine à tréfiler de monocristaux	устройство для вытяжки монокристалла
S 156	**single crystal sputtering**	Einkristallzerstäubung *f*	pulvérisation *f* de monocristaux	монокристаллическое распыление
S 157	**single drum dryer**	Einwalzentrockner *m*	sécheur *m* mono-cylindrique	одновальцовая сушилка
S 158	**single grain source**	Feinbedampfer *m*	vaporisateur *m* en fin, appareil *m* pour la métallisation dans le vide poussé à millitorr	однозерновой испаритель, микроиспаритель
	single-point recorder, dot recorder	Punktschreiber *m*	enregistreur *m* en poin-tillé	точечный самописец
S 159	**single-pole tube**	Einpolröhre *f*	tube *m* unipolaire	лампа с односторонней проводимостью
S 160	**single-rod seal**	Einzelstabeinschmelzung *f*	scellement *m* de barre simple	стержневой впай
S 161	**single-scattering**	Einzelstreuung *f*	dispersion (diffusion) *f* séparée	однократное рассеяние
S 162	**single-screw injector**	Einschneckenpresse *f*	presse *f* par vis sans fin	одношнековый инжектор, винтовой инжектор с одним витком
S 163	**single—twin coating unit**	Einfach—Doppel-überzugsgerät *n*	appareil *m* à couverture simple—double	установка для однократного—двойного покрытия
S 164	**single-V butt weld**	V-Naht *f*	joint *m* en V, soudure *f* en V	стыковое V-образное сварное соединение, сварное соединение со скосом двух кромок
S 165	**single-wire glassed header**	Durchführung *f* mit einem eingeglasten Draht	ensemble *m* de scellement monofilaire	однопроводная ножка, однопроводный проходник
S 165a	**sink**	Senke *f*	dépression *f*, pôle *m* négatif	сток
S 166	**sinter corundum**	Sinterkorund *m*	corindon *m* fritté	спеченный корунд
S 167	**sintered glass**	Sinterglas *n*	verre *m* fritté (sintérisé)	синтерированное стекло
	sintered glass plate, pressed glass plate	Preßglasteller *m*	platine *f* en fritté	штампованная стеклянная пластина
S 168	**sintered metal powder process,** Telefunken process	Telefunken-Verfahren *n*	procédé *m* de Telefunken	процесс металлизации керамики по способу фирмы Телефункен
S 169	**sintered metal process**	Aufsintern *n* eines Metalls auf Keramik, Metalli-sierung *f* von Keramik	métallisation *f* de céra-mique	металлизация керамики
S 170	**sinter fuse**	aufsintern	agglomérer par frittage	спекать обжигом
S 171	**sintering plant**	Sinteranlage *f*	installation *f* d'agglomé-ration, installation par frittage	установка для спекания (синтерирования)
	siphon, lift tube, feed tube, lift pipe	Heber *m*	siphon *m*	сифон
S 172	**skin effect**	Skin-Effekt *m*	effet *m* pelliculaire	скин-эффект
S 173	**skirt**	Düsenmantel *m*	enveloppe *f* de tuyère	рубашка (обтекатель) сопла
S 174	**skull**	Pfannenrest *m*	fond *m* de poche	настыль в ковше
S 175	**skull furnace**	Schalenschmelzofen *m*	four *m* de fusion à coquille	печь для гарниссажной плавки
S 176	**skull melting**	Schalenschmelzen *n*	fusion *f* à fond de moule refroidi, fusion à coquille	литье в вакууме, вакуумная плавка и разливка, гарниссажная плавка
	slab, bloom	Bramme *f*, Walzblock *m*	brame *f*	сляб, листовой (плоский) слиток, болванка
	slanting seat valve, inclined seat valve, bevel seat valve	Schrägsitzventil *n*	vanne *f* (robinet *m*) à siège oblique	вентиль с наклонным шпинделем
S 177	**sleeve**	Buchse *f*, Hülse *f*, Muffe *f*, Manschette *f*	douille *f*	втулка, муфта
	sleeve gasket, sealing sleeve, sealing collar	Dichtungsmanschette *f*	garniture *f* cylindrique, embouti *m* de joint, manchon *m* d'étanchéité	уплотняющая манжета, набивочная манжета
	slide ring seal, face seal, slip-ring seal	Gleitringdichtung *f*, Schleifringdichtung *f*	joint *m* à anneaux glis-sants, garniture *f* étanche à anneau glissant	скользящее кольцевое уплотнение, уплотнение со скользящим кольцом
	slide valve, gate valve, vacuum gate valve	Schieberventil *n*, Vakuumschieber *m*, Torventil *n*	robinet *m*, vanne *f* à tiroir (coulisse), robinet à tiroir, soupape *f* à pas-sage direct à vide	задвижка, заслонка, вакуумная задвижка

S 178	slide-valve recipro- cating vacuum pump	Schieberpumpe f	pompe f glissante (à tiroir)	поршневой (плунжер- ный) вакуумный насос с золотниковым рас- пределением, золотни- ковый насос
S 179	sliding blade, sliding vane, vane	Drehschieber m	palette f, tiroir m, vanne f, coulisse f	пластина, выдвижная лопатка, скользящая лопатка
S 180	sliding bottom	Schiebeboden m	fond m mobile	скользящее (откидное) дно
S 181	sliding discharge	Gleitentladung f	décharge f glissante	скользящий (поверх- ностный) разряд
S 182	sliding flow, slip flow, viscous flow	Schlupfströmung f, Gleitung f, Gleit- strömung f, viskose Strömung f	glissement m, courant m de glissement, écoule- ment m visqueux	течение со скольжением, вязкостное течение
S 183	sliding friction	Gleitreibung f	frottement m de glisse- ment	трение скольжения
S 184	sliding line, slip line	Gleitlinie f	ligne f de glissement	линия скольжения
S 185	sliding socket joint	Gleitmuffe f, Steckmuffe f	manchon m de glissement	соединение со скользя- щей муфтой
	sliding vane	s. sliding blade		
S 186	sliding vane rotary pump, vane type rotary pump	Drehschieberpumpe f	pompe f à palettes, pompe rotative à palettes	пластинчато-роторный насос
	slip flow	s. sliding flow		
	slip line, sliding line	Gleitlinie f	ligne f de glissement	линия скольжения
S 187	slippage	Schlupf m	glissement m	проскальзывание
S 188	slip plane	Gleitebene f	glissière f, plaine f de glissement	плоскость скольжения
S 189	slip ring	Schleifring m	bague f de contact, anneau m glissant	скользящее кольцо
S 190	slip-ring assembly	Schleifringanordnung f	arrangement m à anneaux glissants	устройство со скользя- щим кольцом
	slip-ring seal	s. slide ring seal		
S 191	slip velocity	Gleitgeschwindigkeit f	vitesse f de glissement	скорость скольжения
S 192	slit cathode	Schlitzkatode f	cathode f à fente	щелевой катод
S 193	slitlike tube	Spaltrohr n	tuyau m de clivage	щелевая лампа
S 194	slow freezing	langsames Gefrieren n	congélation f de petite vitesse	медленное заморажи- вание
S 195	sluice	Schleuse f	écluse f	шлюз
S 196	sluice chamber, lock chamber	Schleusenkammer f	chambre f d'écluse	шлюзовая камера
	small-angle scattering	Kleinwinkelstreuung f	dispersion f d'angle miniature (petit), fuite f d'angle miniature	рассеяние под малым углом
S 197	smallest detectable leak rate	kleinste nachweisbare Leckrate f	plus petit débit m déce- lable (détectable) d'une fuite	наименьшая обнаружи- ваемая скорость нате- кания
S 198	small flange connection	Kleinflanschverbindung f	raccord m à petites brides	соединение с малыми фланцами
S 199	small-signal theory	Kleinsignaltheorie f	théorie f de signalisation petite	теория малых сигналов (возмущений)
S 200	snap valve	Kolbenventil n	soupape f à piston	поршневой (цилиндри- ческий) вентиль
	sniffer [probe], sampling probe, pressure probe	Schnüffelsonde f, Schnüff- ler m, Leckschnüffler m	sonde f reniflante, reni- fleur m	всасывающий щуп
S 201	snifting valve	Schnarchventil n	soupape f reniflante	фыркающий клапан, выдувной клапан
S 202	soaking in	Innentränkung f	imprégnation f	пропитывание, импрегнирование
S 203	soap bubble leak detection	Abpreßmethode f <Lecksuche mit Seifenwasser>	méthode f à faire l'épreuve de pression, méthode à éprouver à l'air com- primé	метод опрессовки с ин- дикацией мыльными пузырями, метод мыльных пузырей
S 204	soap-bubble method	Seifenblasenmethode f	méthode f de bulles de savon	метод мыльных пузырей
S 205	socket	Muffenkelch m	prise f de courant	цоколь
S 206	socket head cap screw	Innensechskantschraube f	boulon m à tête à six pans	винт с граненным отвер- стием в головке
S 207	socket pipe	Muffenrohr n	tuyau m à manchon	труба с раструбом
S 208	soda glass	Sodaglas n, Natronglas n	verre m de soude	натриевое стекло
	soda lime glass, lime glass	Natron-Kalk-Glas n	verre m calcaire-natron	известковое стекло
S 209	sodium silicate	Natronwasserglas n	silicate m de soude	натриевое жидкое стекло
S 210	softening point $<\eta$ glass $= 10^{7,65}$ poises>	Erweichungspunkt m, Littleton-Punkt m $<\eta$ Glas $= 10^{7,65}$ Poise>	point m d'amollissement $<\eta$ verre $= 10^{7,65}$ poise>	точка размягчения $<\eta$ стекла $= 10^{7,65}$ пз>
S 211	softening temperature $<\eta$ glass $= 10^{11}...10^{12}$ poises>	Erweichungstemperatur f $<\eta$ Glas $= 10^{11}...10^{12}$ Poise>	température f d'amollisse- ment $<\eta$ verre $= 10^{11}$ $...10^{12}$ poise>	температура размягчения $<\eta$ стекла $= 10^{11}-10^{12}$ пз>
S 212	soft glass	Weichglas n	verre m mou	мягкое стекло
S 213	softness number, softness value	Weichheitszahl f	nombre m de mollesse	мягкость <твердость для мягких материалов>
S 214	soft solder flux	Flußmittel n für Weichlot	fondant m à souder à l'étain	флюс для пайки мягким припоем

S 215	**soft X-ray photo-electric current,** X-ray photocurrent, X-ray current	Röntgenstrom *m*, durch weiche Röntgen-strahlung erzeugter Fotostrom *m*	courant *m* photo-électrique par rayonnement X doux	рентгеновский фототок
S 216	**solar constant**	Solarkonstante *f*	constante *f* solaire	солнечная постоянная
S 217	**solarization**	Solarisation *f*	solarisation *f*	соляризация, выдержка на солнце
S 218	**solder**	weichlöten	souder à l'étain	паять мягким припоем
S 219	**solder**	Weichlot *n*	soudure *f*, étain *m* à souder	припой
S 220	**solder glass**	Glaslot *n*	soudure *f* de verre	стекольный припой
S 221	**soldering**	Weichlöten *n*	soudage *m* à l'étain	пайка мягким припоем
S 222	**soldering flux**	Lötflußmittel *n*	fondant *m* à souder (braser)	флюс для пайки
S 223	**soldering iron**	Lötkolben *m*	fer *m* à souder	паяльник
S 224	**soldering point**	Lötstützpunkt *m*	point *m* d'attache par soudure	точка (место) пайки
S 225	**solder joint, solder seal**	Lötdichtung *f*, Lötver-bindung *f*	joint *m* soudé (brasé)	уплотнение пайкой, соединение пайкой
S 226	**solenoid operated valve**	magnetisch betätigtes Ventil *n*	vanne *f* à commande magnétique	вентиль с соленоидным (электромагнитным) приводом
	solenoid valve, magnetically operated valve, magnetic valve	Magnetventil *n*	vanne *f* magnétique à commande magnétique, robinet *m* solénoïde	электромагнитный клапан
	solid carbon dioxide, dry ice, solidified carbon dioxide gas, carbon dioxide snow	Trockeneis *n*, feste Kohlensäure *f*, Kohlen-säureschnee *m*	glace *f* sèche, carboglace *f*, anhydride *m* carbonique en neige, neige *f* carbonique, acide *m* carbonique solide	сухой лед, твердая углекислота, снег из твердой углекислоты
	solidification, setting	Erstarrung *f*, Solidifika-tion *f*, Verfestigung *f*	solidification *f*	кристаллизация, затвердевание
	solidification point	*s.* setting point		
	solidified carbon dioxide gas, dry ice, carbon dioxide snow, solid carbon dioxide	Trockeneis *n*, feste Kohlensäure *f*, Kohlen-säureschnee *m*	glace *f* sèche, carboglace *f*, anhydride *m* carbonique en neige, neige *f* carbonique, acide *m* carbonique solide	сухой лед, твердая углекислота, снег из твердой углекислоты
	solidifying point	*s.* setting point		
S 227	**solid solution**	feste Lösung *f*	solution *f* solide	твердый раствор
S 228	**solid state physics**	Festkörperphysik *f*	physique *f* des corps solides	физика твердого тела
S 229	**solid surface**	Festkörperoberfläche *f*	surface *f* des corps solides	поверхность твердого тела
S 230	**solidus**	Soliduslinie *f*, Solidus-fläche *f*	solidus *m*	солидус
S 231	**solidus temperature**	Soliduspunkt *m* ‹unterer Schmelzpunkt›	point *m* de solidification	температура солидуса
S 232	**solubility**	Löslichkeit *f*	solubilité *f*	растворимость
S 233	**solvent**	Lösungsmittel *n*	solvant *m*, dissolvant *m*	растворитель
	solvent-free melting adhesive, melting cement	Schmelzkitt *m*	mastic *m* à fusion	плавкая замазка (мастика), расплавляемая замазка
S 234	**solvent-vapour cleaning unit**	Lösungsmitteldampf-reinigungsanlage *f*	appareil *m* de purification de la vapeur des solvants	очистительная установка с парообразным растворителем
S 235	**sorb**	sorbieren	sorber	сорбировать
S 236	**sorbate**	Sorbat *n*, sorbiertes Gas *n*	gaz *m* sorbé (absorbé), sorbat *m*	сорбат, сорбированный газ
S 237	**sorbent,** sorption agent	Sorbens *n*, Sorptions-mittel *n*	sorbant *m*, agent *m* de sorption, absorbant *m*	сорбент, сорбирующее вещество, сорбционное вещество
S 238	**sorption**	Sorption *f*, Gasaufnahme *f*	sorption *f*	сорбция
	sorption agent	*s.* sorbent		
S 239	**sorption loop**	Sorptionsschleife *f*	boucle *f* de sorption	сорбционная петля
S 240	**sorption pump**	Sorptionspumpe *f*	pompe *f* à sorption	сорбционный насос
S 241	**sorption site**	Sorptionsplatz *m*	emplacement *m* de sorption	сорбционная площадка
S 242	**source current stabilizer**	Quellenstromstabilisator *m*	stabilisateur *m* de courant de source	стабилизатор тока источника
S 243	**S.P.** (η glass = $10^{14,5}$ poises›, strain point	Spannungspunkt *m*, 15 h — Entspannungs-temperatur *f*, untere Entspannungstempera-tur *f* ‹η Glas = $10^{14,5}$ Poise›	point *m* de tension thermique	точка натяжения ‹η стекла = $10^{14,5}$ пуаз›
S 244	**space chamber**	Raumkammer *f*	chambre *f* d'espace	камера, моделирующая [космическое] пространство
S 245	**space charge**	Raumladung *f*	charge *f* spatiale	пространственный (объемный) заряд
S 246	**space charge limited**	raumladungsbegrenzt	limité de charge d'espace	ограниченный пространственным зарядом
S 247	**space environmental simulation,** space simulation	Weltraumsimulation *f*	simulation *f* d'espace de l'univers, simulation d'espace	моделирование космического пространства

S 248	**space lattice**	Raumgitter n	grille f spatiale	пространственная решетка
S 249	**spacer**	Abstandsstück n, Zwischenstück n	rondelle f d'épaisseur, pièce f intermédiaire	разделительная прокладка, спейсер
S 250	**space research**	Weltraumforschung f	recherche f d'espace	исследование космического пространства
	space simulation	s. space environmental simulation		
	space simulation chamber, outer space simulator	Weltraumsimulierkammer f	chambre f de simulateur d'espace	камера для имитации условий космического пространства
	space simulator, altitude simulator	Höhensimulator m, Raumsimulator m	simulator m cosmique, chambre f spatiale	имитатор космического пространства
S 251	**space vehicle,** spatial vehicle	Raumschiff n	véhicule m spatial	космический корабль
S 252	**spare filament**	Ersatzglühfaden m	filament m de réserve	резервная спираль накала
S 253	**spare part**	Ersatzteil n	pièce f de rechange	запасная часть
S 254	**spark coil,** sparker, Tesla coil	Induktionsspule f, Tesla-Spule f	bobine f d'induction	катушка Тесла
	spark-coil detector, high-frequency vacuum tester	Hochfrequenzvakuumprüfer m	appareil m à haute fréquence pour le contrôle du vide, contrôleur m à vide à haute fréquence	искровой течеискатель
S 255	**spark coil leak detector**	Hochfrequenzlecksucher m	détecteur m de fuites à haute fréquence	искровой течеискатель
S 256	**spark coil search method**	Lecksuche f mit Hochfrequenz-Vakuumprüfer	détection f de fuites avec indicateur de vide à hautes fréquences	метод обнаружения течи искровым течеискателем
S 257	**spark coil test**	Hochfrequenz- (Leck-, Vakuum-) Prüfung f	essai m pour le contrôle du vide à haute fréquence, épreuve f pour le contrôle du vide à haute fréquence	проверка искровым течеискателем
	sparker	s. spark coil		
S 258	**spark erosion**	Funkenerosion f	érosion f d'étincelles	искровая эрозия
	spatial vehicle, space vehicle	Raumschiff n	véhicule m spatial	космический корабль
S 259	**spatter shield**	Spritzschutzschirm m	écran m protecteur à moulage, écran de protection contre les aspersions	экран для защиты от разбрызгивания
S 260	**specific density**	Dichtezahl f	densité f spécifique	плотность
S 261	**specific heat**	spezifische Wärme f	chaleur f spécifique	удельная теплоемкость
	specific heat ratio, adiabatic exponent	Adiabatenexponent m	indice m adiabatique	показатель степени в уравнении адиабаты, показатель адиабаты
S 262	**specific ionization coefficient**	spezifische differentielle Ionisierung f	ionisation f spécifique	объемный коэффициент ионизации
S 263	**specific speed**	spezifisches Saugvermögen n	débit m spécifique	удельная быстрота откачки
S 264	**spectral emission coefficient, spectral emissivity**	spektraler Emissionskoeffizient m	coefficient m d'émission spectrale	спектральный (монохроматический) коэффициент излучения
S 265	**spectroscopic**	spektralanalytisch	spectroscopique	спектроскопический
S 266	**specular**	spiegelnd	miroitant	зеркальный
	specular layer, mirror coating, reflecting layer	Spiegelbelag m, Reflexbelag m	couche f spéculaire, revêtement m réfléchissant	зеркальное покрытие
S 267	**specular scattering**	spiegelnde Streuung f	dispersion f spéculaire (miroitante), diffusion f spéculaire (miroitante)	зеркальное рассеяние
S 268	**speed curve,** speed-pressure curve	Darstellung f des Saugvermögens als Funktion des Druckes	courbe f de la vitesse d'aspiration et de la pression	зависимость скорости откачки от давления
	speed factor, Ho coefficient, Ho factor	Ho-Faktor m	coefficient m Ho	коэффициент Хо
	speed of a pump, pumping capacity, pumping speed, pump speed, suction speed	Saugvermögen n, Pumpgeschwindigkeit f	débit m de pompage, vitesse f d'aspiration, débit d'une pompe, vitesse de pompage	скорость откачки, быстрота действия насоса
S 269	**speed of evacuation, speed of exhaust, speed of exhaustion**	Auspumpgeschwindigkeit f	vitesse f de faire le vide, vitesse de pompage	понижение давления в единицу времени
S 270	**speed of flow,** volume flow, volumetric flow [rate], volumetric throughput	Volumendurchsatz m, Volumenstrom m	infiltration f de volume, débit m volumétrique, flux m volumétrique	объемный расход, объемная скорость течения
S 271	**speed of response**	Ansprechgeschwindigkeit f	vitesse f de réponse, vitesse de réaction	скорость реагирования
	speed-pressure curve	s. speed curve		
S 272	**speed recording**	Schnellregistrierung f	enregistrement m rapide	быстродействующий регистратор
S 273	**spherical aberration**	sphärische Aberration f	aberration f sphérique	сферическая аберрация
S 274	**spherical seal valve**	Kugelschliffventil n	soupape f à rodage sphérique	вентиль со сферическим шлифом

	English	German	French	Russian
S 275	spherical-type cold trap	Kugelkühlfalle *f*	piège *m* [refroidi] sphérique	шаровая охлаждаемая ловушка
	spherical valve, ball valve, globe valve	Kugelventil *n*	soupape *f* à bille (boulet), soupape sphérique	вентиль со сферическим клапаном, шаровой клапан
S 276	spider	Armkreuz *n*, Zentrierkreuz *n*	croisillon *m* de centrage	крестовина, звездочка
S 277	spin freezing	Schleuderschichtgefrieren *n*	congélation *f* à rotation verticale rapide	центробежное замораживание тонких слоев
S 278	spinwave	Spin-Welle *f*	onde *f* de spin	спиновая волна
	spin welding, friction welding	Reibungsschweißen *n*	soudure *f* par frottement	сварка трением
S 279	spiral agitator	Bandrührer *m*	agitateur *m* en forme de ruban	спиральная мешалка
S 280	splash baffle	Spritzbaffle *n*	baffle *m* à jet	экран-отражатель
S 281	splash lubrication	Tauchschmierung *f*	graissage *m* par immersion (barbotage)	смазка разбрызгиванием
S 282	split-magnetron ionization gauge	Schlitzmagnetron-Ionisationsvakuummeter *n*	manomètre *m* à ionisation à cathode froide	щелевой магнетронный ионизационный манометр
S 283	split ring	geteilter Ring *m*	anneau *m* divisé (partagé)	разделительное кольцо
S 284	sponge rubber	Schwammgummi *m*	éponge *f* de caoutchouc	губчатая резина
S 285	spoon gauge	Röhrenfedervakuummeter *n*	vacuomètre *m* à tube courbé	трубчатый манометр, манометр с трубчатой пружиной
S 286	spot	Brennfleck *m*	foyer *m*	фокусное пятно
S 287	spot size	Brennfleckgröße *f*	grandeur *f* du foyer	размер фокусного пятна
S 288	spot welding	Punktschweißen *n*	soudage *m* par points	точечная электросварка
	spray, atomize	zerstäuben <Flüssigkeiten>	pulvériser	распылять жидкость
S 289	spray, sprinkle	absprühen	asperger, arroser	обдувать
S 290	spray chamber	Diffusionskammer *f*	chambre *f* de diffusion	смесительная камера, диффузионная камера
S 291	spray degassing, stream-droplet degassing, stream-drop degassing	Sprühentgasung *f*, Durchlauftropfenentgasung *f*	dégazage *m* par ruissellement du jet	капельно-струйное обезгаживание, обезгаживание в расплавленной струе, дегазация капель в процессе протекания
S 292	spray dryer	Zerstäubungstrockner *m*	sécheur *m* par pulvérisation	распылительная сушилка
S 293	spray drying	Sprühtrocknung *f*	séchage *m* par ruissellement du jet	сушка распылением
S 294	spray drying plant	Sprühtrocknungsanlage *f*	installation *f* de séchage par ruissellement du jet	распылительная сушильная установка
S 295	spray freezer	Sprühgefrieranlage *f*	congélateur *m* par jaillissement	морозильный аппарат для замораживания орошением
S 296	spray gun	Spritzpistole *f*	pistolet *m* pulvérisateur	пистолет для распыления, пульверизатор
S 297	spray-limiting sleeve, spray sleeve	Spritzschutzmuffe *f*	manchon *m* pour la protection contre les jaillissements, manchon de protection contre les aspersions	муфта для защиты от разбрызгивания
S 298	spray nozzle	Zerstäuberdüse *f*, Sprühdüse *f*	pulvérisateur *m*	распылительное сопло
	spray sleeve	*s.* spray-limiting sleeve		
S 299	spread	spreiten	étendre	распределять по поверхности
S 300	spreading coefficient	Streuungsbeiwert *m*	coefficient *m* de dispersion	коэффициент рассеяния
S 301	Sprengel pump	Sprengel-Pumpe *f*	pompe *f* Sprengel	насос Шпренгеля
S 302	spring contact	Federkontakt *m*	contact *m* élastique	пружинный (пружинящий) контакт
S 303	spring loaded	federbelastet	à charge par ressort, commandé par un ressort	удерживаемый пружиной
S 304	spring-loaded valve	federbelastetes Auspuffventil *n*	soupape *f* d'échappement commandé par un ressort	выпускной клапан, удерживаемый пружиной
S 305	spring steel	Federstahl *m*	acier *m* pour ressorts	пружинная сталь
	sprinkle, spray	absprühen	asperger, arroser	обдувать
S 306	sputter	zerstäuben	pulvériser	распылять
S 307	sputtering	Zerstäubung *f* <eines Metalls>	pulvérisation *f*	распыление
S 308	sputtering bell jar	Zerstäuberglocke *f*	cloche *f* pulvérisateuse	распылительный колпак
S 309	sputtering equipment	Zerstäubungseinrichtung *f*	pulvérisateur *m*, atomiseur *m*	распылительное устройство
S 310	sputtering rate	Zerstäubungsrate *f*, Abstäuberate *f*	taux *m* de pulvérisation	скорость распыления
S 311	sputtering time	Zerstäubungszeit *f*	temps *m* de pulvérisation	время распыления, продолжительность распыления
S 312	sputtering unit	Zerstäubereinheit *f*	unité *f* du pulvérisateur, unité de pulvérisation	установка для распыления
S 313	sputtering voltage	Zerstäubungsspannung *f*	voltage *m* (tension *f*) de pulvérisation	распыляющее напряжение

S 314	**sputtering yield**	Zerstäubungsergiebigkeit *f*	bon rendement *m* de la pulvérisation, richesse *f* de la pulvérisation	выход распыления
	sputter-ion pump, cold-cathode ion pump, penning-type pump	Ionenzerstäuberpumpe *f*	pompe *f* ionique de pulvérisation, pompe ionique à cathode froide, penning-pompe *f*	магнитный электроразрядный насос, ионно-распылительный насос, ионный насос с холодным катодом, электроразрядный насос, насос Пеннинга
	square-butt weld, flat-butt weld	zweiseitige Vollnaht *f*	pleine soudure *f* bilatérale, plein joint *m* bilatéral	стыковый сварной шов без скоса кромок
S 315	**square thread**	Flachgewinde *n*	filetage (filet) *m* plat	квадратная (прямоугольная) резьба
	squealer, audible leak indicator, audio leak indicator	akustischer Leckanzeiger *m*	indicateur *m* acoustique de fuite	звуковой индикатор течи
	squeeze off, pinch off	abquetschen, abpressen, abklemmen	extraire (faire sortir) par pression, écraser, éprouver à l'air comprimé	отделять сжатием, отжимать, отдавливать, отсоединять пережатием
S 316	**squeezing**	Pressung *f*, Quetschen *n*	pression *f*, moulage *m*	прессование, сжатие
	squeezing cock, clip, hose press clamp, pinchcock	Quetschhahn *m*	pince *f* pressante	зажим для шланга
S 317	**stagnation point**	Staupunkt *m*	point *m* de compression	точка застоя, застойная зона
	stagnation pressure, dynamic pressure, velocity pressure, velocity head	Staudruck *m*, dynamischer Druck *m*	pression *f* dynamique	скоростной напор, динамическое давление
S 318	**stagnation temperature**	Haltetemperatur *f*	température *f* d'entretien	температура задержки
	stainless, corrosion-resistant, corrosion-proof	korrosionsbeständig	inoxydable, non-corrosif, résistant à la corrosion	нержавеющий
S 319	**stainless steel**	nichtrostender Stahl *m*	acier *m* inoxydable	нержавеющая сталь
S 320	**stamp**	stanzen	estamper	штамповать
	standard atmosphere, normal atmosphere, atmospheric pressure ‹sea level, °C›, atmosphere, A.P., atm	normaler Atmosphärendruck *m* ‹760 Torr›, physikalische Atmosphäre *f*, Normdruck *m*, Atm	pression *f* atmosphérique normale, atmosphère *f* physique, atmosphère normale	физическая атмосфера, атм
S 321	**standard conditions of temperature and pressure,** standard temperature and pressure, S.T.P., STP	Normalbedingungen *fpl*, Normaldruck *m* und Normaltemperatur *f*	conditions *fpl* normales, pression *f* normale et température *f* normale	нормальные условия температуры и давления
	standard leak, calibrated leak, sensitivity calibrator, reference leak, test leak	Eichleck *n*, Testleck *n*, Vergleichsleck *n*, Leck *n* bekannter Größe, Bezugsleck *n*	fuite *f* calibrée	стандартная течь, калиброванная течь, эталонная течь
S 322	**standard leak rate**	normale Leckrate *f*, normale Ausflußrate *f*	quote-part *m* normal de fuite, quote-part normal d'écoulement	стандартное натекание
S 323	**standard room temperature** ‹20 °C or 25 °C›	normale Zimmertemperatur *f* ‹20 °C oder 25 °C›	température *f* ambiante normale ‹20 °C ou 25 °C›	нормальная комнатная температура ‹20 °C или 25 °C›
	standard temperature and pressure	*s.* standard conditions of temperature and pressure		
S 324	**stand-by plant**	Bereitschaftsanlage *f*	installation *f* en attente	резервная (дежурная) установка
S 325	**stand-by unit**	Reserveeinheit *f*	unité *f* de réserve	запасной агрегат
S 326	**starting [fore-] pressure**	Startdruck *m*, Einschaltdruck *m*	pression *f* d'enclenchement, pression de mise en marche	начальное давление ‹при котором насос начинает работать›, пусковое давление
S 327	**static friction**	Haftreibung *f*	frottement *m* statique (adhérent)	трение покоя
S 328	**static head**	statischer Druck *m* ‹in mm Flüssigkeitssäule›	pression *f* statique	статическое давление
S 329	**static pressure**	statischer Druck *m* (in kp · cm⁻²), Ruhedruck *m*	pression *f* statique	статическое давление
S 330	**static vacuum system**	statisches Vakuumsystem *n*	système *m* statique	статическая вакуумная система
S 331	**stationary exhaust system,** trolley exhaust system ‹US›	stationärer Pumpautomat *m*, Pumpstraße *f* mit stationären Pumpeinrichtungen	banc (carrousel) *m* de pompage stationnaire	откачной автомат
S 332	**stationary flow,** steady state flow	stationäre Strömung *f*	courant (écoulement) *m* stationnaire	стационарное (установившееся) течение
S 333	**stationary state**	stationärer Zustand *m*	état *m* stationnaire	стационарное состояние
S 334	**stationary tank** ‹for liquefied gases›	Standtank *m* ‹für verflüssigte Gase›	réservoir *m* stationnaire ‹à gaz liquéfiés›	стационарный резервуар ‹для сжижаемых газов›

	English	German	French	Russian
S 335	**stator**	Stator *m*, Gehäuse *n*	stator *m*	статор
S 336	**stay bolt**	Stehbolzen *m*	entretoise *f*, tirant *m*	анкерный болт, распорный болт
S 337	**stay-down time**	Stay-down time, Haltezeit *f*	temps *m* de maintien ‹de l'ultra-vide›, durée *f* de non-contamination ‹dans l'ultra-vide›	время до начала повышения давления в сверхвысоко-вакуумной системе
	stay period, adherence time, dwelling time, residence time, retention time, sticking time	Verweilzeit *f*	durée *f*, temps *m* de contact, temps d'exposition	время удерживания (выдержки, контакта, пребывания)
S 338	**steady state**	stationärer Zustand *m*, Dauerzustand *m*	état *m* permanent (stationnaire)	установившееся состояние
	steady state flow, stationary flow	stationäre Strömung *f*	écoulement (courant) *m* stationnaire	стационарное (установившееся) течение
S 339	**steam**, water vapour	Wasserdampf *m*	vapeur *f* d'eau	водяной пар
S 340	**steam chest**	Dampfdom *m*	dôme *m* à vapeur	сухопарник, паровой колпак
S 341	**steam distillation**	Wasserdampfdestillation *f*	distillation *f* de vapeur d'eau	перегонка с водяным паром
	steam ejector [pump], steam ejector [vacuum] pump, jet exhauster	Dampfstrahler *m*, Wasserdampfstrahlsauger *m*, Dampfstrahlsauger *m*, Dampfstrahlvakuumpumpe *f*	éjecteur *m* à vapeur d'eau, aspirateur *m* de vapeur, pompe *f* à vide à vapeur d'eau	паровой эжекторный насос, пароэжекторный насос, паровой (пароструйный) эжектор, пароструйный эжекторный вакуумный насос
S 342	**steam generation**	Dampfentwicklung *f*	génération *f* de vapeur	образование водяного пара
S 343	**steam heating**	Dampfheizung *f*	chauffage *m* à la vapeur	нагрев паром, прогрев паром
S 344	**steam-jet blower**	Dampfstrahlgebläse *n*	soufflante *f* à jet de vapeur	пароструйный инжектор
	steam-jet pump, ejector, ejector pump, vapour-jet pump	Dampfstrahlpumpe *f*	pompe *f* à jet de vapeur, éjecteur *m* [à vapeur]	пароструйный насос, эжектор, эжекторный насос
S 345	**steam-jet refrigeration plant**	Dampfstrahlkälteanlage *f*	installation *f* frigorifique à jet de vapeur	пароэжекторная (пароструйная) холодильная установка
S 346	**steam nozzle**	Dampfdüse *f*	injecteur *m* (tuyère *f*) de vapeur	паровое сопло
S 347	**steam valve**	Dampfventil *n*	valve *f* à vapeur	паровой вентиль
S 348	**steel bell jar**	Stahlglocke *f*	cloche *f* en acier	стальной вакуумный колпак
S 349	**steel degassing**	Stahlentgasung *f*	dégazage *m* de l'acier	обезгаживание стали
S 350	**steel-degassing plant**	Stahlentgasungsanlage *f*	installation *f* de dégazage de l'acier	установка для обезгаживания стали
S 351	**steel strip,** strip steel	Bandstahl *m*	acier *m* feuillard	стальная полоса (лента), ленточная сталь, полосовая сталь
S 352	**Stefan-Boltzmann formula**	Stefan-Boltzmannsches Gesetz *n*	loi *f* de Stefan-Boltzmann	закон Стефана-Больцмана
	stem, pinch, press, pinched base	Quetschfuß *m*	socle *m* pressant, tige *f*	штампованная ножка
S 353	**stem-pressing machine**	Fußquetschmaschine *f*	machine *f* pressante de pied	машина для штамповки ножек
S 354	**stem seal**	Spindeldichtung *f*	joint *m* en tige	стержневое уплотнение
S 355	**step flange**	Stufenflansch *m*	bride *f* étagée, bride à gradins	ступенчатый фланец
S 356	**stepped curve**	Treppenkurve *f*	courbe *f* enchevêtrée (en paliers)	ступенчатая кривая
S 357	**step-seal**	Stufendichtung *f*	joint *m* en cascade	ступенчатое уплотнение
S 358	**sterilize**	sterilisieren, entkeimen	stériliser	стерилизовать
	stick, adhere	stecken, kleben, haften	adhérer	склеивать, прилеплять, прилипать
S 359	**sticking coefficient, sticking factor**	Haftkoeffizient *m*, Haftzahl *f*	coefficient *m* d'adhésion, coefficient d'adhérence	коэффициент прилипания (сцепления)
	sticking probability, trapping coefficient, trapping efficiency, capture coefficient, capture probability	Einfangkoeffizient *m*, Haftwahrscheinlichkeit *f*	probabilité *f* d'adhérence, probabilité d'adhésion, probabilité du pouvoir adhérent, coefficient *m* d'arrêt, coefficient de blocage, probabilité de fixation (collage)	коэффициент прилипания (захвата), вероятность захвата (прилипания)
	sticking time	*s.* stay period		
	sticking vacuum, capillary drag	Klebevakuum *n*	vide *m* adhésif (collant, d'adhésion du mercure)	вакуум прилипания
	still, alembic, distillation still	Destillationskolben *m*, Destillationsblase *f*	ampoule *f* à distiller	перегонная колба
S 360	**stinger rod**	Elektrodenstange *f*	barrette *f* à électrodes	стержень электрода
S 361	**stirring coil**	Rührspule *f*	bobine *f* à agitateur	перемешивающая катушка
S 362	**stirring of molten mass**	Schmelzbadbewegung *f*	brassage *m* du bain de fusion	перемешивание расплавленного металла
S 363	**stirrup**	U-förmiger Bügel *m*	étrier *m* en U	хомут
S 364	**stirrup getter**	Bügelgetter *m*	getter *m* à étrier	хомутообразный поглотитель, бугельный газопоглотитель

	stopcock, plug valve	Absperrhahn m	robinet m à boisseau, robinet d'arrêt	кран, стопорный кран
S 365	stopcock grease	Hahnfett n	graisse f pour robinets	смазка для кранов
	stop-off lacquer, masking lacquer	Abdecklack m	laque f de couverture	лак для защиты при обработке
S 366	stopper	verschließen	boucher, fermer, sceller	запирать, закрывать
S 367	stopper	Stöpsel m, Pfropfen m	bouchon m	пробка
S 368	stopper rod	Stopfenstange f	quenouille f, barre f de bouchage	стопорный стержень
S 369	stop slide valve, stop valve	Absperrschieber m	robinet-vanne m	запорный вентиль, запорная заслонка
S 370	storage ring	Speicherkreis m	circuit m d'accumulation	накопительное кольцо
S 371	storage tank	Vorratsbehälter m	réservoir m	накопительный баллон (бачок)
S 372	storage tube	Speicherröhre f	tube m accumulateur (à mémoire)	запоминающая трубка
	S.T.P., STP	s. standard conditions of temperature and pressure		
S 373	straight-line focus anode	Strichfokusantikatode f	foyer m linéaire-anti-cathode, ligne f focal-anticathode	антикатод со штрихо-вым фокусным пятном
	straight through valve, in-line valve, through-way valve, straight-way valve	Durchgangsventil n	soupape f droite, soupape de traversée, vanne f à passage droit	проходной вентиль
S 374	straight vacuum forming	Vakuumsaugverfahren n	procédé m d'aspiration sous vide	негативное вакуумное формование
	straight-way valve, through-way valve, in-line valve, straight through valve	Durchgangsventil n	soupape f droite, vanne f à passage droit, soupape de traversée	проходной вентиль
S 375	strain gauge	Dehnungsmesser m	dilatomètre m	тензодатчик, тензометр дилатометр
	strain point, S.P. (η glass = $10^{14,5}$ poises)	Spannungspunkt m, 15-h-Entspannungstemperatur f, untere Entspannungs-temperatur f (η Glas = $10^{14,8}$ Poise)	point m de tension thermique	точка натяжения (η стекла = $10^{14,5}$ пуаз)
S 376	stream, flow	Strömung f, Strom m	écoulement m	поток, течение
	stream degassing	Gießstrahlentgasung f, Durchlaufentgasung f	dégazage m pendant la coulée	обезгаживание в раз-ливочной струе
	stream-drop degas-sing, stream-droplet degassing, spray degassing	Sprühentgasung f, Durchlauftropfen-entgasung f	dégazage m par ruisselle-ment du jet	капельно-струйное обезгаживание, обезгаживание в расплавленной струе, дегазация ка-пель в процессе про-текания
S 377	street reducer	Reduzierstück n mit ein-seitiger Muffe	réducteur m avec raccord unilatéral	редуктор с односторон-ней муфтой
	strengthening, hardening	Verfestigung f	solidification f, durcisse-ment m, consolidation f	упрочнение, наклеп
S 378	stress	Beanspruchung f, Spannung f	contrainte f, tension f	напряженное состояние, натяжение
S 379	stress coat method	Reißlackverfahren n	procédé m de vernis de dé-chirure	метод трескающегося покрытия
S 380	stria	Schlieren fpl	stries fpl	свили
S 381	striated column	geschichtete Säule f	colonne f disposée en couches, colonne strati-forme	слоистый столб, столб со стратами
S 382	strike solution	Vorbehandlungsbad n	bain m de traitement préalable (préliminaire)	раствор для предвари-тельной обработки
S 383	striking <of an arc>	Zündung f <eines Licht-bogens>	allumage m, amorçage m	возникновение (зажи-гание) дуги
S 384	strip coating	Abziehlack m	vernis m à enlever, vernis à tirer	снимаемое покрытие
	strip coating, band coating, roll coating	Bandbedampfung f	vaporisation f à bande, métallisation f au dé-roulé	нанесение покрытия на ленту
S 385	strip evaporation plant	Bandbedampfungsanlage f	installation f de métallisa-tion au déroulé	установка для нанесения покрытия на ленту
S 386	strip heating	Bandheizung f	chauffage m de bande	ленточный нагрев
S 387	strip line	Bandstraße f	train m à feuillards, voie f à bande	конвейер
	strip recorder, continuous-line recorder	Linienschreiber m	enregistreur m en trait continu	ленточный самописец
	strip steel, steel strip	Bandstahl m	acier m feuillard	стальная полоса (лента), ленточная сталь, полосовая сталь
S 388	stroke-speed of the piston	Kolbenhubgeschwindig-keit f	vitesse f élévatoire du piston	скорость хода поршня
S 389	strong focusing principle	Prinzip n der starken Fokussierung	principe m de la focalisa-tion intense	принцип сильной фокусировки
	structural member, constructional element	Bauteil n	élément m de construction	структурный элемент

S 390	**stud, stud bolt,** trunnion	Stiftschraube *f*	goujon *m*, vis *f* à cheville	шпилька, палец, цапфа
S 391	**stud welding**	Stiftschweißung *f*	soudure *f* par pointe (goujon)	точечная (штифтовая) сварка
	stuffing box, packing gland	Stopfbuchse *f*	presse-étoupe *m*, boîte *f* à bourrage	сальник, сальниковая коробка
	stuffing-box packing, gland packing	Stopfbuchsenpackung *f*	garniture *f* de presse-étoupe	набивка сальника
S 392	**stuffing-box sealing**	Stopfbuchsendichtung *f*	joint *m* par presse-étoupe	сальниковое уплотнение
S 393	**sublimate,** sublime	sublimieren	sublimer	возгонять
S 394	**sublimate separator**	Sublimatabscheider *m*	séparateur *m* de sublimé (sublimat)	сублимационный сепаратор
S 395	**sublimation**	Sublimation *f*	sublimation *f*	сублимация, возгонка
	sublimation from the frozen state, cryo-drying, freeze-drying, lyophilization	Gefriertrocknung *f*, Lyophilisation *f*	lyophilisation *f*, dessiccation *f* par congélation, cryodessiccation *f*	сублимационная сушка из замороженного состояния, лиофилизация
	sublimation heat, heat of sublimation	Sublimationswärme *f*	chaleur *f* de sublimation	теплота сублимации
S 396	**sublimation plant**	Sublimationsanlage *f*	installation *f* de sublimation	сублимационная установка
S 397	**sublimation pump**	Sublimationspumpe *f*	pompe *f* de sublimation	сублимационный насос
S 398	**sublimation rate**	Sublimationsrate *f*	taux *m* de la sublimation	скорость возгонки (сублимации)
S 399	**sublimation technique**	Sublimationstechnik *f*	technique *f* de sublimation	сублимационная (возгонная) технология
	sublime, sublimate	sublimieren	sublimer	возгонять
S 400	**subliming**	Sublimieren *n*	sublimation *f*	сублимация
	subsequent drying, afterdrying, final drying, secondary drying	Nachtrocknung *f*, Endtrocknung *f*	dessiccation *f* secondaire (finale), séchage *m* final (secondaire)	досушивание, досушка, вторичная осушка, конечная (окончательная) сушка
S 401	**subsonic**	Unterschall-	sous-sonore	дозвуковой, инфразвуковой
S 402	**substrate**	Schichtträger *m*, Unterlage *f*, Basis[platte] *f*	base *f*, lit *m*	подложка
S 403	**substrate heater**	Substratheizer *m*	réchauffeur *m* de substratum	нагреватель подложки
S 404	**substrate holder**	Substrathalter *m*	support *m* de substratum	держатель подложки
S 405	**substrate magazin**	Schichtträgermagazin *n*	magasin *m* à support d'émulsion	кассета для подложек
S 406	**substrate-mask changer**	Substrat-Maskenwechsler *m*	inverseur *m* de masque à substrat	механизм для смены масок-подложек
S 407	**substrate temperature control**	Schichtträgertemperatursteuerung *f*	commande *f* thermique du côté sensible du film	регулирование температуры подложки
S 408	**successive dilution**	sukzessive Verdünnung *f*	dilution *f* successive	последовательное разбавление
	suck, pump	ansaugen, saugen, pumpen	sucer, aspirer, pomper	всасывать, засасывать, качать
S 409	**suck-back**	zurücksaugen	respirer	всасывать обратно
	sucking, suction, inspiration	Ansaugen *n*, Saugen *n*, Einsaugen *n*	succion *f*, aspiration *f*	всасывание, засасывание
	sucking and forcing pump, double acting pump, lift and force pump, suction and pressure pump	Saug- und Druckpumpe *f*	pompe *f* aspirante et refoulante	всасывающий и нагнетательный насос, насос двойного действия
	suck off, draw off	absaugen	évacuer, pomper	отсасывать
	suction, sucking, inspiration	Saugen *n*, Ansaugen *n*, Einsaugen *n*	succion *f*, aspiration *f*	всасывание, засасывание
	suction and pressure pump	*s.* sucking and forcing pump		
	suction capacity, exhausting power, throughput	Saugleistung *f*	puissance *f* d'une pompe, puissance d'admission, débit *m* massique d'une pompe, rendement *m* d'aspiration	производительность по всасыванию
S 410	**suction casting,** vacuum casting	Vakuumgießen *n*	coulée *f* sous vide, coulage *m* sous vide	вакуумное литье, литье в вакууме
S 411	**suction chamber**	Schöpfraum *m*	chambre *f* d'aspiration, chambre de la pompe	рабочая камера
S 412	**suction device**	Saugvorrichtung *f*	aspirateur *m*	всасывающее устройство
S 413	**suction filter**	Saugfilter *n*	essoreuse *f* à vide, filtre *m* d'aspiration	вакуумфильтр
	suction line, intake line	Saugleitung *f*	conduite (canalisation) *f* d'aspiration, tuyau *m* de pompage	впускной (всасывающий) трубопровод
S 414	**suction mould**	Saugform *f*	mode *m* (forme *f*) d'aspiration	форма с отсосом
	suction port, pump inlet, intake port	Ansaugöffnung *f*	fente *f* d'aspiration, ouverture *f* d'aspiration	впускной патрубок
	suction pressure, inlet pressure, intake pressure, fine pressure, head pressure	Ansaugdruck *m*	pression *f* d'aspiration, pression d'admission	впускное давление
	suction side, collecting side, inlet side, intake side, vacuum side	Saugseite *f*, Vakuumseite *f*	côté *m* vide, côté d'aspiration	собирающая сторона, сторона впуска

	English	German	French	Russian
	suction speed, pump speed, pumping speed, pumping capacity, speed of a pump	Saugvermögen n, Pumpgeschwindigkeit f	débit m de pompage, vitesse f de pompage, débit d'une pompe, vitesse d'aspiration	быстрота действия насоса, скорость откачки
S 415	**supercharge**	vorverdichten	précondenser, précomprimer	нагнетать, создавать избыточное давление
S 416	**supercharging**	Vorverdichtung f	précompression f, précondensation f	наддув, создание избыточного давления
S 417	**superconductivity,** supraconductivity	Supraleitfähigkeit f	supraconductibilité f	сверхпроводимость
S 418	**superconductor**	Supraleiter m	supraconducteur m	сверхпроводник
S 419	**supercooling**	Unterkühlung f	sous-refroidissement m	переохлаждение
	superheated vapour, dry vapour	überhitzter Dampf m	vapeur f surchauffée	перегретый пар, сухой пар
S 420	**superheater**	Überhitzer m	surchauffeur m	пароперегреватель
S 421	**superheating**	Überhitzung f	surchauffage m	перегрев
	supersaturated vapour, oversaturated vapour	übersättigter Dampf m	vapeur f sursaturée	пересыщенный пар
S 422	**supersaturation**	Übersättigung f	sursaturation f	пересыщение, перенасыщение
S 423	**supersonic driving jet,** supersonic speed jet	Überschalltreibstrahl m	jet m supersonique	сверхзвуковая струя
S 424	**supersonic flow**	Überschallströmung f	courant (écoulement) m ultrasonique	сверхзвуковое течение
S 425	**supersonic molecular beam**	Überschallmolekularstrahl m	jet m moléculaire ultrasonique	сверхзвуковой молекулярный пучок, сверхзвуковая молекулярная струя
	supersonic speed jet	s. supersonic driving jet		
S 426	**supersonic vapour jet**	Überschalldampfstrahl m	jet m supersonique	сверхзвуковая паровая струя
S 427	**supported flange joint**	Klemmflansch m	bride f de serrage	зажимное фланцевое соединение
S 428	**supported screwed joint**	Klemmverschraubung f	vissage m de serrage	зажимное винтовое (болтовое) соединение
	supporting gas dryer, airborne dryer, suspended particle dryer	Schwebegastrockner m	sécheur m de gaz pendant	газовая сушилка суспендированного вещества
	supporting ring, bearing ring	Tragring m	anneau m portatif, bague-support f	опорное кольцо
S 429	**suppressor electrode**	Bremselektrode f	électrode f de frein	защитный электрод
S 430	**suppressor ion gauge**	Ionisationsvakuummeter n mit Bremselektrode	manomètre m à ionisation avec électrode de frein	ионизационный манометр с электродом для подавления фототока
S 431	**supraconduction**	Supraleitung f	supraconduction f	сверхпроводимость
	supraconductivity, superconductivity	Supraleitfähigkeit f	supraconductibilité f	сверхпроводимость ‹явление›
S 432	**surface atom**	Oberflächenatom n	atome m de la surface	поверхностный атом
	surface coefficient of heat transfer, film coefficient of heat transfer, surface film conductance	Wärmeübergangszahl f	coefficient m d'écoulement thermique, nombre m de la transmission thermique, coefficient de transmission thermique	коэффициент теплоотдачи (теплопередачи)
S 433	**surface condenser**	Oberflächenkondensator m	condensateur (condenseur) m à surface	конденсатор поверхностного типа
S 434	**surface coverage**	Oberflächenbedeckung f	couvercle m superficiel, couverture f superficielle	поверхностное покрытие
S 435	**surface diffusion**	Oberflächendiffusion f	diffusion f superficielle	поверхностная диффузия
S 436	**surface etching**	Oberflächenätzung f	corrosion f superficielle	поверхностное травление
S 437	**surface evaporation**	Oberflächenverdampfung f	évaporation f de la surface	испарение поверхности
	surface film conductance, film coefficient of heat transfer, surface coefficient of heat transfer	Wärmeübergangszahl f	coefficient m d'écoulement thermique, nombre m de la transmission thermique, coefficient de transmission thermique	коэффициент теплоотдачи (теплопередачи)
	surface finish, finish, surface quality	Oberflächenbeschaffenheit f, Aussehen n, Oberflächengüte f	fini m de la surface, qualité f de surface, constitution f de la surface	чистота (качество, состояние) поверхности, отделка
S 438	**surface flow**	Oberflächenströmung f	courant m superficiel	поверхностное течение
S 439	**surface ionization**	Oberflächenionisation f	ionisation f superficielle	поверхностная ионизация
S 440	**surface leakage,** tracking	Kriechwegbildung f	formation f d'un courant vagabond	образование следа от утечки
S 441	**surface migration**	Oberflächenwanderung f	migration f superficielle (de la surface)	поверхностная миграция
	surface quality	s. surface finish		
S 442	**surfacer**	Oberflächenkitt m, Porenfüller m	mastic m superficiel	шпаклевка
S 443	**surface refining**	Oberflächenveredelung f	raffinage m de surface	поверхностная очистка
S 444	**surface replica,** surface replication	Oberflächenabdruck m	impression (copie) f superficielle, calque m superficiel	оттиск поверхностных микронеровностей, поверхностная реплика

S 445	**surface roughness**	Oberflächenrauhigkeit *f*	rugosité *f* superficielle (de la surface)	поверхностная шероховатость
S 446	**surface site**	Oberflächenplatz *m*	place *f* de la surface	участок (площадка) поверхности
S 447	**surface state**	Oberflächenzustand *m*	état *m* de surface, condition *f* superficielle	состояние поверхности
S 448	**surface tension**	Oberflächenspannung *f*	tension *f* superficielle	поверхностное натяжение
S 449	**surge**	Spannungsdurchschlag *m*	percement *m* de tension	перенапряжение
	surge voltage, impulse voltage	Stoßspannung *f*	tension *f* de choc, surtension *f* impulsionnelle	импульсное напряжение
S 450	**susceptor**	Sekundärzylinder *m*	cylindre *m* secondaire	вторичный цилиндр
	suspended particle dryer, airborne dryer, supporting gas dryer	Schwebegastrockner *m*	sécheur *m* de gaz pendant	газовая сушилка суспендированного вещества
S 451	**Sutherland's constant**	Sutherland-Konstante *f*	constante *f* de Sutherland	постоянная Сезерленда
S 452	**sweating**	Schwitzwasserbildung *f*, Beschlagen *n*	formation *f* d'eau de condensation	запотевание, образование конденсата
S 453	**swelling isotherm**	Quellungsisotherme *f*	isotherme *f* de soufflage	изотерма набухания
S 454	**swept volume,** volumetric displacement	Hubvolumen *n*, Schöpfvolumen *n*	volume *m* engendré	объем рабочей камеры, описываемый объем
	switch desk, control desk	Schaltpult *n*	pupitre *m* de commande	пульт управления
S 455	**switching pressure**	Schaltdruck *m*	pression *f* de distribution	выключающее давление
S 456	**synthetic resin paste**	Kunstharzmasse *f*	pâte *f* de résine synthétique	паста из синтетической смолы
	syphon degassing, circulation degassing, continuous degassing by the syphon method	Umlaufentgasung *f*	dégazage *m* par circulation, dégagement *m* par circulation	циркуляционное (сифонное) обезгаживание

T

T 1	**tack-welding**	Heftschweißen *n*	queue-soudure *f*	сварка прихваточными швами
T 2	**tail piece**	Endstück *n*	morceau *m* final, bout *m*	наконечник
T 3	**tail pipe**	Saugrohr *n*	tuyau *m* d'aspiration	всасывающая труба
	take-hold pressure, holding vacuum	Haltevakuum *n*	vide *m* d'entretien, vide à maintenir	давление, созданное поддерживающим насосом, поддерживающий вакуум
	take-hold pressure, cross-over forepressure	Vorvakuumdruck *m* <bei dem der Ansaugdruck gleich dem Vorvakuumdruck ist>	pression *f* du vide préliminaire	противодавление полного срыва <при котором впускное давление становится равным выпускному>
T 4	**tamping material**	Stampfmasse *f*	masse *f* de damage	масса для набивки, набивная масса
	tank, container, vessel	Rezipient *m*, Behälter *m*	récipient *m*	контейнер, вместилище
	tank gas, bomb gas	Flaschengas *n*	gaz *m* en bouteille	баллонный газ
T 5	**tap degassing**	Abstichentgasung *f*	dégazage *m* dans la poche de coulée	обезгаживание, дегазация при разливке
T 6	**tap hole**	Stichloch *n*	trou *m* de coulée	летка, выпускное отверстие
T 7	**tapping bar**	Abstichstange *f*	sonde *f* de perçage	стержень для разбивания (пробивания)
T 8	**tapping ladle**	Abstichpfanne *f*	poche *f* de coulée	разливочный ковш
T 9	**tapping side**	Abstichseite *f*	côté *m* de coulée	выпускная сторона <печи>
T 10	**tap water**	Leitungswasser *n*	eau *f* de conduite	водопроводная вода
	target, ion collector, plate	Ionenkollektor *m*, Ionenauffänger *m*	collecteur *m*, collecteur *m* ionique	ионный коллектор
T 11	**tear resistance**	Reißfestigkeit *f*	résistance *f* à la rupture	предел прочности на разрыв
T 12	**technique**	Technik *f* <im Sinne der Verfahrenstechnik>, Methode *f*, Verfahren *n*	technique *f*	технология, техника, метод
T 13	**technology of materials**	Werkstofftechnik *f*	technique *f* des matières premières	технология материалов
T 14	**telefocus system**	Fernfokussystem *n*	système *m* de foyer à distance	длиннофокусная система
	Telefunken process, sintered metal powder process	Telefunken-Verfahren *n*	procédé *m* de Telefunken	процесс металлизации керамики по способу фирмы Телефункен
T 15	**temperature control**	Temperaturregelung *f*	réglage *m* de température	терморегулировка
T 16	**temperature controller**	Temperaturwächter *m*	contrôleur *m* de la température, régulateur *m* thermique, thermostat *m*	терморегулятор
T 17	**temperature cycling**	Temperaturwechsel *m*	changement *m* de température	температурное колебание
T 18	**temperature sensor**	Temperaturfühler *m*	sonde *f* thermique (de température), thermosonde *f*	температурный датчик
T 19	**tempering**	Anlassen *n*	malléabilisation *f*, revenu *m*	отпуск, отжиг для снятия натяжений

	English	German	French	Russian
T 20	**tensile strength**	Zugfestigkeit *f*, Reißfestigkeit *f*	résistance *f* à la traction	предел прочности при растяжении, прочность при растяжении
	terminal box, end box	Kabelendverschluß *m*	boîte *f* d'extrémité du câble	присоединительная коробка, кабельный бокс
	Tesla coil, sparker, spark coil	Induktionsspule *f*, Tesla-Spule *f*	bobine *f* d'induction	катушка Тесла
	test dome, measuring dome	Meßdom *m*	dôme *m* de mesure	испытательный колпак
	test gas, probe gas, search gas, tracer gas	Testgas *n*, Prüfgas *n*	gaz *m* témoin (traceur, d'épreuve), gaz-sonde *m*	пробный газ
	test leak, calibrated leak, sensitivity calibrator, standard leak, reference leak	Eichleck *n*, Testleck *n*, Vergleichsleck *n*, Leck *n* bekannter Größe, Bezugsleck *n*	fuite *f* calibrée	стандартная течь, калиброванная течь, эталонная течь
T 21	**test sieve**	Prüfsieb *n*	blutoir *m* d'essai	лабораторное сито
T 22	**tetrode ionization gauge**	Tetroden-Ionisations-vakuummeter *n*	tétrode-vacuomètre *m* d'ionisation	тетродный ионизационный манометр
T 23	**tetrode sputtering**	Tetrodenzerstäubung *f*	pulvérisation *f* par tétrodes	тетродное распыление
	thaw, defrost, thaw-off	auftauen, abtauen, entfrosten	dégeler, dégivrer	оттаивать, размораживать
	thawing, defrosting	Auftauen *n*, Abtauen *n*	dégel *m*, dégivrage *m*	оттаивание
	thaw-off, thaw, defrost	abtauen, auftauen, entfrosten	dégivrer, dégeler	оттаивать, размораживать
	THEED	*s.* transmission high energy electron diffraction		
	theoretical speed, intrinsic speed	Eigensaugvermögen *n*	débit *m* intrinsèque	теоретическая быстрота откачки
T 24	**thermal agitation noise**	Wärmerauschen *n*	bruit *m* thermique	тепловой шум
T 25	**thermal arrest**	thermischer Haltepunkt *m*	point *m* thermique d'arrêt	площадка на кривой изменения температуры
T 26	**thermal breakdown**	Wärmedurchschlag *m*	infiltration *f* thermique	тепловой пробой
T 27	**thermal breakdown range**	Temperaturbereich *m* des Wärmedurchschlages	zone *f* de température de la décharge thermique	температурный диапазон теплового пробоя
	thermal conductance, heat conduction power	Wärmeleitfähigkeit *f*, Wärmeleitvermögen *n*	conductibilité *f* thermique, conductivité *f* calorifique	теплопроводность, коэффициент теплопроводности
	thermal conduction, heat conduction	Wärmeleitung *f*	conductibilité *f* thermique	теплопроводность
T 28	**thermal conductivity**	Wärmeleitvermögen *n*	conductibilité *f* thermique	теплопроводность
	thermal conductivity vacuum gauge, heat-conduction gauge, thermal gauge	Wärmeleitungs-vakuummeter *n*	jauge *f* thermique, vacuomètre *m* à conductibilité thermique, manomètre *m* thermique	тепловой манометр, теплоэлектрический манометр
T 29	**thermal decomposition trap**	Zersetzungsfalle *f*	piège *m* à décomposition thermique	термическая ловушка
T 30	**thermal desorption**	thermische Desorption *f*	désorption *f* thermique	термическая (тепловая) десорбция
T 31	**thermal diffusion,** thermal transpiration	thermische Transpiration *f*, thermische Effusion *f*, Thermodiffusion *f*, thermomolekulare Strömung *f*	transpiration *f* thermique	температурная (тепловая) транспирация
T 32	**thermal diffusion**	Thermodiffusion *f*	diffusion *f* thermique	термодиффузия, тепловая диффузия
T 33	**thermal diffusivity**	Temperaturleitfähigkeit *f*	conductibilité *f* de température	температуропроводность, коэффициент температуропроводности
T 34	**thermal electron energy**	thermische Elektronen-energie *f*	énergie *f* thermique des électrons	тепловая энергия электронов
	thermal emissivity, radiation emittance	thermisches Emissions-vermögen *n*	pouvoir *m* d'émission thermique	термическая эмиссионная способность
T 35	**thermal equilibrium**	thermisches Gleich-gewicht *n*	équilibre *m* thermique	термическое равновесие
T 36	**thermal equivalent**	Wärmeäquivalent *n*	équivalent *m* calorifique	тепловой эквивалент
T 37	**thermal expansion coefficient**	thermischer Aus-dehnungskoeffizient *m*	coefficient *m* thermique de dilatation	коэффициент теплового расширения
	thermal gauge	*s.* thermal conductivity vacuum gauge		
T 38	**thermal impulse welding**	Wärmeimpulsschweißen *n*	soudure *f* par impulsion thermique	термоимпульсная сварка
	thermal insulation, heat insulation	Wärmeisolation *f*	isolation *f* thermique (calorifuge)	теплоизоляция
T 39	**thermal ionization**	thermische Ionisierung *f*	ionisation *f* thermique	термическая ионизация
T 40	**thermal oxidation**	thermische Oxydation *f*	oxydation *f* thermique	термическое окисление
T 41	**thermal radiation**	Wärmestrahlung *f*	rayonnement *m* thermique	тепловое излучение
T 42	**thermal radiation protection glass**	Wärmestrahlungs-schutzglas *n*	verre *m* de protection à rayonnement de chaleur	стекло для защиты от теплоизлучения
T 43	**thermal release**	thermische Auslösung *f*	déclenchement *m* thermique	термический отпуск

T 44	thermal resistivity	Temperaturwechsel-beständigkeit *f*	résistance *f* à changement de température	стойкость к температурным колебаниям
T 45	thermal space chamber	thermische Raumkammer *f*	chambre *f* thermique au volume	тепловая камера для имитации космического пространства
T 46	thermal stress resistance	Wärmespannungs-widerstand *m*	résistance *f* d'effort thermique	прочность по отношению к термическим напряжениям, сопротивление термическим напряжениям
	thermal transmission, heat transfer, heat transmission	Wärmeübergang *m*, Wärmeübertragung *f*	écoulement *m* thermique, transmission *f* de chaleur, transmission calorifique (thermique), conduction *f* thermique	теплопередача, перенос тепла
	thermal transmittance, overall coefficient of heat transfer	Wärmedurchgangszahl *f*	coefficient *m* du passage de chaleur	коэффициент теплопередачи
	thermal transpiration, thermal diffusion	thermische Transpiration *f*, thermische Effusion *f*, Thermodiffusion *f*, thermomolekulare Strömung *f*	transpiration *f* thermique	температурная (тепловая) транспирация
T 47	thermal transpiration ratio	Verhältniszahl *f* der thermischen Transpiration	coefficient *m* proportionnel de la transpiration thermique	коэффициент термической транспирации
T 48	thermal transpiration vacuum pump	Thermomolekularpumpe *f*	pompe *f* thermomoléculaire	термомолекулярный насос
T 49	thermal velocity	thermische Geschwindigkeit *f*	vitesse *f* thermique	тепловая скорость
T 50	thermic cathode thermionic cathode, hot cathode	thermische Katode *f* Glühkatode *f*	cathode *f* thermique cathode *f* chaude (chauffée, incandescente, à incandescence)	термокатод термоэлектронный катод, термокатод, накаленный катод
T 51	thermionic cathode ionization gauge	Ionisationsvakuummeter *n* mit geheizter Katode	vacuomètre *m* ionique à cathode chaude, manomètre *m* à ionisation à cathode chaude	ионизационный манометр с накаленным катодом
T 52	thermionic emission	Glühelektronenemission *f*	émission *f* thermoélectrique (d'incandescence)	термоэлектронная эмиссия
T 53	thermionic emissivity	Elektronenemissions-vermögen *n*	pouvoir *m* émissif d'électrons	электронная эмиссионная способность
	thermionic tube, hot-cathode vacuum tube	Glühkatodenröhre *f*	tube *m* à cathode chaude, tube thermionique	электронная лампа с накаленным катодом
	thermionic valve, electronic valve	Elektronenröhre *f*	tube *m* électronique	электронная лампа, радиолампа
T 54	thermistor ‹NTC resistor›	Thermistor *m* ‹NTC-Widerstand›	thermistor *m* ‹NTC-résistance›	термистор
T 55	thermistor gauge	Thermistor-Vakuummeter *n*	thermistor-vacuomètre *m*	термисторный манометр
T 56	thermocouple	Thermoelement *n*	couple *m* thermoélectrique	термопара
T 57	thermocouple vacuum gauge	thermoelektrisches Vakuummeter *n*, Thermokreuz *n*	jauge *f* à vide thermoélectrique	термопарный манометр
T 58	thermodiffusion column	Trennrohr *n*	tuyau *m* séparateur (de séparation)	термодиффузионная разделительная колонна
T 59	thermodynamics	Thermodynamik *f*, Wärmelehre *f*	thermodynamique *f*	термодинамика
T 60	thermoelectrically cooled baffle	Dampfsperre *f* mit Peltierkühlung	baffle *m* refroidi thermoélectrique (Peltier-refroidissement)	ловушка с термоэлектрическим охлаждением
T 61	thermoelectric assembly	Anordnung *f* thermoelektrischer Elemente ‹Peltier-Effekt›	arrangement *m* de couples thermoélectriques	батарея термоэлементов, термобатарея
	thermoelectric baffle, Peltier trap	Peltier-Falle *f*, Peltier-Baffle *n*	piège *m* à effet Peltier, piège (baffle *m*) thermoélectrique	термоэлектрическая ловушка
T 62	thermoelectric refrigeration	thermoelektrische Kälteerzeugung *f*	réfrigération *f* thermoélectrique	термоэлектрическое охлаждение
T 63	thermopile	Thermosäule *f*	pile *f* thermoélectrique	термостолбик
T 64	thermoplast, thermoplastic	Thermoplast *m*	matière *f* thermoplastique	термопласты, термопластичные пластмассы
T 65	thermostatic expansion valve	thermostatisches Expansionsventil *n*	soupape *f* thermostatique à expansion	термостатированный расширительный вентиль
T 66	thickness gauge	Dickenmeßgerät *n*	micromètre *m*	толщиномер
T 67	thickness-shear oscillation	Dickenscherschwingung *f*	vibration *f* de cisaillement d'épaisseur	колебания толщины среза
	thimble trap, cold finger, cooling finger	Kühlfinger *m*	doigt *m* refroidisseur, sonde *f* de refroidissement	пальцевый холодильник
	thimble type cold trap, reservoir-type freezing trap	Behälterkühlfalle *f*, Kühlfinger *m*	réservoir-piège *m* à refroidissement, sonde *f* de refroidissement, doigt *m* refroidisseur	охлаждаемая ловушка с резервуаром для хладоагента
T 68	thin-film circuit	Dünnfilmschaltkreis *m*	connexion *f* circulaire de couches minces	тонкопленочная схема

T 69	**thin-film degassing**	Dünnschichtentgasung f	émission f de gaz de filets minces	тонкопленочное обезгаживание
T 70	**thin-film degassing column**	Dünnschichtentgasungskolonne f	colonne f de dégazage de couches minces	колонка для тонкопленочного обезгаживания
T 71	**thin-film distillation,** thin-layer distillation	Dünnschichtdestillation f	distillation f pelliculaire	тонкопленочная дистилляция
T 72	**thin-film dryer**	Dünnschichttrockner m	sécheur m à couche fine, installation f de séchage à couche fine	пленочная сушилка, сушилка для высушивания в тонком слое
	thin-film evaporator, falling-film evaporator	Dünnschichtverdampfer m	vaporisateur m à couche fine	испаритель с падающей пленкой
T 73	**thin-film light filter**	Dünnschichtlichtfilter n	filtration f de la lumière par couches minces	тонкопленочный светофильтр
T 74	**thin film lubrication**	Grenzschmierung f	graissage m limite, lubrification f limite	смазка тонким слоем
T 75	**thin-film memory element**	Dünnschichtspeicherelement n	élément m de mémoire de couches minces	тонкопленочный запоминающий элемент, тонкопленочный элемент памяти
T 76	**thin-film monolithic circuit**	integrierte Dünnschichtschaltung f	circuit m intégré des couches minces	тонкопленочная интегральная схема
T 77	**thin-film physics**	Physik f dünner Schichten	physique f de couches minces	физика тонких пленок
T 78	**thin-film resistance thermometer**	Dünnschichtwiderstandsthermometer n	thermomètre m à résistance à couches minces	тонкопленочный термометр сопротивления
T 79	**thin-film short-path distillation**	Dünnschicht-Kurzwegdestillation f	distillation f pelliculaire à court trajet	тонкопленочный дистиллятор с коротким путем
T 80	**thin-film technology**	Dünnschichttechnik f	technique f des couches minces	технология тонких пленок
T 81	**thin-film thermocouple**	Dünnschichtthermoelement n	élément m thermoélectrique à couches minces, thermocouple m à couches minces	тонкопленочная термопара
	thin-layer distillation, thin-film distillation	Dünnschichtdestillation f	distillation f pelliculaire	тонкопленочная дистилляция
T 82	**thinly rolled sheet of fine silver**	dünngewalztes Feinsilberblech n	feuille f mince d'argent fin laminé	тонко прокатанный лист чистого серебра
T 83	**thin-wall leak**	Leck n in dünner Wand	fuite f de paroi mince	течь в тонкой стенке
T 84	**thoriated filament**	thorierte Katode f	cathode f thoriée	торированный катод
T 85	**threaded sleeve**	Rohrverschraubung f	raccord m à vis des tuyaux, tuyau m raccordé par vis	муфта с резьбой, резьбовое соединение
T 86	**thread reducer**	Reduzierstück n ‹mit beiderseitigem Gewinde›, Gewindereduzierstück n	raccord m de réduction	редуктор с двухсторонней резьбой, нарезной редукторный переход
T 87	**three-body collision**	Dreierstoß m	choc m triple	тройное соударение
T 88	**three-electrode sputter-ion pump,** triode type pump, triode ion pump	Triodenionengetterpumpe f, Ionenzerstäuberpumpe f vom Triodentyp	pompe f ionique à getter du type Penning	триодный магнитоэлектроразрядный насос
T 89	**three-way cock,** three-way valve	Dreiwegehahn m	robinet m à trois voies, robinet tournant à trois lumières	трехходовой кран
T 90	**threshold energy**	Schwellenenergie f	énergie f de seuil	пороговая энергия
T 91	**threshold region**	Schwellenbereich m	région f de seuil	пороговая область
	throat, throat area, intake area, inlet area	Ansaugquerschnitt m	section f d'aspiration	площадь сечения горловины сопла
	throat of diffuser, diffuser throat	Diffusorhals m ‹engste Stelle einer Venturi-Düse›	col m de diffuseur	горловина диффузора
	throat of nozzle, nozzle throat	kleinste Düsenspaltfläche f, Düsenhals m	buse f (col m) de diffuseur	горловина сопла
T 92	**throat of vapour pump**	Pumpenhals m	face f de fente de diffusion	горловина диффузионного насоса
	throttle valve, butterfly valve, quarter swing valve	Drosselventil n, Drosselklappe f	papillon m d'obturation, robinet m d'étranglement	дроссельный вентиль
T 93	**throttling**	Drosselung f	étranglement m	дросселирование
T 94	**through-hole coating**	Bedampfung f durch Löcher	métallisation f par trou	напыление сквозь отверстие
	throughput, exhausting power, suction capacity	Saugleistung f	puissance f d'une pompe, puissance d'admission, débit m massique d'une pompe, rendement m d'aspiration	производительность по всасыванию
	throughput, load, capacity	Förderleistung f, Saugleistung f, Fördermenge f, Durchsatz m, Durchsatzmenge f	flux m massique, débit m massique	скорость откачку
T 95	**through-type furnace**	Durchlaufofen m	four m continu	проходная печь
	through-way valve, straight-way valve, in-line valve, straight-through valve	Durchgangsventil n	soupape f droite, vanne f à passage droit, soupape de traversée	проходной вентиль

T 96	**thrust**	Schub *m*	poussée *f*	сдвиг, скалывание, срез. скольжение, тяга
	thrust leadthrough, linear motion drive, linear motion feedthrough	Schiebedurchführung *f*	traversée *f* entrée, douille *f* glissante	ввод для передачи линейного перемещения
	tie rod, centre rod	Verbindungsstab *m*, Zentralstab *m*	bâton *m* de connexion, bâton central	стержень для крепления и центровки
T 97	**tie wire**	Bindedraht *m*	fil *m* à lier, fil d'attache	проволока для связки
	tighten, seal, seal off, tip off	abdichten, dichten, abschmelzen	étancher, calfeutrer, boucher, rendre étanche, assurer l'étanchéité, séparer par fusion	уплотнять, герметизировать, заделывать, изолировать, отпаивать
T 98	**tightening torque**	Anzugsdrehmoment *n*	couple *m* moteur de démarrage	затягивающий момент
	tightness control, leak proving, leak testing	Dichtigkeitsprüfung *f*	test *m* (épreuve *f*, essai *m*) d'étanchéité	испытание (проверка) на герметичность
T 99	**tiltable crucible**	kippbarer Schmelztiegel *m*, Kipptiegel *m*	creuset *m* basculable (basculant)	опрокидывающийся тигель
T 100	**tiltable vacuum bell jar**	kippbare Vakuumglocke *f*	cloche *f* basculable à vide	опрокидывающаяся вакуумная камера
T 101	**tilted-cylinder dryer,** tumbler, tumbling dryer	Taumeltrockner *m*	sécheur *m* chancelant	сушилка с качающимся цилиндром, сушилка типа «пьяной бочки»
T 102	**tilted-cylinder mixer**	schrägstehender Trommelmischer *m*	tambour *m* sécheur en position inclinée	смеситель с наклонным барабаном
T 103	**tilting-crucible vacuum furnace**	Vakuumschmelzofen *m* mit kippbarem Schmelztiegel	four *m* de fusion à vide à basculant creuset fusible	вакуумная печь с опрокидывающимся тиглем
T 104	**tilting furnace**	Kippkesselofen *m*, Schaukelofen *m*	four *m* basculant	опрокидывающаяся печь
T 105	**tilting vacuum furnace**	Kippkessel-Vakuumschmelzofen *m*	four *m* basculant de fusion à vide	поворотная печь для вакуумной плавки
T 106	**time delay**	Zeitverzögerung *f*	retard *m* de temps	временная задержка
T 107	**time delay relay**	Verzögerungsrelais *n*	relais *m* temporisé (à retardement)	реле времени
	time lag, deceleration, delay, retardation	Verzögerung *f*	retard *m*, retardation *f*, ralentissement *m*, temporisation *f*	задержка, запаздывание
	time of evacuation, time of exhaust, pump-down time	Pumpzeit *f*, Evakuierungszeit *f*, Auspumpzeit *f*	durée *f* d'évacuation, temps *m* de pompage	время откачки ‹до заданного давления›
T 108	**time-of-flight mass spectrometer**	Laufzeitmassenspektrometer *n*	spectromètre *m* de masse de temps de propagation	времяпролетный масс-спектрометр, хронотрон
T 109	**time-of-flight method**	Flugzeitmethode *f*	méthode *f* de temps de vol	времяпролетный метод
	time of response, response time	Ansprechzeit *f*	durée *f* de réponse, temps *m* de réaction (réponse)	время срабатывания (запаздывания)
T 110	**timer**	Zeitmesser *m*, Zeitschalter *m*, Zeitgeber *m*	compteur (mesureur) *m* de temps, interrupteur *m* à minuterie	таймер, отметчик времени
T 111	**time switch**	Zeitschalter *m*	interrupteur *m* à minuterie, commutateur *m* à temps	механизм, программный выключатель
	timing, setting	Einstellung *f*	réglage *m*, ajustage *m*	регулировка, установка
	tin foil coating, layer of tin foil	Stanniolbelag *m*	couche *f* de feuilles d'étain	станиолевое покрытие
T 112	**tin-fusion gas analysis**	Gasanalyse *f* bei der Zinnschmelze, Heißextraktionsanalyse *f* mit Zinn als Badmetall	analyse *f* d'extraction chaude à l'étain	газовый анализ методом экстракции из расплава
	tip off, seal off, seal, tighten	abschmelzen, abdichten, dichten	séparer par fusion, étancher, calfeutrer, boucher, rendre étanche, assurer l'étanchéité	отпаивать, уплотнять, герметизировать, заделывать, изолировать
T 113	**tip-off,** tipping	Abschmelzen *n*, Abklemmen *n* ‹Pumpenstengel bei Elektronenröhren›	séparation *f* par fusion	отпаивание
T 114	**tissue freeze drying**	Gewebegefriertrocknung *f*	lyophilisation *f* de tissu, dessiccation *f* par congélation de tissu	сушка замораживанием тканей ‹клеток›
T 114 a	**titanium booster, titanium evaporator**	Titanverdampfer *m*	évaporateur *m* de titane	испаритель титана
T 115	**titanium evaporator pump**	Titanverdampferpumpe *f*	pompe *f* à évaporation de titane, pompe getter à évaporation de titane, pompe à getter par évaporation de titane	титановый испарительный насос
T 116	**titanium hydride sealing method**	Titanhydridverfahren *n*	méthode *f* à l'hydrure de titane, technique *f* à l'hydrure de titane, procédé *m* à l'hydrure de titane	титано-гидридная пайка ‹технология›
T 117	**titanium ionization pump, titanium ion pump**	Titanionenpumpe *f*	pompe *f* ionique de titane	титановый геттерно-ионный насос
T 118	**titanium sponge**	Titanschwamm *m*	éponge *f* de titane	губчатый титан

T 119	**titanium sublimation pump**	Titansublimationspumpe *f*	pompe *f* de sublimation de titane	титановый сублимационный насос
T 120	**titrate**	abtitrieren	titrer	титровать
T 121	**Toepler pump**	Toepler-Pumpe *f*	pompe *f* de Toepler	насос Теплера
	tolerable forepressure, maximum forepressure, maximum backing pressure, forepressure tolerance	„zulässiger" Vorvakuumdruck *m* ‹der einem um 10% größeren Ansaugdruck als bei normalem Vorvakuum entspricht›	pression *f* tolérée du vide préliminaire	максимально-допустимое выпускное давление ‹при превышении которого впускное давление возрастает более, чем на 10%›
T 122	**tombac**	Tombak *m* ‹Legierung›	tombac *m*	томпак
T 123	**tombac**	Metallbalg *m* aus Tombak	tombac *m*	сильфон из томпака
	top cap	s. top jet cap		
	top coat, top layer, overcoat	Deckschicht *f*	couche *f* de recouvrement	наружный слой
	top jet cap, jet cap, top cap	Düsenhut *m*	baffle *m* à chapeau, baffle-chapeau *m*	колпачок сопла
	top layer, top coat, overcoat	Deckschicht *f*	couche *f* de recouvrement	наружный слой
T 124	**torch brazing**	Hartlöten *n* mittels Flamme, Flammenlöten *n*	brasage *m* par flamme	газовая пайка
T 125	**torching**	Entgasen *n* einer Vakuumanlage während des Pumpprozesses mittels Gasflamme, Ausheizen *n* mittels Gasflamme, Flammen *n*	flambage *m*, chauffage *m* avec un brûleur à gaz	обезгаживание во время откачки с помощью газовой горелки
T 126	**torch soldering**	Weichlöten *n* mittels Flamme	soudure *f* à l'étain par flamme	пайка горелкой
T 127	**torque**	Drehmoment *n*	moment (couple) *m* de rotation	вращательный момент, крутящий момент
T 128	**torque wrench**	Drehmomentenschlüssel *m*	clef *f* dynamométrique	динамометрический ключ
T 129	**torr**	Torr *n*	torr *m*	торр
T 130	**Torricellian vacuum**	Torricelli-Vakuum *n*	vide *m* de Torricelli	вакуум Торричелли
T 131	**torsional strength**	Torsionsfestigkeit *f*	résistance *f* à la torsion	сопротивление при кручении, предел прочности при кручении
T 132	**torsional vibration vacuum gauge**	Drehschwingungsvakuummeter *n*	vacuomètre *m* à vibration rotative	динамический манометр с крутильными колебаниями
T 133	**torsion-microbalance**	Torsionsmikrowaage *f*	microbalance *f* de torsion	торсионные микровесы
T 134	**total emission coefficient, total emissive power, total emissivity**	Gesamtemissionsvermögen *n*	émissivité *f* totale, émission *f* totale, pouvoir *m* d'émission totale	полная излучательная способность, полная лучеиспускательная способность
	total extension, fracture strain, breaking elongation	Bruchdehnung *f*	dilatation *f* (allongement *m*) de rupture	растяжение (удлинение) при разрыве
T 135	**total flux**	Gesamtfluß *m*	flux *m* total	полный (интегральный) поток
	total heat, enthalpy, heat content	Enthalpie *f*	enthalpie *f*	энтальпия, теплосодержание
	total leakage, total leaks, integral leakage	Gesamtundichtigkeit *f*, Leckrate *f*	fuite *f* totale	общее натекание
T 136	**total pressure**	Gesamtdruck *m*	pression *f* totale	полное давление
T 137	**total pressure gauge**	Totaldruckmesser *m*	manomètre *m* total	манометр для измерения полного давления
T 138	**total radiation pyrometer**	Gesamtstrahlungspyrometer *n*	pyromètre *m* à radiation totale	пирометр суммарного излучения
T 139	**total speed**	Gesamtsaugvermögen *n*	débit *m* total d'une pompe, débit global	полная быстрота разрежения, полная скорость откачки
T 140	**town gas**	Stadtgas *n*, Leuchtgas *n*	gaz *m* de ville, gaz d'éclairage	светильный газ
T 141	**T-piece**	T-Stück *n*	pièce *f* en T	тройник
T 142	**tracer**	Indikator *m*, Leitisotop *n*	indicateur *m*	изотопный индикатор, меченый атом
	tracer gas, probe gas, search gas, test gas	Prüfgas *n*, Testgas *n*	gaz *m* traceur (témoin, d'épreuve), gaz-sonde *m*	пробный газ
	tracking, surface leakage	Kriechwegbildung *f*	formation *f* d'un courant vagabond	образование следа от утечки
T 143	**train of pumping units**	Pumpstraße *f* mit Pumpwagen	train *m* de pompage ‹comportant des organes de pompage mobiles›	«железная дорога»
T 144	**transformation temperature,** Tr.T.	Transformationspunkt *m*, Tr. P.	point *m* de transformation	точка превращения (перехода)
T 145	**transition flange**	Übergangsflansch *m*	bride *f* transitoire	переходный фланец
	transition flow	s. Knudsen flow		
T 146	**transition fusing glass**	Übergangsglas *n*	verre *m* de transition	переходное стекло
T 147	**transition metal**	Übergangsmetall *n*	métal *m* transitoire (de transition)	переходный металл
	transition piece, adapter, intermediate piece	Zwischenstück *n*	raccord *m* (pièce *f*) intermédiaire, adapteur *m*	промежуточный элемент, переходник

T 148	**transition range (region)**	Übergangsbereich *m*	étendue (zone) *f* transitoire	переходная область
T 149	**transition temperature**	Umwandlungstemperatur *f*	température *f* de transformation	температура превращения
T 150	**transit time**	Laufzeit *f*	durée *f* de fonctionnement (parcours), temps *m* de transit	время пролета
T 151	**translational degree of freedom**	Translationsfreiheitsgrad *m*	degré *m* de liberté de la translation	поступательная степень свободы
T 152	**translational energy**	Translationsenergie *f*	énergie *f* de translation	энергия поступательного движения
T 153	**translucent,** transparent	optisch dünn	translucide	просвечивающий, прозрачный для излучения
T 154	**transmission**	Transmission *f*	transmission *f*	передача
T 155	**transmission coefficient**	Transmissionskoeffizient *m*	coefficient *m* de transmission	коэффициент пропускания
T 156	**transmission electron diffraction**	Elektronenbeugung *f* in Transmission, Elektronenbeugung im Durchstrahlungsverfahren	diffraction *f* des électrons par transmission	дифракция электронов в проходящем пучке
T 157	**transmission high energy electron diffraction,** THEED	Beugung *f* schneller Elektronen in Durchstrahlung	diffraction *f* des électrons à grande vitesse	дифракция быстрых электронов в проходящем пучке
T 158	**transmission probability**	Übergangswahrscheinlichkeit *f*	probabilité *f* de changement (passage)	вероятность перехода
	transmission probability, Clausing's correction factor, Clausing's factor	Clausing-Faktor *m*	facteur *m* de Clausing	поправочный коэффициент Клаузинга
T 159	**transonic flow in pumps**	Überschallströmung *f* in Pumpen	écoulement *m* ultrasonique dans pompes	сверхзвуковое течение в насосах
T 160	**transparency**	Lichtdurchlässigkeit *f*	transparence *f*, translucidité *f*	прозрачность
	transparent, translucent	optisch dünn	translucide	просвечивающий, прозрачный для излучения
T 161	**transpiration**	Transpiration *f*	transpiration *f*	транспирация
T 162	**transport efficiency**	Transportwirkungsgrad *m*	rendement *m* de transport	эффективность переноса
T 163	**transverse electron gun, transverse gun**	Querfeld-Elektronenstrahler *m*, Querfeld-Katode *f*, Flachstrahlkanone *f*	cathode *f* transversale, canon *m* transversal à électrons	электронная пушка с наклонным пучком, электронная пушка с поворотом пучка
	transverse strength, flexural strength, maximum surface stress in bend, modulus of rupture	Biegefestigkeit *f*	résistance *f* à la flexion	предел прочности при изгибе, прочность при изгибе
T 164	**trap**	Falle *f*, Kühlfalle *f*	trappe *f*, piège *m*	ловушка
T 165	**trapezoidal groove**	Trapeznut *f*	encoche *f* trapézoïdale	трапецеидальный паз
T 166	**trapped gauge**	Vakuummeterröhre *f* mit vorgeschalteter Kühlfalle	tube *m* de manomètre	защищенный манометр, манометр с охлаждаемой ловушкой
	trapping, freezing	Ausfrieren *n*	congélation *f*, refroidissement *m*	вымораживание
	trapping coefficient (efficiency), sticking probability, capture coefficient, capture probability	Einfangkoeffizient *m*, Haftwahrscheinlichkeit *f*	probabilité *f* d'adhérence, probabilité d'adhésion, probabilité du pouvoir adhérent, coefficient *m* d'arrêt, coefficient de blocage, probabilité de fixation (collage)	коэффициент прилипания (захвата), вероятность захвата (прилипания)
T 167	**travelling-wave magnetron**	„Travelling-Wave"-Magnetron *n*	magnétron *m* à ondes progressives	магнетрон типа бегущей волны
T 168	**treatment of impregnant**	Imprägniermittelaufbereitung *f*	traitement *m* d'agent d'imprégnation	обработка импрегнирующим составом
T 169	**triatomic**	dreiatomig	triatomique	трехатомный
T 170	**trigger discharge gauge**	Trigger-Ionisationsvakuummeter *n*, getriggerte Entladungsmeßröhre *f*	tube *m* déclenché de mesure à décharge	электроразрядный манометр
T 171	**trimmability**	Justierbarkeit *f*	ajustabilité *f*	способность подстраиваться, подстраиваемость
	triode ion pump	s. three-electrode sputter-ion pump		
T 172	**triode sputtering**	Triodenzerstäubung *f*	pulvérisation *f* de triode	триодное распыление
	triode type pump	s. three-electrode sputter-ion pump		
T 173	**triple point**	Tripelpunkt *m*	point *m* triple	тройная точка
T 174	**tripolar**	dreipolig	tripolaire, à trois pôles	трехполюсный
T 175	**trolley**	Pumpstraßenwagen *m*, fahrbarer Pumpstand *m* einer Pumpstraße	poste *m* roulant de pompage, station *f* roulante de pompage	передвижная откачная позиция
	trolley exhaust system <US>, stationary exhaust system	stationärer Pumpautomat *m*, Pumpstraße *f* mit stationären Pumpeinrichtungen	banc (carrousel) *m* de pompage stationnaire	откачной автомат

T 176	trouble-free	störungsfrei	sans trouble, libre de para-sites, sans dérangement, sans perturbation	безаварийный
	trough dryer, screw dryer	Muldentrockner m	sécheur m en auge	лотошная сушилка
	Tr.T.	s. transformation temperature		
T 177	truck	Hubwagen m	chariot m avec dispositif de levage	подъемная тележка
	true speed, operational speed, effective speed, net speed	wirksames Saugvermögen n	débit m utile, débit effectif	эффективная скорость откачки, быстрота раз-режения объекта
T 178	true temperature	wahre Temperatur f	température f véritable	истинная температура
	trunnion, stud, stud bolt	Stiftschraube f	goujon m, vis f à cheville	шпилька, палец, цапфа
T 179	tube basing cement	Röhrensockelkitt m	mastic m à culot de tube, ciment m à support de tube	мастика для цоколевки
	tube furnace, retort furnace	Retortenofen m	four m tubulaire (à cornue), alambic-four m	реторная печь
	tubing, duct, line, pipe	Rohrleitung f	canalisation f, tuyau m, conduite f	трубопровод
	tubular cooler, coil condenser	Röhrenkühler m	radiateur (réfrigérant) m tubulaire	змеевиковый конденса-тор (охладитель)
T 180	tubular heating element	Heizleiterrohr n	élément m tubulaire de chauffage	трубчатый нагреватель-ный элемент
T 181	tubular seal	Rohranschmelzung f, Endanglasung f	fondage m de tuyau	трубчатый спай
T 182	tubulation <of a tube>	Röhrenpumpstengel m, Verbindungsleitung f, Rohransatz m	conduite f de raccorde-ment, tuyau m de raccordement, raccord m tubulaire	штенгель, соединитель-ный трубопровод
T 183	tumble drying, tumbling freeze drying	Taumelgefriertrocknung f	lyophilisation f chancelante	сушка вымораживанием в качающейся сушил-ке
	tumbler, tumbling dryer, tilted-cylinder dryer	Taumeltrockner m	sécheur m chancelant	сушилка с качающимся цилиндром, сушилка типа «пьяной бочки»
	tumbling freeze drying	s. tumble drying		
T 184	tungsten, wolfram	Wolfram n	tungstène m, wolfram m	вольфрам
T 185	tungsten coil	Wolframwendel f	hélice f (filament m) de tungstène	вольфрамовая спираль
T 186	tungsten disk	Wolframscheibe f, Wolframpastille f	disque m de tungstène	вольфрамовый диск
T 187	tungsten inert-gas arc welding	Wolfram-Inertgas-Schweißen n, WIG-Schweißen n	soudage m avec tungstène et gaz inerte	дуговая сварка вольфра-мовым электродом в среде инертного газа
T 188	tungsten tip	Wolframfinger m	bout m tungstique (de tungstène)	вольфрамовый наконечник
	tuning scale, degree scale	Einstellskale f	échelle f graduée, cadran m de réglage	шкала настройки, гра-дуированная шкала
T 189	tunnel action	Tunnelvorgang m	action f tunnel	туннельное действие
T 190	tunnel dryer	Durchlauftrockner m, Kammertrockner m, Kanaltrockner m	sécheur m continu, sécheur m tunnel	туннельная сушилка
T 191	tunnel effect	Tunneleffekt m	tunnel effet m, effet m tunnel	туннельный эффект
	tunnel freezer, freezing tunnel	Gefriertunnel m	tunnel m de congélation	туннельный морозиль-ный аппарат, моро-зильный туннель
	tunnel furnace, continuous pusher-type furnace	Tunnelofen m, Durch-satzofen m	four m tunnel	туннельная печь, печь с проталкиванием деталей
T 192	tunnelling	Durchtunnelung f	tunnel m	прохождение туннеля
T 193	tunnel oven	Durchlaufofen m, Tunnelofen m	four m continu, four à tunnel	печь непрерывного действия, туннельная печь
T 194	turbocompressor	Turboverdichter m	turbocompresseur m, turbosurpresseur m	турбокомпрессор
T 195	turbodryer	Turbinentrockner m, Ringetagentrockner m	turboisécheur m	турбосушилка
T 196	turbo-molecular pump	Turbomolekularpumpe f	pompe f turbomolé-culaire	турбомолекулярный насос
	turbo-pump, non-positive displacement pump	Turbopumpe f	turbopompe f, pompe f mécanique non volu-métrique	турбонасос
T 197	turbulent flow	turbulente Strömung f	écoulement (courant) m turbulent	турбулентное течение
T 198	turbulent layer dryer	Wirbelschichttrockner m	sécheur m par couches tourbillonnaires	сушилка с псевдоожи-женным слоем
T 199	turnable compression manometer	drehbares Kompressions-vakuummeter n	manomètre m rotatif à compression	поворотный компрес-сионный манометр
T 200	turntable	Drehtisch m	table f pivotante	поворотный стол
T 201	turret	Revolverkopf m	tête-révolver f, tourelle f révolver	револьверная головка
T 202	tweezers	Pinzette f, Federzange f	pincette f	пинцет
T 203	twin crystal	Zwillingskristall m	macle m	двойниковый кристалл

T 204	**twinning**	Zwillingsausbildung f	structure f double	двойная (дублированная) конструкция
T 205	**twisted**	verdrillt	torsadé	витой, скрученный
T 206	**twist welding**	Rührschweißen n	soudage m d'agitation	сварка с перемешиванием
T 207	**two-chamber ionization gauge**	Zweikammerionisationsvakuummeter n	vacuomètre m d'ionisation à deux chambres	двухкамерный ионизационный манометр
T 208	**two-chamber plant**	Zweikammeranlage f	installation f à deux salles	двухкамерная установка
	two-electrode sputter-ion pump, diode sputter-ion pump, diode type pump	Ionenzerstäuberpumpe f vom Diodentyp	pompe f ionique à getter par pulvérisation cathodique	диодный магнито-электроразрядный насос
T 209	**two-gauge calibrated conductance method**	Leitwertmethode f <zur Bestimmung des Saugvermögens>	méthode f de conductance	метод определения скорости откачки по пропускной способности
	two-gauge method, calibrated conductance method	Leitwertmethode f	méthode f de conductance	метод определения быстроты действия насоса по известной пропускной способности, метод двух манометров
T 210	**two-roll extruder**	Zweiwalzenextruder m	extrudeuse f à deux cylindres	двухвальцовая шприцмашина
T 211	**two-way stopcock**	Zweiwegehahn m	robinet m à deux voies	двухходовой кран

U

	uhv, UHV	s. ultrahigh vacuum		
U 1	**ultimate density**	Enddichte f	densité f finale	предельная плотность
	ultimate pressure, ultimate vacuum, blank-off (limiting, final) pressure	Enddruck m	pression f finale	предельный вакуум, предельное (остаточное) давление
U 2	**ultracentrifuge**	Ultrazentrifuge f	ultracentrifugateur m	ультрацентрифуга
U 3	**ultrahigh-strength steel**	ultrafester Stahl m	acier m ultradur	сверхтвёрдая сталь
U 4	**ultrahigh vacuum,** uhv, UHV	Ultrahochvakuum n, UHV	ultravide m, UHV, vide m ultrapoussé	сверхвысокий вакуум
U 5	**ultrahigh vacuum flange**	Ultrahochvakuumflansch m	bride f à ultravide	сверхвысоковакуумный фланец
U 6	**ultrahigh vacuum oil diffusion pump**	Ultrahochvakuum-Öldiffusionspumpe f	pompe f à diffusion d'huile à ultravide	сверхвысоковакуумный масляный диффузионный насос
U 7	**ultrahigh vacuum system**	Ultrahochvakuumanlage f	système m à ultravide, appareil m à ultravide	сверхвысоковакуумная система
U 8	**ultrahigh vacuum valve**	Ultrahochvakuumventil n	soupape f à ultravide	сверхвысоковакуумный вентиль
U 9	**ultra-low temperature**	ultratiefe Temperatur f	température f ultrabasse	сверхнизкая температура
U 10	**ultrasonic cleaning**	Ultraschallreinigung f	purification f ultrasonore (supersonique)	ультразвуковая очистка, очистка ультразвуком
U 11	**ultrasonic erosion**	Ultraschallerosion f	érosion f supersonique (ultrasonore)	ультразвуковая эрозия
U 12	**ultrasonic soldering**	Ultraschallöten n	soudage m ultrasonore	пайка ультразвуком
	umbrella jet, inverted jet nozzle, annular jet nozzle	ringförmige Strahlumlenkdüse f	diffuseur m, diffuseurdéflecteur m, tuyère f à jet inversé	зонтичная ступень, зонтичное сопло, обращенное сопло Лаваля
U 13	**unbaffled speed**	Saugvermögen n ohne Baffle	débit m d'une pompe sans baffle	скорость откачки в отсутствии ловушки
U 14	**unballasted pumping speed**	Saugvermögen n ohne Gasballast	débit m d'une pompe sans lest d'air	быстрота действия насоса без подачи балласта
	undercoat, base coat of lacquer	Grundlackschicht f	couche f de laque d'accrochage, couche de fond de vernis	лаковая основа
U 15	**under pressure**	spannungsführend	sous tension	под напряжением
U 16	**uniflow furnace**	Einwegofen m	four m à une seule direction	прямоточная печь
U 17	**uniformity of density**	Dichtehomogenität f	homogénéité f de densité	однородность по плотности
	unit-assembly principle, building block principle	Baukastenprinzip n	principe m des unités de montage pour l'agencement rapide <des appareillages à vide>, principe de boîte de construction	принцип сборки из готовых узлов
U 18	**unitized column**	Bauteilkolonne f	colonne f de sous-ensemble	составная колонка
U 19	**unit process**	Grundverfahren n	procédé m fondamental, technique f fondamentale	основной метод обработки
	universal gas constant, ideal gas constant	universelle Gaskonstante f	constante f universelle des gaz	универсальная газовая постоянная

	English	German	French	Russian
	universal gas constant related to one molecule, Boltzmann constant	Boltzmann-Konstante f	constante f universelle des gaz, constante de Boltzmann	постоянная Больцмана
U 20	**unlocking**	Entriegelung f	déverrouillage m, déblocage m	размыкание, деблокирование
	unobstructed path still, open path still, short path still	Kurzwegdestillations-anlage f	distillerie f à court trajet, installation f de distilla-tion à court trajet	установка для моле-кулярной перегонки
U 21	**unsaturated vapour**	ungesättigter Dampf m	vapeur f non saturée	ненасыщенный пар
U 22	**unsolder**	ablöten	dessouder	распаивать
U 23	**unsteady state flow**	instationäre Strömung f	courant m instationnaire	неустановившееся течение
U 24	**up-evaporation,** vertical evaporation	Aufwärtsverdampfung f	vaporisation f ascendante	вертикальное испарение, испарение вверх
U 25	**upstream**	stromauf	courant montant	вверх по течению
U 26	**uranium**	Uran n	uranium m	уран
U 27	**useful refrigerating effect**	Nutzkälteleistung f	effet (débit) m utile du froid	полезная хладопроиз-водительность
U 28	**U-shaped clamp**	U-Schellenverbindung f	assemblage (raccorde-ment) m par collier en U	соединение с U-образ-ным хомутом
	U-shaped manom-eter, U-tube manometer, mercury level manometer	U-Rohr-Manometer n	manomètre m en U tronqué	U-образный манометр
U 29	**U-type mercury manometer**	Quecksilber-U-Rohr-Manometer n	manomètre m à mercure en U	U-образный ртутный манометр

V

	English	German	French	Russian
V 1	**vacancy**	Leerstelle f	lacune f	вакансия
V 2	**vacancy flow**	Leerstellenfluß m	flux m des lacunes	поток дырок
V 3	**vacuo / in**	im Vakuum	dans le vide, sous vide, à vide	в вакууме
V 4	**vacustat**	Vakustat m	vacustat m	вакустат
V 5	**vacuum**	Vakuum n	vide m	вакуум
V 6	**vacuum accessories**	Vakuumzubehör n	accessoires mpl pour le vide, utilisés mpl en technique du vide	вакуумные принадлеж-ности
V 7	**vacuum annealing**	Blankglühen n im Vakuum, Vakuum-glühen n	recuit m brillant sous vide	вакуумный отжиг
V 8	**vacuum annealing furnace,** vacuum heat treatment furnace	Vakuumglühofen m	four m à recuire à vide	печь вакуумного отжига, печь для вакуумного отжига (термообра-ботки)
V 9	**vacuum arc**	Vakuumlichtbogen m	arc m électrique de vide	дуга в вакууме, вакуумная дуга
V 10	**vacuum-arc deposition**	Vakuumbogenbedamp-fung f	métallisation f à l'arc sous vide, traitement m par métallisation à l'arc sous vide	напыление в вакуумной дуге
V 11	**vacuum-arc furnace**	Vakuumlichtbogenofen m	four m à arc sous (à) vide	вакуумная дуговая печь
V 12	**vacuum-arc melting**	Vakuumlichtbogen-schmelzen n	fusion f à l'arc sous vide	вакуумная дуговая плавка
	vacuum bag moulding, deflatable bag moulding	Vakuumgummisack-verfahren n	procédé m par le sac en caoutchouc	вакуумная формовка в резиновую форму
V 13	**vacuum balance**	Vakuumwaage f	balance f à vide	вакуумные весы
V 14	**vacuum bell jar**	Vakuumglocke f	cloche f à vide	вакуумный колпак
V 15	**vacuum brazing**	Hartlöten n im Vakuum	brasage m sous vide	вакуумная пайка твердым припоем
V 16	**vacuum brazing furnace**	Vakuumlötofen m	four m de brasure à vide	вакуумная печь для пайки
V 17	**vacuum breakdown**	elektrischer Überschlag m im Vakuum	claquage (éclat, flash) m sous vide	электрический пробой в вакууме, вакуум-ный пробой
V 18	**vacuum cabinet**	Vakuumschrank m	armoire f à vide	вакуумный шкаф
V 19	**vacuum calcination**	Vakuumkalzination f	calcination f sous vide	вакуумный обжиг, вакуумное кальцини-рование
	vacuum casting, suction casting	Vakuumgießen n	coulée f sous vide, coulage m sous vide	вакуумное литье, литье в вакууме
V 20	**vacuum casting plant,** vacuum moulding plant	Vakuumgießanlage f	installation f (poste m, appareillage m) de cou-lée à vide	установка для литья под вакуумом
V 21	**vacuum cement,** vacuum putty	Vakuumkitt m	mastic m pour le vide, ciment m à vide	вакуумный цемент
V 22	**vacuum chamber**	Vakuumkammer f	chambre f à vide	вакуумная камера
V 23	**vacuum chromatog-raphy**	Vakuumchromatografie f	chromatographie f sous vide	вакуумная хромато-графия
V 24	**vacuum circulating evaporator**	Vakuumumlaufver-dampfer m	évaporateur m de circulation à vide	циркуляционный вакуумный испаритель
V 25	**vacuum cleaner**	Staubsauger m	aspirateur m de poussière	пылесос
V 26	**vacuum coater,** vacuum coating equipment (plant)	Vakuumaufdampfanlage f, Vakuum-bedampfungsanlage f	installation f d'évapora-tion (de métallisation) sous vide	установка вакуумного напыления

	vacuum coating, evaporative coating, evaporation coating	Aufdampfen *n*, Vakuum-bedampfung *f*	métallisation *f* sous vide	вакуумное напыление, нанесение в вакууме
	vacuum coating equipment (plant)	*s.* vacuum coater		
V 27	**vacuum coil**	Wendel *f* für Vakuum-lampen	spirale *f* pour lampes à vide	спираль для вакуумной лампы
V 28	**vacuum cold-wall furnace**	Vakuumkaltwandofen *m*	four *m* de paroi froide à vide	вакуумная печь с холод-ными стенками
V 29	**vacuum concentration**	Vakuumkonzentration *f*	concentration *f* du vide	вакуумная концентрация, вакуумное обогащение
V 30	**vacuum concentrator**	Vakuumkonzentrations-anlage *f*	installation *f* de con-centration à vide	вакуумный концентра-тор, вакуумный обогатитель
V 31	**vacuum concrete**	Vakuumbeton *m*	béton *m* de vide	вакуумированный бетон
V 32	**vacuum condensation**	Vakuumkondensation *f*	condensation *f* sous vide	вакуумная конденсация
V 33	**vacuum condition**	Vakuumbedingung *f*	condition *f* sous vide	вакуумные условия
V 34	**vacuum container,** vacuum vessel, vacuum tank	Vakuumbehälter *m*	récipient *m* à vide, enceinte *f* à vide	вакуумная камера, вакуумный сосуд, резервуар
	vacuum cooking appliance, cold boiler	Vakuumkochapparat *m*, Vakuumkocher *m*	marmite *f* à vide	вакуумный кипятиль-ник (бойлер)
V 35	**vacuum cooling**	Vakuumkühlung *f*	refroidissement *m* sous vide	вакуумное охлаждение
V 36	**vacuum cryostat**	Vakuumkryostat *m*	cryostat *m* de vide	вакуумный криостат
V 37	**vacuum crystallization**	Vakuumkristallisieren *n*	cristallisation *f* sous vide	вакуумная кристаллиза-ция
V 38	**vacuum cup stopcock,** vacuum stopcock	Vakuumhahn *m*, Hahn *m* mit offenem Küken	robinet *m* à vide	вакуумный кран
V 39	**vacuum decarboniza-tion (decarburiza-tion)**	Entkohlung *f* im Vakuum	décarburation (décar-bonisation) *f* sous vide	вакуумное обезуглерожи-вание
	vacuum deep drawing, deep drawn vacuum forming	Tiefziehansaugen *n*	pompage *m* d'embou-tissage	вакуумная формовка с глубокой вытяжкой
V 40	**vacuum degassing furnace**	Vakuumentgasungsofen *m*	four *m* de dégazage sous vide	печь для вакуумного обезгаживания
V 41	**vacuum dehydration**	Vakuumentwässerung *f*	déshydratation *f* sous vide	вакуумное обезвоживание
V 42	**vacuum deposition**	Beschichtung *f* im Vakuum	revêtement (dépôt) *m* sous vide	вакуумное нанесение (покрытие)
V 43	**vacuum desiccator,** vacuum exsiccator	Vakuumexsikkator *m*	dessiccateur *m* à vide	вакуумный эксикатор
	vacuum device, vacuum sealed device, sealed-off vacuum system	abgeschlossenes Vakuumsystem *n*	installation *f* scellée de vide, système *m* séparé par fusion	отпаянный вакуумный прибор, отпаянная вакуумная система
V 44	**vacuum die-casting,** vacuum injection moulding	Vakuumspritzguß *m*	moulage *m* par injection sous vide, coulage *m* par injection sous vide	литье под давлением в вакууме
V 45	**vacuum diffusion welding**	Vakuumdiffusions-schweißen *n*	soudure *f* de diffusion sous vide	вакуумная диффузион-ная сварка
V 46	**vacuum dilatometer**	Vakuumdilatometer *n*	dilatomètre *m* à vide	вакуумный дилатометр
V 47	**vacuum distillation**	Vakuumdestillation *f*	distillation *f* sous (dans le) vide	вакуумная перегонка, вакуумная дистилляция
V 48	**vacuum distillation camera**	Vakuumdestillations-kamera *f*	appareil *m* photogra-phique de distillation sous vide	вакуумная дистилля-ционная камера
V 49	**vacuum distillator**	Vakuumdestillations-anlage *f*	appareil *m* distillatoire à vide, installation *f* de distillation sous vide	вакуумный перегонный аппарат, вакуумный дистиллятор
V 50	**vacuum drip melting**	Vakuumabtropf-schmelzen *n*	fusion *f* d'égouttage sous vide	вакуумная капельная плавка
V 51	**vacuum drum dryer**	Vakuumwalzentrockner *m*	sécheur *m* cylindrique à vide	вакуумная вальцовая сушилка
V 52	**vacuum drum filter**	Vakuumtrommelfilter *m*	filtre *m* à tambour à vide	барабанный вакуум-фильтр
V 53	**vacuum dryer**	Vakuumtrockner *m*	dessiccateur *m* à vide, appareil *m* de séchage à vide, sécheur *m* à vide	вакуумная сушилка
V 54	**vacuum drying**	Vakuumtrocknung *f*	séchage *m* à vide, dessicca-tion *f* sous vide	вакуумная сушка, сушка в вакууме
V 55	**vacuum drying cham-ber (cupboard),** vacuum shelf dryer	Vakuumtrockenschrank *m*	armoire *f* de séchage sous vide, séchoir *m* à vide, étuve *f* à vide	вакуумная полочная сушилка, вакуумный сушильный шкаф
V 56	**vacuum drying plant**	Vakuumtrocknungs-anlage *f*	installation *f* de dessicca-tion sous vide, installa-tion de séchage à vide	вакуум-сушилка
V 57	**vacuum eccentric tumbling dryer**	Vakuumtaumeltrockner *m*	sécheur *m* chancelant à vide	вакуумная качающаяся сушилка
V 58	**vacuum electron beam melting furnace**	Vakuumelektronenstrahl-schmelzanlage *f*	installation *f* de fusion à rayons électronique à vide	вакуумная электронно-лучевая плавильная печь
V 59	**vacuum encapsulation,** vacuum potting	Vergießen (Verkapseln) *n* unter Vakuum	coulée *f* sous vide	вакуумная инкапсуля-ция, заливка под вакуумом

	English	German	French	Russian
	vacuum etching, cathodic etching	katodisches Ätzen n, Glimmen n, Abglimmen n, Beglimmen n	gravure f cathodique	катодное травление
V 60	vacuum evaporation	Verdampfung f im Vakuum	vaporisation (évaporation) f sous vide	вакуумное испарение
	vacuum exsiccator	s. vacuum desiccator		
V 61	vacuum extraction	Vakuumextraktion f	extraction f sous vide	вакуумная экстракция
V 62	vacuum extrusion	Vakuumstrangpressen n	extrusion f sous vide	вакуумная экструзия
V 63	vacuum factor	Vakuumfaktor m	facteur (coefficient, degré) m du vide	вакуумфактор
V 64	vacuum filter	Nutsche f	entonnoir m filtre	вакуум-фильтр
V 65	vacuum filtration	Nutschen n ‹Filtrieren unter Vakuum›	filtration f sous vide, essorage m, filtrage m par succion	вакуумное фильтрование
V 66	vacuum firing, vacuum furnacing, vacuum heat treatment	Wärmebehandlung f im Vakuumofen	traitement m thermique (à chaud) au four à vide	термообработка в вакуумной печи
V 67	vacuum forming, vacuum moulding	Vakuumformverfahren n, Vakuumverformung f	déformation f sous vide	вакуумная формовка
V 68	vacuum freeze dryer	Vakuumgefriertrockner m	sécheur m de lyophilisation de vide	вакуумная сублимационная сушилка
V 69	vacuum freeze-drying equipment	Vakuumgefriertrocknungsanlage f	installation f de lyophilisation de vide	установка для вакуумной сублимационный сушки
V 70	vacuum freezing	Vakuumgefrieren n	congélation f sous vide	вакуумное замораживание
V 71	vacuum fumigation	Räuchern n im Vakuum	enfumage m sous vide	вакуумное обкуривание
V 72	vacuum furnace	Vakuumofen m	four m à vide	вакуумная печь
	vacuum furnacing, vacuum heat treatment, vacuum firing	Wärmebehandlung f im Vakuumofen	traitement m thermique (à chaud) au four à vide	термообработка в вакуумной печи
V 73	vacuum fusion apparatus	Heißextraktionsapparatur f	appareillage m d'extraction à chaude	установка для горячей экстракции
	vacuum fusion gas analysis, hot extraction gas analysis	Heißextraktionsanalyse f	analyse f d'extraction à chaude	экстракционный анализ газов путем вакуумного расплавления
	vacuum fusion gas extraction, hot extraction process	Heißextraktionsverfahren n	procédé m d'extraction à chaude	процесс горячей экстракции
	vacuum gate valve, gate valve, slide valve	Vakuumschieber m, Schieberventil n, Torventil n	vanne f à coulisse (tiroir), soupape f à passage direct à vide, robinet m [à tiroir]	[вакуумная] задвижка, заслонка
V 74	vacuum gauge	Vakuummeter n, Vakuummeßgerät n	manomètre m à vide, vacuomètre m	вакуумметр
V 75	vacuum gauge control box (circuit), vacuum gauge power supply, vacuum gauge power rack	Stromversorgungsgerät n für Vakuummeter	appareil m pour l'alimentation de courant d'un manomètre à vide	блок питания манометра, схема питания манометра
V 76	vacuum gauge head, vacuum gauge tube	Vakuummeßröhre f, Vakuummeßkopf m	burette f à vide	датчик вакуумметра
	vacuum gauge power rack (supply)	s. vacuum gauge control box		
	vacuum gauge tube, vacuum gauge head	Vakuummeßröhre f, Vakuummeßkopf m	burette f à vide	датчик вакуумметра
V 77	vacuum grease	Vakuumfett n	graisse f à vide	вакуумная смазка
V 78	vacuum heat treatment	Wärmebehandlung f im Vakuum	traitement m thermique sous vide	термическая обработка в вакууме
	vacuum heat treatment	s. vacuum furnacing		
	vacuum heat treatment furnace, vacuum annealing furnace	Vakuumglühofen m	four m à recuire à vide	печь для вакуумного отжига (термообработки), печь вакуумного отжига
V 79	vacuum-hot-press and quenching furnace	Vakuum-Heißpreß- und Abschreckofen m	four m à vide à pression chaude et trempe	вакуумная печь для горячей прессовки и закалки
V 80	vacuum hot-pressing	Heißpressen n im Vakuum	compression f à chaud sous vide	горячая прессовка в вакууме
V 81	vacuum hot-wall furnace	Vakuumheißwandofen m	four m de paroi surchauffée à vide	вакуумная печь с нагретыми стенками
V 82	vacuum-impregnated	vakuumgetränkt	imprégné sous vide	пропитанный под вакуумом, вакуумно импрегнированный
V 83	vacuum impregnation	Vakuumimprägnierung f	imprégnation f sous vide	вакуумная пропитка
V 84	vacuum indicator, vacuum tester	Vakuumprüfer m	indicateur m de vide, appareil m pour le contrôle de vide	индикатор вакуума
V 85	vacuum induction casting	Vakuuminduktionsgießen n	coulage m d'induction sous vide	вакуумное индукционное литье
V 86	vacuum induction furnace	Vakuuminduktionsofen m	four m à induction à vide	вакуумная индукционная печь
V 87	vacuum induction melting	Vakuuminduktionsschmelzen n	fusion f d'induction sous vide	вакуумная индукционная плавка
V 88	vacuum induction melting furnace	Vakuuminduktionsschmelzofen m	four m de fusion à induction sous vide	индукционная печь вакуумной плавки

V 89	vacuum ingot casting	Vakuumblockguß *m*	fonte *f* de lingots sous vide	вакуумная плавка слитка, вакуумная отливка
	vacuum injection moulding	*s.* vacuum die-casting		
V 90	vacuum insulation	Vakuumisolation *f*	isolement *m* à vide	вакуумная изоляция
V 91	vacuum interrupter	Vakuumunterbrecher *m*	interrupteur *m* à vide	вакуумный прерыватель
V 92	vacuum jacket	Vakuummantel *m*	enveloppe *f* à vide	вакуумная рубашка
V 93	vacuum-jacketed feed tube	Vakuummantelheber *m*	tube *m* d'ascension à double paroi à vide	питающее устройство <трубка> с вакуумной рубашкой
V 94	vacuum jet exhauster	Vakuumstrahlsauger *m*	aspirateur *m* du rayon sous vide	вакуумный струйный аспиратор
V 95	vacuum laminating	Vakuumkaschieren *n*, Kaschieren *n* eines Formkörpers unter Vakuum	laminage *m* sous vide	нанесение слоев под вакуумом
V 96	vacuum leakproofness, vacuum tightness	Vakuumdichtigkeit *f*	étanchéité *f* au vide	вакуумплотность
V 97	vacuum line, vacuum pipe, vacuum tubing	Vakuumleitung *f*	conduite *f* à vide	вакуумная коммуникация, вакуум-провод
V 98	vacuum lock	Vakuumschleuse *f*	écluse *f* (sas *m*) à vide	вакуумный шлюз
V 99	vacuum measurement, vacuum measuring	Vakuummessung *f*	mesure *f* du vide	измерение вакуума
V 100	vacuum measuring technique	Vakuummeßtechnik *f*	technique *f* de mesure du vide	техника вакуумных измерений
V 101	vacuum-melted metals	vakuumgeschmolzene Metalle *npl*	métaux *mpl* fondus sous vide	металлы вакуумной плавки
V 102	vacuum melting	Vakuumschmelzen *n*	fusion *f* sous vide	вакуумная плавка
V 103	vacuum melting furnace	Vakuumschmelzofen *m*	four *m* de fusion à vide	вакуумная печь для плавки
V 104	vacuum metallizing	Metallisierung *f* im Vakuum	métallisation *f* sous vide	вакуумная металлизация
V 105	vacuum metallizing plant	Vakuummetallisierungsanlage *f*	installation *f* de métallisation sous vide	установка для вакуумной металлизации
V 106	vacuum metallurgy	Vakuummetallurgie *f*	métallurgie *f* du vide	вакуумная металлургия
V 107	vacuum method, vacuum process	Vakuumverfahren *n*	procédé *m* à vide	вакуумный процесс (прием)
V 108	vacuum microbalance	Vakuummikrowaage *f*	microbalance *f* à vide	вакуумные микровесы
V 109	vacuum mixer	Vakuummischer *m*	mélangeur *m* à vide	вакуумная мешалка
V 110	vacuum monochromator	Vakuummonochromator *m*	monochromateur *m* à vide	вакуумный монохроматр
	vacuum moulding	*s.* vacuum forming		
	vacuum moulding plant, vacuum casting plant	Vakuumgießanlage *f*	installation *f* (poste *m*, appareillage *m*) de coulée à vide	установка для литья под вакуумом
V 111	vacuum packaging	Vakuumverpackung *f*	emballage *m* sous vide, conditionnement *m* sous vide	упаковка в вакууме, вакуум-упаковка
V 112	vacuum paddle dryer	Vakuumschaufeltrockner *m*	sécheur *m* en pale à vide	вакуумная сушилка с перелопачивающей мешалкой
V 113	vacuum paddle mixer	Vakuumtrockner *m* mit Rührwerk	sécheur *m* à vide avec agitateur mécanique	вакуумная сушилка с лопастной мешалкой
	vacuum paint, sealing paint	Dichtungslack *m*, Vakuumlack *m*	laques *fpl* d'étanchéifier, laques d'étanchéité	вакуумный лак
V 114	vacuum path X-ray diffractometer	Vakuumröntgenstrahldiffraktometer *n*	diffractomètre *m* à rayons X à vide	вакуумный рентгеновский дифрактометр
V 115	vacuum physics	Vakuumphysik *f*	physique *f* du vide	вакуумная физика
	vacuum pipe, vacuum line, vacuum tubing	Vakuumleitung *f*	conduite *f* à vide	вакуумная коммуникация, вакуумпровод
V 116	vacuum pipe joining	Vakuumrohranschluß *f*, Vakuumrohrverbindung *f*	assemblage *m* de tuyaux à vide	вакуумное трубчатое соединение
V 117	vacuum pipework	Vakuumpumpleitung *f*	tuyau *m* de pompe à vide	вакуумный трубопровод, вакуумпровод
V 118	vacuum plant, vacuum system <US>, vacuum unit	Vakuumanlage *f*	système *m* à vide, installation *f* de vide	вакуумная система, вакуумная установка
V 119	vacuum plumbing	Vakuuminstallation *f*	installation *f* à vide	система вакуумпроводов, вакуумная арматура
V 120	vacuum polarization	Vakuumpolarisation *f*	polarisation *f* de vide	поляризация вакуума
	vacuum port, evacuation port, vacuum tubulation	Vakuumstutzen *m*	raccord *m* pour le vide, tubulaire *f* de vide, tubulaire d'évacuation	впускной патрубок
	vacuum potting, vacuum encapsulation	Vergießen (Verkapseln) *n* unter Vakuum	coulée *f* sous vide	вакуумная инкапсуляция, заливка под вакуумом
V 121	vacuum powder filling equipment	Vakuumpulverabfüllbehälter *m*	installation *f* pour le chargement dosé sous vide de poudre	вакуумная установка для наполнения порошком
V 122	vacuum powder insulation	Vakuumpulverisolation *f*	isolement *m* pulvérulent à vide	вакуумная порошковая изоляция
V 123	vacuum pressure plant	Vakuumdruckanlage *f*	installation *f* de pression à vide	вакуумная давильная установка
V 124	vacuum pressure sintering	Vakuumdrucksintern *n*	frittage *m* sous pression sous vide	вакуумное спекание под давлением
	vacuum process	*s.* vacuum method		

V 125	**vacuum processing technique**	Vakuumverfahrenstechnik *f*	technique (technologie) *f* des procédés sous vide	технология вакуумной обработки
V 126	**vacuum pump**	Vakuumpumpe *f*	pompe *f* à vide	вакуумный насос
V 127	**vacuum pumping**	Erzeugung *f* von Vakuum	pompage *m*, évacuation *f*	откачка, создание вакуума
V 128	**vacuum pumping system**	Vakuumpumpsatz *m*	groupe *m* de pompage à vide	вакуумная насосная система
V 129	**vacuum purification**	Vakuumreinigung *f*	purification *f* du vide	вакуумная очистка
	vacuum putty, vacuum cement	Vakuumkitt *m*	mastic *m* pour le vide, ciment *m* à vide	вакуумный цемент
V 130	**vacuum reaction**	Vakuumreaktion *f*	réaction *f* sous vide	реакция в вакууме, вакуумная реакция
V 131	**vacuum reactor**	Vakuumreaktor *m*	pile *f* atomique à vide	вакуумный реактор
V 132	**vacuum rectification**	Vakuumrektifikation *f*	rectification *f* de vide	вакуумная ректификация
V 133	**vacuum rectification plant**	Vakuumrektifizieranlage *f*	installation *f* de rectification à vide	вакуумная ректификационная установка
V 134	**vacuum rectification tower**	Vakuumrektifizierkolonne *f*	colonne *f* de rectification à vide	вакуумная ректификационная колонна
V 135	**vacuum relay**	Vakuumrelais *n*	relais *m* à vide	вакуумное реле
V 136	**vacuum reservoir**	Zusatzvakuum *n*	réservoir *m* à vide, vide *m* auxiliaire	вакуумный резервуар
V 137	**vacuum resin casting plant**	Vakuumgießharzanlage *f*	installation *f* de résine moulable à vide	вакуумная установка для литья смолы
V 138	**vacuum resistance furnace**	widerstandsbeheizter Vakuumofen *m*, Vakuumwiderstandsofen *m*	four *m* à résistance de (à) vide	вакуумная печь сопротивления
V 139	**vacuum retort furnace**	Vakuumretortenofen *m*	four *m* à cornue à vide	вакуумная реторная печь
V 140	**vacuum rotary dryer**	Vakuumtrommeltrockner *m*	tambour *m* sécheur à vide	вакуумная сушилка с вращающимся барабаном
V 141	**vacuum rotary evaporator**	Vakuumrotationsverdampfer *m*	évaporateur *m* rotatif à vide	вращающийся вакуумный испаритель
V 142	**vacuum science**	Vakuumwissenschaft *f*	science *f* du vide, vacuologie *f*	наука о вакууме
	vacuum sealed device, vacuum device, sealed-off vacuum system	abgeschlossenes Vakuumsystem *n*	installation *f* scellée de vide, système *m* séparé par fusion	отпаянный вакуумный прибор, отпаянная вакуумная система
V 143	**vacuum sealing disk**	Vakuumdichtscheibe *f*	rondelle *f* joint au vide	уплотнительная шайба
V 144	**vacuum-sealing unit**	Vakuumverschließeinrichtung *f*	fermeture *f* à vide	устройство для вакуумного уплотнения
V 145	**vacuum sheathing** <of power cables>	Vakuumverkleidung *f* <von Starkstromkabeln>	revêtement *m* à vide <des câbles à courant fort>	вакуумная оболочка <мощных кабелей>
	vacuum shelf dryer	*s.* vacuum drying chamber		
V 146	**vacuum shell**	Vakuumgehäuse *n*	boîte *f* à vide	вакуумная оболочка, вакуумный кожух
V 147	**vacuum shut-off valve**	Vakuumabsperrventil *n*	vanne *f* d'arrêt à vide	вакуумный отсечной клапан
	vacuum side, collecting side, inlet side, intake side, suction side	Saugseite *f*, Vakuumseite *f*	côté *m* vide, côté d'aspiration	собирающая сторона, сторона впуска
V 148	**vacuum sintering**	Vakuumsintern *n*	frittage *m* (agglomération *f*) sous vide	спекание в вакууме
V 149	**vacuum-sintering plant**	Vakuumsinteranlage *f*	installation *f* de frittage sous vide	установка для вакуумного спекания (синтерирования)
	vacuum-siphon degassing, pipette degassing by lifting	Vakuumheberentgasung *f*	dégagement *m* par siphon sous vide	сифонное обезгаживание в вакууме
V 150	**vacuum sluice valve**	Vakuumschieberventil *n*	soupape *f* à vanne à vide	вакуумная задвижка
V 151	**vacuum spark gap**	Vakuumfunkenstrecke *f*	espace *m* à étincelle sous vide, éclateur *m* sous vide, distance *f* explosive sous vide	вакуумный искровой разрядник, вакуумный искровой промежуток
V 152	**vacuum spectral region**	Vakuumultraviolett *n*	vide *m* ultraviolet	вакуумный ультрафиолет
V 153	**vacuum spectrograph**	Vakuumspektrograf *m*	spectrographe *m* à vide	вакуумный спектрограф
V 154	**vacuum-stable**	vakuumbeständig	avec stabilité du vide	вакуумноустойчивый
V 155	**vacuum steam**	Vakuumdampf *m*	vapeur *f* sous dépression, vapeur sous vide	разреженный пар, вакуумный пар
V 156	**vacuum-steam generator**	Vakuumdampferzeuger *m*	générateur *m* de vapeur à vide	вакуумный парогенератор
V 157	**vacuum steam heating**	Vakuumdampfheizung *f*	chauffage *m* à la vapeur sous vide	вакуумно-паровой нагрев
V 158	**vacuum steam installation**	Vakuumdampferzeugung *f*	production *f* de vapeur sous vide (dépression)	вакуум-паровая установка
V 159	**vacuum steel degassing**	Vakuumstahlentgasung *f*	dégagement *m* de l'acier sous vide	вакуумное обезгаживание стали
	vacuum stopcock	*s.* vacuum cup stopcock		
V 160	**vacuum stream degassing**	Vakuumstrahlentgasung *f*	dégagement *m* gazeux à air comprimé sous vide	вакуумное струйное обезгаживание
V 161	**vacuum stream-degassing system**	Vakuumstrahlentgaser *m*	dégazeur *m* de courant dans le vide	система для вакуумного струйного обезгаживания

V 162	**vacuum stream-droplet process**	Vakuumgießstrahlverfahren *n*	méthode *f* de rayon de coulée à vide	вакуумное капельно-струйное рафинирование
V 163	**vacuum sublimation**	Vakuumsublimation *f*	sublimation *f* sous vide	вакуумная сублимация, вакуумная возгонка
V 164	**vacuum switch**	Vakuumschalter *m*	interrupteur *m* à vide	вакуумный выключатель
V 165	**vacuum system** **vacuum system** ‹US›	Vakuumsystem *n* *s.* vacuum plant	système *m* de vide	вакуумная система
V 166	**vacuum system and evaporator**	Vakuum-Pump-und-Aufdampfanlage *f*	installation *f* de pompage et installation d'évaporation sous vide	установка для откачки и вакуумного испарения
	vacuum tank, vacuum vessel, vacuum container	Vakuumbehälter *m*	récipient *m* à vide, enceinte *f* à vide	вакуумная камера, вакуумный сосуд, резервуар
V 167	**vacuum technology**	Vakuumtechnik *f*	technique *f* du vide	вакуумная техника
	vacuum tester, vacuum indicator	Vakuumprüfer *m*	indicateur *m* de vide, appareil *m* pour le contrôle de vide	индикатор вакуума
V 168	**vacuum testing, vacuum testing method**	Absprühmethode *f*, Vakuumprüfung *f*	contrôle *m* de vide	вакуумные испытания
V 169	**vacuum thermal insulation**	Vakuum-Wärme-Isolation *f*	isolation *f* calorifique du vide	вакуумно-тепловая изоляция
V 170	**vacuum thermo-balance**	Vakuumthermowaage *f*	thermobalance *f* à vide	вакуумные термовесы
V 171	**vacuum thermocouple**	Vakuumthermoelement *n*	thermo-élément (thermocouple) *m* du vide	вакуумная термопара
V 172	**vacuum thermogravi-metry**	Vakuumthermogravimetrie *f*	thermogravimétrie *f* de vide	вакуумная термогравиметрия
V 173	**vacuumtight**	vakuumdicht	étanche au vide	вакуумно-плотный, вакуумплотный, вакуумный
	vacuum tightness, vacuum leakproofness	Vakuumdichtigkeit *f*	étanchéité *f* au vide	вакуумплотность
V 174	**vacuum trap**	Vakuumfalle *f*	piège *m*, trappe *f* à vide	вакуумная ловушка
V 175	**vacuum treatment**	Behandlung *f* im Vakuum	traitement *m* sous vide	вакуумная обработка
V 176	**vacuum tube,** vacuum valve	Vakuumröhre *f*	tube *m* à vide, kénotron *m*	вакуумная лампа
V 177	**vacuum tube**	Vakuumschlauch *m*	tuyau *m* à vide	вакуумный шланг
	vacuum tubing, vacuum line, vacuum pipe	Vakuumleitung *f*	conduite *f* à vide	вакуумная коммуникация, вакуум-провод
	vacuum tubulation	*s.* vacuum port		
V 178	**vacuum ultraviolet source**	Vakuum-UV-Quelle *f*	UV-source *f* de vide	источник вакуумного ультрафиолета
	vacuum unit	*s.* vacuum plant		
	vacuum valve, vacuum tube	Vakuumröhre *f*	tube *m* à vide, kénotron *m*	вакуумная лампа
V 179	**vacuum valve**	Vakuumventil *n*	soupape *f* à vide	вакуумный вентиль
V 180	**vacuum-vapour drying,** vapour stream drying	Dampfstromtrocknung *f*	séchage *m* par courant de vapeur, dessiccation *f* par courant de vapeur	вакуумная сушка струей пара
	vacuum vessel	*s.* vacuum container		
	vacuum wax, sealing wax	Dichtungswachs *n*, Vakuumwachs *n*	cire *f* pour joint étanche, cire d'étanchéité, cire-laque *f*	уплотняющий воск, вакуумная замазка
V 181	**vacuum wettability**	Vakuumbenetzbarkeit *f*, Benetzbarkeit *f* im Vakuum	mouillure *f* dans le vide	смачиваемость в вакууме
V 182	**vacuum zone melting**	Vakuumzonenschmelzen *n*	fusion *f* de zone sous vide, fusion à zones sous vide	вакуумная зонная плавка
V 183	**vacuum zone purification**	Zonenreinigungsverfahren *n* im Vakuum	raffinage *m* en zones progressives sous vide	вакуумная зонная очистка
V 184	**vagabond cathode rays**	vagabundierende Katodenstrahlen *mpl*	rayons *mpl* cathodiques aberrants	блуждающие электронные лучи
V 185	**value of the full scale**	Skalenendwert *m*	déviation *f* totale (maxima)	полный отсчет шкалы
	valve, blow off, bleed, release	abblasen	souffler, laisser échapper, lâcher, évacuer	выпускать
V 186	**valve**	Ventil *n*	valve *f*, soupape *f*	вентиль, клапан
V 187	**valve area**	Ventilquerschnitt *m*	section *f* de soupape	проходное сечение вентиля
V 188	**valve block**	Ventilblock *m*	bloc *m* de soupapes couplées, ensemble *m* mécanique de vannes	вентильная коробка
	valve body, body of valve	Ventilkörper *m*	corps *m* de soupape	корпус вентиля
V 189	**valve bonnet, valve cap, valve cover,** valve hood	Ventildeckel *m* ‹mit Spindelführung›, Ventilkappe *f*	couvercle *m* (hotte *f*) de la soupape, chapeau *m* de la vanne	крышка вентиля, колпачок вентиля
V 190	**valve disk,** valve plate, valve head	Ventilteller *m*, Ventilplatte *f*	plateau *m* de soupape, disque *m* de vanne, obturateur *m* d'un robinet, tête *f* soupape	тарелка (диск) вентиля, тарелка клапана
V 191	**valve disk gasket**	Ventiltellerdichtung *f*, Scheibendichtung *f*	joint *m* du disque de vanne, joint de l'obturateur d'un robinet	уплотнитель тарелки вентиля

V 192	**valve face**	Ventildichtungsfläche f	surface f d'obturation de soupape	уплотнительная поверхность вентиля
V 193	**valve flap**	Ventilklappe f	clapet m	заслонка клапана, заслонка вентиля
V 194	**valve gear**	Ventilsteuerung f	commande f de vanne (robinet), conduite f de soupape	вентильное (клапанное) распределение, клапанный (золотниковый) привод
V 195	**valve guard,** valve stop	Ventilanschlag m, Ventilhubbegrenzung f	dispositif m de sécurité de soupape, butée f de soupape, arrêt m de vanne	ограничитель хода клапана, клапанный упор
	valve head	s. valve disk		
	valve hood	s. valve bonnet		
V 196	**valve interlocking combination**	Ventilkombination f	combinaison f de soupape	комбинация вентилей
V 197	**valveless**	ventillos	sans soupape	бесклапанный, безвентильный
V 198	**valve lift**	Ventilhub m	levée f de soupape, course f de soupape	ход клапана
V 199	**valve piston**	Ventilkolben m	piston m de soupape	клапанный поршень
	valve plate	s. valve disk		
	valve plate gasket, disk gasket	Ventiltellerdichtung f, Scheibendichtung f	joint m du disque de vanne, joint de l'obturateur d'un robinet	уплотнение тарелки вентиля (затвора)
V 200	**valve plug**	Ventilstempel m, Dichtstempel m	piston m de vanne	пробка вентиля, запорная часть вентиля
V 201	**valve rod,** valve spindle, valve stem	Ventilspindel f, Ventilschaft m	broche f de soupape	шток вентиля
V 202	**valve seat**	Ventilsitz m	siège m de soupape	седло вентиля (клапана)
V 203	**valve seat gasket**	Ventilsitzdichtung f	joint m du siège d'un robinet, joint du siège d'une soupape	уплотнитель седла вентиля
	valve spindle, valve stem	s. valve rod		
V 204	**valve stem guide**	Ventilspindelführung f	guide m de vis (tige) à soupape	направляющий шток вентиля
	valve stop	s. valve guard		
V 205	**valve tongue**	Schieberzunge f	changement m à aiguille de tiroir	язычок клапана
V 206	**valving**	Ventilanordnung f	arrangement m (disposition f) de soupapes	размещение вентилей
	vane, sliding blade, sliding vane	Drehschieber m	palette f, tiroir m, vanne f, coulisse f	пластина, выдвижная лопатка, скользящая лопатка
V 207	**vane pump**	Flügelpumpe f	pompe f à palettes	пластинчатый насос
	vane type rotary pump, sliding vane rotary pump	Drehschieberpumpe f	pompe f à palettes, pompe rotative à palettes	пластинчато-роторный насос
	vaporization, evaporation	Verdampfung f, Verdampfungsprozeß m, Verdunstung f	évaporation f, vaporisation f	испарение
V 208	**vaporization limit**	Verdampfungsgrenze f	limite f de vaporisation	предел парообразования
	vaporization rate, evaporation rate	Verdampfungsgeschwindigkeit f, Verdampfungsrate f	vitesse (rapidité) f de vaporisation, taux m d'évaporation	скорость испарения (парообразования)
	vaporization temperature, evaporation temperature	Verdampfungstemperatur f	température f d'évaporation	температура испарения
	vaporize, evaporate	verdampfen	s'évaporer, se vaporiser	испарять
	vaporizer, evaporator	Verdampfer m	évaporateur m, vaporisateur m	испаритель, выпарной аппарат
V 209	**vapour**	Dampf m	vapeur f	пар
V 210	**vapour beam,** vapour jet	Dampfstrahl m	jet m de vapeur	струя пара
	vapour booster pump, diffusion-ejector pump	Diffusionsejektorpumpe f <Pumpe mit Strahl- und Diffusionsstufen>	éjecteur m à diffusion	диффузионный эжекторный насос <насос с диффузионной и эжекторной ступенями>
	vapour-catching cone cup, cooled cover, cold cap <baffle>	Düsenhutdampfsperre f	baffle m à chapeau, baffle-chapeau m	охлаждаемый колпачковый отражатель, охлаждаемая коническая насадка
V 211	**vapour-cooled**	siedegekühlt	réfrigéré par ébullition	охлаждаемый паром
V 212	**vapour-degrease,** vapour [phase] degreasing	Dampfentfetten n	dégraissage m de vapeur	очистка пара от масел
V 213	**vapour degreaser**	Dampfentfettungsanlage f	installation f à dégraissage de vapeur	установка для обезжиривания паров
	vapour degreasing	s. vapour-degrease		
V 214	**vapour density**	Dampfdichte f	densité f de vapeur	плотность паров
V 215	**vapour-deposited**	aufgedampft	déposé par vaporisation, métallisé sous vide	нанесенный испарением
V 216	**vapour-deposited metal film**	metallische Aufdampfschicht f	couche f métallique par évaporation	металлическая пленка, нанесенная испарением
V 217	**vapour-ejecting pump**	Dampfstrahlejektorpumpe f	injecteur m à vapeur	пароэжекторный насос

V 218	**vapour exhaustion**	Dampfabsaugung f	aspiration f (échappement m) de vapeur	отсос паров, выпуск пара
V 219	**vapour hood**	Brüdenhaube f, Wrasenhaube f	capotage m de buée, calotte f de buée	паровой (пароулавливающий, пароотводящий) колпак
	vapour jet, vapour beam	Dampfstrahl m	jet m de vapeur	струя пара
V 220	**vapour-jet pump,** vapour pump	Treibdampfpumpe f	pompe f à flux de vapeur	пароструйный насос
	vapour-jet pump, ejector, ejector pump, steam-jet pump	Dampfstrahlpumpe f	pompe f à jet de vapeur, éjecteur m [à vapeur]	пароструйный насос, эжектор
V 221	**vapour lock**	Dampfsperre f, Dampfabschluß m	baffle m, chicane f, piège m à vapeur	паровой затвор
V 222	**vapour particle**	Dampfteilchen n	particule f de vapeur	частицы пара
	vapour phase degreasing	*s.* vapour-degrease		
	vapour pipe, jet stack	Dampfsteigrohr n	tuyau m de vapeur, colonne f de vapeur, tube m d'amenée de vapeur	паропровод
V 223	**vapour pressure**	Dampfdruck m	tension f de vapeur, pression f de vapeur	давление паров
V 224	**vapour-pressure thermometer**	Dampfdruckthermometer n	thermomètre m à pression de vapeur	парогазовый термометр
	vapour pump, diffusion pump	Diffusionspumpe f	pompe f à diffusion	диффузионный насос
	vapour pump	*s.* vapour-jet pump		
V 225	**vapour pump oil**	Treibdampfpumpenöl n	huile f d'une pompe à vapeur	масло для высоковакуумного насоса
V 226	**vapour source**	Dampfquelle f	source f de vapeur	источник пара
V 227	**vapour source turret**	Verdampferquellenrevolver m	révolver m de source d'évaporation	револьверная головка с испарителями
	vapour stream drying	*s.* vacuum-vapour drying		
V 228	**vapour superheating**	Treibdampfüberhitzung f	surchauffe f de vapeur du fluide moteur, surchauffage m de vapeur du fluide moteur	перегрев паров, перегрев рабочей жидкости
	variable leak, adjustable leak	einstellbares Leck n	fuite f réglable	регулируемая течь
	variable leak valve, dosing valve, leak valve	Gaseinlaßventil n, Dosierventil n	soupape f réglable, vanne f de dosage, fuite f réglable	дозировочный вентиль, вентиль впуска газа
	varnish, lacquer	lackieren	vernir, laquer	лакировать
V 229	**varnish**	Firnis m	vernis m	олифа
V 230	**varnish impregnating plant**	Tränklackimprägnieranlage f	installation f d'imprégnation au vernis	установка для пропитки лаком
V 231	**V-belt, Vee belt**	Keilriemen m	courroie f trapézoïdale (conique, en forme de coin)	клиновидный ремень
V 232	**Vee-ring,** V-ring gasket	V-Dichtungsring m, Rundring m mit V-förmigem Querschnitt, Trapezring m, V-Ringdichtung f, Trapezringdichtung f	anneau m d'obturation, bague f d'obturation, anneau joint, joint m torique trapézoïdal, anneau à section trapézoïdal	уплотняющее кольцо с V-образным поперечным сечением, трапецеидальное кольцо
V 233	**velocity distribution,** velocity spread	Geschwindigkeitsverteilung f	répartition f de vitesse, distribution f de vitesse	разброс скоростей, распределение по скоростям
	velocity head, dynamic pressure, velocity pressure, stagnation pressure	dynamischer Druck m, Staudruck m	pression f dynamique	динамическое давление, скоростной напор
	velocity of migration, migration velocity	Wanderungsgeschwindigkeit f	vitesse f de migration	скорость миграции
	velocity pressure, dynamic pressure, stagnation pressure, velocity head	Staudruck m, dynamischer Druck m	pression f dynamique	скоростной напор, динамическое давление
	velocity spread	*s.* velocity distribution		
	vent, admit air, break vacuum, aerate	belüften	respirer, aérer	аэрировать, впускать воздух, продувать
	vented exhaust, gas ballast	Gasballast m	lest m de gaz, lest (injection f) d'air	газовый балласт
	vented-exhaust [mechanical] pump, gas ballast pump	Gasballastpumpe f	pompe f à injection d'air, pompe ballast à gaz	газобалластный насос
	ventilation, venting, deairing, deaerating	Entlüftung f	ventilation f, évacuation f de l'air, respiration f, désaération f	выпуск воздуха, удаление воздуха, деаэрация, деаэрирование
	venting, airing, aeration, air inlet	Belüftung f	aération f, entrée f d'air	аэрация, впуск воздуха, обдувание
	venting screw, air release plug	Entlüftungsschraube f	vis f d'évacuation d'air, vis de sortie d'air	пробка (винт) для выпуска воздуха
V 234	**venting time**	Belüftungszeit f	temps m d'aération, temps d'entrée d'air	время повышения давления до атмосферного

	venting valve, air admittance valve, air release valve, airing valve, bleed valve, air inlet valve	Belüftungsventil n, Flut-ventil n, Lufteinlaß-ventil n	soupape f à admettre de l'air, robinet m d'entrée d'air	вентиль пуска воздуха, напускной вентиль
V 235	venturi jet	Venturi-Düse f	tube m de Venturi	трубка Вентури, сопло Вентури
	vertical evaporation, up evaporation	Aufwärtsverdampfung f	vaporisation f ascendante	вертикальное испарение, испарение вверх
V 236	vessel, container, tank	Rezipient m, Behälter m	récipient m	контейнер, вместилище
	vial	Phiole f, Ampulle f	fiole f	пузырек, ампула
	vibrating crystal thick-ness monitor, quartz crystal thin film monitor	Schwingquarzschicht-dickenmeßgerät n	appareil m de mesure des couches minces à quartz oscillatoire	прибор с кварцевым датчиком для измерения толщин пленок
	vibrating quartz crystal, high-frequency quartz crystal, quartz crystal oscillator, crystal transducer, oscillating quartz crystal	Schwingquarz m	quartz m oscillatoire (de résonance)	кристалл кварцевого генератора, генери-рующий кварцевый кристалл, генерирую-щий кварц
V 237	vibrating read con-denser electrometer	Schwingkondensator-elektrometer n	électromètre m à conden-sateur oscillant	электрометр с динамиче-ским конденсатором
V 238	vibration	Schwingung f, Schwingen n	vibration f, oscillation f	вибрация
V 239	vibratory feed	Schwingförderer m	transporteur m oscillatoire	качающийся конвейер
V 240	Vickers hardness	Vickershärte f	dureté f Vickers	твердость по Виккерсу
	viewing glass, viewing port, viewing win-dow, sight glass, inspection glass, obser-vation glass, sight port, sight window	Einblickfenster n, Schau-glas n, Beobachtungs-fenster n	regard m de contrôle, fenêtre f de coup d'œil, fenêtre de contrôle, regard, fenêtre d'obser-vation, hublot m	смотровое отверстие (окно), наблюдатель-ное окно, контрольное (смотровое) стекло, глазок
V 241	virtual leak	virtuelles Leck n, scheinbares Leck	fuite f virtuelle (apparente)	ложная течь, кажущаяся течь
V 242	viscosity, viscosity coefficient	dynamische Zähigkeit f, Viskosität f, Koeffizient m der inneren Reibung, Zähigkeit	viscosité f	вязкость
	viscosity manometer, viscosity type of gauge, viscosity vacuum gauge, decre-ment gauge, decrement viscosity gauge, friction type vacuum gauge	Reibungsvakuummeter n	manomètre m à amor-tissement (viscosité), jauge f à vide à frotte-ment, vacuomètre m d'adhérence, vacuo-mètre de frottement, tube m gradué à viscosité	динамический (вязкост-ный) манометр
	viscous flow, sliding flow, slip flow	Schlupfströmung f, Gleitung f, Gleit-strömung f, viskose Strömung f	glissement m, courant m de glissement, écoule-ment m visqueux	течение со скольжением, вязкостное течение
V 243	viscous leak	viskoses Leck n	fuite f visqueuse	течь с вязкостным потоком
	vitreous silica, opaque quartz, quartz ware	Quarzgut n	silice f vitreuse (fondue)	кварцевая порода, непрозрачный кварц
V 244	vitrification	Glasfluß m	flux m de verre	переход в стекловидное состояние
V 245	vitrify	beglasen	vitrer, vitrifier	остекловывать
V 246	volatilized getter	Verdampfungsgetter m	getter m à évaporation	распыляемый (испаряе-мый) газопоглотитель
	volume flow, volu-metric flow [rate], volumetric throughput, speed of flow	Volumendurchsatz m, Volumenstrom m	infiltration f de volume, débit m volumétrique, flux m volumétrique	объемный расход, объем-ная скорость течения
V 247	volume of gram molecule of a gas	Volumen n einer Gramm-molekel eines Gases	volume m du gramme-molécule d'un gaz	объем грамм-молекулы газа
V 248	volume-ratio calibra-tion system	volumetrisches Kalibrier-system n	système m de calibration volumétrique	волюметрическая кали-бровочная система
V 249	volumetric capacity	volumetrische Förder-leistung f	débit m volumétrique, vitesse f d'aspiration	быстрота откачки
	volumetric displace-ment, swept volume	Hubvolumen n, Schöpf-volumen n	volume m engendré	объем рабочей камеры, описываемый объем
V 250	volumetric displace-ment method	Bürettenmethode f	méthode f de burette	метод «бюретки»
V 251	volumetric efficiency	volumetrischer Wirkungs-grad m	rendement m volu-métrique	объемный коэффициент полезного действия
	volumetric flow, volu-metric flow rate, volumetric throuput	s. volume flow		
V 252	vortex flow, vortical flow	Wirbelströmung f	écoulement m tourbillon-naire, écoulement tourbillonnant	вихревое течение
	V-ring gasket, Vee-ring	V-Dichtungsring m, Rundring m mit V-förmigem Querschnitt, Trapezring m, V-Ring-dichtung f, Trapezring-dichtung f	anneau m d'obturation, bague f d'obturation, anneau joint, joint m torique trapézoïdal, anneau à section trapézoïdal	уплотняющее кольцо с V-образным попереч-ным сечением, трапецеидальное кольцо

W

	English	German	French	Russian
W 1	**wall-collision**	Wandstoß m	collision f murale	соударение со стенкой
W 2	**warm-up time**	Anheizzeit f	temps m de chauffage	время разогрева
W 3	**wash-gas method**	Spülgasmethode f	méthode f du gaz de balayage	метод промывки газом
W 4	**waste**	Abfall m, Ausschuß m	déchet m	отходы, потери
W 5	**waste heat**	Abwärme f	chaleur f perdue	отходящее тепло, использованное тепло
	waste space, dead space	schädlicher Raum m	espace m mort (nuisible)	мертвое пространство, вредное пространство
W 6	**water absorption**	Wasseraufnahme f	absorption f de l'eau	водопоглощенне
	water blast pump, water jet pump, aspirator pump	Wasserstrahlpumpe f	pompe f à jet d'eau	водоструйный насос
W 7	**waterbolt**	Wasserriegel m	verrou m d'eau	водяной затвор
W 8	**water coat,** water film	Wasserhaut f	peau (pellicule, couche) f d'eau	водяная пленка
W 9	**water content**	Wassergehalt m	teneur f en eau	водосодержание
W 10	**water-cooled copper mould**	wassergekühlte Kokille f	lingotière f refroidie par eau	изложница с водяным охлаждением
W 11	**water-cooled ladle**	wassergekühlte Gieß-schale f	poche f de coulée refroidie par eau	ковш с водяным охлаждением
	water-cooling connec-tion, cooling-water connection	Kühlwasseranschluß m	raccord m de l'eau de refroidissement	охлаждаемое водой сое-динение
W 12	**water defrosting**	Abtauen n durch Sprüh-wasser	dégel m par eau fine	размораживание ороше-нием водой
	water film	s. water coat		
W 13	**water hammer**	Wasserschlag m	coup m de bélier	гидравлический удар
W 14	**water inlet**	Wassereinlaß m	entrée f d'eau	ввод воды
W 15	**water jet condenser**	Wasserstrahlkondensator m	condensateur m à jet d'eau	водоструйный конденсатор
	water jet pump, water blast pump, aspirator pump	Wasserstrahlpumpe f	pompe f à jet d'eau	водоструйный насос
W 16	**water resistance**	Wasserbeständigkeit f	résistance f à l'eau	гидродинамическое сопротивление, гидравлическое сопротивление
W 17	**water-ring pump,** water-ring sealed fan pump	Wasserringpumpe f	pompe f annulaire d'eau, pompe à anneau d'eau	водокольцевой насос
	water-ring sealed fan pump	s. water-ring pump		
	water vapour, steam	Wasserdampf m	vapeur f d'eau	водяной пар
	water vapour handling capacity, maximum rate of handling water vapour, maximum capacity of treated water	Wasserdampfkapazität f	capacité f de vapeur d'eau, capacité de pompage de vapeur d'eau	наибольшая производи-тельность по воде ‹газобалластного насоса›
	water vapour toler-ance pressure, maxi-mum safe inlet pressure of water vapour, maxi-mum admissible water vapour pressure	Wasserdampfverträglich-keit f	compatibilité f à vapeur d'eau	максимальное допусти-мое впускное давление паров воды
W 18	**wax impregnation plant**	Wachsimprägnieranlage f	installation f d'imprégna-tion à la cire	установка для пропитки воском
W 19	**wear,** wear and tear	Verschleiß m, Abnutzung f	usure f	износ, истирание
W 20	**wear hardening**	Kaltverfestigung f	durcissement m à froid	холодное упрочнение, холодный наклеп
W 21	**wear index**	Verschleißzahl f	nombre m d'usure, indice m d'usure	коэффициент износа
W 22	**wearing part**	Verschleißteil n	partie (pièce) f d'usure	изнашивающаяся деталь, изнашивающийся узел
W 23	**weathering**	Verwitterung f	désagrégation f, efflorescence f	выветривание
	web dryer, sheeting dryer	Bahnentrockner m	sécheur m en lés	листовая сушилка
W 24	**weldability**	Schweißbarkeit f	soudabilité f	свариваемость
W 25	**welded flange**	Schweißflansch m	bride f soudée	приваренный фланец
W 26	**welded sleeve**	Schweißmuffe f	manchon m de soudage	сварная муфта
W 27	**welding current**	Schweißstrom m	courant m de soudure	сварочный ток
W 28	**welding neck flange**	Vorschweißflansch m	bride f à épaulement de soudure, bride à arête de soudure	фланец подготовленный к сварке
W 29	**welding neck quick flange**	Vorschweißkleinflansch m	bride f petite à épaulement de soudure	малый фланец подготов-ленный к сварке
W 30	**welding speed**	Schweißgeschwindigkeit f	vitesse f de soudage	скорость сварки
W 31	**welding torch**	Schweißbrenner m	soudeur m, brûleur m de soudure, chalumeau m à souder	сварочная горелка
W 32	**weldment**	Schweißkonstruktion f	construction f de soudure	сварная конструкция
W 33	**weld width,** width of bead	Schweißnahtbreite f	largeur f du joint soudé, largeur de joint de soudure	ширина сварного шва, ширина кромки

W 34	weld zone	Schweißzone f	zone f de soudage	зона сварки
	wettability angle, contact angle	Randwinkel m, Benetzungswinkel m	angle m mouillant	краевой угол, угол смачивания
W 35	wetting agent	Benetzungsmittel n, Netzmittel n	agent m mouillant	смачивающий агент
W 36	wetting power	Benetzungsfähigkeit f, Benetzungsvermögen n	mouillabilité f	смачиваемость
W 37	wet vapour	Naßdampf m	vapeur f humide	влажный пар
W 38	Wheatstone bridge	Wheatstonesche Brücke f	pont m de Wheatstone	мост Уитстона
W 39	whirlpool	Wirbel m	tourbillon m	вихрь, завихрение
W 40	whispering pump	Flüsterpumpe f	pompe f silencieuse	бесшумно-работающий насос
	width of bead, weld width	Schweißnahtbreite f	largeur f de joint de soudure, largeur du joint soudé	ширина кромки, ширина сварного шва
W 41	Wiedemann-Franz's law	Wiedemann-Franzsches Gesetz n	loi f de Wiedemann-Franz	закон Видемана-Франца
W 42	Wien's law of displacement	Wiensches Verschiebungs-gesetz n	loi f de translation de Wien	закон смещения Вина
W 43	Wilson seal	Wilsondichtung f	scellement m de Wilson, joint m Wilson	вилсоновское уплот-нение
W 44	wind channel, wind tunnel	Windkanal m	tunnel m aérodynamique, soufflerie f	вентиляционный тун-нель, аэродинамиче-ская труба
W 45	wiped-film molecular still	Fraktionierbürsten-Molekulardestillations-anlage f	installation f de distillation moléculaire à brosse tournante (fraction-nante)	дистилляционная уста-новка со щетками
W 46	wire cloth mesh	Drahtsiebmaschenweite f	largeur f de maille d'un tamis métallique	меш проволочной сетки
W 47	wire evaporation type getter-ion pump	Ionengetterpumpe f mit Drahtverdampfung	pompe f ionique à getter avec vaporisation de fil	геттеро-ионный насос с проволочным испари-телем, сорбционно-ионный насос с про-волочным испарителем
	wire gauze electrode, net-shaped electrode	Netzelektrode f	électrode f réticulaire (de réseau), électrode en toile métallique	электрод из проволоч-ной сетки
W 48	wire seal	Drahtdichtung f	garniture f étanche de fil, joint m de fil	проволочное уплотнение
	wiring, binding	Abbinden	câblage m	схватывание, связывание
W 49	wobble drive shaft	Wackelschwanzdreh-durchführung f Vakuumkurbel f	sortie f d'arbre tournant	вакуумный курбель
	wolfram, tungsten	Wolfram n	tungstène m, wolfram m	вольфрам
W 50	Wood's alloy, Wood's metal	Wood-Metall n	métal m de Wood	сплав Вуда
W 51	workability	Formänderungs-vermögen n	pouvoir m de déformation	обрабатываемость
W 52	work-accelerated electron gun	Elektronenstrahler m mit Fremdbeschleunigung	canon m à électrons d'ac-célération étrangère	электронная пушка с не-зависимым ускоре-нием
W 53	workholder	Werkstückhalter m	porte-outil m	держатель заготовки
	working fluid, working medium, operating fluid, pump fluid, pump oil	Pumpentreibmittel n, Pumpenöl n	agent m moteur, fluide m moteur d'une pompe, huile f à pompe	вакуумное масло, масло для высоковакуум-ного насоса
W 54	working piston	Arbeitskolben m	piston m moteur	рабочий поршень
W 55	working point	Verarbeitungstemperatur f	température f de façonne-ment	точка (температура) об-работки
	working pressure, operating pressure	Arbeitsdruck m, Betriebsdruck m	pression f de service, pression de travail	рабочее (эксплуатацион-ное) давление
W 56	working range	Verarbeitungsbereich m	zone f d'usinage	диапазон переработки
	worm, coiled cooling pipe	Kühlschlange f	serpentin m réfrigérant (refroidisseur, conden-seur)	охлаждающий змеевик
	wrought alloy, forgeable alloy	Knetlegierung f	alliage m baratté (de pétrissage, à pétrir)	ковкий (пластичный, прокованный) сплав

X

X 1	xenon	Xenon n	xénon m	ксенон
	X-ray current, soft X-ray photoelectric current, X-ray photo-current	durch weiche Röntgen-strahlung erzeugter Fotostrom m, Röntgenstrom m	courant m photo-élec-trique par rayonnement X doux	рентгеновский фототок
X 2	X-ray diffraction	Röntgenbeugung f	diffraction f des rayons X	дифракция рентгенов-ских лучей
X 3	X-ray effect <in ionization gauge>	Röntgenstrahleffekt m <im Ionisations-manometer>	effet m des rayons X <dans un manomètre à ionisation>	влияние собственного рентгеновского излу-чения на показания <ионизационного манометра>
X 4	X-ray limit, X-ray photocurrent pressure limit	Röntgengrenze f	rayonnement m X limite	рентгеновская граница

	X-ray photocurrent	s. X-ray current		
	X-ray photocurrent pressure limit	s. X-ray limit		
X 5	**X-rays**	Röntgenstrahlen *mpl*	rayons *mpl* X, rayons de Röntgen	рентгеновские лучи
X 6	**X-ray wavelength**	Röntgenwellenlänge *f*	longueur *f* d'onde des rayons X	длина волны рентгеновских лучей
X 7	**X-ring**	X-Ring *m*, Dichtungsring *m* mit x-förmigem Querschnitt	anneau *m* en X	уплотняющее кольцо с X-образным сечением

Y

	Y-branch, angle tee	Abzweigung *f*	dérivation *f*, embranchement *m*, ramification *f*	тройник в виде буквы «Y»
Y 1	**yield curve**	Ausbeutekurve *f*	courbe *f* de rendement, courbe de débit	кривая выхода
Y 2	**yield point,** yield value	Fließgrenze *f*	limite *f* d'allongement	предел текучести
	yield point, maximum tolerable tensile strength	Streckgrenze *f*	limite *f* d'allongement	предел текучести
Y 3	**yield pressure**	Fließdruck *m*	pression *f* continue	давление истечения
	yield value	s. yield point		
Y 4	**yoke**	Joch *n*, Magnetjoch *n*	culasse *f* de l'aimant	ярмо ‹магнитопровода›
	Young's modulus, modulus of elasticity	Elastizitätsmodul *m*	module *m* d'élasticité	модуль упругости при растяжении, модуль Юнга

Z

Z 1	**zeolite**	Zeolith *m*	zéolite *f*	цеолит
Z 2	**zeolite adsorption pump, zeolite molecular sieve pump**	Zeolith-Sorptionspumpe *f*	pompe *f* à sorption par zéolite	цеолитовый адсорбционный насос
Z 3	**zeolite trap**	Zeolith-Falle *f*	trappe *f* (piège *m*) à zéolite	цеолитовая ловушка
Z 4	**zero-adjusting device**	Nullpunkteinstellvorrichtung *f*	dispositif *m* de mise à zéro	прибор с установкой на нуль
Z 5	**zero drift**	Nullpunktsdrift *f*	mobilité *f* du point zéro	дрейф нуля
Z 6	**zero-point constancy**	Nullpunktkonstanz *f*	constance (stabilité) *f* du zéro	стабильность нулевой точки
Z 7	**zero-point setting**	Nullpunkteinstellung *f*	mise *f* à zéro, réglage *m* du zéro	установка на нуль
Z 8	**zero potential**	Nullpotential *n*	potentiel *m* zéro	нулевой потенциал
	zero resetting, resetting of zero	Nullabgleich *m*, Rückstellung *f* auf den Nullpunkt	réglage *m* du point zéro	сброс на нуль
Z 9	**zero suppression**	Nullunterdrückung *f*	suppression *f* du zéro	подавление нулевого значения измеряемой величины
Z 10	**zinc**	verzinken	zinquer, galvaniser au zinc	оцинковывать
Z 11	**zinc-bearing**	zinkhaltig	zincifère	содержащий цинк
Z 12	**zinc-cadmium sulfate**	Zink-Kadmium-Sulfat *n*	sulfate *m* de cadmium et zinc	цинко-кадмиевый сульфат
Z 13	**zone levelling**	waagerechtes Tiegelverfahren *n*	procédé *m* de creuset horizontal	зонное выращивание
Z 14	**zone levelling technique**	Zonenschmelzkristallziehverfahren *n*	procédé *m* de fusion à zones par tirage des cristaux	техника зонной плавки для выращивания кристалла
Z 15	**zone melting,** zone segregation	Zonenschmelzen *n*, Zonenseigerung *f*	fusion *f* de (à) zone	зонная плавка
Z 16	**zone melting plant**	Zonenschmelzanlage *f*	installation *f* de fusion à zones	установка для зонной плавки
Z 17	**zone purification,** zone refining	Zonenreinigen *n*, Zonenreinigung *f*	rectification (purification) *f* des zones	зонная очистка
Z 18	**zone refiner**	Zonenreinigungsapparatur *f*	installation *f* de purification à zones	установка для зонной очистки
	zone refining	s. zone purification		
	zone segregation	s. zone melting		

DEUTSCH

INHALT

1.   Vakuumphysik
1.1   Materie im Gaszustand
1.1.1 Reale und ideale Gase
1.1.2 Gasgesetze
1.2   Kinetische Gastheorie
1.3   Gasströmung

2.   Vakuumtechnik
2.1   Größen und Maßeinheiten
2.2   Vakuummeter
2.2.1 Totaldruckmeßgeräte
2.2.2 Partialdruckmeßgeräte
2.3   Vakuumerzeuger
2.3.1 Verdrängerpumpen
2.3.2 Molekularpumpen
2.3.3 Dampfstrahlpumpen
2.3.4 Diffusionspumpen
2.3.5 Gasaufzehrungspumpen
2.3.6 Kryopumpen
2.3.7 Pumpenzubehör
2.4   Dampfsperren
2.5   Ventile und Hähne
2.5.1 Absperrventile
2.5.2 Belüftungsventile
2.5.3 Gaseinlaßventile
2.5.4 Ventilkombinationen
2.5.5 Hähne
2.6   Vakuumleitungen
2.6.1 Starre Leitungen
2.6.2 Bewegliche Leitungen
2.7   Verbindungen
2.7.1 Untrennbare Verbindungen
2.7.2 Trennbare Verbindungen

2.8   Vakuumbehälter
2.9   Durchführungen
2.10   Fenster
2.11   Lecksuche
2.11.1 Lecksuchgeräte
2.11.2 Lecksuchmethoden
2.12   Vakuumsysteme
2.12.1 Statische Vakuumsysteme
2.12.2 Dynamische Vakuumsysteme
2.13   UHV-Technik

3.   Anwendungen der Vakuumtechnik
3.1   Vakuumtrocknung
3.2   Vakuumentgasung
3.3   Vakuumimprägnierung
3.4   Vakuumlötung
3.5   Vakuumverpackung
3.6   Vakuumdestillation
3.7   Vakuumsublimation
3.8   Vakuumkondensation
3.9   Vakuumfiltration
3.10   Beschichtungstechnik
3.11   Optische Schichten
3.12   Schutzschichten
3.13   Metallisieren
3.14   Dünnschichttechnik
3.15   Vakuumschmelzen
3.16   Vakuumentgasen
3.17   Vakuumgießen
3.18   Vakuumglühen
3.19   Vakuumsintern
3.20   Gefriertrocknung
3.21   Raumsimulation

A

Abbindekitt S 92
Abbinden B 121
abblasen B 134
Abblasventil B 149
abblättern C 132
Abbrechspitze B 197
Abbrennschweißen F 80
Abdampf E 200
Abdeckblende S 137
abdecken M 24
Abdecken M 27
Abdecklack M 28
Abdeckmaske S 98
Abdeckmaske für Mikro-
 schaltung M 112
Abdeckschicht R 140
abdichten S 36
Abdichtung G 45
Abdruck R 112
abdrücken P 182
Abdrücken P 181
Abdrückschraube F 154
Abfall W 4
Abfallzeit R 103
Abfüllvorrichtung F 49
Abgasturbine E 198
abgeschlossenes Vakuum-
 system S 39
abgestuft G 104
Abglimmeinrichtung G 102
Abglimmen C 73, D 158
Abguß C 52
abklemmen P 61
Abklemmen T 113
abkühlen C 127
Abkühlungsgeschwindig-
 keit C 328
Abkühlzeit C 319
ablagern / sich S 97
Ablagerungserscheinung
 S 19
Ablaßhahn B 135
Ablaßschraube D 59
Ablaßstopfen D 59
Ablaßventil P 223
Ablenkmagnet D 32
Ablenksystem D 80
ablöten U 22
abnutschen F 58
Abnutzung W 19
Abnutzungswiderstand A 4
abpressen P 61
Abpressen P 181
Abpreßmethode S 203
abpumpen E 158
Abquetschdichtung C 342
abquetschen P 61
Abrieb A 2
Abriebtest E 150
Absaugeanlage E 197
absaugen D 237
Absaugvorrichtung E 202
Abscheider S 86
Abscherfestigkeit S 107
Abschmelzdauer F 236
Abschmelzelektrode C 280
abschmelzen S 36
Abschmelzen T 113
Abschmelzgeschwindigkeit
 M 68
Abschmelzkapillare S 52
Abschmelzkatode M 72
Abschmelzkopf M 62
Abschmelzschweißen F 240
Abschmelzschweißung F 92
Abschmelzstab M 69
Abschrecken Q 11
Abschreckfestigkeit H 47
Abschreckung Q 11
absenkbarer Boden L 110
absetzen S 97
absolute Atmosphäre B 44
absoluter Druck A 7
absoluter Nullpunkt A 13
absoluter Nullpunkt
 der Temperatur A 14

absolutes Manometer A 6
absolutes Vakuum A 10
absolutes Vakuummeter
 A 11
absolute Temperatur A 8
absolute Temperaturskale
 A 5, A 9
absolute Verdampfungs-
 rate K 15
Absorbens A 17
absorbieren A 15
Absorption A 20
Absorptionsfaktor A 21
Absorptionsfalle A 19
Absorptionskältemaschine
 A 23
Absorptionskältesatz A 25
Absorptionsmittel A 17
Absorptionspumpe A 22
Absorptionsspektrum A 24
Absorptionsvermögen A 18
Absperrhahn P 90
Absperrhahn mit
 Quecksilberabdichtung
 M 81
Absperrschieber S 369
Absperrventil C 420, S 136
absprühen S 289
Absprühmethode V 168
Absprühsonde P 189
Abstandsstück S 249
Abstäuberate S 310
Abstichentgasung T 5
Abstichpfanne T 8
Abstichseite T 9
Abstichstange T 7
Abtastelektronenbeugung
 S 22
Abtastelektronenmikroskop
 S 23
abtasten S 21
Abtastgeschwindigkeit S 24
abtauen D 34
Abtauen D 36
Abtauen durch Sprühwasser
 W 12
Abtauen mit Heißgas
 H 131
Abtauwanne D 242
abtitrieren T 120
Abtropfbrett D 230
Abtropfgestell D 231
Abtropfschmelze D 241
abwalzen R 168
Abwärme W 5
Abwärtsverdampfung
 D 222
Abweichung D 79
Abziehlack S 384
Abzugskanal D 224
Abzweigung A 145
Adatom A 56
Additionspolymerisation
 P 102
Adhäsionsvermögen A 66
Adiabatenexponent A 69
adiabatisch A 68
Adsorbat A 75
Adsorbens A 77
adsorbiertes Gas A 75
Adsorption A 78
Adsorption-Desorptions-
 Technik A 79
Adsorption in Poren P 36
Adsorptionsisotherme A 81
Adsorptionsmittel A 77
Adsorptionspumpe A 83
Adsorptionsschicht A 76
Adsorptionsspektrometer
 A 84
Adsorptionsvakuummeter
 A 87
Adsorptionsvakuumprüf-
 gerät A 87
Adsorptionsvermögen A 82
Adsorptionszeit A 85
Adsorptiv A 16
Aircomatic-Verfahren I 50

Akkommodationskoeffi-
 zient A 32
Akkommodationszahl A 32
Aktinium A 39
Aktivator A 46
Aktivgetter A 49
aktiviertes Aluminiumoxid
 A 41
Aktivierungsanalyse A 43
Aktivierungsenergie A 44
Aktivierung von
 Oxidkatoden A 45
Aktivität A 50
Aktivkohle A 42
Aktivmetallverfahren A 47
akustischer Leckanzeiger
 A 209
akustisches Signal A 210
Akzeptor A 29
Alkylphenolharze A 121
Allglasionisationsmeßzelle
 A 122
Alpha-Strahler A 128
Alpha-Teilchen A 129
Alphatron A 130
Alterungshärtung A 92
Alterung thorierter
 Katoden A 93
Aluminiumbedampfung
 A 136
Aluminiumfolie A 134
Aluminiumoxid A 133
Amalgamation A 138
amalgamieren A 137
Ammoniaklecksuche A 141
Ampulle V 236
Analysatorröhre A 142
Aneinanderfügung J 17
Anfangsdruck I 65
Anfangssuszeptibilität I 66
angepaßte Verbindung M 40
Angström-Einheit A 147
Anheizzeit W 2
Anlage I 81
Anlage mit linearem
 Durchlauf L 63
anlassen A 150
Anlassen I 19
Anodenabstand A 164
Anodendunkelraum A 162
Anodendurchgriff P 27
Anodeneinzelzelle S 152
Anodengitter A 165
Anodenglimmhaut A 166
Anodenverlustleistung
 A 163
anodische Oxydation A 167
anodischer Dunkelraum
 A 162
anodisches Glimmlicht
 A 166
Anordnung thermo-
 elektrischer Elemente
 T 61
anormale Glimmentladung
 A 1
anpassen A 52
Anpassungsflansch A 55
Anregung E 194
Anregungsfrequenz E 195
Anregungsspannung E 196
Anreicherungsmethode
 B 215
Anreicherungsverfahren
 A 34
Ansaugdruck F 63
ansaugen P 208
Ansaugen I 80
Ansaugöffnung I 88
Ansaugquerschnitt I 69
anschichten P 55
Anschluß C 273, J 17
Anschlußflansch C 269
Anschlußklemme C 271
Anschlußleistung P 127
Anschmelzstelle S 47
Ansprechempfindlichkeit
 R 149

ansprechen R 146
Ansprechgeschwindigkeit
 S 271
Ansprechwahrscheinlich-
 keit S 80
Ansprechzeit R 148
Antikatode A 169
Antimon A 174
Antireflexbelag A 176
Antireflexschicht A 176
antireflexvergütet B 145
Antrieb P 197
Antriebsritzel D 243
Antriebsrolle D 244
Anwärmen C 172
Anzeigeverzögerung R 147
Anzugsdrehmoment T 98
aperiodisch D 5
Aquadag-Belag A 181
Arbeitsdruck O 51
Arbeitskolben W 54
Arbeitsplanung S 25
Arbeitstakt A 40
Arcatom-Schweißen A 182
Arcatom-Schweißung A 182
Arcatom-Schweißver-
 fahren A 182
Ardometer A 187
Argonarc-Schweißver-
 fahren A 188
Argonbehandlung A 192
Argoninstabilität A 189
Argonreinigung A 190
Argonstabilität A 191
Armkreuz S 276
Asbest A 194
Asbestpappe A 195
Asbestschnur A 196
Assoziationswahrschein-
 lichkeit P 188
Astonscher Dunkelraum
 A 198
Astrotorusbaffle A 199
Atm A 200
atmende Folie B 198
Atmosphärendruck A 202
Atmosphärenüberdruck
 A 201
atomar sauber A 203
Atomisieren A 207
Atomstrahl A 204
Atomvolumen A 206
Aufdampfen E 170
Aufdampfkatode E 167
Aufdampfrate D 67
Aufdampfschicht C 176
Aufdampftechnologie D 68
aufgedampft V 215
Aufkohlung C 37, R 49
Auflösungsvermögen R 142
Aufschlagventil B 189
aufsintern S 170
Aufsintern eines Metalls
 auf Keramik S 169
Aufspannvorrichtung J 16
Aufspreizwinkel O 4
Aufsprudeln E 4
auftauen D 34
Auftauen D 36
auftragen C 173
Auftragschweißen H 14
auftretende Zer-
 setzungen / an den
 Elektroden D 18
Auftriebskraft L 56
Aufwachsverfahren
 G 53
Aufwallen E 4
Aufwärtsverdampfung U 24
Aufzehrungszeit C 156
Ausbeute E 16
Ausbeutekurve Y 1
Ausdehnungskoeffizient
 C 180
Ausfallzeit D 225
ausfiltern F 60
Ausfluß E 120
Ausflußrate R 34

Ausflußziffer D 159
Ausgangssignal-
Rauschverhältnis O 69
Ausgleichsdruck B 35
Ausgleichskörper C 231
Ausgleichsventil E 148
Ausglühen F 223
Aushärtung C 414
ausheizbar B 26
ausheizbarer Rezipient H 22
ausheizbares Vakuumdurch-
gangsventil großer Nenn-
weite L 15
Ausheizen B 33
Ausheizen mittels
Gasflamme T 125
Ausheizmantel B 30
Ausheizofen B 29
Ausheizsteuerung B 28
Ausheiztemperatur B 32
Ausheiztisch B 31
Ausheizzyklus B 27
Auskleidung L 69
auskristallisierte Gläser
G 85
Auspuff D 166
Auspuffdruck D 167
Auspuffilter D 162
Auspuffleitung D 165
Auspufföffnung D 166
Auspuffprozeß E 160
auspuffseitiger Ölfänger
O 36
Auspuffventil D 171
auspumpen E 158
Auspumpgeschwindigkeit
S 269
Auspumpkurve P 211
Auspumpzeit P 212
Ausschleusvorrichtung
D 161
Ausschuß W 4
ausschwemmen F 140
Aussehen F 65
Außenanglasung O 71
Außendurchmesser E 219
Außengewinde E 221
Außenkondensator E 218
Außenmetallisierung O 66
Außenzylinderlötung O 70
äußerer Fotoeffekt P 44
äußerer lichtelektrischer
Effekt P 44
Aussiebung F 59
ausspülen F 140, R 163
Ausstoßdruck D 167
ausströmen F 124
Ausströmung E 120
austauschbar I 92
austenitischer rostfreier
Stahl A 211
Austrittsblende E 203
Austrittszahl D 159
Austrocknung D 72
Automat A 212
automatische Blende A 214
Autoradiografie A 215
Avogadrosches Gesetz
A 223
Avogadrosche Zahl L 105
Axialdichtring A 228
Axialgebläse A 225
Axialkolbenpumpe A 229
Axialpumpe A 227
Axialverdichter A 226

B

Badmetall B 65
Badschmelzen C 282
Baffle B 23
Baffleventil B 25
Bahnentrockner S 112
Bajonettfassung B 70
Bajonettsockel B 68
Bajonettverschluß B 69
Bandbedampfung B 42

Bandbedampfungsanlage
S 385
Bandenspektrum B 43
Bandförderer B 100
Bandheizung S 386
Bandrührer S 279
Bandstahl S 351
Bandstraße S 387
Bandtrockner B 101
Bar B 44
Bariumkarbonat B 48
Bariumoxid B 49
Barometer B 50
Barometerkompensation
B 52
Barometerstand B 51
barometrische Höhenformel
B 53
Barren B 113
Barrenform I 59
Basis S 402
Basisfläche B 58
Basisplatte S 402
Baukastenprinzip B 214
Bauteil C 279
Bauteilkolonne U 18
Bayard-Alpert-Röhre B 66
Beanspruchung S 378
Bearbeitungsgrad B 51
Beaufort-Skale B 88
Bedampfung durch Löcher
T 94
Bedampfungskammer D 65
Bedampfungsmaske D 66
bedecken C 173
Bedeckungsgrad C 360
Bedienungspult C 312
Bedienungsventil S 88
Befilmen F 50
beglasen V 245
Beglimmen C 73
Begrenzungswiderstand L 58
Behälter C 293
Behälterkühlfalle R 114
Behandlung im Vakuum
V 175
Beiluft S 54
Belegungsdichte C 359
belüften A 73
Belüftung A 89
Belüftungsventil A 95
Belüftungsventil
der Vorpumpe P 221
Belüftungszeit V 234
Benetzbarkeit im Vakuum
V 181
Benetzungsfähigkeit W 36
Benetzungsmittel W 35
Benetzungsvermögen W 36
Benetzungswinkel C 283
Beobachtungsfenster I 78
Bereitschaftsanlage S 324
Bernoullische Gleichung
B 105
Berstscheibe B 232
beruhigter Stahl K 6
Berylliumfenster B 106
Beschatten O 3
Beschattung O 3
beschichten Z 42
Beschichten durch
Eintauchen D 145
Beschichtung im Vakuum
V 42
Beschichtungstechnik
C 177
Beschichtungstechnologie
D 68
Beschickung C 114
Beschickungsvorrichtung
C 115
Beschießen B 162
Beschleunigungselektrode
A 26
Beschleunigungsfeld A 27
Beschleunigungsspannung
A 28
Bestrahlungsstation I 171

Bestrahlungszeit E 216
Beta-Teilchen B 107
Betatron B 108
BET-Isotherme B 109
Betriebsdruck O 51
betriebsfertiger Pumpstand
C 229
Betriebsmeßinstrument S 87
betriebssicher F 7
Betriebssicherheit F 151
Beugung langsamer
Elektronen D 123
Beugung schneller
Elektronen H 72
Beugung schneller Elektro-
nen in Durchstrahlung
T 157
Beugung schneller
Elektronen in Reflexion
R 86
Beugungsebene D 124
bewegliche Kupplung F 99
Bezugsdruck R 75
Bezugsleck C 11
Bezugspotential R 78
Bezugspunkt R 74
Bezugsspannung R 78
Bezugsstromkreis R 73
Bezugstemperatur R 76
Biegefestigkeit F 103
Biegesteifigkeit F 102
Bifilarwicklung B 112
bildsame Formgebung durch
Schmieden, Walzen,
Pressen und Ziehen F 1
Biluxlampe B 114
Bimetalldruckschalter B 119
Bimetallkontakt B 115
Bimetallschalter B 117
Bimetallthermometer B 118
Bimetallvakuummeter
B 116
binärer Thermodiffusions-
faktor B 120
Bindedraht T 97
Bindemittel C 80
Bindung B 121
Bindungsenergie B 122
Bindungskraft B 165
Bindungsstärke B 165
Blankglühen B 202
Blankglühen im Vakuum
V 7
Blankziehen B 203
Blasenbildung B 139
Blasenentwicklung P 207
Blasenkammer B 211
Blasfolie B 148
Blattventil L 35
Blears-Effekt B 133
Bleidichtung L 34
Blende C 62
Blendenmethode O 63
Blende über der Ver-
dampferquelle B 183
Blindflansch B 128
blindflanschen B 129
Blindversuch B 131
blitzartiges Hochheizen
F 88
Blochsche Wand B 140
Bloch-Wand B 140
Blockabzug I 60
Blockabzugsvorrichtung
I 63
Blockform I 59
Blockkatode B 141
Bodenabgußtiegel B 174
Bodenabstich B 177
Bodengießmethode B 175
Bogenentladung A 183
Boltzmann-Konstante B 161
Bolzenkatode B 160
Booster-Pumpe B 170
Bördelmaschine B 72
bördeln B 71
Bördelrohr F 74
Bördelverbindung F 78

Borsilikatglas B 172
Bourdon-Rohr B 183
Bourdon-Vakuummeter
B 183
Boyle-Mariottesches Gesetz
B 186
Bramme B 144
Brechungsindex R 90
Bremselektrode S 429
Bremsfeld R 153
Brennebene F 145
Brenner B 227
Brennermund B 228
Brennfleck S 286
Brennfleckgröße S 287
Brenngas F 221
Brennstoffelement F 220
Brinellhärte B 205
Brownsche Bewegung
B 209
Bruchdehnung B 193
Bruchfestigkeit B 195
Bruchplatte B 232
Brückenbrecher B 200
Brückenschaltung B 201
Brüdenhaube V 219
Buchse S 177
Bügelgetter S 364
Bürettenmethode V 250
Burnauer-Emmet-Teller-
Isotherme B 109
Bürstendestillieranlage
B 210
Büschelentladung B 225

C

Carnotscher Kreisprozeß
C 38
Carnotscher Wirkungsgrad
C 39
Celsius-Skale C 83
Chargenbetrieb B 63
Chargierventil C 118
Chargiervorrichtung L 94
chemisch rein C 122
Chemisorption C 123
Chemisorptionsbindung
C 124
Chevron-Baffle C 125
Chevron-Dichtung C 126
Chlorkobalt C 133
Chlorplatin C 134
Clausing-Faktor C 152
Clausius-Clapayronsche
Gleichung C 151
„corner"-Golddraht-
Dichtung C 344
Coulombsche
Abstoßungskraft C 355
Crookesscher Dunkelraum
C 375
Crookessche Röhre C 376
Curie-Punkt C 415
Curie-Temperatur C 416
Czochralski-Verfahren
C 433

D

Daltonsches Gesetz D 1
Dampf V 209
Dampfabsaugung V 218
Dampfabschluß V 221
Dampfdichte V 214
Dampfdom S 340
Dampfdruck V 223
Dampfdruckthermometer
V 224
Dampfdüse S 346
Dampfentfetten V 212
Dampfentfettungsanlage
V 213
Dampfentwicklung S 342
Dampfgefäß B 154
Dampfheizung S 343

Dampfkessel B 154
Dampfquelle V 226
Dampfsperre B 23, V 221
Dampfsperre mit Kriech-
  schutz N 10
Dampfsperre mit Peltier-
  kühlung T 60
Dampfsperre mit Ring-
  blechen M 198
Dampfsperre ohne axialen
  Durchblick O 52
Dampfsteigrohr J 13
Dampfstrahl V 210
Dampfstrahlejektorpumpe
  V 217
Dampfstrahler J 9
Dampfstrahlgebläse S 344
Dampfstrahlkälteanlage
  S 345
Dampfstrahlpumpe E 20
Dampfstrahlsauger J 9
Dampfstrahlstufe E 22
Dampfstrahlvakuumpumpe
  J 9
Dampfstromtrocknung
  V 180
Dampfteilchen V 222
Dämpfungsdiode D 3
Dämpfungskapillare D 2
Dampfventil S 347
Darstellung des Saugver-
  mögens als Funktion des
  Druckes S 268
Dauerbetrieb C 299
Dauerbiegefestigkeit B 104
Dauerbruch F 15
Dauerelektrode N 19
Dauerfestigkeit F 16
Dauerfestigkeitsverhältnis
  F 17
Dauergußform P 31
Dauerhaltbarkeit R 31
Dauerleistung C 300
Dauerprüfung E 137
Dauerstandfestigkeit C 367,
  P 32
Dauerstrichmagnetron
  C 305
Dauerversuch E 137
Dauerzustand S 338
Deckgetter C 175
Deckschicht O 19
Deckungssphäre P 198
Deckwalzen S 109
Deformationstemperatur
  D 33
Dehnbarkeit M 15
Dehngrenze O 7
Dehnungsfuge E 209
Dehnungsgeschwindigkeit
  C 365
Dehnungsmesser S 375
Dehnungsmeßstreifen
  E 32
dekantieren D 9
Dekantieren D 10
demontierbare Hoch-
  vakuumanlage D 61
Depressionskonstante D 69
Desorption D 73
Desorption durch
  Elektronen E 90
Desorptionsspektrometer
  D 74
Desorptionsspektrometrie
  D 75
Desorptionswärme H 37
Desoxydationsmittel D 64
Destillationsanlage D 196
Destillationsapparat D 194
Destillationsblase A 118
Destillationskolben A 118
Destillatpumpe D 76
Destillierapparat D 194
Destillierbehälter D 197
Destillierblase P 117
destilliertes Wasser D 195
Dewar-Gefäß D 85

dialysieren D 91
Dichtegradient C 251
Dichtehomogenität U 17
dichten S 36
Dichtezahl S 260
Dichtfläche J 19
Dichtflüssigkeit S 43
Dichtigkeitsprüfung L 45
Dichtkraft S 44
Dichtscheibe S 42
Dichtstempel V 200
Dichtung G 45
Dichtung mit abgesetztem
  Rechteckring C 188
Dichtungsdruck S 48
Dichtungselement G 45
Dichtungsfett P 5
Dichtungsfläche J 19
Dichtungskitt S 40
Dichtungslack S 46
dichtungslos G 47
Dichtungsmanschette S 41
Dichtungsmaterial S 37
Dichtungsnut G 46
Dichtungsring G 48
Dichtungsring aus
  Plattenmaterial F 95
Dichtungsring mit
  x-förmigem Querschnitt
  X 7
Dichtungstragring R 151
Dichtungswachs S 40, S 50
Dichtungswulst S 49
Dickenmeßgerät T 66
Dickenscherschwingung
  T 67
dielektrische Erwärmung
  D 104
dielektrische Hochfrequenz-
  erwärmung D 105
dielektrischer Verlustfaktor
  D 191, L 106
dielektrische Trocknung
  D 103
Dielektrizitätskonstante
  D 102
Differentialgetriebe D 107
Differentialgetterpumpe
  D 108
Differentiallecksuchgerät
  D 113
Differentialmanometer
  D 115
Differentialmikromano-
  meter D 116
Differential-Pirani-
  Lecksuchgerät D 117
Differentialpumpstufe
  D 121
differentielle Ionisierung
  D 110
differentieller Stoßquer-
  schnitt D 106
Differenzdruck D 118
Differenzdruckmessung
  D 119
Differenzdruckvakuum-
  meter D 115
diffus D 125
Diffusion D 128
Diffusionsaktivität D 129
Diffusionsdüse D 134
Diffusionsejektorpumpe
  D 131
Diffusionsfenster D 133
Diffusionsgeschwindigkeit
  D 137
Diffusionskammer S 290
Diffusionskoeffizient D 130
Diffusionskonstante D 130
Diffusionspumpe D 135
Diffusionspumpenöl D 136
Diffusionspumpe zwischen
  Haupt- und Vorpumpe
  B 168
Diffusionsspaltbreite J 7
Diffusionsspaltfläche A 74
Diffusionsstufe D 139

Diffusionsverbindung
  D 138
Diffusionsverdampfung
  D 132
Diffusor D 126
Diffusorhals D 127
Diodenmeßgerät C 402
Diodenzerstäubung D 141
Diode-Pentode D 140
Diode-Triode D 143
direkte Verdampfung
  D 149
direkt geheizt D 155
dispergieren D 183
Dispersion D 184
Dispersionsgetter D 181
Dissoziationsdruck D 192
Dissoziationsgrad D 49
Divergenz A 149
Donator D 200
Doppeleinschmelzung
  D 216
Doppelfadenpendel D 215
doppelfokussierender
  Massenspektrograf D 211
Doppelglockenanlage
  D 206
Doppelglockenanordnung
  D 209
Doppelglockenmethode
  D 208
Doppelglockenprinzip
  D 207
Doppelschneidendichtung
  D 213
Doppelsitzdichtung B 20
Doppelwandglocke D 218
doppelwandiges System
  D 209
doppelwandige Vakuum-
  kammer D 219
Doppelwendel D 185
Doppelwendelheizer C 186
Dorn M 17
Dosierpumpe D 202
Dosierventil D 203
Dosiervorrichtung D 201
Drahtdichtung W 48
Drahtsiebmaschenweite
  W 46
Drehanode R 190
drehbares Kompressions-
  vakuummeter T 199
drehbare Vorrichtung
  R 174
Drehdurchführung R 178
Drehfeld R 191
Drehflansch L 104
Drehkolben I 18
Drehkolbengebläse R 181
Drehkolbenpumpe R 182
Drehkolbenverdichter
  R 172
Drehkorb R 171
Drehmoment T 127
Drehmomentenschlüssel
  T 128
Drehschieber S 179
Drehschieberbauart R 186
Drehschieberpumpe
  R 176, S 186
Drehschwingungs-
  vakuummeter T 132
Drehspulgalvanometer
  M 184
Drehtisch T 200
Drehtrommel P 78, R 174
dreiatomig T 169
Dreierstoß T 87
dreipolig T 174
Dreiwegehahn T 89
Drift D 79
Drosselklappe B 233
Drosselung T 93
Drosselventil B 233
Drosselzeit C 135
Druck P 146
Druckabfall P 158

Druckabfallmethode P 159
druckabhängig P 154
Druckablesung P 166
Druckänderung P 150
Druckanlage P 184
Druckanstieg P 169
Druckanstiegsgeschwindig-
  keit R 35
Druckanstiegsmethode
  I 176
Druckausgleich P 151
Druckausgleichsleitung
  P 168
Druckausgleichsventil
  P 160
Druckbereich P 165
Druckdifferenz P 155
Druckdifferenzlecksucher
  D 113
Druckdifferenzlecksuch-
  methode D 112
Druckdiffusion P 156
Druckeinheit P 178
Druckeinstellung P 171
druckelektrisch P 53
Druckempfindlichkeit
  P 170
Druckfenster P 180
Druckfestigkeit C 249
Druckglaseinschmelzung
  C 247
Druckguß P 149
Druckhöhe P 161
Druckimpulsvakuum-
  messung F 85
Druckleitung D 164
Druckluft P 183
druckluftbetätigtes Ventil
  A 109
Druckluftstrom A 97
Druckmeßgerät M 19
Druckminderventil P 167
Druckprüfung P 175
Druckreduzierventil P 167
Druckregler P 152
Druckschalter P 174
Druckschmierung F 153
Druckschreiber M 18
Drucksinterofen P 145
Druckstoß P 147
Druckstufe P 172
Druckstufendichtung
  P 173
Druckstufenschleuse
  D 114
Druckstufensystem D 122
Druckteilung P 157
Drucktest P 175
Druckumlaufschmierung
  F 116
Druckumrechnungs-
  konstante P 153
Druckumwerter P 177
Druck unter 10⁻³ Torr
  B 127
Druckverdüsung A 207
Druckverhältnis C 243
Druckverhältnisse in Pum-
  pen P 179
Druckverlust P 162
Druck-Zeit-Kurve P 176
Duktilität D 267
Dunkelentladung D 4
Dünnfilmschaltkreis
  T 68
dünngewalztes Feinsilber-
  blech T 82
Dünnschichtdestillation
  T 71
Dünnschichtentgasung
  T 69
Dünnschichtentgasungs-
  kolonne T 70
Dünnschicht-Kurzweg-
  destillation T 79
Dünnschichtlichtfilter
  T 73

Dünnschichtspeicher-
  element T 75
Dünnschichttechnik T 80
Dünnschichtthermo-
  element T 81
Dünnschichttrockner T 72
Dünnschichtverdampfer
  F 8
Dünnschichtwiderstands-
  thermometer T 78
Duoplasmatron-
  Ionenquelle D 270
Duplexpumpe C 237
Durchbrennen B 231
Durchbrennen des
  Glühfadens F 41
„Durchbruchs"-Vor-
  vakuumdruck B 194
Durchflußmenge M 31
Durchflußmengen-
  verstärker F 128
Durchflußmesser F 122
Durchführung L 30
Durchführung für Ein-
  tauch-Ionisations-
  vakuummeter-System
  N 41
Durchführung mit einem
  eingeglasten Draht
  S 165
Durchführung mit mehre-
  ren eingeglasten Drähten
  M 203
Durchführungsring E 40
Durchgangsventil I 72
Durchlässigkeit P 33
Durchlaufchargen-
  destillationsanlage C 423
Durchlaufentgasung S 376
Durchlaufentgasung von
  Pfanne zu Pfanne L 7
Durchlaufglühofen C 295
Durchlaufofen T 95, T 193
Durchlaufsinterofen C 302
Durchlauftrockner T 190
Durchlauftropfenent-
  gasung S 291
Durchsatz C 20
Durchsatzmenge C 20
Durchsatzofen C 301
Durchschlagfestigkeit
  B 190
Durchschlagsspannung
  B 191
Durchschnittstemperatur
  A 222
Durchtunnelung T 192
Düse N 31
Düsenantrieb J 12
Düsenhals N 35
Düsenhut J 5, C 191
Düsenhutdampfsperre
  C 192
Düsenhutheizer J 6
Düsenkopf N 32
Düsenmantel S 173
Düsenmund N 34
Düsenmündung N 34
Düsensatz J 4
Düsenschirm A 180
Düsenstock J 4
Düsenvorgang J 10
Dushmansches Reibungs-
  vakuummeter D 271
Dyn D 282
dynamische Druckstufe
  D 280
dynamische Leckprüfung
  D 278
dynamischer Druck D 279
dynamisches Vakuum-
  system D 281
dynamische Vakuumanlage
  D 281
dynamische Viskosität
  A 12
dynamische Zähigkeit
  A 12, V 242

E

Ebonit E 3
Ecken-Golddrahtdichtung
  C 344
Ecknaht C 345
Eckventil A 146
Edelgas I 47
Edelgasröhre R 27
effektive Ordnungszahl
  E 11
effektives Kompressions-
  verhältnis E 12
effektives Saugvermögen
  N 6
effektive Verdampfungsrate
  E 14
Effusion E 17
Effusionsgesetz E 18
Eichgas C 15
Eichleck C 11
Eichleckdimensionierung
  C 12
Eichmarke R 74
Eichung C 13
Eigendiffusion S 64
Eigenfrequenz F 225
Eigensaugvermögen I 119
Eigenstreuung S 73
Eigenstruktur S 75
Eigenverdampfung S 66
einadrig S 154
einatomare Schicht M 168
Einbau-Omegatron-
  Massenspektrometer N 42
Einbausystem B 46
einbetten E 121
Einbetten E 122
Einblickfenster I 78
Einblockpumpe M 169
einbrennen B 229
Eindringhärte I 38
Eindringtiefe D 70
Einfach-Doppel-Überzugs-
  Gerät S 163
Einfachwendel S 153
einfachwirkend S 150/1
Einfallswinkel A 144
Einfangkoeffizient C 31
Einfangwahrscheinlichkeit
  C 31
Einfriergeschwindigkeit
  F 205
Einfriervorrichtung F 201
Einführungsdraht L 32
Einführungskabel L 31
Eingangsschleuse S 144
Einhebelventilblock O 41
einkapseln E 130
Einkapselung E 131
Einklemmen P 59
Einkristalldraht M 171
Einkristallzerstäubung
  S 156
Einkristallziehmaschine
  S 155
Einlassen von Luft A 98
Einlaßrate I 52
Einpolröhre S 159
Einsatzhärten C 49
Einsaugen I 80
Einschaltdruck S 326
Einschmelzdraht S 51
Einschmelzglas S 45
Einschmelzstelle G 45
Einschmelzung G 45
Einschneckenpresse S 162
Einspannvorrichtung
  G 125, J 16
Einspritzkondensator I 67
einstellbares Leck A 71
Einstellbereich S 95
Einstellgenauigkeit A 36
Einstellskale D 53
Einstellung S 90
Einstellzeit S 96
einströmen F 121
Eintauchionenquelle I 74

Eintauchkühlfinger I 12
Eintauchmeßröhre D 150
Eintauchsystem B 46, N 43
Eintrittsstutzen I 70
Einwalzentrockner S 157
Einwegofen U 16
Einzelstabeinschmelzung
  S 160
Einzelstreuung S 161
Eisbildung I 2
Eisenschlacke U 170
Eiskern I 3
Eiskondensator I 1
Ejektorpumpe E 20
Ejektorstufe E 22
elastischer Wirkungsgrad
  R 127
Elastizitätsgrenze E 24
Elastizitätsmodul M 135
Elastizitätszahl E 23
Elastomer E 26
Elastomerdichtung E 27
elektrisch ausglühen A 151
elektrische Doppelschicht
  E 30
elektrische Durchschlags-
  festigkeit E 36
elektrische Gasaufzehrung
  E 29
elektrischer Überschlag im
  Vakuum V 17
elektrisches Massenfilter
  E 31
elektrisches Stumpf-
  schweißen F 80
Elektrodenabbrand E 39
Elektrodenhalter E 41
Elektrodenmetall F 45
Elektrodenstange S 360
Elektrodenvorschub E 37
Elektrodenvorspannung
  E 38
Elektroerosion E 42
elektroerosive Bearbeitung
  E 43
Elektrolumineszenz-
  manometer E 44
Elektrolyse E 45
elektrolytisches Metalli-
  sieren E 113
elektrolytisches Polieren
  E 46
elektrolytisch polieren
  E 115
Elektrolytsilber E 47
elektromagnetisch betätigt
  E 48
Elektrometerröhre E 49
Elektronenanlagerung
  E 84
Elektronenanregung E 91
Elektronenausbeute C 65
Elektronenbahn E 101
Elektronenbahnlänge E 102
Elektronenbeschuß E 81
Elektronenbeugung im
  Durchstrahlungs-
  verfahren T 156
Elektronenbeugung in
  Reflexion R 85
Elektronenbeugung in
  Transmission T 156
Elektronenbeweglichkeit
  E 99
Elektronendruck E 103
Elektronendurchtunnelung
  E 111
Elektroneneinfang E 84
Elektronenemission E 87
Elektronenemissions-
  vermögen T 53
Elektronenflachstrahl E 109
Elektronengas E 88
Elektronenhülle E 107
Elektronenkanone E 89
Elektronenlawine E 50
Elektronenlinse E 96
Elektronenoptik E 100

elektronenoptische
  Abbildung E 97
Elektronenrekombination
  E 105
Elektronenröhre E 92
Elektronenröhren-
  herstellung E 110
Elektronenschale E 108
Elektronensonde E 104
Elektronenspiegel-
  mikroskop E 98
Elektronenstoß E 86, E 93
elektronenstoßgeheizte
  Verdampferquelle
  E 80
Elektronenstoßheizung E 94
Elektronenstoßionenquelle
  E 83
Elektronenstoßionisation
  E 95
Elektronenstoßverdampfer-
  quelle E 82
Elektronenstrahl E 52
Elektronenstrahlabgleich
  E 55
Elektronenstrahlbearbeitung
  E 70
Elektronenstrahlbohren
  E 61
Elektronenstrahlbohrgerät
  E 62
Elektronenstrahler E 89
Elektronenstrahler mit
  Eigenbeschleunigung
  S 60
Elektronenstrahler mit
  Fremdbeschleunigung
  W 52
Elektronen-Strahler-
  schleuse G 141
Elektronenstrahlerzeuger
  E 67
Elektronenstrahlerzeugung
  E 58
Elektronenstrahlgerät E 69
Elektronenstrahlgießofen
  E 57
Elektronenstrahlglühanlage
  E 54
Elektronenstrahlglühen E 53
Elektronenstrahlheizung
  E 68
Elektronenstrahl hoher
  Perveanz H 76
Elektronenstrahllöten E 56
Elektronenstrahlmehr-
  kammerofen E 72
Elektronenstrahl-Mehr-
  kanonen-Schmelzofen
  E 73
Elektronenstrahlofen E 66
Elektronenstrahlpolymeri-
  sation E 74
Elektronenstrahlschmelzen
  E 71
Elektronenstrahlschmelz-
  ofen mit Fernkanone
  R 109
Elektronenstrahlschneiden
  E 60
Elektronenstrahlschreiben
  E 75
Elektronenstrahlschweißen
  E 79
Elektronenstrahlsintern
  E 76
Elektronenstrahlsonde E 104
Elektronenstrahl-Stahl-
  bandbedampfungsanlage
  E 77
Elektronenstrahlstromdichte
  E 59
Elektronenstrahltechnologie
  E 78
Elektronenstrahlverdampfer
  E 65
Elektronenstrahl-
  verdampfung E 63

Elektronenstrahl-
verdampfungsanlage E 64
Elektronenstrahlverdamp-
fungseinrichtung E 64
Elektronenwolke C 168,
E 85
Elektroplattieren E 113
elektropneumatisch
betätigt E 114
Elektropolieren E 116
elektrostatische
Abschirmung E 117
Elektrovalenz E 118
eloxieren A 168
Emanation E 120
Emissionsausbeute E 127
Emissionsfläche E 126
Emissionsmaß E 125
Emissionsstromdichte E 124
Emissionsvermögen E 128
Empfindlichkeit S 80
Empfindlichkeits-
einstellung S 81
Empfindlichkeitsstufe S 81
emulgieren E 129
Endanglasung T 181
Endausschalter L 59
Enddichte U 1
Enddruck B 130
Endgitter E 134
Endmaß E 132
endotherm E 135
Endröhre E 136
Endstück T 2
Endtrocknung A 90
Energiedichte P 123
Energieniveaudiagramm
E 139
Energiesprung E 138
Energiestreuung E 140
engster Strahlquerschnitt
C 381
Entdröhnen N 11
enteisen D 56
Enteisungsanlage D 35
Entfernen einer Aufdampf-
schicht D 15
entfetten D 47
Entfeuchtung D 54
Entflammung I 9
entfrosten D 34
entgasen D 37
Entgasen D 39, G 54
Entgasen einer Vakuum-
anlage während des
Pumpprozesses mittels
Gasflamme T 125
Entgasung D 39
Entgasung durch Bom-
bardement D 41
Entgasungsanlage D 44
Entgasungsgefäß D 8
Entgasungsglocke D 40
Entgasungskolonne D 42
Entgasungsrate D 45
Entgasungsregler D 38
Entgasungsturm D 46
entglasen D 84
Entglasung D 83
Enthalpie E 141
entkeimen S 358
entkohlen D 12
Entkohlung im Vakuum
V 39
Entladungsrohr D 169
Entladungssperre D 168
Entladungsstrecke D 163
Entladungsstrom D 160
Entleerung D 229
Entleerungshahn D 228
Entleerungsventil D 232
Entlüftung D 7
Entlüftungsloch A 113
Entlüftungsschraube A 112
Entlüftungsventil A 101
Entmagnetisierung D 60
Entmischungseffekt S 84
Entriegelung U 20

Entropie E 143
Entspannungstemperatur
H 102
Entwachsungseinrichtung
D 86
entzundern D 71
Epitaxie E 146
Epoxydharz E 147
Erdpotential E 2
Erdung E 1
Erg E 151
Erhöhung des Röhren-
vakuums I 36
Erholungszeit C 157
Erinnerungseffekt M 74
Ermüdung F 14
Ermüdungsbruch F 15
Erneuerung R 111
Erniedrigung des Röhren-
vakuums D 22
Ersatzglühfaden S 252
Ersatzteil S 253
Erstarrung S 91
Erstarrungsbereich F 208
Erstschmelze F 67
Erweichungspunkt S 210
Erweichungstemperatur
S 211
Erzaufbereitung O 59
Erzeugung von Vakuum
V 127
erzwungene Konvektion
F 152
eutektischer Punkt E 157
evakuieren E 158
evakuierter Zwischen-
kanal D 120
Evakuierungsprozeß E 160
Evakuierungszeit P 212
Exhaustor A 197
Exoelektronenemission
E 204
exotherm E 205
expandieren E 206
Expansionsdüse E 207
Expansionskühlung E 211
Expansionsmaschine D 19
Expansionsraum E 208
Expansionsschieber E 212
Expansionsventil E 213
Expansionsverhältnis E 210
Explosionsgemisch E 215
Explosivverdampfung E 214
extrahieren E 222
Extraktionsmittel E 223
extrapolieren E 225
Extruder E 227

F

fahrbar P 108
fahrbarer Pumpstand einer
Pumpstraße T 175
fahrbares Pult M 130
Fahrenheit-Skale F 5
Fallbügelregler H 115
Falle T 164
Faltenbalg B 93
falzen F 149
Falzverbindung L 14
Faradayscher Dunkelraum
F 12
Faradayscher Käfig F 11
Farbschicht D 274
Farbtemperatur C 227
Fasertextur F 32
Federbalg B 93
Federbalgdifferential-
manometer mit Null-
ablesung B 121
Federbalghochvakuum-
ventil B 94
Federbalgventil B 95/6
federbelastet S 303
federkraftbelastetes Auspuff-
ventil S 304
Federkeil F 19

Federkontakt S 302
Federrohrvakuummeter
B 183
Federstahl S 305
Federungskörper B 93
Federventil F 20
Federzange T 202
Fehlbedienung F 18
Feinbedampfer S 158
Feinbeizen F 62
feines Loch P 62
Feinfühligkeit S 80
Feingoldkatode R 80
Feinvakuum F 64
Feldelektronenmikroskop
F 34
Feldemissions-Adsorptions-
spektrometrie F 33
Feldemissionsmikroskop
F 36
Feldionenmikroskop F 35
Feldionisation F 38
Feldionisationsvakuum-
meter F 39
Feldionisierung F 38
Feldverdampfung F 37
Fermi-Kante F 27
Fermi-Niveau F 28
Fernanzeige R 110
Fernfokussystem T 14
Fernkanone R 108
Fernkatode D 193
Fernsteuerung R 107
Ferritschaltkern F 29
Fertigmaß F 66
feste Kohlensäure C 33
feste Lösung S 227
feste Verbindung F 68
Festfressen S 58
fest haftend A 62
festklemmen C 147
Festkörpermikroelektronik
I 90
Festkörperoberfläche S 229
Festkörperphysik S 228
Fettverbindung A 120
fettwachsartig A 70
Feuchtigkeit H 144
Feuchtigkeitsentzug D 54
Feuchtigkeitsgehalt M 139
Feuchtigkeitswaage M 138
filtern F 57
Filtration F 61
Firnis V 229
Fixierstift A 119
Fixpunkt F 69
Flachdichtung F 95
flachdrücken F 98
Flächenkapazität S 111
Flächenstoßhäufigkeit I 21
flächenzentriert F 2
Flachflansch P 75
Flachfolie F 94
Flachgewinde S 315
Flachquetschfuß F 97
Flachstrahlkanone T 163
Flammen T 125
Flammenlöten T 124
Flammenschweißen F 71
Flammofen A 104
Flammpunkt F 91
Flammspritzen F 70
Flansch F 72
Flanschanschluß F 73
Flanschdichtung F 75
Flanschkerbe F 76
Flansch mit zwei konzen-
trischen Golddraht-
Eckdichtungen D 210
Flanschverbindung F 73
Flaschengas B 163
Flash-filament-Methode
F 85
Flash-filament-Technik F 86
Fleckseigerung F 181
Fliehkrafteinlaß C 86
Fließdruck Y 3
Fließgrenze Y 2

Fließkunde R 156
Fließpunkt F 126
Fließtemperatur F 126
Fließverhalten F 134
Flip-Flop-Schaltung F 106
Flockenbildung F 112
Flockpunkt F 113
Flockung F 112
Flotation F 117
Flüchtigkeit F 222
Flügelpumpe V 207
Flugzeitmethode T 109
Fluidität F 134
Fluoreszenzschicht F 137
Fluoreszenzschirm F 138
fluoridfest F 139
flüssiger Stickstoff L 85
Flüssigkeit geringen
Dampfdruckes L 118
Flüssigkeitsdichtung L 88
flüssigkeitsgedichtete
mechanische Pumpe
L 89
flüssigkeitsgekühlt L 78
Flüssigkeitsmanometer L 81
Flüssigkeitsringpumpe
F 136
Flüssigkeitsstand L 79
Flüssigkeitsstandanzeiger
L 80
Flüssigkeitsstrahlpumpe
F 135
Flüssigkeitsströmung F 131
Flußmittel für Hartlot H 18
Flußmittel für Weichlot
S 214
Flußstahl L 108
Flüsterpumpe W 40
Fluten F 115
Flutventil A 95
Fokussierung F 146
Folienblasen S 110
Foliendichtung F 147
Folienelektronik P 24
Folienfenster F 148
Fördergas C 41
Förderhöhe P 161
Förderleistung C 20, D 185,
P 215
Fördermenge C 20
Förderpumpe P 24
Förderrutsche C 316
Formänderungsvermögen
W 51
Formbeständigkeit F 172
Formenschleuse L 96
Formfaktor S 102
Formgebung S 103
Formguß S 101
Formiergas F 170
Formierspannung F 171
Formmaskenguß S 119
Formmaskenverfahren S 13
Formschluß C 50
Fortpflanzungsrichtung
D 154
Fotoätzen P 46
Fotoelektronenkonstante
P 45
Fotoionisation P 47
Fotokatode P 42
Fotolackmaske L 4
Fotolumineszenz P 48
Fotometer mit modulier-
tem Lichtstrahl M 133
Fotostrom / durch weiche
Röntgenstrahlung
erzeugter S 215
Fotozelle P 40
Fraktionierbürste F 178
Fraktionierbürsten-
Molekulardestillations-
anlage W 45
fraktionierende Pumpe
F 180
Fraktionierkolonne F 179
fraktionierte Destillation
F 176

fraktionierte Hochvakuum-
destillation H 87
freie Energie F 184
freie Expansion F 185
freie Konvektion N 1
Freifallverdampfer F 186
Freiflußventil F 187
Freistrahlen mit Stahlsand
S 132
Freiwegdestillation F 189
Fremdatom F 156
Fremdfeld E 220
Fremdschicht F 157
Frequenzverschiebung F 209
Fritte F 216
Fugazität F 222
Fühler S 79
Führungslager G 137
Führungsstange G 140
Führungsstift G 139
Fülldruck F 48
Füllkörperkolonne P 1
Füllmittel F 44
Füllstandanzeiger L 52
Füllstandsfühler L 82
Füllstandwächter L 53
Fülltablett C 116
Funkelrauschen F 104
Funkenerosion S 258
Funkenüberschlag F 89
Fußquetschmaschine S 353
Futter L 69

G

Gaedesches Molekular-
vakuummeter G 3
Galvanostegie E 113
Gammastrahlen G 4
Ganzmetallausführung
A 123
Ganzmetallventil A 124
garantiertes Vakuum G 133
Gasabgabe O 68
Gasadsorption G 25
Gas-Adsorptionschromato-
graphie A 80
Gasanalyse G 7
Gasanalyse bei der Zinn-
schmelze T 112
Gasanalyse durch fraktio-
nierte Adsorption F 173
Gasanalyse durch fraktio-
nierte Desorption F 175
Gasanalyse durch fraktio-
nierte Kondensation F 174
Gasanalyse durch fraktio-
nierte Verdampfung
F 177
Gasaufnahme S 238
Gasaufzehrung C 154
Gasaufzehrungspumpe
C 155
Gasausbruch G 17
Gasballast G 10
Gasballastbetrieb A 96
Gasballastpumpe G 11
Gasballastpumpe nach
Gaede G 1
Gasballastventil G 13
Gasblasenpumpe G 16
Gaschromatografie G 18
gasdicht G 61
Gasdichte D 63
Gasdichtheit G 62
Gasdosierleck G 23
Gasdosierung G 22
Gasdruckanstieg I 35
Gasdurchlässigkeit G 51
gasdynamische Strömung
D 275
Gaseinfüllventil G 32
Gaseinlaßventil D 203, G 6
Gaselektronik G 27
Gasentladung G 26
Gasentwicklung G 30
Gaserzeugung G 38

Gasfaktor G 31
Gasfokussierung G 36
gasförmig G 24
Gasführungsflansch G 34
Gasgehalt G 20
Gashaut G 33
Gasinjektion G 40
Gasion G 28
Gasionenkonstante G 41
Gaskältemaschine C 392
Gaskinetik G 50
Gaslöslichkeit G 58
Gas-Luft-Gemisch A 105
Gasmenge Q 4
Gasquelle G 59
Gasreinigung G 55
Gasrückströmung G 9
Gasschicht G 33
Gasschmelzschweißung
G 37a
Gasstrahl G 43
Gasstrahlpumpe G 44
Gasstrom F 123
Gasströmung F 123
Gasströmungsmesser G 35
gasundurchlässig G 39
Gasverflüssigung L 75
Gasvolumeter G 63
Gasvorrat G 60
Gaszerlegungsanlage G 56
Gaszusammensetzung G 19
Gaszutritt G 5
Gay-Lussacsches Gesetz
C 119
geätzte Kupfermaske E 152
Gebläse B 147
Gebläsebrenner B 150/1
gebündelt B 76
gefederte Ventilplatte S 2
geflutetes System F 114
Gefrierätztechnik F 198
Gefrierätzung F 197
gefrieren F 190
Gefrieren F 199
gefriergetrocknetes Gut
F 192
Gefrierkonzentration F 191
Gefrierleistung F 200
Gefrierpunkt F 203
Gefrierpunktserniedrigung
F 204
gefriertrocknen F 193
Gefriertrocknung C 391
Gefriertrocknungsanlage
F 195
Gefriertrocknungsgeschwin-
digkeit F 196
Gefriertrocknungskammer
F 194
Gefriertunnel F 207
Gefrierzeit F 206
Gegenfeldelektrode für
Elektronen E 106
Gegenfeldelektrode für
Ionen I 160
Gegenfeldzerstäubung B 111
Gegenflansch C 357
Gegenstrom C 358
Gegenstromaustauscher
C 356
Gegenvakuum A 177
Gegenwendel D 212
Gehäuse S 335
Geiger-Müller-Zählrohr
G 70
Geiger-Zähler G 70
Geißlerrohr G 71
Geisterlinie G 77
geklebte Steckmuffe C 81
gekühltes Wasser C 128
gemessene Rückströmungs-
rate M 50
gemessenes Saugvermögen
M 51
Genauigkeit A 35
gepreßte Katode M 182
gerichtet B 76
geringes Vakuum L 116

Germanium G 72
Gesamtdruck T 136
Gesamtemissionsvermögen
T 134
Gesamtfluß T 135
Gesamtsaugvermögen
T 139
Gesamtstrahlungs-
pyrometer A 187, T 138
Gesamtundichtigkeit I 89
gesättigter Dampf S 16
geschichtete Säule S 381
geschlossenes Quecksilber-
manometer C 166
geschmiedeter Flansch
F 167
Geschwindigkeitsverteilung
V 233
Gesenk D 100
Gestaltfaktor S 102
Gestaltfestigkeit R 31
gestrahlt B 76
geteilter Ring S 283
getriggerte Entladungs-
meßröhre T 170
Getter G 73
Gettermaterial G 73
Getterplättchen G 75
Getterpumpe G 76
Getterung C 154
Getterung mittels ionisieren-
der Gasentladung E 35
Gewebegefriertrocknung
T 114
gewellte Membran C 350
Gewindereduzierstück T 86
gezogenes Metall D 236
GFPH-Kupfer G 37
Gießbereich M 67
Gießen C 55
Gießform M 181
Gießharz C 54
Gießharzanlage R 129
Gießharzmantel C 178
Gießpfanne C 53
Gießpfanne mit Gieß-
schnauze L 73
Gießrinne P 119
Gießschnauze L 72
Gießschweißen C 57
Gießstrahlentgasung S 376
Girlandentrockner F 30
Gitterebene L 22
Gitterelektron L 20
Gitterendabdeckung G 122
Gitterfehler L 18
Gitterfehlstelle L 21
Gitterfehlstellendichte L 19
Gitterkonstante L 17
Gitterkühlung C 327
Gitterleerstelle L 23
Gitterschwingung L 24
Gitterstegung G 123
Glasanschluß G 95
Glasbalg G 82
Glasfaser G 87
Glasfasertuch F 31
glasfaserverstärkt G 88
Glasfilter G 89
Glasfluß V 244
Glasfuß G 84
Glasglocke G 81
Glaskitt G 84
Glas-Kovar-Verschmelzung
K 17
Glas-Kupfer-Verschmelzung
C 341
Glaslot S 220
Glasmantel G 86
Glas-Metall-Einschmelzung
M 97
Glas-Metall-Verschmelzung
M 97
Glasperle G 80
Glasquetschfuß G 94
Glasrohr G 97
Glasröhre G 97
Glasschliff G 93

Glasschliffverbindung
G 127
Glastinte E 154
Glasur G 98
Gleichung des idealen
Gases G 29
Gleitebene S 188
Gleitentladung S 181
Gleitgeschwindigkeit S 191
Gleitlagermetall A 172
Gleitlinie S 184
Gleitmuffe S 185
Gleitreibung S 183
Gleitringdichtung F 4
Gleitströmung S 182
Gleitung S 182
Glimmeinrichtung G 100
Glimmen C 73
Glimmentladung G 99
Glimmentladungsrohr G 101
Glimmer M 101
Glimmerabblätterung
M 102
Glimmerfenster M 108
Glimmerisolation M 103
Glimmerplatte M 105
Glimmerscheibe M 106
Glimmreinigung D 158
Glockenanlage B 91
Glockenventil B 98
Glühdraht F 40
Glühdraht mit Wolfram-
karbidbekleidung C 35
Glühdrahtzange H 140
Glühelektronenemission
T 52
glühen A 150
Glühen von rostfreiem
Stahl A 153
Glühfaden F 40
Glühfadenlebensdauer F 42
Glühfadenpyrometer D 157
Glühfadentemperatur F 43
Glühkatode H 121
Glühkatodenionisations-
vakuummeter H 122
Glühkatodenionisations-
vakuummeter nach
Bayard und Alpert B 67
Glühkatodenröhre H 125
Glühkatodenventil H 126
Glühlampe I 34
Glühofen A 152
Glutfestigkeit I 10
Glyzerin-Bleiglätte L 92
Glyzerin-Bleiglätte-Kitt
L 33
Golddrahtdichtung G 103
Grad Celsius D 48
Grad Rankine D 52
Grahamsches Gesetz E 18
Grammatom G 110
Grammolekül G 111
Granulat G 112
Graphitelektrode G 115
Graphitschicht A 181
Graphittiegel G 114
Graphitüberzug G 113
grauer Körper G 120
Grenzdruck der Vor-
vakuumbeständigkeit
C 370
Grenzfläche I 96
Grenzflächenenergie I 97
Grenzflächenfilm I 98
Grenzflächenreibung B 179
Grenzflächenschmierung
B 182
Grenzschicht B 180
Grenzschichtströmung
B 181
Grenzschmierung T 74
Grenztemperatur M 45
Grenzwellenlänge C 374
Grobevakuieren C 171
grobpumpen P 133
Grobpumpzeit F 161
Grobvakuum R 200

Grobvakuumleitung R 197
Grobvakuumpumpe R 198
Grobvakuum- und Halte-
pumpe mit Gasballast-
einrichtung G 12
Grobvakuumventil B 240,
R 199
Grübchenbildung P 73
Grundanstrich P 186
Grundfrequenz F 225
Grundgröße F 224
Grundlack B 57
Grundlackschicht B 56
Grundlagenuntersuchung
B 61
Grundplatte B 59
Grundschicht P 186
Grundverfahren U 19
Grundviskosität I 121
Grünfuttertrockner G 119
Gummidichtung R 203
gummielastischer Werk-
stoff R 205
Gummifingerling R 202
Gummischlauch R 204
Gummiverbindungsstück
R 201
Guß C 51
Gußblase B 138
Gußeisen C 55
Gußverbindung C 56
Gütegrad C 181
Gütezahl C 181
Guttapercha G 142

H

Haarkristall C 412
Haarnadelkatode H 1
Haarriß C 25
Hafenofen P 116
haftbeständig A 62
haften A 59
haftfeste Bedampfung A 64
Haftfestigkeit A 60
Haftkoeffizient S 359
Haftreibung S 327
Haftwahrscheinlichkeit C 31
Haftzahl S 359
Hahnfassung K 5
Hahnfett S 365
Hahnküken K 4
Hahn mit offenem Küken
V 38
Halbleiterbauelement S 77
Halbleiterkatode S 76
Halbschattenbereich P 30
Halbwertszeit H 2
Halbzeug S 78
Halogen H 3
Halogenanzeigegerät H 4
Halogen-Diode H 5
Halogen-Dioden-Meßzelle
H 6
Halogenkleinschnüffler H 9
Halogenlecksucher H 7
Halogenlecksuchgerät H 7
Haltepumpe H 101, P 194
Haltetemperatur S 318
Haltevakuum H 103
Haltezeit H 102, S 337
handbetätigt M 22
handbetätigtes Ventil H 10
Handgebläse H 11
Handschuhkasten D 250
Hängetrockner F 30
Härte H 16
härten H 12
Härteprüfung H 17
Hartgummi E 3
hartlöten B 187
Hartlöten B 188
Hartlöten im Vakuum V 15
Hartlöten mittels Flamme
T 124
Hartlöten mittels Ofens
F 226

Hartlöten unter Wasser-
stoffatmosphäre H 149
Harzhärter R 130
Haubenleckprüfung C 105
Haubenlecksuchverfahren
C 105
Hauptkondensator M 11
Haupttrocknung M 12
Hebelsystem M 57
Hebelventil L 54
Heber F 25
Heftschweißen T 1
Heißextraktionsanalyse mit
Zinn als Badmetall T 112
Heißextraktionsanalysen-
anlage F 238
Heißextraktionsapparatur
V 73
Heißextraktionsgasanalyse
H 129
Heißextraktionsverfahren
H 130
Heißluftdusche D 43
Heißpressen H 134
Heißpressen im Vakuum
V 80
Heißvulkanisation H 139
Heißwalzen H 135
Heizblech J 2
Heizdraht F 40
Heizereinschub H 32
Heizleistung H 31
Heizleistung bei Netz-
spannungsanschluß R 32
Heizleiterrohr T 180
Heizmantel H 35
Heizpatrone C 43
Heizplatte J 2
Heizplattenmagnetrührer
H 132
Heizplattentrockner J 3
Heizwendel H 57
Heliarc-Verfahren H 54
Heliumdurchlässigkeit
H 60
Heliumdurchlässigkeit von
Glas H 61
Heliumgaskälteanlage H 58
Heliumlecksuchgerät H 59
Heliumrückgewinnungs-
anlage H 62
Helligkeitspyrometer D 157
Helmholtz-Spule H 64
Herdofen O 46
hermetisch A 110
hermetisch abgedichtet
H 65
hermetische Abdichtung
H 67
hermetischer Verschluß
H 67
Hilfselektrode A 217
Hilfskatode A 216
hintereinanderschalten
C 272
hinzutropfen D 245
Hittorfscher Dunkelraum
C 375
Hochdruck-Ionisations-
vakuummeter H 77
Hochfrequenzionenquelle
H 73
Hochfrequenzlecksucher
S 255
Hochfrequenzmassen-
spektrometer H 74
Hochfrequenzprüfung S 257
Hochfrequenzvakuum-
prüfer H 75
Hochfrequenzzerstäubung
R 20
Hochglühen zum Ver-
dampfen F 87
Hochkantfilter E 10
Hochleistungsdiffusions-
pumpe H 79
Hochleistungspumpstand
H 80

Hochleistungstiefkühlfalle
H 71
hochsiedend H 70
Hochspannungsschweißen
H 99
Hochstromdurchführung
H 53
Hochvakuum H 81
Hochvakuumanlage H 90
Hochvakuumanlage für
Wachsimprägnierung
H 98
Hochvakuumaufdampf-
anlage H 82
Hochvakuumdestillations-
kolonne H 83
Hochvakuumdiagnostik-
röhre H 86
Hochvakuumgleichrichter
H 93
Hochvakuumhahn H 95
Hochvakuumpumpe H 92
Hochvakuumpräzisionsguß
H 91
Hochvakuumschmelz-
verfahren H 88
Hochvakuumsinterofen
H 94
Hochvakuumsperre H 85
Hochvakuumtherapieröhre
H 96
Hochvakuumventil H 97
Hochvakuumverbindung
H 84
Hochvakuumzüchtung
H 89
Ho-Faktor H 100
Höhenkammer A 131
Höhensimulator A 132
Hohlkatode H 106
Hohlraum C 77
Hohlstrahl H 105
homogenisieren H 110
Homogenisierung H 109
Homogenität H 108
homöopolar H 111
Hordengefrieranlage S 117
Hordentrockner S 116
horizontaler Kammerofen
H 117
Houstonsches Ionisations-
vakuummeter H 143
Hub-Kolbenpumpe R 51
Hubvolumen S 454
Hubwagen T 177
Hüllabdruck E 145
Hülle E 131
Hüllenelektron S 108
Hüllenmethode H 114
Hüllentest C 105
Hülse S 177
Hutmanschette C 21
hydraulischer Radius H 145
Hydridverfahren H 146
hydrolytische Klassifikation
H 150

I

ideale Pumpe I 8
ideales Gas I 6
Idealkristall I 5
Ikonoskop I 4
Impfling C 406
implodieren I 22
Implosion I 23
Implosionsschutz I 24
Imprägnieranlage I 27
imprägnieren I 25
imprägnieren I 28
Imprägniermittel-
aufbereitung T 168
imprägnierte Katode I 26
Imprägnierung I 28
Impuls I 29
Impulsdesorption F 86
Impulsübertragung I 30

I-Naht P 74
Inbus-Schlüssel I 33
Indikator T 142
indirekt geheizt I 40
indischer Glimmer M 2
Indium-Drahtdichtung I 41
Induktionsheizung I 44
Induktionsplasmabrenner
I 42
Induktionsschmelzen I 45
Induktionsschmelzofen I 43
Induktionsspule S 254
induktives Rühren I 46
induktive Wärmebehand-
lung I 44
Inertgasschweißung I 49
Infrarotabsorption I 53
Infrarotadsorptionsspektro-
skopie I 54
Infrarotheizung I 55
Infrarotlecksuchgerät I 56
inhomogen I 64
Innenanglasung I 77
Innendurchmesser I 73
Innengewinde F 26
Innenheizer I 114
innenmattiert I 75
Innenraum I 106
Innensechskantschraube
S 206
Innentränkung S 202
Innen- und Außen-
anglasung E 8
Innenzylinderlötung I 76
innere Reibung I 113
innerer Fotoeffekt P 41
Inselschicht I 173
Inselstruktur I 174
instationäre Strömung
U 23
integrierte Dünnschicht-
schaltung T 76
Interferenzfilter I 102
Interferenzröhre I 103
Interferenzschicht I 100
Interferenzschichtsystem
I 101
Interferenzvakuummeter
D 109
Interferenz-Wärme-
reflexionsfilter H 44
interferometrisches Diffe-
rentialvakuummeter
D 109
interferometrisches
Ölvakuummeter I 105
interferometrisches
Vakuummeter I 104
interkristalliner Bruch I 95
intermittierend B 64
Internationale Union für
Forschung, Technik und
Anwendung des Vakuums
I 115
Ionenätzen E 153
Ionenauffänger I 133
Ionenaufzehrung I 130
Ionenausbeute I 167
Ionenaustauscher I 136
Ionenbaffle I 126
Ionenbeschuß I 129
Ioneneinfang I 131
Ioneneinfangwahrschein-
lichkeit I 132
Ionenform / in I 140
Ionengetterpumpe G 74
Ionengetterpumpe mit
Drahtverdampfung W 47
Ionenkollektor I 133
Ionenkonzentration C 252
Ionenleitfähigkeit I 141
Ionenmikrosonden-
Massenspektrometer
I 155
Ionenpumpe I 150
Ionenpumpgeschwindigkeit
I 158
Ionenquelle I 166

Ionenrakete I 163
Ionenresonanz I 161
Ionenresonanzspektrometer
I 162
Ionepsorption I 164
Ionensorptionspumpe I 165
Ionensorptionsröhre I 156
Ionenstoß I 142
Ionenstrahlbedampfung
I 127
Ionenstrahltechnik I 128
Ionenstrom, bedingt durch
den Dampfdruck des
Verdampfungsgutes
E 163
Ionenstrom, bedingt durch
den Restgasdruck G 42
Ionenstromtransmission
I 134
Ionenverarmung D 21
Ionenverdampferpumpe
E 177
Ionenwirkungsgrad I 135
Ionenzerstäuberpumpe
C 197
Ionenzerstäuberpumpe vom
Diodentyp D 142
Ionenzerstäuberpumpe vom
Magnetrontyp M 10
Ionenzerstäuberpumpe vom
Triodentyp T 88
Ionenzerstäuberpumpe vom
Typ des umgekehrten
Magnetrons I 125
Ionisationsdetektor I 145
Ionisationskammer I 144
Ionisationsmanometer I 137
Ionisationsmeßgerät I 138
Ionisationsratemonitor
I 139
Ionisationsvakuummeter
I 137, I 152
Ionisationsvakuummeter
mit Bremselektrode
S 430
Ionisationsvakuummeter
mit Extraktor E 224
Ionisationsvakuummeter
mit geheizter Katode
T 51
Ionisationsvakuummeter
mit radioaktivem Prä-
parat R 16
Ionisationsvakuummeter
ohne Magnet N 21
ionisch I 140
ionisch hergestellte Schicht
I 157
ionisieren I 154
Ionisierung im starken
elektrischen Feld F 38
Ionisierungsarbeit I 147
Ionisierungseinrichtung
I 146
Ionisierungsraum I 151
Ionisierungsspannung I 148
Ionisierungsvolumen I 153
Ionisierungswahrschein-
lichkeit I 149
Iridium-Glühfaden I 168
Irisblende I 169
isentrop I 172
isobar I 175
Isolierfähigkeit I 86
Isolierkeramik C 99
Isoliermasse I 85
Isoliermaterial I 83
Isolierstoff I 83
Isoplethe I 177
isotherm I 178
Isotopengasanalyse I 182
Isotopenhäufigkeits-
bestimmung I 182
Isotopenhäufigkeits-
verhältnis I 179
Isotopentrennung I 181
Isotopenverhältnis I 180
IUFTAV I 115

J

Joch Y 4
Justierbarkeit T 171
Justiervorrichtung S 93

K

Kabelendverschluß E 133
Kabelkessel C 2
Kadmiumwolframat C 3
kalandern C 5
kalibrieren C 8
Kalibrierung C 13
Kalibrierungskurve C 14
Kalibriervorrichtung
C 16/7
kaltabbindend C 210
Kaltbrüchigkeits-
temperatur B 208
Kälteaggregat R 98
Kälteanlage L 115
Kältebedarf C 326
Kälteerzeugung R 99
Kälteleistung R 95
Kältemaschine R 97
Kältemischung F 202
Kältemittel C 327
Kältetechnik R 96
Kaltfließen C 202
Kaltfließpressen C 200
Kaltgaskältemaschine
C 189
kaltgereckter Draht C 212
kaltgewalzt C 208
kalthärtend C 210
Kaltkatode C 193
Kaltkatodengasentladung
C 194
Kaltkatodenionenquelle
C 198
Kaltkatodenvakuummeter
C 196
Kaltlichtspiegel C 205
Kaltluftmaschine A 111
Kaltprägen C 211
Kaltpreßschweißen C 207
Kaltschweißen C 217
Kaltstauchen C 204
Kaltverfestigung W 20
kaltverformt C 208
Kaltverformung C 219
Kaltvergoldung C 203
Kaltvulkanisation C 215
Kaltwalzen C 209
Kaltwandvakuum-
hochtemperaturlötofen
mit Innenheizer C 216
Kaltwasser C 128
Kalzinierung C 6
Kalziumwolframat C 7
Kammerofen C 104
Kammertrockner T 190
Kanalstrahlen C 18
Kanaltrockner T 190
Kantenfilter E 10
Kantenschweißung E 9
Kapillardepression C 26
Kapillare C 23
Kapillarkondensation C 24
Kapillarmethode C 9
Kapillarröhre C 23
Kapillarwirkung C 22
Kapselfedervakuummeter
C 30
Kapselmeßgerät C 30
kapseln E 130
Kapselpumpe B 184
Karburieren C 37
Kardanaufhängung G 78
Karussell C 40
Kaschieren eines Form-
körpers unter Vakuum
V 95
Kaskadenanordnung C 48
Kaskadenelektron C 47
Kaskadenkühler C 44

Kaskadenschaltung C 45
Kastenofen B 185
Katalysator C 58
Kataphorese C 59
Kathetometer C 60
Katodenaktivierung C 61
Katodenalterung C 62
Katodenanheizzeit C 67
Katodendunkelraum C 375
Katodenfall C 64
Katodenglimmlicht C 66
Katodenstrahl E 51
Katodenstrahlablenkung
C 69
Katodenstrahlbündel C 70
Katodenumsetzung C 63
Katodenvergiftung C 68
Katodenzerstäubung C 71
katodisches Ätzen C 73
katodisches Reinigen C 72
Katodolumineszenz C 74
Kavitation C 75
Kegelbedampfung C 264
Kegelbeschattung C 265
Kegeldichtung C 263
Kehlnaht F 47
Keilriemen V 231
Keim N 40
Keimbildung N 38
Keramik C 101
Keramikabschmelzung
C 102
Keramiken mit hohem
Aluminiumoxidgehalt
H 69
Keramik-Metall-
Verbindung C 103
Keramikscheibe C 98
Keramiktiegel C 97
Kerbempfindlichkeit N 30
Kerbschlagversuch N 28
Kerbschlagzähigkeit N 29
Kern N 40
Kernbildung N 38
Kernfusion N 37
Kernladungszahl A 205
Kernschliff M 14
Kernspaltung N 36
Kernteilchen N 39
kinematische Zähigkeit
K 7
kinetische Gastheorie
K 8
kippbarer Schmelztiegel
T 99
kippbare Vakuumglocke
T 100
Kippkesselofen T 104
Kippkessel-Vakuum-
schmelzofen T 105
Kippmoment P 71
Kirchhoffsches Gesetz K 9
Klammerflansch C 148
Klappventil F 77
Klebefolie A 65
Klebemittel A 63
kleben A 59
Kleben B 164
Kleber A 63, C 80
Klebevakuum C 27
Klebwachs A 67
Kleinflansch Q 15
Kleinflanschverbindung
S 198
Kleinsignaltheorie S 199
kleinste Düsenspaltfläche
N 35
kleinste nachweisbare
Druckänderung M 122
kleinste nachweisbare
Leckrate S 197
kleinstes nachweisbares
Leck M 121
Kleinwinkelstreuung S 196

Klemmflansch S 427
Klemmring C 150
Klemmverbindung P 59
Klemmverschraubung
S 428
Klinkwerk R 30
klopffest A 173
Knetlegierung F 166
Knetmaschine K 10
Knickfestigkeit B 212
Knickpunkttemperatur I 51
Knopfprobe B 235
Knopfschmelze B 235
Knopfschmelztiegel B 234
Knudsensches Gesetz K 16
Knudsensches Radiometer-
Vakuummeter K 13
Knudsen-Strömung K 12
Knudsen-Vakuummeter
K 13
Knudsen-Zahl K 14
Knüppel B 113
koaxiale Durchführung
C 179
Koeffizient der inneren
Reibung V 242
Kohärenzlänge C 182
Kohäsion C 183
Kohlebogenverdampfung
C 32
Kohlefalle C 110
Kohlensäureschnee C 33
Kohlenwasserstoff H 147
kohlenwasserstoffreies
Vakuum H 148
Kohlespitzenverdampfung
C 34
Kokillenrand M 183
Kokillentisch I 61
Kolbenanschmelzmaschine
B 217
Kolbenblasmaschine G 83
Kolbenhubgeschwindigkeit
S 388
Kolbenpumpe P 67
Kolbenring P 68
Kolbenventil S 200
Kolbenverdichter R 50
Kollodium C 224
kolloidal C 225
Kolophonium C 226
Komponentenkammer
C 233
Kompressionsfaktor C 240
Kompressionskapillare
C 246
Kompressionsraum C 239
Kompressionsvakuum-
meter C 241
Kompressionsverhältnis
C 243
Kompressor C 250
Kondensatabscheider C 260
Kondensation C 256
Kondensationsfläche B 23
Kondensations-
geschwindigkeit C 258
Kondensationskoeffizient
A 33
Kondensationspumpe C 257
Kondensationsrate C 258
Kondensationszahl A 33
Kondensator C 260
Kondensatorkapazität C 261
Kondensatorschichtdicken-
meßgerät C 19
kondensierbares Gas C 255
Kondensor C 260
Kondenswasser C 259
Konservierungsmethode
P 141
Konstanz C 275
Konstitutionskitt C 278
Kontaktgefrieranlage C 288
Kontaktgefrieren C 289
Kontaktgetterung C 290
Kontaktpotential C 291
Kontaktschmierstoff C 149

Kontaktsubstanz C 58
Kontaktthermometer E 34
Kontakttrockner C 286
Kontakttrocknung C 287
Kontaktunstetigkeit C 285
Kontaktverfahren C 292
kontinuierliche Anlage
  A 117
Kontinuitätsgleichung
  C 294
Kontinuum C 306
Kontrollfenster für den
  Ölstand I 79
Kontrollkristall M 166
Kontrollventil C 121
Kontrollwiderstandsabgriff
  M 167
Konvektion C 314
Konzentrationsgefälle C 251
Konzentrationsverhältnis
  C 253
Konzentrator C 254
Konzentrieren durch
  Gefrieren F 191
Kopie R 112
Korbspule B 62
Korngrenze G 106
Korngrenzenstreuung
  G 107
Korngröße G 109
Kornwachstum G 108
Korpuskularstrahl C 347
korrosionsbeständig C 348
korrosionshindernde
  Schicht C 349
Kratzer auf der Dichtung
  G 49
Kratzprüfgerät S 28
Kreisbeschleuniger C 138
Kreisblende C 139
Kreiselgebläse C 86
Kreiselpumpe C 94
Kreiselverdichter C 88
kreisförmige Hochvakuum-
  röhre des Betatrons
  D 220
Kreislaufkühlung C 165
Kreuzfeuerbrenner C 378
Kreuzquetschfuß C 140
Kreuzstrombrenner N 33
Kreuzstück C 377
Kreuzstück mit Reduzier-
  flanschen R 67
Kriechbarriere C 363
Kriechen M 116
Kriechfestigkeit C 367
Kriechschutz M 117
Kriechsperre A 170
Kriechstrom L 37
Kriechweg C 366
Kriechwegbildung S 440
Kristalldruse C 170
Kristallfehler C 405
Kristallfläche C 403
Kristallisationskeim C 406
Kristallisationskern C 406,
  N 40
Kristallisationsschwelle R 63
Kristallkeim C 406
Kristallplättchen C 411
Kristallscheibchen C 411
Kristallstörstelle C 405
Kristallstörung C 405
Kristallversetzung C 407
Kristallziehanlage C 404
Kristallziehen C 409
Kristallziehofen C 404
kritischer Druck C 372
kritischer Punkt C 371
kritische Temperatur C 373
Krümmungsradius B 103
Krustenbildung I 37
Kryofläche C 397
Kryogetterpumpe C 396
Kryopumpe C 393
Kryopumpen C 394
Kryosorptionspumpe C 398
Kryostatenprinzip C 399

Kryosublimationsfalle
  C 400
Kryotrapping C 395
Kryotrappingeffekt C 401
kubisch-flächenzentriert F 3
Kugelgelenk B 36
Kugelkühlfalle S 275
Kugellager B 39
Kugelrückschlagventil B 40
Kugelschliff G 128
Kugelschliffmagnetventil
  M 4
Kugelschliffventil S 274
Kugelventil B 41
Kühlaggregat C 329
Kühldauer H 102
Kühlfalle C 213, T 164
Kühlfalle mit geringem
  Kühlmittelverlust L 111
Kühlfinger C 201, R 114
Kühlflüssigkeit C 317
Kühlkappe C 191
Kühlmantel C 325
Kühlmittel C 317
Kühlmittelverbrauch C 318
Kühlpunkt A 154
Kühlrippe C 324
Kühlschlange C 187
Kühlsole C 321
Kühlspirale C 323
Kühlung C 129
Kühlwasseranschluß C 330
Kühlwasserkontrollschalter
  C 331
Kühlwassermantel C 332
Kühlwasserpumpe C 333
Kühlzeit C 130
kunstharzgetränkt R 128
Kunstharzmasse S 456
künstlicher Zeolith A 193
Kunststoffmetallisierung
  P 86
Kupferfoliendichtung C 338
Kupferfolienfalle C 339
Kupferkristallisator C 337
Kupfermanteldraht C 335
Kupfertiegel C 336
kurzes Glas S 129
Kurzwegdestillation O 47
Kurzwegdestillations-
  anlage O 48
Kurzwegdestillationsgerät
  mit mechanischer
  Wandreinigung S 130
Kurzzeit-Restgasmessung
  S 131

L

Labyrinthdichtung L 1
lackieren L 2
Lackschicht L 3
Lacküberzug L 3
Ladungsträger C 111
Ladungsträgerdichte C 112
Ladungsverteilung C 113
Lafferty-Ionisations-
  vakuummeter H 123
Lagermetall B 1
Lagerspiel B 85
Lambertsche Kosinus-
  verteilung C 353
Lambertsches Gesetz C 354
Lambertsches Kosinus-
  gesetz C 354
laminare Strömung L 8
Laminarströmung L 8
langes Glas L 100
Langlebensdauerröhre L 101
Langmuir-Dushmansches
  Molekularvakuummeter
  L 11
langsames Gefrieren S 194
Langzeitstabilität L 102
läppen L 12
Laser-Spiegel L 16
Laufrad M 187

Laufzeit T 150
Laufzeitmassen-
  spektrometer T 108
Laval-Düse L 26
Leck L 36
Leckanzeigevorrichtung
  L 42
Leck bekannter Größe C 11
Leckdichtungsmaterial L 46
leckfrei L 38
Leck in dünner Wand
  T 83
Leckprüfung L 40, S 257
Leckrate I 68, I 89
Leckschnüffler P 164
Leckstrom L 39
Leckströmung L 39
Lecksuche L 38
Lecksuche mit Ammoniak
  A 141
Lecksuche mit Anschluß des
  Lecksuchers an das Vor-
  vakuum B 16
Lecksuche mit Geißlerrohr
  G 21
Lecksuche mit Hoch-
  frequenz-Vakuumprüfer
  S 256
Lecksuche mit Petroleum
  K 1
Lecksuche mit radioaktivem
  Indikator R 19
Lecksucher L 43
Lecksuchgerät mit Ionen-
  pumpe I 159
Lecksuchmassen-
  spektrometer L 47
Lecksuchröhre L 41
Lecksuchtechnik mit
  Absprühsonde P 191
Lederdichtung L 48
Leerlaufzeit D 225
Leerstelle V 1
Leerstellenfluß V 2
Legierung A 126
Legierungszusatz A 127
Lehre G 65
leichtsiedend L 109
Leistungsaufnahme P 127
Leistungsdichte P 124
Leistungsverdampfer L 93
Leistungsziffer C 181
Leitelektrode G 138
Leitisotop T 142
Leitungswasser T 10
Leitung zur Grob-
  evakuierung R 197
Leitwert C 262
Leitwert bei Molekular-
  strömung M 148
Leitwert des geöffneten
  Ventils O 45
Leitwert des geschlossenen
  Ventils M 147
Leitwert einer Verbindung
  S 38
Leitwertmethode C 10,
  T 209
Leuchtgas T 140
Leuchts altbild M 119
Lichtbogenofen A 184
Lichtbogenschmelzen A 185
Lichtbogenschweißung
  A 186
Lichtdurchlässigkeit T 160
Lichtpunktschreiber F 142
Lichtzählrohr P 43
Lindescher Kreisprozeß
  L 61
Linearbeschleuniger L 60
Linienbohrung L 64
Linienschreiber C 298
Linienspektrum L 66
Lippendichtung L 71
Liquiduslinie L 90
Liquiduspunkt L 91
Littleton-Punkt S 210

L-Katode L 29
Loch C 76
Löchereinfang H 104
Lochkreis P 69
Lochkreisdurchmesser P 70
logarithmische Druckskale
  L 99
lösbare Verbindung D 62
Löschen Q 11
Loschmidtsche Zahl A 224
Löslichkeit S 232
Lösungsmittel S 233
Lösungsmitteldampf-
  reinigungsanlage S 234
Lösungswärme H 40
Lötdichtung S 225
Lötflußmittel S 222
Lötkolben S 223
Lötstützpunkt S 224
Lötung mit flexiblem
  Ausgleichsteil F 101
Lötverbindung S 225
Luftabscheidegefäß D 8
luftdicht A 110
luftdicht abgeschlossen H 65
luftdicht eingeschmolzene
  Elektrode H 66
Lufteinbruch A 106
Lufteinlaßventil A 95
Luftfilter A 102
luftgekühlt A 100
Luftklappe A 103
Luftschleuse A 108
Luftspaltsäule A 115
Luftverflüssigung A 107
Luftwäscher A 114
Luftwiderstand D 226
Luftwiderstandswaage
  D 227
Luftzerlegungsanlage A 115
Luft-zu-Luft-Anlage A 117
Lunker C 76
Luppenstahl B 146
Lyophilisation C 391

M

Mach-Zahl M 1
Madenschraube G 132
magnetisch betätigtes
  Ventil S 226
magnetische Feldstärke M 5
magnetische Induktion M 6
Magnetjoch M 8, Y 4
Magnetron M 9
Magnetron-Ionisations-
  vakuummeter mit Glüh-
  katode H 123
Magnetronvakuummeter
  C 199
Magnetronvakuummeter-
  röhre mit heißer Katode
  und Unterdrückung des
  Foto-Elektronenstromes
  H 124
Magnetspeicherschicht M 7
Magnetventil M 3
Mammutpumpe G 16
Manometer M 19
Manometeräquivalent M 21
Manometerkonstante G 67
Manometerschenkel M 20
Manschette S 177
Manteldüse C 28
Mantelheber J 1
Mantelthermoelement
  S 120
Mantelwalzen S 109
Marmorkitt M 23
Maschinenkühlung M 59
Maskenkarussell M 30
Maskenrahmen M 26
Maskenspeicherung M 29
Maskenwechsler M 25
Masseleitfähigkeit B 219
Massenbereich M 35
Massendurchlauf M 36

Massendurchsatz M 32
Massendurchsatzmeßgerät M 33
Massenfilter E 31, Q 2
Massenlinie M 34
Massenspektrometer M 38
Massenspektrometer-lecksuchgerät M 39
Massentrennung M 37
Massenwirkungsgesetz L 27
Masseteilchen C 346
massiver Stoff B 223
Massivgetter B 222
Materialzuführung F 22
Materialzugabe F 22
Matrixkatode M 41
Mattbeizen P 52
mattieren F 218
Mattieren R 196
Mattierpulver E 155
Mattiersalz E 156
maximale Verdampfungs-rate K 15
Maxwell-Boltzmannsche Verteilungsfunktion M 46
Mc Leod M 47
Mc Leodsches Kompres-sionsvakuummeter M 47
mechanische Pumpe M 58
mechanische Vorbehand-lung P 139
Mehrfach-Dünnschicht-Kondensator M 191
Mehrfachdüse M 195
Mehrfach-Punkt-Schreiber M 204
Mehrfachschicht M 192
Mehrfachumlaufband-vakuumtrockner M 197
mehrgängiges Gewinde M 201
Mehrschichtadsorption M 190
Mehrschichtenspiegel M 193
Mehrschichtkondensator M 191
mehrstufig C 235
mehrstufige mechanische Pumpe C 237
mehrstufige Pumpe M 207
mehrteilige Ventilröhre M 202
Mehrzweckofen M 205
Meißner-Falle M 61
Membran D 92
Membrandose D 93
Membrandruckschalter D 96
Membranleck M 73
Membranmanometer D 95
Membranpumpe D 97
Membranschalter D 96
Membranvakuummeßgerät D 98
Membranvakuummeter D 98
Membranventil D 99
Mengenstrom R 34
Meßbereichsumschaltung R 22
Meßbürette M 53
Meßdom M 54
Meßfühler S 79
Meßgenauigkeit M 52
Meßgerät G 66
Meßkopf G 66
Meßröhre G 66
Meßumformer M 55
Meßwandler M 56
Metallanglasung G 96
Metallanschluß M 96
Metallbalg B 82
Metallbalg aus Tombak T 123
Metallbalgdichtung M 83
Metallbearbeitung M 98
Metallbeschattung M 94
Metalldichtung M 90

Metallentgasungsanlage M 86
metallgedichtet M 93
Metallgewebe M 92
Metallgewinnung M 89
Metall-Glasverschmelzung G 96
Metall-Glimmer-Verbindung M 107
Metall-Inertgas-Schweißen I 48
metallische Aufdampf-schicht V 216
metallischer Schutzüberzug P 205
Metallisieren M 85
Metallisierung M 85
Metallisierung im Vakuum V 104
Metallisierung von Keramik S 169
Metallkeramik M 84
Metallkeramikanschluß C 100
Metall-Keramik-Verbindung C 103
Metall-Keramik-Verbin-dung nach dem Molyb-dän-Mangan-Verfahren M 164
Metallmembrandichtung M 88
Metallmembranfilter M 87
Metallphysik M 91
Metall-Saphir-Verbindung S 14
Metallschlauch F 100
Metallspritzen M 95
Metallversilberung D 268
metastabil M 99
metastabiler Zustand M 100
Methode T 12
Methode des konstanten Druckes C 276
Methode des konstanten Volumens C 277
Mikanit M 104
Mikrobestimmung M 114
Mikrokapillare M 110
Mikrominiaturschaltung M 115
Mikroschaltkreisschablone M 113
Mikroschaltung M 111
Mikrowaage M 109
Miniaturmeßröhre M 120
15-Minuten-Entspannungs-temperatur A 154
Mischdüse M 127
Mischfilmdestillation M 131
Mischkondensator C 284
Mischrohr M 129
Mischungslücke M 125
Mischungswärme H 39
mittlere Anregungsenergie A 219
mittlere freie Weglänge M 48
mittlere Gesamtstoßzahl pro Zeit- und Volumen-einheit C 222
mittlere Geschwindigkeit M 49
mittlere Molekular-geschwindigkeit A 221
mittlerer linearer Aus-dehnungskoeffizient A 220
mittleres Geschwindig-keitsquadrat R 169
mittlere spezifische Wandstoßrate I 21
mittlere Teilchen-geschwindigkeit A 221
mittlere Temperatur A 222
mittlere Wandstoßzahl pro Zeit- und Flächeneinheit I 21

Modulator M 134
Mohs-Härte M 137
Mol G 111
Moleküladhäsion M 142
Molekularanziehung M 144
Molekulardestillation M 150
Molekulardestillations-anlage M 162
Molekulardestillations-anlage mit fallendem Film F 9
Molekulareffusion M 153
molekulare Leitfähigkeit M 149
Molekularelektronik M 154
Molekulargeschwindigkeit M 161
Molekularkraft M 156
Molekularleck M 157
Molekularluftpumpe M 143
Molekularpumpe M 152
molekular rauh M 158
Molekularsieb M 159
Molekularstrahl M 145
Molekularstrahlquelle M 146
Molekularströmung M 155
Molekularvakuummeter M 151
Moleküldichte N 45
Molekülsenke M 160
Molekülstoß M 147
Molenbruch M 140
Molvolumen M 141
Molybdänglas G 90
Molybdän-Mangan-Verfahren M 165
Molybdänschiffchen M 163
monomolekular M 176
monomolekulare Adsorption M 177
Monopol-Partialdruck-analysator M 178
Monopolspektrometer M 179
Monoschicht M 172
Monoschichtbedeckung M 173
Monoschichtverdampfung M 174
Monozelle M 170
Montageplatte B 59
Muffe S 177
Muffelofen M 186
Muffenkelch S 205
Muffenkupplung M 185
Muffenrohr S 207
Muffenverbindung B 89
Muldentrockner S 33
Muster S 8
Musternehmen S 10

N

Nachfüllvorrichtung R 79
Nachfüllvorrichtung für flüssigen Stickstoff L 87
Nachglimmen A 91
Nachkalibrierung R 48
Nachleuchten A 91
Nachschwitzmethode G 52
Nachtrocknung A 90
Nachverdampfer R 45
Nachweisempfindlichkeits-grenze D 78
Nachweisgrenze D 77
Nadelventil N 2
Nahkatode R 159
Nahtschweißen S 53
Nahtschweißung L 68
narrensicher F 150
Naßdampf W 37
Naßluftpumpe R 54
Natronglas S 208
Natron-Kalk-Glas L 57a

Natronwasserglas S 209
natürliches Abtauen O 6
Naturspeckstein B 143
negatives Glimmlicht N 3
Nenndruck N 16
Nennsaugleistung N 17
Nennsaugvermögen R 33
Nennweite N 15
Netzebene C 408, L 22
Netzebenenabstand I 116
Netzelektrode N 7
Netzgerät P 128
Netzmittel W 35
Netzspannungsschwankung L 67
Netzteil P 128
Neutralglasfilter G 121
nicht abschmelzbare Elektrode N 19
nichtflüchtig N 25
nichtkondensierbares Gas N 18
Niederdruckquecksilber-lampe L 117
Niederdruckstufe L 114
Niederspannungsschweißen L 119
Nippel N 8
Niveaukonstanthalter L 51
Normalbedingungen S 321
Normaldruck und Normal-temperatur S 321
normale Ausflußrate S 322
normale Glimmentladung N 27
normale Leckrate S 322
normaler Atmosphären-druck A 200
,,normaler Durchbruch''-Vorvakuumdruck D 276
normale Zimmertemperatur S 323
Normdruck A 200
NTC-Widerstand N 4
Nukleon N 39
Nullabgleich R 115
Nullabgleichgerät N 44
Nullabgleichverfahren B 34
Null einstellen / auf A 72
Nullpotential Z 8
Nullpunkteinstellung Z 7
Nullpunkteinstellvorrich-tung Z 4
Nullpunktkonstanz Z 6
Nullpunktsdrift Z 5
Nullunterdrückung Z 9
Nusselt-Zahl N 47
Nutsche V 64
Nutschen V 52
Nutzkälteleistung U 27
Nutzraum E 13
NW N 15

O

obere Entspannungs-temperatur A 154
Oberflächenabdruck S 444
Oberflächenatom S 432
Oberflächenätzung S 436
Oberflächenbedeckung S 434
Oberflächenbeschaffenheit F 65
Oberflächendiffusion S 435
Oberflächengüte F 65
Oberflächenionisation S 439
Oberflächenkitt S 442
Oberflächenkondensator S 433
Oberflächenplatz S 446
Oberflächenrauhigkeit S 445
Oberflächenspannung S 448
Oberflächenspiegel F 217
Oberflächenströmung S 438
Oberflächenverdampfung S 437

Oberflächenveredelung S 443
Oberflächenvergütung L 50
Oberflächenwanderung S 441
Oberflächenzustand S 447
Objekthalter O 1
Objektträgertrommel O 2
Ofendurchsatz F 230
Ofenkapazität F 227
Ofenkessel F 229
Ofenlöten F 226
Ofenpfanne F 228
Öffnung O 62
OFHC-Kupfer O 83
Okklusion O 5
Ölablaß O 19
Ölablaßschraube O 20
Ölablaßventil O 21
Ölabscheider O 34
Ölaufbereitungsanlage O 15
Öldampfstrahlpumpe O 22
Öldichtung O 33
Öldiffusionspumpe O 17
Öleinfüllschraube O 24
Öleinfüllventil O 12
Ölejektorpumpe O 22
Ölfangblech O 9
Ölfänger O 37
Ölfilter O 25
Ölfiltereinrichtung O 25
Ölfilter mit Ölrücklauf O 30
Ölfüllung O 11
ölgedichtete Rotations-pumpe R 180
Ölkriechen O 16
Ölkriechsperre A 170
Ölmeßstab O 18
Ölnebelfilter O 29
Ölregenerieranlage O 31
Ölreiniger O 14
Ölrücklauffilter O 30
Ölrücklaufleitung O 32
Ölrückströmung O 8
Ölrücktritt B 4
Ölstand O 27
Ölstandsauge O 35
Ölstandschraube O 28
Ölstandsschauglas O 35
Öltreibdampfpumpe O 38
ölüberlagert O 26
Ölumlauf O 23
Ölumlaufheizung O 13
Ölwechsel O 10
Omegatron O 39
Opazität O 42
optisch aktiver Körper O 53
optisch dicht O 43
optisch dichte Dampf-sperre O 52
optisch dünn T 153
optisches Schichtdicken-meßgerät O 55
Orbitronionenpumpe O 56
Orbitron-Ionisations-vakuummeterröhre O 58
Orbitronvakuumpumpe O 56
Ordnungszahl A 205
orientiertes Aufwachsen O 61
O-Ring O 64
O-Ring-Doppeldichtung D 214
Oxidkatode O 81
oxidkeramische Stampf-masse P 82
Ozalidpapier O 85

P

Packungsdichte P 3
Palladiumlecksuchgerät P 13
Palladiumschwarz P 11
Palladium-Wasserstoff-Ionisationsvakuummeter P 9
Palladium-Wasserstoff-Leck P 12
Palladium-Wasserstoff-Lecksucher P 10
Partialdruck P 15
Partialdruckanalysator P 16
Partialdruckempfindlich-keit P 17
Partialdruckmeßgerät P 14
Partialdruckvakuummeter P 18
Paschensches Gesetz P 20
Paßstift A 119
Pechblende P 23
Peltier-Baffle P 25
Peltier-Falle P 25
Pendelschieberventil P 26
Penetrationsmesser P 28
Penning-Vakuummeter C 196
periodische Elektronen-zusammenballung B 226
Permanentelektrode N 19
permanentes Gas N 18
Permeation P 34
Persorption P 36
Pfannenentgasung L 5
Pfannenrest S 174
Pfannentrocknungsanlage L 6
Pfropfen S 367
PG-Draht C 343
Phasendiagramm P 37
Phasenregel P 39
Phasenverschiebung P 38
Philips-Vakuummeter C 196
Phiole V 236
Photon P 49
physikalische Atmosphäre A 200
Physik dünner Schichten T 77
Physisorption P 51
piezoelektrisch P 53
Pilgerschrittverfahren R 52
Pincheffekt P 60
Pinzette T 202
Pipettenmethode P 65
Pirani-Vakuummeter P 66
Plancksches Strahlungs-gesetz P 77
Plancksches Wirkungs-quantum P 76
Planetenkäfig P 78
Plasmabrenner P 83
Plasmagas P 80
Plasmaionenquelle P 81
Plasmaspritzen P 82
Plaste P 85
Platinglas G 91
Platin-Halogen-Lecksuch-gerät H 133
plattdrücken F 98
Plattenabstand C 158
Plattendampfsperre P 87
Plattengefrieranlage P 89
Plattenglimmer S 113
Plattentrommeltemperatur S 115
Plattenventil D 177
Poiseuillesches Gesetz P 94
Poiseuille-Strömung P 93
Poissonsche Konstante P 97
Polardiagramm P 98
Polaroidfolie P 100
Polaroidfolie als Analysator A 143

Polaroidfolie als Polarisator P 99
Polschuh P 101
Polymerisation P 103
Polymorphie P 104
polytrop P 105
Pore P 62
Porenfüller S 442
Porengetter P 107
Porigkeit P 106
Porosität P 106
Positioniervorrichtung P 109
positive Säule P 110
Potentialberg P 113
Potentialschwelle P 113
Potentialtopf P 115
Potentialverteilung P 114
Prallfläche B 23
Prallplatte B 24
Prandtl-Zahl P 129
Präzisionsmessung H 68
Preßbernstein P 142
Preßdichtung C 245
Preßglasteller P 143
Preßglimmer M 104
Preßguß P 149
Preßstumpfschweißen P 148
Pressung S 316
Prinzip der starken Fokussierung S 389
Prinzipschaltung B 60
Probe S 8
Probenahme S 10
Probenehmen S 10
Probenwechsler S 9
Profildichtung P 195
Programmschalter C 424
Programmventil P 196
Prüfgas P 190
Prüfling S 8
Prüfsieb T 21
pulvergespeiste Flash-Verdampfungsquelle P 121
Pulvermetallurgie P 122
Pumpanlage mit Sorptionspumpe D 263
Pumpautomat A 213, E 202
Pumpcharakteristik P 211
Pumpe P 209
pumpen P 208
Pumpenbelüftungsventil P 221
Pumpengehäuse P 210
Pumpenhals P 92
Pumpenkopf H 20
Pumpenöl O 50
Pumpenstiefel B 54
Pumpentreibmittel O 50
Pumpgeschwindigkeit P 215
Pumpkarussell R 175
Pumpkombination P 216
Pumpsatz G 130
Pumpstandsteuerung P 220
Pumpstengel E 201
Pumpstraße I 71
Pumpstraße mit Pump-wagen T 143
Pumpstraße mit stationären Pumpeinrichtungen S 331
Pumpstraßenwagen T 175
Pumpsystem P 218
Pumpverfahren P 217
Pumpwirkung P 213
Pumpzeit P 212
Pumpzeitkonstante P 219
Punktquelle P 92
Punktschreiber D 204
Punktschweißen S 288
Putzstrahlen B 132
Pyrolyse P 227
pyrolytische Beschichtung C 174

Q

Quadrupolfeld Q 1
Quadrupol-Hochfrequenz-massenspektrometer Q 2
Quadrupol-Ionisations-vakuummeter Q 3
Quadrupol-Massenspektro-meter nach Paul P 21
Quantenausbeute Q 6
quantenmechanisch Q 5
quarzähnliches Glas H 78
Quarzauskleidung S 143
Quarzglas F 232
Quarzgut O 44
Quarzoszillator R 144
Quarzröhre Q 9
Quarz-Vakuum-Mikrowaage Q 10
Quecksilber M 75
Quecksilberdampfstrahl-pumpe M 79
Quecksilberdiffusionspumpe M 78
Quecksilbersäule M 76
Quecksilber-U-Rohr-Manometer U 29
Quecksilberverschlußhahn M 81
Quecksilberverschlußventil M 77
Quellenstromstabilisator S 242
Quellungsisotherme S 453
Querdehnungsziffer P 97
Querfeld-Elektronen-strahlen T 163
Querfeld-Katode T 163
Querstromgebläse C 379
Quetschen S 316
Quetschfuß P 58
Quetschfußeinschmelzung F 96
Quetschfußzange P 144
Quetschhahn C 164

R

0R A 5
Radialdichtring für Welle R 5
Radialgebläse R 173
Radialpumpe R 2
Radialverdichter C 88
Radialwellendichtring mit Manschette R 4
radioaktive Quelle R 17
radioaktiver Indikator R 18
Radiometer-Vakuummeter K 13
Raffinieren R 81
Randbedingung B 178
Randfeld F 214
Randwinkel C 283
Rankinescher Wirkungs-grad R 23
Raschig-Ring R 28
Rasterdampfsperre C 125
Rastgesperre R 29
Räuchern im Vakuum V 71
Rauhwerden R 196
Rauhwert P 22
Raumgitter S 248
Raumkammer S 244
Raumladung S 245
raumladungsbegrenzt S 246
Raumschiff S 251
Raumsimulator A 132
raumzentriert B 152
Rauschpegel N 12
Rauschtest N 14
Re R 155
reagieren R 146
reaktionsfähig R 40
Reaktionskitt R 38
reaktiver Getter R 42
reaktives Verdampfen R 41

reaktive Zerstäubung R 43
Reaktivierung R 39
reales Gas I 20
Realkristall I 19
Réaumur-Skale R 44
Rechen für Ampullen-
  trocknung D 260
Rechtecknut R 59
Reduktionsofen R 68
Reduzier-Kreuzstück R 67
Reduzierkupplung R 66
Reduzierstück R 64, T 86
Reduzierstück mit
  einseitiger Muffe S 377
Reduzierverbindungsstück
  R 65
Reemission R 70
Reflexbelag M 124
Reflexionsbeugung R 84
Reflexionskoeffizient R 83
Reflexionsverminderung
  A 175
Reflexionsvermögen R 82
reflexmindernde Schicht
  A 176
regelbares Gasballast-
  ventil C 310
Regelgerät C 311
Regelventil C 313
Regelzeit A 38
Registriergerät P 29
Registriervorrichtung
  R 57
Registrierzeit R 58
Reibung F 210
Reibungsbremse F 211
reibungsfreie Strömung
  E 142
Reibungsschweißen F 213
Reibungsströmung F 212
Reibungsvakuummeter
  D 23, O 65
Reibungsvakuummeter mit
  Quarzfadenpendel Q 8
Reifbildung F 219
Reindichte N 5
Reinheitsgrad D 50
Reinigung C 153
Reinigungsanlage P 225
Reinigungshahn P 224
Reinmetallkatode P 222
Reinstkupfer O 83
Reißfestigkeit T 11, T 20
Reißlack B 206
Reißlackverfahren S 379
Rekombination R 56
Rekristallisation R 61
Rekristallisations-
  temperatur R 62
rektifizieren R 60
relative Feuchtigkeit R 101
Relaxationszeit R 102
Reserveeinheit S 325
Resonanzfrequenz R 143
Respirationsmassen-
  spektrometer R 145
Restaktivität R 116
Restdampfdruck R 125
Restfeuchtigkeit R 123
Restgas R 119
Restgasanalysator R 120
Restgasdruck R 122
Restgaszusammensetzung
  R 121
Restleitwert R 117
Reststrom R 118
Retortenofen R 154
Revolverkopf T 201
Reynoldssche Zahl R 155
Rezipient C 293
Rezipientenglocke B 90
Rezipiententeller B 92
Richtstrahlwirkung B 79
Richtungseffekt D 151,
  O 60
Richtungsfokussierung
  D 153
Richtungskosinus D 152

Richtwirkung D 151
Riemen B 99
Rieselentgasung O 49
Rieseltrockner C 46
Rieseltrocknung F 105
Riffelung C 352
Rillung C 352
Ringanschmelzung A 161
Ringbrenner A 155
Ringdichtung R 161
Ringdüse A 159
Ringdüsendampfstrahl-
  pumpe A 158
Ringetagentrockner T 195
ringförmige Nahkatode
  R 159
ringförmige Strahl-
  umlenkdüse A 157
Ringgetter R 162
Ringkatode R 159
Ringkatoden-Elektronen-
  stoßquelle R 160
Ringkolbenschieber A 160
Ringspaltdichtung A 156
Ringspannung H 116
Ringstrahlnahkatode R 159
Ringstrahlsystem R 158
Riß C 361, L 36
Ritzhärte A 4
Rockwell-Härte R 165
Rohdichte G 126
Roheisen P 54
Rohmetall C 390
Rohransatz T 182
Rohranschmelzung T 181
Rohrbogen B 102
Rohrdraht I 84
Röhrenempfindlichkeit
  G 68
Röhrenfedermanometer
  B 183
Röhrenfedervakuummeter
  S 285
Röhrenkonstante G 41,
  G 67
Röhrenkühler C 184
Röhrenpumpstengel T 182
Röhrensockelkitt T 179
Röhrentrockner C 431
Röhrenwandpotential
  B 216
Rohrkrümmer B 102
Rohrleitung D 266
Rohrtrockner F 82
Rohrverschraubung T 85
Rollmembran C 432
Röntgenbeugung X 2
Röntgengrenze X 4
Röntgenstrahleffekt X 3
Röntgenstrahlen X 5
Röntgenstrom S 215
Röntgenwellenlänge X 6
Roots-Pumpe R 170
rostfreier Stahl S 319
Rotationsgefrieren R 194
Rotationsölluftpumpe
  R 179
Rotationspumpe R 183
Rotationsquecksilber-
  luftpumpe R 177
Rotationsquecksilber-
  pumpe nach Gaede G 2
Rotationsvakuumpumpe
  R 188
Rotationswasserstrahl-
  pumpe R 193
rotierender Mischkon-
  densator R 192
rotierendes Pumpsystem
  R 175
Rotor R 195
Rückdiffusion B 2
Rückfluß R 87
Rückfüllung B 3
Rückgewinnung R 55
Rückkriechen B 18
Rücklaufpumpe R 89
Rücklaufsperre R 88

Rückprallelastizität R 127
Rückprallhärte R 47
Rückschlagventil B 19
Rücksieder R 46
Rücksprunghärte R 47
Rückstand R 126
Rückstellmoment R 150
Rückstellung auf den
  Nullpunkt R 115
Rückstrahlung R 113
Rückstrahlverfahren
  nach Laue L 25
Rückströmrate B 22
Rückströmung B 21
Rückverdampfung
  infolge Kriechens B 18
Rückzündung B 7
Ruhedruck S 329
ruhendes Gas G 8
Rühreinrichtung A 94
Rührschweißen T 206
Rührspule S 361
Rührwerksmischer M 126
Rundring O 64
Rundring mit
  V-förmigem Querschnitt
  V 232
Rundschnurring O 64
Runzelbildung R 196

S

Sackloch B 137
Sammelstück H 21
Sanden S 12
Sandstrahlen S 12
Sattdampfdruck S 17
sättigen S 15
Sättigungsdampfdruck S 17
Sättigungsniveau S 18
Sauerstofflecksuchgerät
  O 84
saugen P 208
Saugen I 80
Saugfilter S 413
Saugform S 414
Saugleistung C 20, E 199
Saugleistungsmaximum
  C 109
Saugleitung I 87
Saugpumpe D 235
Saugrohr T 3
Saugseite C 220
Saugstrahlpumpe E 20
Saug- und Druckpumpe
  D 205
Saugvermögen D 185,
  P 215
Saugvermögen bei
  Gasballast B 37
Saugvermögen für Luft
  A 116
Saugvermögen ohne Baffle
  U 13
Saugvermögen ohne
  Gasballast U 14
Saugvorrichtung S 412
Saugwirkung P 214
Säulenstruktur C 228
säurebeständig A 37
Schaber S 27
Schachtelhalm G 105
Schachtofen P 72
schädlicher Raum D 6
Schalenschmelzen S 176
Schalenschmelzofen S 175
Schalldämpfer N 13
Schalldämpfung N 11
Schaltdruck S 455
Schalter für flüssige Luft
  L 77
Schaltleistung B 192
Schaltpult C 307
Schaltschrank C 312
Schalttafel C 312
Schaltventil P 57
Schaltwerk R 30

Schaufeltrockner P 8
Schauglas I 78
Schaukelofen T 104
Schaum F 144
schäumen F 143
Schaumglas C 78
Schaumstoff A 88
Scheibenanschmelzung
  D 175
Scheibendichtung D 174,
  V 191
Scheibenkatode D 173
scheinbares Leck V 241
Schellack S 118
Schenkel L 57
Scherdichtung S 106
Scherfestigkeit S 107
Scherströmung S 105
Schichtdickenmeßgerät
  F 56
Schichtgetter C 175
Schichtgitter F 54
Schichtkontinuität F 53
Schichtkunststoff L 9
Schicht niedriger
  Austrittsarbeit L 120
Schichtreinheit F 55
Schichtspaltung P 91
Schichtträger S 402
Schichtträgermagazin
  S 405
Schichtträgertemperatur-
  steuerung S 407
Schichtung L 10
Schichtwiderstands-
  meßgerät R 135
Schichtzusammensetzung
  F 52
Schiebeboden S 180
Schiebedurchführung
  L 62, R 53
Schieberpumpe S 178
Schieberventil G 64
Schieberzunge V 205
Schirmgitter S 30
Schlagbiegefestigkeit I 16
Schlagbiegeversuch I 17
Schlagzähigkeit I 16
Schlauch H 118
Schlauchklemme H 119
Schlauchtülle H 120
Schlauchwelle H 120
Schleifentrockner F 30
Schleifhärte G 124
Schleifkorn A 3
Schleifring S 189
Schleifringanordnung
  S 190
Schleifringdichtung F 4
Schleudergefriertrockner
  C 91
Schleudergefriertrocknung
  C 92
Schleuderguß C 87
Schleuderschichtgefrieren
  S 277
Schleuse L 96, S 195
Schleusenkammer L 97
Schleusenventil L 98
Schlieren S 380
Schliffdichtung C 263
Schliffverbindung G 129
Schlitzkatode S 192
Schlitzmagnetron-
  Ionisationsvakuummeter
  S 282
Schlupf S 187
Schlupfströmung S 182
Schöpfraum S 411
Schöpfvolumen S 454
Schmelzbad M 63
Schmelzbadbewegung
  S 362
schmelzbar F 234
Schmelzbarkeit F 233
Schmelzblock I 57
Schmelzblockgewicht I 62
Schmelzblockschleuse I 58

Schmelzeinsatz F 235
Schmelzen F 237
Schmelzen mit flüssiger
    Katode C 282
Schmelzfluß C 232
Schmelzflußschmierung
    M 71
Schmelzgut M 69
Schmelzkessel M 70
Schmelzkitt M 64
Schmelz-, Legierungs- und
    Gießverfahren F 1
Schmelzleistung M 65
Schmelzmetall-Vakuum-
    ventil L 84
Schmelzofen M 66
Schmelzprozeß F 237
Schmelzpunkt F 239
Schmelzschweißen F 240
Schmelzschweißung F 240
Schmelzsumpf M 63
Schmelztiegel C 383
Schmelzwärme H 38
Schmelzzone F 111
Schmiedbarkeit F 165,
    M 15
Schmiedeeisen L 108
Schmierfähigkeit L 124
Schmiermittel L 121
Schmierung L 123
Schmucküberzug D 20
Schmutzfänger D 156
Schnarchventil S 201
Schneckenpresse S 35
Schneide 422
Schneidenanglasung E 8
Schneidendichtung K 11
Schnellabtastung R 24
Schnelldurchlauf R 24
Schnellentleerung Q 14
Schnellentspannungs-
    temperatur I 82
Schnellgefrieren Q 16
Schnellkühlung Q 13
Schnellregistrierung S 272
Schnellschlußhandventil
    Q 12
Schnellschlußventil F 13
Schnelltrockner F 82
Schnellverschluß Q 17
Schnittwerkzeug D 100
Schnüffelmethode S 11
Schnüffelsonde P 164
Schnüffeltest S 11
Schnüffler S 11
Schnüfflermethode S 11
Schrägbedampfung O 3
schräger Verdichtungsstoß
    O 4
Schrägsitzventil B 110
schrägstehender
    Trommelmischer T 102
Schranktrockner C 1
schraubenförmig H 55
Schraubenversetzung S 32
Schraubmuffenverbindung
    S 34
Schraubverbindung S 31
Schreiber P 29
Schrotrauschen S 133
Schrott S 26
Schrumpfen S 134
Schrumpfmuffe S 135
Schrumpfriß C 120
Schub T 96
Schubmodul M 136
Schüttdichte B 220
Schutzgasatmosphäre P 202
Schutzgaskontakt P 203
Schutzgaslöten C 308
Schutzgasofen C 309
Schutzlack P 204
Schutzring I 82
Schutzringelektroden-
    system G 135
Schutzschaltung P 199
Schutzschicht P 200
Schutzüberzug E 217

Schutzvakuum G 136
Schutzverriegelung S 4
Schutzverriegelungssystem
    S 5
Schutzvorrichtung P 201
Schwabbelscheibe B 213
Schwalbenschwanznut
    D 221
Schwammgummi S 284
schwarzer Körper B 124
Schwarzverchromen B 125
Schwebegastrockner A 99
Schwebepotential F 107
Schwebeschmelzen F 188
Schwebeverdampfen L 55
Schwebezonenschmelz-
    verfahren F 108
Schweißbarkeit W 24
Schweißbrenner W 31
Schweißflansch W 25
Schweißgeschwindigkeit
    W 30
Schweißkonstruktion W 32
Schweißmuffe W 26
Schweißnahtbreite W 33
Schweißstab F 46
Schweißstrom W 27
Schweißtiefe-Breite-
    Verhältnis R 36
Schweißzone W 34
Schwellenbereich T 91
Schwellenenergie T 90
Schwellfestigkeit I 117
Schwere G 116
Schwerkraft G 116
Schwerkraftölablaß-
    schraube G 117
schwerschmelzende Metalle
    R 93
Schwimmaufbereitung
    F 117
Schwimmerschalter F 109
Schwimmerventil F 110
Schwingen V 238
Schwingförderer V 239
Schwingkondensator-
    elektrometer V 237
Schwingquarz C 410
Schwingquarzschicht-
    dickenmeßgerät Q 7
Schwingung V 238
Schwingungsdämpfer A 178
Schwingungsfreiheit F 183
Schwitzwasserbildung
    S 452
sedimentieren S 97
Seelendraht C 343
Segmentbogen L 95
Seifenblasenmethode S 204
Seigerung E 119
Seitendampfstrahlstufe
    S 138
Sektorgerät S 57
Sekundärzylinder S 450
selbständige Entladung
    S 70
selbstansaugend S 71
Selbstevakuierung
    der Röntgenröhre S 65
Selbstfraktionierungs-
    einrichtung S 67
Selbsthaftkoeffizient S 74
selbstheilend S 68
Selbstreinigung S 61
Selbstsperrung S 69
selbstverzehrende Elektrode
    C 280, S 62
selektiver Getter S 59
Senke S 165a
Sensibilisator S 82
Shorehärte S 128
Sicherheitsdruck-
    auslösung S 1
Sicherheitshebevor-
    richtung S 3
Sicherheitsventil S 7
Sicherheitsverriegelung
    S 4

Sicherheitsvorrichtung S 1
Sickenmaschine B 72
Siebbandtrockner S 29
Siedegefäß B 155
siedegekühlt V 211
Siedekühlung H 127
Sieden B 157
Siedepunkt B 159
Siedeverzug B 158
Siemens-Martin-Ofen
    O 46
Signal-Rauschverhältnis
    S 139
Sikkativ D 239
Silber-Sauerstoff-Leck
    S 148
Silikagel S 141
Silikagel-Lecksuchgerät
    S 142
Silikonfett S 144
Silikongummi S 140
Silikonöl S 145
Siliziumstahl S 146
Sinteranlage S 171
Sinterglas S 167
Sinterkorund S 166
Sintermetallurgie P 122
Skalenendwert V 185
Skin-Effekt S 172
Sodaglas S 208
Solarisation S 217
Solarkonstante S 216
Solidifikation S 91
Solidusfläche S 230
Soliduslinie S 230
Soliduspunkt S 231
Sonde P 189
Sorbat S 236
Sorbens S 237
sorbieren S 235
sorbiertes Gas S 236
Sorption S 238
Sorptionsfalle A 86, C 110
Sorptionsfalle mit
    Aktivkohle A 48
Sorptionsgetter N 26
Sorptionsmittel S 237
Sorptionsplatz S 241
Sorptionspumpe S 240
Sorptionsschleife S 239
Spachtel S 27
Spaltbarkeit C 160
Spaltbreite R 3
Spaltdichtung C 159, D 94
spalten C 163
Spaltfläche C 161
Spaltprodukt C 162
Spaltrohr S 193
Span C 131
Spannring C 150
Spannung S 378
Spannungsdurchschlag
    S 449
spannungsführend U 15
Spannungspunkt S 243
spannungsunabhängig
    I 39
Spardiode B 169
Speicherkreis S 370
Speicherröhre S 372
Speisepumpe F 23
spektralanalytisch S 265
spektraler Emissions-
    koeffizient S 264
Spektrometer mit
    magnetischem Sektorfeld
    S 56
Sperrflüssigkeit C 267
Sperrgitter C 266
Sperrichtung I 122
Sperrschicht B 55
Sperrschichtfotozelle P 50
Sperrschieber H 107
Sperrschieberpumpe R 182
Sperrvorrichtung R 29
Sperrwirkung R 142
spezifische differentielle
    Ionisierung S 262

spezifische Ionisierung
    D 111
spezifische Kondensationsge-
    schwindigkeit C 258
spezifischer elektrischer
    Widerstand E 33
spezifisches Saugvermögen
    S 263
spezifische Wärme S 261
sphärische Aberration
    S 273
Spiegelbelag M 124
spiegelnd S 266
spiegelnde Streuung S 267
Spindeldichtung S 354
Spin-Welle S 278
spreiten S 299
Sprengel-Pumpe S 301
Sprengring R 152
Spritzbaffle S 280
Spritzgießen D 101
Spritzpistole S 296
Spritzschutzmuffe S 297
Spritzschutzschirm S 259
Spritzzündung J 11
Sprödigkeit B 207
Sprühdüse S 298
Sprühentgasung S 291
Sprühgefrieranlage S 295
Sprühtrocknung S 293
Sprühtrocknungsanlage
    S 294
Sprung C 361
Spülelektrode R 164
spülen R 163
Spülen F 141
Spülgas B 136
Spülgasflansch G 14
Spülgasmethode W 3
Spülgasventil G 15
Spülung F 141
Stabschmelzen R 166
Stabsintereinrichtung R 167
Stadtgas T 140
Stahlentgasung S 349
Stahlentgasungsanlage
    S 350
Stahlglocke S 348
Stampfmasse T 4
Standtank S 334
Stangenvorschubschleuse
    B 47
Stanniolbelag L 28
stanzen S 320
Starrkugelmodell H 19
Startdruck I 65, S 326
stationärer Pumpautomat
    S 331
stationärer Zustand
    S 333, S 338
stationäre Strömung S 332
statischer Druck S 328,
    S 329
statischer Druckverlust
    L 107
statisches Vakuumsystem
    S 330
statistische Verteilung R 21
Stator S 335
Staubabscheider D 273
Staubentwicklung F 168
Staubfilter D 272
Staubsauger V 25
Staudruck D 279
Staudüse D 126
Staupunkt S 317
Stay-down time S 337
stecken S 320
Steckmuffe S 185
Stefan-Boltzmannsches
    Gesetz S 352
Stehbolzen S 336
Stellfläche S 114
Stellschraube S 89
Stempel D 100
sterilisieren S 358
Steuerpult C 307
Steuerwelle A 51

Stichloch T 6
Stickstoffäquivalent N 9
Stifteinschmelzung P 63
Stiftschraube S 390
Stiftschweißung S 391
Stirnstoß E 7
Stockpunkt P 120
Stopfbuchse P 4
Stopfbuchsendichtung S 392
Stopfbüchsenpackung G 79
stopfbuchsloses Vakuumventil P 6
stopfbuchsloses Ventil P 7
Stopfbuchsventil P 2
Stopfen K 4
Stopfengußpfanne B 176
Stopfenstange S 368
Stopfenzug B 45
Stöpsel S 367
Störfaktor S 139
Störpegel N 12
Störstellenplatz D 30
störungsfrei T 176
Stößel P 226
stoßempfindlich S 124
Stoßfestigkeit I 16
Stoßfront S 122
Stoßhäufigkeit C 222
Stoßionisation I 14
Stoßionisierung I 143
Stoßpolarendiagramm S 123
Stoßquerschnitt C 221
Stoßschweißung E 112
Stoßspannung I 31
Stoßverdampfung F 84
Stoßwahrscheinlichkeit C 223, I 15
stoßweises Sieden B 224
Stoßwelle S 126
Stoßwellenrohr S 125
Stoßzahl C 223
Stoßzahlverhältnis I 21
Strahlantrieb J 12
Strahlapertur B 73
Strahldichte B 75
Strahldruck B 82
Strahldüse E 21
Strahleindringtiefe B 80
strahlen R 9
Strahlendurchlässigkeit R 13
Strahlenschutz R 37
Strahlerschleuse G 141
Strahlerzeugung B 77
Strahlfleck B 84
Strahlleistung B 81
Strahlleitrohr B 78
Strahlstrom B 74
Strahlsystem J 14
Strahlteiler B 83
Strahltriebwerk J 15
Strahlung R 10
Strahlungsaustausch R 15
Strahlungsdruck R 14
Strahlungsgasbrenner R 6
Strahlungsheizer R 7
Strahlungsheizung R 8
Strahlungshitzemesser O 54
Strahlungsofen R 12
Strahlungsschirm H 46
Strahlungsschutzschirm H 46
Strahlverdichtung E 19
Strahlwirkungsgrad J 8
Strangabzugsvorrichtung D 81
Stranggußverfahren C 296
strangpressen E 226
Strangpressen E 228
Strangpreßrohling E 229
Strangschmelzen E 230
Streckbarkeit M 15
Streckformen D 234
Streckgrenze E 24, M 44

Streckgrenzenverhältnis E 25
streifende Inzidenz G 118
streifender Einfall G 118
Streufeld F 214
Streuflußbildung F 215
Streuung D 184
Streuungsbeiwert S 300
Strichfokusantikatode S 373
Strichfokusröhre L 65
Strom F 118
stromab D 223
stromauf U 25
strombegrenzend C 418
Stromdurchführung C 417
Stromtrocknungsanlage F 83
Strömung F 118
Strömungsgeschwindigkeit F 127
Strömungskontinuum C 297
Strömungsmesser F 122
Strömungspumpe F 130
Strömungsschalter F 129
Strömungsweg F 125
Strömungswiderstand F 120
Strömungswiderstand einer Öffnung A 179
Stromversorgungsgerät für Vakuummeter V 75
Stufendichtung S 357
Stufenflansch S 355
Stufenversetzung E 6
stufenweise G 104
Stufenzahl N 46
Stumpfanglasung D 175
Stumpfanschmelzung B 236
Stumpfschweißen B 237
Stundendurchsatz H 142
15-Stunden-Entspannungstemperatur S 243
Stundenschmelzleistung M 68
Sublimatabscheider S 394
Sublimation S 395
Sublimationsanlage S 396
Sublimationshorde C 117
Sublimationspumpe S 397
Sublimationsrate S 398
Sublimationstechnik S 399
Sublimationstrocknung D 255
Sublimationswärme H 41
sublimieren S 393
Sublimieren S 400
Substrathalter S 404
Substratheizer S 403
Substrat-Maskenwechsler S 406
Suchlampe H 8
sukzessive Verdünnung S 408
Supraleiter S 418
Supraleitfähigkeit S 417
Supraleitung S 431
Sutherland-Konstante S 451

T

Taktfrequenz B 87
Taktzeit C 425
Talk B 143
tatsächliches Saugvermögen I 118
Tauchgefrieranlage I 13
Tauchgettern D 146
Tauch-Hartlöten D 144
Tauchlöten D 148
Tauchschmierung S 281
Tauchveredelung H 128
Tauchverkupferung C 340
Tauchzündung D 147
Taumelgefriertrocknung T 183

Taumeltrockner T 101
Taupunkt D 87
Technik T 12
Teilchendichte P 19
teilfraktionierende Pumpe S 72
Telefunken-Verfahren S 168
Tellerdrehmaschine F 79
Tellerfeder C 413
Tellerfedernsäule D 176
Tellertrockner P 88
Tellerventil D 177
Temperaturbereich des rein elektrischen Durchschlages D 190
Temperaturbereich des Wärmedurchschlages T 27
Temperaturfühler T 18
Temperaturleitfähigkeit T 33
Temperaturregelung T 15
Temperaturskale nach Celsius C 84
Temperaturskale nach Fahrenheit F 6
Temperaturwächter T 16
Temperaturwechsel T 17
Temperaturwechselbeständigkeit H 47, T 44
Temperguß M 16
tempern A 150
Temperpunkt S 94
Termschema E 139
Tesla-Spule S 254
Testgas P 190
Testleck C 11
Tetroden-Ionisationsvakuummeter T 22
Tetrodenzerstäubung T 23
theoretischer Drucksatz I 120
thermische Auslösung T 43
thermische Desorption T 30
thermische Effusion T 31
thermische Elektronenenergie T 34
thermische Geschwindigkeit T 49
thermische Ionisierung T 39
thermische Katode T 50
thermische Oxydation T 40
thermische Raumkammer T 45
thermischer Ausdehnungskoeffizient T 37
thermischer Haltepunkt T 25
thermisches Emissionsvermögen R 11
thermisches Gleichgewicht T 35
thermische Transpiration T 31
Thermistor T 54
Thermistor-Vakuummeter T 55
Thermodiffusion T 31, T 32
Thermodynamik T 59
thermoelektrische Kälteerzeugung T 62
thermoelektrisches Vakuummeter T 57
Thermoelement T 56
Thermokreuz T 57
thermomolekulare Strömung T 31
Thermomolekularpumpe T 48
Thermoplast T 64
Thermosäule T 63

thermostatisches Expansionsventil T 65
thorierte Katode T 84
Tiefgefrieranlage S 104
Tiefgefrieren D 27
Tiefkühltruhe D 26
Tiefkühlung D 24
Tiefschweißeffekt D 28
Tiefschweißen D 29
tiefsiedend L 109
Tiefziehansaugen D 25
tiegelfreies Zonenschmelzen C 385
Tiegelhalter C 388
Tiegelheizer C 387
Tiegelkippvorrichtung D 82
Tiegelofen C 386
Tiegelreaktion C 389
Tiegelverdampfung C 384
Titanhydridverfahren T 116
Titanionenpumpe T 117
Titanschwamm T 118
Titansublimationspumpe T 119
Titanverdampfer T 114a
Titanverdampferpumpe T 115
Toepler-Pumpe T 121
Tombak T 122
Tonerdehydrat A 135
Topfglühofen P 112
Torr T 129
Torricelli-Vakuum T 130
Torsionsfestigkeit T 131
Torsionsmikrowaage T 133
Torventil G 64
Totaldruckmesser T 137
Totzeit D 225
Townsend-Entladung D 4
Trägerdraht C 42
Trägergas C 41
Tragring B 86
Tränklackimprägnieranlage V 230
Transformationspunkt T 144
Translationsenergie T 152
Translationsfreiheitsgrad T 151
Transmission T 154
Transmissionsfaktor E 171
Transmissionskoeffizient T 155
Transpiration T 161
transportabel P 108
Transporttank M 132
Transportwirkungsgrad T 162
Trapeznut T 165
Trapezring V 232
Trapezringdichtung V 232
„Travelling-Wave" Magnetron T 167
Treibdampf M 128
Treibdampfpumpe B 171, V 220
Treibdampfpumpenöl V 225
Treibdampfüberhitzung V 228
Treibgas P 126
Treibmittel P 197
Treibmitteldampfdruck im Siedegefäß B 156
treibmittelfrei F 132
treibmittelfreies Vakuum F 133
Trennahtschweißen H 141
Trennfaktor S 83
Trennrohr T 58
Trennsäule F 179
Trennscheibe C 421
Trennwirkung S 85
Treppenkurve S 356
Trigger-Ionisationsvakuummeter T 170

Triodenionengetterpumpe T 88
Triodenzerstäubung T 172
Tripelpunkt T 173
Trockendichtung D 264
Trockeneis C 33
Trockengestell D 231
Trockenkammer D 256
Trockenmittel D 55
Trockenprodukt mit Eiskern D 258
Trockenrechen D 257
Trockenreibung D 254
Trockenschleuder C 89
Trockenschmiermittel D 253
Trockenstoff D 239
Trockenschrank S 116
Trockenzeit D 262
Trockner D 251
Trocknung D 54
Trocknungsanlage D 259
Trocknungsgeschwindigkeit D 261
Trocknungsrate D 261
Trommelgefriertrockner D 249
Trommeltrockner C 431
Tröpfchenwachstum D 246
Tropfendüse D 248
Tropföler D 240
Tropfpunkt D 247
Tropfschale D 242
Tr. P. T 144
Trübungspunkt C 169
Trümmerventil B 196
T-Stück T 141
Tunneleffekt T 191
Tunnelofen C 301, T 193
Tunnelvorgang T 189
Turbinentrockner T 195
Turbomolekularpumpe T 196
Turbopumpe N 22
Turboverdichter T 194
turbulente Störung T 197

U

überbrücken B 199
Überdruck O 77
Überdruckdichtigkeitsprüfung O 79
Überdruckventil R 104
Überdruckverfahren O 78
Überexpansion O 74
Übergangsbereich T 148
Übergangsflansch T 145
Übergangsglas T 146
Übergangsglasrohr G 105
Übergangsmetall T 147
Übergangsströmung I 110, K 12
Übergangswahrscheinlichkeit T 158
Übergang vom festen in den flüssigen Aggregatzustand C 106
Überhitzer S 420
überhitzter Dampf D 265
Überhitzung S 421
Überlappstoß L 13
Überlast O 75
Überlastungsschutz O 76
übersättigter Dampf O 80
Übersättigung S 422
Überschalldampfstrahl S 426
Überschallmolekularstrahl S 425
Überschallströmung S 424
Überschallströmung in Pumpen T 159
Überschalltreibstrahl S 423
Überschlagsfestigkeit F 90
Überstromrelais C 419
Überwurfmutter C 29

überziehen C 173
U-Eisen C 108
U-förmiger Bügel S 363
UHV U 4
ultrafester Stahl U 3
Ultrahochvakuum U 4
Ultrahochvakuumanlage U 7
Ultrahochvakuumflansch U 5
Ultrahochvakuum-Öldiffusionspumpe U 6
Ultrahochvakuumventil U 8
Ultrahochvakuumventil mit Schmelzmetalldichtung L 83
Ultraschallerosion U 11
Ultraschallöten U 12
Ultraschallreinigung U 10
ultratiefe Temperatur U 9
Ultrazentrifuge U 2
Umfangsgeschwindigkeit C 146
Umformfaktor C 315
Umgebungsdruck A 139
Umgebungstemperatur A 140
umgehen B 238
Umgehungsleitung B 239
umgekehrtes Magnetronvakuummeter C 195
Umhüllung E 131
Umhüllungstechnik H 114
Umkehrstrahl I 123
Umlaufbahn O 57
Umlaufentgasung C 144
Umlaufheizthermostat C 142
umleiten B 238
Umpumpleitung C 143
Umrechnungsfaktor C 315
umschmelzen R 105
Umschmelzhärte R 106
Umwälzpumpe C 145
Umwandlungstemperatur T 149
Umwegleitung B 239
unberuhigter Stahl R 157
Undichtigkeit I 68, L 36
Undichtigkeitsnachweis mittels Entladungsrohres D 170
Undurchsichtigkeit O 42
unempfindlich gegen Fehlbedienung F 150
ungekühlte Dampfsperre N 23
ungesättigter Dampf U 21
Universalmeßinstrument M 194
universelle Gasgleichung E 149
universelle Gaskonstante I 7
unlegierter Stahl C 36
unlösbare Verbindung F 68
unselbständige Entladung N 24
unsichtbare Entladung B 126
Untenentleerung B 173
Unterbrecherblende S 137
Unterdruck L 112
Unterdruckkammer A 131
untere Druckmeßgrenze L 113
untere Entspannungstemperatur S 243
Untergrundgas B 6
Untergrundstrom B 5
Unterkühlung S 419
Unterlage V 402
Unterschall- S 401
Uran U 26
U-Rohr-Manometer M 80
U-Schellenverbindung U 28

V

vagabundierende Katodenstrahlen V 184
Vakustat V 4
Vakuum V 5
Vakuum / im V 3
Vakuumabsperrventil V 147
Vakuumabtropfschmelzen V 50
Vakuumanlage V 118
Vakuumaufdampfanlage V 26
Vakuumbedampfung E 170
Vakuumbedampfungsanlage V 26
Vakuumbedingung V 33
Vakuumbehälter V 34
Vakuumbenetzbarkeit V 181
vakuumbeständig V 154
Vakuumbeton V 31
Vakuumblockguß V 89
Vakuumbogenbedampfung V 10
Vakuumbogenverdampfung mit selbstverzehrender Katode C 281
Vakuumchromatografie V 23
Vakuumdampf V 155
Vakuumdampferzeuger V 156
Vakuumdampferzeugung V 158
Vakuumdampfheizung V 157
Vakuumdestillation V 47
Vakuumdestillationsanlage V 49
Vakuumdestillationskamera V 48
vakuumdicht L 44, V 173
Vakuumdichtigkeit V 96
Vakuumdichtscheibe V 143
Vakuumdiffusionsschweißen V 45
Vakuumdilatometer V 46
Vakuumdrehdichtung R 189
Vakuumdrehfilter R 187
Vakuumdruckanlage V 123
Vakuumdrucksintern V 124
Vakuumdurchlaufbandglühofen C 303
Vakuumdurchlaufmetallbedampfer C 304
Vakuumelektronenstrahlschmelzanlage V 58
Vakuumentgasungsofen V 40
Vakuumentwässerung V 41
Vakuumerzeugung C 362
Vakuumexsiccator V 43
Vakuumextraktion V 61
Vakuumfaktor V 63
Vakuumfalle V 174
Vakuumfett V 77
Vakuumfließbettsublimation E 159
Vakuumformofen F 182
Vakuumformverfahren V 67
Vakuumfunkenstrecke V 151
Vakuumgefrieren V 70
Vakuumgefriertrockner V 68
Vakuumgefriertrocknungsanlage V 69
Vakuumgehäuse V 146
vakuumgeschmolzene Metalle V 101
vakuumgetränkt V 82
Vakuumgießanlage V 20
Vakuumgießen S 410

Vakuumgießharzanlage V 137
Vakuumgießstrahlverfahren V 162
Vakuumglocke B 90, V 14
Vakuumglühen V 7
Vakuumglühofen V 8
Vakuumgummisackverfahren D 31
Vakuumhahn V 38
Vakuumheberentgasung P 64
Vakuum-Heißpreß- und Abschreckofen V 79
Vakuumheißwandofen V 81
Vakuumimprägnierung V 83
Vakuuminduktionsgießen V 85
Vakuuminduktionsofen V 86
Vakuuminduktionsschmelzen V 87
Vakuuminduktionsschmelzofen V 88
Vakuuminstallation V 119
Vakuumisolation V 90
Vakuumkaltwandofen V 28
Vakuumkalzination V 19
Vakuumkammer B 90, V 22
Vakuumkaschieren V 95
Vakuumkitt V 21
Vakuumkochapparat C 190
Vakuumkocher C 190
Vakuumkondensation V 32
Vakuumkonzentration V 29
Vakuumkonzentrationsanlage V 30
Vakuumkristallisieren V 37
Vakuumkryostat V 36
Vakuumkühler F 81
Vakuumkühlung V 35
Vakuumkurbel W 49
Vakuumlack S 46
Vakuumleitung V 97
Vakuumlichtbogen V 9
Vakuumlichtbogenofen V 11
Vakuumlichtbogenofen mit selbstverzehrender Elektrode C 206
Vakuumlichtbogenschmelzen V 12
Vakuumlötofen V 16
Vakuummantel G 136, V 92
Vakuummantelheber V 93
Vakuummeßgerät V 74
Vakuummeßkopf V 76
Vakuummeßröhre V 76
Vakuummeßtechnik V 100
Vakuummessung V 99
Vakuummetallisierungsanlage V 105
Vakuummetallurgie V 106
Vakuummeter V 74
Vakuummeter mit optischer Refraktionsmessung R 91
Vakuummeter nach dem Omegatronprinzip O 40
Vakuummeterröhre mit vorgeschalteter Kühlfalle T 166
Vakuummeter vom Typ des umgekehrten Magnetrons I 124
Vakuummikrowaage V 108
Vakuummischer V 109
Vakuummonochromator V 110

Vakuumofen V 72
Vakuumphysik V 115
Vakuumpolarisation V 120
Vakuumprüfer V 84
Vakuumprüfung S 257,
　V 168
Vakuumpulverabfüll-
　behälter V 121
Vakuumpulverisolation
　V 122
Vakuumpumpe V 126
Vakuumpumpleitung V 117
Vakuumpumpsatz V 128
Vakuum-Pump-und-
　Aufdampfanlage V 166
Vakuumreaktion V 130
Vakuumreaktor V 131
Vakuumreinigung V 129
Vakuumrektifikation
　V 132
Vakuumrektifizieranlage
　V 133
Vakuumrektifizierkolonne
　V 134
Vakuumrelais V 135
Vakuumretortenofen V 39
Vakuumrohranschluß
　V 116
Vakuumröhre V 176
Vakuumrohrverbindung
　V 116
Vakuumröntgenstrahl-
　diffraktometer V 114
Vakuumrotationsver-
　dampfer V 141
Vakuumsaugverfahren
　S 374
Vakuumschalter V 164
Vakuumschaufeltrockner
　V 112
Vakuumschieber G 64
Vakuumschieberventil
　V 150
Vakuumschlauch V 177
Vakuumschleuse V 98
Vakuumschmelzen V 102
Vakuumschmelzofen V 103
Vakuumschmelzofen mit
　kippbarem Schmelztiegel
　T 103
Vakuumschmiermittel
　L 122
Vakuumschrank V 18
Vakuumseite C 220
Vakuumsinteranlage V 149
Vakuumsintern V 148
Vakuumspektrograf V 153
Vakuumspritzguß V 44
Vakuumstahlentgasung
　V 159
Vakuumstrahlentgaser
　V 161
Vakuumstrahlentgasung
　V 160
Vakuumstrahlsauger V 94
Vakuumstrangpressen
　V 62
Vakuumstreckform-
　verfahren D 233
Vakuumstutzen E 161,
　E 201
Vakuumsublimation V 163
Vakuumsystem V 165
Vakuumtaumeltrockner
　V 57
Vakuumtechnik V 167
Vakuumthermoelement
　V 171
Vakuumthermogravimetrie
　V 172
Vakuumthermowaage
　V 170
Vakuumtrockenschrank
　V 55
Vakuumtrockner V 53
Vakuumtrockner mit
　Rührwerk V 113
Vakuumtrocknung V 54

Vakuumtrocknungsanlage
　V 56
Vakuumtrommelfilter
　V 52
Vakuumtrommeltrockner
　V 140
Vakuumultraviolett V 152
Vakuumumlaufverdampfer
　V 24
Vakuumunterbrecher
　V 91
Vakuum-UV-Quelle
　V 178
Vakuumventil V 179
Vakuumverfahren V 107
Vakuumverfahrenstechnik
　V 125
Vakuumverformung V 67
Vakuumverkleidung V 145
Vakuumverpackung V 111
Vakuumverschließ-
　einrichtung V 144
Vakuumwaage V 13
Vakuumwachs S 50
Vakuumwalzentrockner
　V 51
Vakuum-Wärme-
　Isolation V 169
Vakuumwiderstandsofen
　V 138
Vakuumwissenschaft
　V 142
Vakuumzonenschmelzen
　V 182
Vakuumzubehör V 6
V-Dichtungsring V 232
Ventil V 186
Ventilanordnung V 206
Ventilanschlag V 195
Ventilator F 10
Ventilblock V 188
Ventildeckel V 189
Ventildeckeldichtung
　B 167
Ventildeckel-Ventilkörper-
　Dichtung B 166
Ventildichtungsfläche
　V 192
Ventilhub V 198
Ventilhubbegrenzung
　V 195
Ventil in der
　Umgehungsleitung B 240
Ventil in der Umweg-
　leitung B 240
Ventilkappe V 189
Ventilklappe V 193
Ventilkolben V 199
Ventilkombination V 196
Ventilkörper B 153
ventillos V 197
Ventil mit Federbalg-
　dichtung B 95/6
Ventil mit Stopfbuchse
　P 2
Ventilplatte V 190
Ventilquerschnitt V 187
Ventilschaft V 201
Ventilsitz V 202
Ventilsitzdichtung V 203
Ventilspindel V 201
Ventilspindelführung V 204
Ventilstempel V 200
Ventilsteuerung V 194
Ventilteller V 190
Ventiltellerdichtung
　D 174, V 191
Venturi-Düse V 235
Verarbeitungsbereich
　W 56
Verarbeitungstemperatur
　W 55
Verbindung B 164, C 273
Verbindungsflansch C 269
Verbindungsglas C 270
Verbindungsleitung T 182
Verbindungsschweißen
　J 18

Verbindungsstab C 85
Verbindungsstück C 274
Verbindungtechnik I 94
Verblocken I 108
Verblockung I 108
Verbrauch an flüssigem
　Stickstoff L 86
Verbrennung B 230
Verbundglas C 236
Verbundmetall C 234
verchromen C 137
verchromt C 136
verdampfen E 164
Verdampfer E 190
Verdampferkarussell E 184
Verdampferpumpe E 192
Verdampferquelle E 193
Verdampferquellenrevolver
　V 227
Verdampferschiffchen
　E 191
Verdampferwendel H 56
Verdampfung E 165
Verdampfung im Vakuum
　V 60
Verdampfungsanalyse
　E 166
Verdampfungscharakteri-
　stik E 169
Verdampfungsgeschwin-
　digkeit E 180
Verdampfungsgetter V 246
Verdampfungsgetterung
　D 182
Verdampfungsglühen
　F 87
Verdampfungsgrenze
　V 208
Verdampfungsgut E 162
Verdampfungskammer
　E 168
Verdampfungskoeffizient
　E 171
Verdampfungskühlung
　E 174
Verdampfungsmaske E 179
Verdampfungsprozeß E 165
Verdampfungsrate E 180
Verdampfungsraten-
　meßanordnung von
　geheizten Elektroden
　H 30
Verdampfungsraten-
　meßgerät E 182
Verdampfungsraten-
　regelsystem E 181
Verdampfungsregler E 173
Verdampfungssteuerung
　E 172
Verdampfungssynthese
　E 185
Verdampfungstechnik
　E 186
Verdampfungstemperatur
　E 187
Verdampfungstrocknung
　E 175
Verdampfungsverlust
　E 178
Verdampfungswärme E 176
Verdichter C 250
Verdichterkälteanlage
　C 244
Verdichterkältemaschine
　C 248
Verdichtung C 238
Verdichtungsdruck D 167
Verdichtungsenddruck
　C 242
Verdichtungsstoß S 121
Verdichtungsstoßsystem
　S 127
Verdrängerkolben D 188
Verdrängermanometer
　D 187
Verdrängerpumpe D 189
Verdrängungsverdichter
　P 111

verdrillt T 205
verdünnen R 26
verdünntes Gas R 25
Verdunstung E 165
Verdunstungskühlung E 188
Verdunstungstrocknung
　E 189
Verfahren T 12
Verfahrenstechnik P 193
Verfestigung H 13, S 91
Verflüssiger L 76
Verflüssigung L 74
Verformung F 169
Vergießen unter Vakuum
　V 59
vergiften P 95
Vergiften P 96
Vergiftung P 96
Vergleichsdruck R 75
Vergleichsleck C 11
Vergleichsvakuum R 77
Vergleichungskapillare
　C 230
Vergußmasse P 118
Vergütung optischer Gläser
　L 49
Verhältnis von Partial- zu
　Totaldruck C 253
Verhältniszahl der thermi-
　schen Transpiration T 47
Verkapseln unter Vakuum
　V 59
Verkapselung E 131
Verkettung L 70
Verklebeanlage C 82
verkupfern C 334
Verlust durch Auslaufen
　L 36
Verlustwinkel P 125
Vernetzung C 380
verriegeln I 107
Verriegeln I 108
Verriegelung I 108
Verschiebungsstrom D 186
Verschleiß W 19
Verschleißfestigkeit R 138
Verschleißteil W 22
Verschleißzahl W 21
verschließen S 366
Verschlußventil C 420
Verschmelzung F 231, G 45
Verschmelzung zwischen
　Dumet-Draht und
　Weichglas D 269
Verschmelzung zwischen
　rostfreiem Stahl und
　Hartglas H 15
Versetzung D 179
Versetzungsdichte D 178
versilbern S 147
Versilberung S 149
Verspiegelung M 123
Versprödung E 132
Versuchsanlage P 56
Verteiler D 199
Verteilerrohr D 199
Verteilerstück D 199
vertikales Zonen-
　schmelzen F 108
Vertikalrohrverdampfer
　L 103
Verunreinigung I 32
Verweilzeit A 61
Verwitterung W 23
verzinken Z 10
verzögerte Kondensation
　D 57
Verzögerung D 14
Verzögerungsrelais T 107
Verzögerungszeit D 58
Verzugszeit D 58
Vibrationsrinne D 198
Vickershärte V 240
Vielfachstreuung M 199
Vielschieberpumpe M 200
Vielstrahlbrenner M 189
Vielstrahlinterferometrie
　M 196

Vielstufenentgasung M 206
virtuelles Leck V 241
viskoses Leck V 243
viskose Strömung S 182
Viskosität V 242
Viskositätszahl I 121
V-Naht S 164
Volumenabsorption B 218
Volumen der Anlage P 79
Volumendiffusion B 221
Volumendurchsatz S 270
Volumen einer Gramm-
  molekel eines Gases V 247
Volumenleitfähigkeit B 219
Volumenstoßhäufigkeit
  C 222
Volumenstrom S 270
volumetrische Förder-
  leistung V 249
volumetrischer Wirkungs-
  grad V 251
volumetrisches Kalibrier-
  system V 248
Vorbehandlung P 185
Vorbehandlungsbad S 382
vorevakuieren P 133
Vorglühen P 135
vorheizen P 134
Vorkühlen P 131
Vorkühler P 130
Vorpumpe B 14
vorpumpen F 160
Vorpumpen C 171
Vorpumpenöl R 184
Vorpumpenventil B 12
Vorpumpzeit F 161
Vorratsbehälter S 371
Vorratskatode D 180
Vorrichtung zum Kaltpreß-
  schweißen und Ab-
  klemmen C 218
Vorschubregelung F 21
Vorschweißflansch W 28
Vorschweißkleinflansch
  W 29
Vorsinterung P 140
Vorspannung A 218
vortrocknen P 132
Vortrocknung P 138
Vorvakuum F 162
Vorvakuumanschluß B 10
Vorvakuumbauteil F 163
Vorvakuumbehälter B 38,
  F 164
Vorvakuumbeständigkeit
  C 370
Vorvakuumdruck B 13,
  C 382, D 277
Vorvakuumfalle B 11
Vorvakuumkessel B 38
Vorvakuumleitung B 8
Vorvakuumpumpe B 14
Vorvakuumraum B 15
Vorvakuumseite F 159
vorvakuumseitiger
  Abscheider B 9
Vorvakuumstutzen F 155
Vorvakuumtechnik B 17
Vorvakuumventil B 12
vorverdichten S 415
Vorverdichtung S 416
vorwärmen P 134
Vorwärmen C 172
Vorwärmer P 136
Vorwärmung P 137
V-Ringdichtung V 232

W

waagerechtes Tiegel-
  verfahren Z 13
Wabe H 112
wabenförmige Anode
  M 188
Wabenstruktur H 113
Wachsimprägnieranlage
  W 18

Wachstumsgeschwindigkeit
  G 131
Wachstumsrate G 131
Wackelschwanzdrehdurch-
  führung W 49
wahres Saugvermögen
  I 118
wahre Temperatur T 178
wahrscheinlichste
  Geschwindigkeit M 180
Walzblock B 144
Walzentrockner C 431
Wälzkolbenpumpe R 170
Wälzlager A 171
Wanderungsgeschwindig-
  keit M 118
Wandstoß W 1
Warmbrüchigkeit H 136
Wärmeäquivalent T 36
Wärmeaustausch H 33
Wärmeaustauscher H 34
Wärmebehandlung H 52
Wärmebehandlung im
  Vakuum H 78
Wärmebehandlung im
  Vakuumofen V 66
Wärmebeständigkeit R 92
Wärmedurchgang H 51
Wärmedurchgangszahl
  O 72
Wärmedurchschlag T 26
Wärmefestigkeitsgrenze
  H 29
Wärmeimpulsschweißen
  T 38
Wärmeisolation H 36
Wärmelehre T 59
Wärmeleitfähigkeit H 26
Wärmeleitung H 23, H 24
Wärmeleitungsvakuum-
  meter H 25
Wärmeleitvermögen H 26,
  T 28
Wärmepumpe H 42
Wärmerauschen T 24
Wärmereflektor H 45
Wärmeriß H 27
Wärmeschutzglas H 43
Wärmespannungs-
  widerstand T 46
Wärmestau H 48
Wärmestrahlung T 41
Wärmestrahlungsschutzglas
  T 42
Wärmeübergang H 49
Wärmeübergangszahl
  F 51, H 50
Wärmeübertragung H 49
Wärmeverteilung H 28
Warmfestigkeit H 137
Warmwalzen H 135
Warmzerreißfestigkeit
  H 138
wartungsfrei M 13
Waschzeit C 156
wasserabstoßende Schicht
  D 252
Wasseraufnahme W 6
Wasserbeständigkeit W 16
Wasserdampf S 339
Wasserdampfdestillation
  S 341
Wasserdampfkapazität M 43
Wasserdampfstrahlsauger
  J 9
Wasserdampfverträglichkeit
  M 42
Wassereinlaß W 14
Wassergehalt W 9
wassergekühlte Gießschale
  W 11
wassergekühlte Kokille
  W 10
Wasserhaut W 8
Wasserkreislauf C 167
Wasserriegel W 7
Wasserringpumpe W 17
Wasserschlag W 13

wasserstoffgekühlt C 320
Wasserstrahlkondensator
  W 15
Wasserstrahlpumpe A 197
Wasserströmungswächter
  P 206
Wasserumlaufkühlung
  C 322
Wechselverdampfer I 93
Wechselwirkung I 91
Wechselwirkung Gas-Fest-
  körper G 57
Weichglas S 212
Weichheitszahl S 213
Weichlot S 219
weichlöten S 218
Weichlöten S 221
Weichlöten mittels
  Flamme T 126
Weichmacher P 84
Wellendichtring R 5
Wellendichtung R 5,
  S 100
Wellendurchführung R 185
Wellen-Öltropfkragen S 99
Wellrohr C 351
Weltraumforschung S 250
Weltraumsimulation S 247
Weltraumsimulierkammer
  O 67
Wendel für Vakuum-
  lampen V 27
Wendelsteigungsfehler H 63
Werkstofftechnik T 13
Werkstückhalter W 53
Wheatstonesche Brücke
  W 38
widerstandsbeheizter
  Vakuumofen V 138
Widerstandserwärmung
  R 134
Widerstandsheizung R 133
Widerstandsnetz R 141
Widerstandsofen R 134
Widerstandsschichtdicken-
  meßgerät R 131
Widerstandsschweißen
  R 139
Widerstandsthermometer
  R 137
Widerstandsverhältnis
  R 136
Wiedemann-Franzsches
  Gesetz W 41
Wiederansprechzeit C 157
Wiederbedeckung S 63
Wiederbedeckungszeit für
  monomolekulare
  Bedeckung M 175
Wiederherstellung R 100
Wiedereinfrieren R 94
Wiedergewinnungsrate
  R 71
wiederverdampfen R 72
Wiedervereinigung R 56
Wiensches Verschiebungs-
  gesetz W 42
WIG-Schweißen T 187
Wilsondichtung W 43
Windkanal W 44
Winkelbereich A 149
Winkelfalle E 28
Winkel-Golddrahtdichtung
  C 344
Winkelverteilung A 148
Wirbel W 39
Wirbelschichttrockner
  T 198
Wirbelstromheizung E 5
Wirbelströmung V 252
wirksames Saugvermögen
  E 15, N 6
Wirksamkeit der Kühlfalle
  C 214
Wirkungsgrad E 16
Wolfram T 184
Wolfram-Einschmelzglas
  G 92

Wolframfaden entkarboni-
  sieren / den D 11
Wolframfinger T 188
Wolfram-Inertgas-
  Schweißen T 187
Wolfram-Inertgas-
  Schweißen mit Helium
  als Schutzgas H 54
Wolframpastille T 186
Wolframscheibe T 186
Wolframwendel T 185
Wood-Metall W 50
Wrasenhaube V 219

X

Xenon X 1
X-Naht D 217
X-Ring X 7

Z

Zähigkeit V 242
Zahnradpumpe G 69
Zahnstangentrieb R 1
Zaponlack C 79
Zäsium C 4
Zeigerbarometer D 88
Zeigerthermometer D 90
Zeigervakuummeter D 89
Zeitdehngrenze C 368
Zeitgeber T 110
Zeitmesser T 110
Zeitschalter T 110, T 111
Zeitstandfestigkeit C 369
Zeitstandsverhalten C 364
Zeitstauchgrenze C 368
Zeitverzögerung T 106
Zentralstab C 85
Zentrierring C 96
Zentriervorrichtung C 95
Zentrierkreuz S 276
Zentrifugalgebläse C 86
Zentrifugal-Molekular-
  destillationsanlage C 93
Zentrifugaltrocknung
  C 90
Zentrifugiertrockner C 89
Zeolith Z 1
Zeolith-Falle Z 3
Zeolith-Sorptionspumpe
  Z 2
Zerfallsgeschwindigkeit
  D 172
Zerfallskonstante D 13
zerlegen S 21
Zersetzung D 16
Zersetzungsanfälligkeit
  D 17
Zersetzungsfalle T 29
zerstäuben A 208, S 306
Zerstäuberdüse S 298
Zerstäubereinheit S 312
Zerstäuberglocke S 308
Zerstäubung A 207, S 307
Zerstäubungseinrichtung
  S 309
Zerstäubungsergiebigkeit
  S 314
Zerstäubungsrate S 310
Zerstäubungsspannung
  S 313
Zerstäubungstrockner
  S 292
Zerstäubungszeit S 311
zerstörungsfrei N 20
Zerstreuung D 184
Ziehdüse D 100
Zieheisen D 100
Ziehfeld D 238
zinkhaltig Z 11
Zink-Kadmium-Sulfat Z 12
Zirkularbeschleuniger
  C 138
Zonenreinigen Z 17
Zonenreinigung Z 17

Zonenreinigungsapparatur
Z 18
Zonenreinigungsverfahren
im Vakuum V 183
Zonenschmelzanlage Z 16
Zonenschmelzen Z 15
Zonenschmelzkristallzieh-
verfahren Z 14
Zonenseigerung Z 15
Zubehör A 30
zufälliger Fehler A 31
Zugfestigkeit T 20
zulässiger Fehler A 125
„zulässiger" Vorvakuum-
druck F 158
Zulässigkeitsgrenze der
Verunreinigungen P 35

Zunderbeizen S 20
Zündspannung I 11
Zündung S 383
Zungenventil R 69
zurückfließen F 119
zurücksaugen S 409
zurückströmen F 119
Zusatz A 57
Zusatzschicht A 58
Zusatzvakuum V 136
Zusatzwerkstoff F 45
Zuschlag A 57
Zustandsänderung C 107
Zweikammeranlage
T 208
Zweikammerionisations-
vakuummeter T 207

zweiseitige Vollnaht F 93
Zweitschmelze S 55
Zweiwalzenextruder
T 210
Zweiwegehahn T 211
zweiwertig B 123
Zwillingsbildung T 204
Zwillingsdoppelverbindung
C 268
Zwillingskristall T 203
Zwischenglas A 54
Zwischenglühen P 192
Zwischenkondensator
I 109
Zwischenkondensor I 109
zwischenmolekular I 112
Zwischenpfanne M 60

Zwischenprobenentnahme
I 111
Zwischenpumpe B 170
Zwischenschicht I 99
Zwischenstück A 53,
S 249
Zwischenvakuum G 136
Zykloiden-Massen-
spektrometer C 426
Zyklotron C 427
Zyklotronfrequenz C 428
Zyklotronresonanz C 429
Zyklotronresonanzfrequenz
C 430
Zykluszeit C 425
Zylindertrockner
C 431

FRANÇAIS

CONTENU

1.   Physique du vide
1.1   Matière en état gazeux
1.1.1   Gaz réel et idéal
1.1.2   Lois des gaz
1.2   Théorie cinétique des gaz
1.3   Flux de gaz

2.   Technique du vide
2.1   Valeurs et unités
2.2   Manomètres à vide
2.2.1   Manomètres totals
2.2.2   Appareils de mesure de pression partielle
2.3   Pompes à vide et getters
2.3.1   Pompes de déplacement
2.3.2   Pompes moléculaires
2.3.3   Ejecteurs à vapeur
2.3.4   Pompes à diffusion
2.3.5   Pompes à getter
2.3.6   Pompes cryostatiques
2.3.7   Accessoires de pompes
2.4   Chicanes
2.5   Valves et robinets d'arrêt
2.5.1   Soupapes de fermeture
2.5.2   Soupapes à admettre de l'air
2.5.3   Soupapes d'admission des gaz
2.5.4   Combinaisons des soupapes
2.5.5   Robinets d'arrêt
2.6   Conduites à vide
2.7   Jonctions
2.7.1   Jonctions fixes
2.7.2   Connexions détachables
2.8   Récipients à vide

2.9   Traversées
2.10   Fenêtres
2.11   Détection de fuites
2.11.1 Détecteurs de fuites
2.11.2 Méthodes de recherche de fuites
2.12   Systèmes de vide
2.12.1 Systèmes de vide scellés
2.12.2 Systèmes dynamiques à vide
2.13   Technique UHV

3.   Applications de la technique du vide
3.1   Dessiccation sous vide
3.2   Dégazage sous vide
3.3   Imprégnation sous vide
3.4   Brasage sous vide
3.5   Emballage sous vide
3.6   Distillation sous vide
3.7   Sublimation sous vide
3.8   Condensation sous vide
3.9   Filtration sous vide
3.10   Technique de métallisation sous vide
3.11   Trempes superficielles
3.12   Couches protectrices
3.13   Métallisation
3.14   Technique des couches minces
3.15   Fusion sous vide
3.16   Dégazage sous vide
3.17   Coulée sous vide
3.18   Recuit brillant sous vide
3.19   Frittage sous vide
3.20   Dessiccation par congélation
3.21   Simulation d'espace

A

abaissement du point de
  congélation F 204
abaissement réfléchissant
  A 175
aberration D 79
aberration sphérique S 273
abrasion A 2
absorbabilité A 18
absorbant A 17
absorbant S 237
absorber A 15
absorption A 20
absorption de gaz C 154
absorption de l'eau W 6
absorption d'ions I 130
absorption infrarouge I 53
absorption volumétrique
  B 218
absorptivité A 18
accélérateur circulaire C 138
accélérateur linéaire L 60
accepteur A 29
accès du gaz G 5
accessoires A 30
accessoires pour le vide V 6
accouplement L 70
accouplement de réduction
  R 66
accouplement par manchon
  M 185
accroissance orientée O 61
accroissement de pression
  P 169
accroissement de pression
  gazeuse I 35
accroissement du vide d'un
  tube I 36
accumulation de masque
  M 29
accumulation thermique
  H 48
acide carbonique solide
  C 33
acier au silicium S 146
acier austénitique
  inoxydable A 211
acier calciné à mort K 6
acier carboné C 36
acier doux L 108
acier en loupes B 146
acier feuillard S 351
acier fondu L 108
acier inoxydable S 319
acier non calmé R 157
acier pour ressorts S 305
acier ultradur U 3
actinium A 39
action de noyer E 122
action réciproque de gaz et
  solide S 7
action tunnel T 189
activateur A 46
activation cathodique C 61
activation des cathodes à
  oxydes A 45
activité A 50
activité de diffusion D 129
activité résiduelle R 116
adapter A 52
adapteur A 53
adatome A 56
addition A 57
adhérence A 60
adhérer A 59
adhésif A 62
adhésion moléculaire M 142
adhésivité A 66
adiabatique A 68
adipocéreux A 70
admission du lest d'air A 96
adoucir A 150
adsorbant A 77
adsorbat A 75
adsorption A 78
adsorption de plusieurs
  couches M 190

adsorption gazeuse G 25
adsorption monomoléculaire
  M 177
aération A 89
aérer A 73
affluer F 121
agent de séchage D 55
agent de sorption S 237
agent frigorifique C 317
agent moteur O 50
agent mouillant W 35
agent siccatif D 55
agglomération périodique
  des électrons B 226
agglomération sous vide
  V 148
agglomérer par frittage
  S 170
agglutinant A 63
agitateur A 94
agitateur en forme de ruban
  S 279
agitateur magnétique à
  plaque de chauffage
  H 132
agitation inductive I 46
agrégat de pompage
  G 130
aigrette B 225
ailette de refroidissement
  C 324
aimant de déviation D 32
air comprimé P 183
air secondaire S 54
ajoutage A 57
ajouter goutte à goutte
  D 245
ajustabilité T 171
ajustage S 90
alambic D 194
alambic-four R 154
alliage A 126
alliage à pétrir F 166
alliage baratté F 166
alliage d'addition A 127
alliage de pétrissage F 166
allongement de rupture
  B 193
allumage S 383
allumage de retour B 7
allumage par immersion
  D 147
allumage par injection J 11
alpha-radiateur A 128
alphatron A 130
alvéole H 112
amalgamation A 138
amalgamer A 137
ambre pressé P 142
amenée de matériaux F 22
amener au zéro A 72
amiante A 194
amorçage S 383
amorce de cristallisation
  C 406
amortissement conique
  C 264
amortissement de son N 11
amortisseur de son N 13
amortisseur de vibrations
  A 178
amplificateur du débit
  volumétrique F 128
ampoule à distiller A 118
analyse d'activation A 43
analyse de l'évaporation
  E 166
analyse des gaz G 7
analyse des gaz d'isotopes
  I 182
analyse des gaz par adsorption fractionnée F 173
analyse des gaz par condensation fractionnée F 174
analyse des gaz par désorption fractionnée F 175
analyse des gaz par évaporation fractionnée F 177

analyse d'extraction à
  chaude H 129
analyse d'extraction chaude
  à l'étain T 112
analyse fondamentale B 61
analyseur de feuille
  polaroïdale A 143
analyseur de gaz résiduel
  R 120
analyseur de pressions
  partielles P 16
analyseur d'extraction à
  chaude F 238
analyseur monopole de
  pressions partielles M 178
angle de pertes P 125
angle d'incidence A 144
angle mouillant C 283
anhydride carbonique en
  neige C 33
anneau à lèvres L 71
anneau à section circulaire
  D 214
anneau à section ronde O 64
anneau à section trapézoïdal
  V 232
anneau de centrage C 96
anneau de piston P 68
anneau de retenue R 152
anneau de serrage C 150
anneau d'étanchéité G 48
anneau d'étanchéité axial
  A 228
anneau divisé S 283
anneau d'obturation V 232
anneau en X X 7
anneau glissant S 189
anneau joint V 232
anneau-joint radial R 5
anneau-joint radial avec
  embouti R 4
anneau partagé S 283
anneau portatif B 86
anneau porteur d'étanchéité
  R 151
anneau profilé P 195
anneau Raschig R 28
anode multicellulaire M 188
anode rotative R 190
anodiser A 168
anti-acide A 37
anticathode A 169
antidépression A 177
antidétonant A 173
antimoine A 174
apériodique D 5
aplatir F 98
appareil à charger C 115
appareil à couverture
  simple-double S 163
appareil à haute fréquence
  pour le contrôle du vide
  H 75
appareil à rayons d'électrons
  E 69
appareil à séparer l'air D 8
appareil à soutirer R 29
appareil à ultra-vide U 7
appareil automatique A 212
appareil à vider F 49
appareil basculant à creuset
  D 82
appareil d'ajustage S 93
appareil d'alimentation
  C 115
appareil d'arrêt R 29
appareil d'échappement de
  lingots I 63
appareil de chargement
  C 115, L 94
appareil de chauffage par
  rayonnement R 7
appareil de commande
  C 311
appareil de congélation
  F 201
appareil de contrôle du
  courant F 122

appareil de dépoussiérage
  D 273
appareil de levage de
  sécurité S 3
appareil de lyophilisation à
  tambour D 249
appareil de mesure de la
  résistance de couche
  R 135
appareil de mesure de la
  résistance des couches
  minces R 131
appareil de mesure de
  l'épaisseur de couche F 56
appareil de mesure de
  pression partielle P 14
appareil de mesure des
  couches minces à condensateur C 19
appareil de mesure des couches minces à quartz
  oscillatoire Q 7
appareil de mesure de
  service S 87
appareil de mesure du débit
  de masses M 33
appareil de mesure pour
  diodes C 402
appareil de purification de la
  vapeur des solvants S 234
appareil de réglage C 311
appareil de réglage du zéro
  N 44
appareil de remplissage
  R 79
appareil de remplissage
  d'azote liquide L 87
appareil de séchage à vide
  V 53
appareil distillatoire D 194
appareil distillatoire à vide
  V 49
appareil enregistreur P 29,
  R 57
appareillage I 81
appareillage de coulée à vide
  V 20
appareillage d'extraction à
  chaude V 73
appareil optique de mesure
  de l'épaisseur de couches
  O 55
appareil photographique de
  distillation sous vide
  V 48
appareil pour la détection de
  fuites L 43
appareil pour l'alimentation
  de courant d'un mano-
  mètre à vide V 75
appareil pour la métalli-
  sation dans le vide poussé
  à millitorr S 158
appareil pour la métalli-
  sation de bande en acier
  par faisceaux électroniques
  E 77
appareil pour le contrôle de
  vide V 84
appareil réglable D 201
appareil rotatif R 174
appauvrissement d'ions
  D 21
application de couches
  antiréfléchissantes L 49
apport par soudure H 14
appui de lingotière I 61
arbre de commande A 51
arc à segment L 95
arcatome-soudage A 182
arc électrique de vide V 9
argent électrolytique E 47
argenter S 147
argenture S 149
argenture matte D 268
armoire à vide V 18
armoire de commande
  C 312

armoire de séchage sous vide V 55
armoire frigorifique D 26
arrangement à anneaux glissants S 190
arrangement à double cloche D 206
arrangement de couples thermoélectriques T 61
arrangement de soupapes V 206
arrêt de vanne V 195
arroser S 289
asbeste A 194
asperger S 289
aspirateur S 412
aspirateur de poussière V 25
aspirateur de vapeur J 9
aspirateur du rayon sous vide V 94
aspiration I 80, P 214
aspiration de vapeur V 218
aspiration mesurée M 51
aspirer P 208
assemblage B 164
assemblage à filet S 31
assemblage à manchons B 89
assemblage à vide poussé H 84
assemblage de tuyaux à vide V 116
assemblage par collier en U U 28
assemblage par manchon à vis S 34
assemblage par pinces P 59
assurer l'étanchéité S 36
atmosphère absolue A 200
atmosphère de gaz protecteur P 202
atmosphère gazeuse protectrice P 202
atmosphère normale A 200
atmosphère physique A 200
atome d'autre origine F 156
atome de la surface S 432
atome étranger F 156
atome-gramme G 110
atomisation A 207
atomisation réactif R 43
atomiseur S 309
attachement des électrons E 84
attraction moléculaire M 144
augmentation de pression P 169
auto-aspirant S 71
auto-diffusion S 64
auto-durcissement du tube pour rayons X S 65
auto-épuration S 61
autoradiographie A 215
avalanche des électrons E 50
avance d'électrode E 37
azote liquide L 85

B

baffle B 23, V 221
baffle à chapeau C 192, J 5
baffle à chevron C 125
baffle à disque P 87
baffle à écran plan P 87
baffle à écrans angulaires multiples C 125
baffle à jet S 280
baffle à plateau P 87
baffle astrotorus A 199
baffle-chapeau C 192, J 5
baffle cryogénique de sublimation C 400
baffle ionique I 126
baffle refroidi thermo-électrique T 60

baffle thermoélectrique P 25
bague de contact S 189
bague d'obturation V 232
bague protectrice G 134
bague-support S 86
baguette de soudure F 46
baguette fusible M 69
baignant dans l'huile O 26
bain de fusion M 63
bain de traitement préalable S 382
bain de traitement préliminaire S 382
balance à vide V 13
balance de résistance de l'air D 227
balance d'humidité M 138
balayage F 141
balayer F 140, S 21
banc de pompage C 229, E 202
banc de pompage rotatoire A 213, R 175
banc de pompage stationnaire S 331
bande de mesure de dilatation E 32
bande transporteuse B 100
bar B 44
baromètre B 50
baromètre à aiguille D 88
barre de bouchage S 368
barre de soudure F 46
barre fusible I 57
barre-support O 1
barrette à électrodes S 360
barrière de fuites d'huile A 170
barrière de grimpement C 363
barrière de potentiel P 113
barrière-écran C 363
base S 402
bas point d'ébullition / à L 109
bâti de masque M 26
bâti de pompage C 229
bâti de séchage D 231
bâton central C 85
bâton de connexion C 85
bec de coulée L 72
besoin frigorifique C 326
bétatron B 108
BET-isotherme B 109
béton de vide V 31
bilame enrobé S 120
bivalent B 123
blindage E 131
blocage I 108
blocage automatique S 69
blocage de protection S 4
blocage de retour R 88
blocage de sécurité S 4
blocage des trous H 104
bloc de soupape à un levier O 41
bloc de soupapes couplées V 188
bloquer I 107
blutoir d'essai T 21
bobine à agitateur S 361
bobine clissée B 62
bobine de Helmholtz H 64
bobine d'induction S 254
boîte à bourrage P 4
boîte à gants D 250
boîte à membrane D 93
boîte à vide V 146
boîte d'extrémité du câble E 133
bombardement B 162
bombardement cathodique E 81
bombardement électronique E 81
bombardement ionique I 129

bon rendement de la pulvérisation S 314
booster pompe à diffusion B 168
bord de Fermi F 27
bord de la coquille M 183
border B 71
boucher S 36, S 366
bouchon S 367
bouchon de décharge D 59
bouchon de vidange d'huile O 20
boucle de sorption S 239
bouilleur B 154, B 155
bouilli à température élevée H 70
boulon à tête à six pans S 206
bourrelet joint S 49
bout T 2
bout de tungstène T 188
bouton-fusion B 235
bout tungstique T 188
brame B 144
branche L 57
branche du manomètre M 20
brasage B 188
brasage de cylindre extérieur O 70
brasage en atmosphère gazeuse protectrice C 308
brasage par bombardement électronique E 56
brasage par flamme T 124
brasage par four F 226
brasage par immersion D 144
brasage sous vide V 15
braser B 187
brassage du bain de fusion S 362
brasure avec compensation flexible F 101
brasure de cylindre intérieur I 76
brasure sous hydrogène H 149
break-seal B 196
bride F 72
bride à arête de soudure W 28
bride à conduite de gaz G 34
bride à épaulement de soudure W 28
bride à gradins S 355
bride à griffe C 148
bride à ultravide U 5
bride avec deux joints d'équerre de fil d'or D 210
bride aveugle B 128
bride d'adaptation A 55
bride d'assemblage C 269
bride de gaz de balayage G 14
bride de serrage S 427
bride d'obturation B 128
bride étagée S 355
bride forgée F 167
bride miniature Q 15
bride obturatrice B 128
bride petite à épaulement de soudure W 29
bride plane P 75
bride rotative L 104
bride soudée W 25
bride tournante L 104
bride transitoire T 145
broche de soupape V 201
brosse fractionnante F 178
brosse tournante F 178
broyeur de pont B 200
bruit de grenaille S 133
bruit de scintillation F 104
bruit thermique T 24
brûlage B 230

brûler B 229
brûleur B 227
brûleur à courant en croix N 33
brûleur à gaz à rayonnement R 6
brûleur annulaire A 155
brûleur à plusieurs rayons M 189
brûleur circulaire A 155
brûleur de plasma P 83
brûleur de soudure W 31
brûleur d'induction à plasma I 42
brûleur en croix C 378
brûlure B 231
bulle de distillation P 117
burette G 66
burette à vide V 76
burette de mesure M 53
burette d'immersion D 150
buse E 21, M 129, N 31
buse cuirassée C 28
buse de diffuseur N 35
buse d'expansion E 207
buse du chalumeau B 228
buse multiple M 195
butée de soupape V 195

C

câblage B 121
câble d'entrée L 31
cadran de réglage D 53
cage de Faraday F 11
cage planétaire P 78
caisson de commande C 312
calamine-décapage S 20
calandrer C 5, R 168
calcination C 6
calcination du filament F 41
calcination haute F 87
calcination préliminaire P 135
calcination sous vide V 19
calfeutrer S 36
calibrage C 13
calibrateur C 16/7
calibre G 65
calibrer C 8
calotte de buée V 219
calque R 112
calque superficiel S 444
canal d'échappement D 224
canalisation D 266
canalisation d'aspiration I 87
canalisation de prévidage R 197
canalisation pour vide primaire B 8
cannelage C 352
cannelure de vibration D 198
canon à distance R 108
canon à électrons E 89
canon à électrons d'accélération étrangère W 52
canon à électrons d'accélération propre S 60
canon électronique E 89
canon ionique de duoplasmatron D 270
canon transversal à électrons I 163
caoutchouc de silicone S 140
caoutchouc durci E 3
capacité d'absorption mesurée M 51
capacité d'aspiration nominale R 33
capacité de congélation F 200
capacité de face S 111
capacité d'émission E 128
capacité de pompage D 185

capacité de pompage de vapeur d'eau M 43
capacité de rupture B 192
capacité de vapeur d'eau M 43
capacité du condensateur C 261
capacité d'une pompe D 185
capacité d'un four F 227
capacité effective d'absorption N 6
capillaire C 23
capillaire à comparaison C 230
capillaire à l'atténuation D 2
capillaire de compression C 246
capillaire séparé par fusion S 52
capotage de buée V 219
capsule de décongélation D 242
capsuler E 130
captage cryogénique C 395
captage d'ions I 130, I 131
capture des électrons E 84
capuchon en caoutchouc R 202
caractéristique de pompage P 211
caractéristique d'évaporation E 169
carboglace C 33
carbonate de baryum B 48
carburant gazeux P 126
carburation C 37
carcasse de pompe P 210
carrousel C 40
carrousel de masque M 30
carrousel de pompage A 213, R 175
carrousel de pompage stationnaire S 331
carter de pompe P 210
carton d'amiante A 195
cartouche chauffante C 43
catalyseur C 58
cataphorèse C 59
cathétomètre C 60
cathode à disque D 173
cathode à distance D 193
cathode à fente D 192
cathode à incandescence H 121
cathode annulaire R 159
cathode à semi-conducteur S 76
cathode à thermistance S 76
cathode auxiliaire A 216
cathode bloc B 141
cathode chaude H 121
cathode chauffée H 121
cathode cyclique R 159
cathode de boulon B 160
cathode de fusion M 72
cathode de métallisation sous vide E 167
cathode de métal pur P 222
cathode de réserve D 180
cathode d'oxyde O 81
cathode en or fin R 80
cathode en U H 1
cathode évidée H 106
cathode froide C 193
cathode imprégnée I 26
cathode incandescente H 121
cathode matrice M 41
cathode moulée M 182
cathode photoélectrique P 42
cathode pressée M 182
cathode thermique T 50
cathode thoriée T 84
cathode transversale T 163
cathodoluminescence C 74

cavitation C 75, P 73
cavité C 77
cellule de mesure G 66
cellule de mesure à ionisation en verre A 122
cellule de mesure insérée B 46
cellule photoélectrique P 40
cellule photoélectrique de couche d'arrêt P 50
Celsius-graduation de température C 84
cémentation C 49
centré B 152
céramique C 101
céramique isolante C 99
céramique métallique M 84
céramiques à teneur haut d'oxyde d'aluminium H 69
cercle des trous P 69
césium C 4
chaleur de désorption H 37
chaleur de dissolution H 40
chaleur de fusion H 38
chaleur de mélange H 39
chaleur de sublimation H 41
chaleur d'évaporation E 176
chaleur perdue W 5
chaleur spécifique S 261
chalumeau B 150/1
chalumeau à souder W 31
chambre à sécher C 1
chambre à vide B 90, V 22
chambre à vide à double paroi D 219
chambre d'aspiration S 411
chambre de bulles B 211
chambre d'écluse L 97
chambre de composante C 233
chambre de compression C 239
chambre de dessiccation D 256
chambre de détente E 208
chambre de diffusion S 290
chambre de la pompe S 411
chambre de métallisation D 65
chambre de séchage D 256
chambre de séchage par congélation F 194
chambre de simulateur d'espace O 67
chambre d'espace S 244
chambre d'évaporation E 168
chambre d'ionisation I 144
chambre lyophilisation F 194
chambre noire cathodique C 375
chambre noire d'Aston A 198
chambre noire de Crooke C 375
chambre noire de Hittorf C 375
chambre spatiale A 132
chambre thermique au volume T 45
champ d'accélération A 27
champ de dispersion F 214
champ de freinage R 153
champ de tirage D 238
champ étrangeur E 220
champ perturbateur E 220
champ quadrupôle Q 1
champ rotatif R 191
champ tournant R 191
changement à aiguille de tiroir V 205
changement de pression P 150

changement d'état C 107
changement de température T 17
changement d'huile O 10
changeur d'échantillons S 9
changeur de masque M 25
changeur de pression P 177
chapeau de la vanne V 189
charbon actif A 42
charbon adsorbant A 42
charge d'huile O 11
chargement C 114
charge par ressort / à S 303
charge spatiale S 245
chargeur mécanique C 115
chariot avec dispositif de levage T 177
châssis de distribution C 312
châssis de séchage D 231
chaudière B 155
chaudière à câble C 2
chaudière à fusion M 70
chaudière de four F 229
chauffage B 33
chauffage à circulation d'huile O 13
chauffage à la vapeur S 343
chauffage à la vapeur sous vide V 157
chauffage à rayonnement R 8
chauffage avec un brûleur à gaz T 125
chauffage de bande S 386
chauffage diélectrique D 104
chauffage direct / à D 155
chauffage indirect / à I 40
chauffage infrarouge I 55
chauffage par choc d'électrons E 94
chauffage par induction I 44
chauffage par rayonnement R 8
chauffage par rayons d'électrons E 68
chauffage par rayons électroniques E 68
chauffage par résistance R 133
chauffage préliminaire C 172
chauffeur à creuset C 387
chauffeur à doubles spirales C 186
chauffeur à rayonnement R 7
chauffeur du baffle à chapeau J 6
chauffeur intérieur I 114
chemin parcouru des électrons E 102
chemise à circulation d'eau de refroidissement C 332
chemise chauffante H 35
chemise d'eau de refroidissement C 332
chemise de chauffage H 35
chemise de refroidissement C 325
chemise d'étuvage B 30
cheville d'ajustage A 119
chevron-garniture étanche C 126
chicane B 23, B 24, V 221
chicane à décharge D 168
chicane non rampante N 10
chicane non réfrigérée N 23
chicane pour retenue d'huile O 9
chimiquement pur C 122

chlorure de cobalt C 133
chlorure platinique C 134
choc S 121
choc de pression P 147
choc de recouvrement L 13
choc d'ions I 142
choc électronique E 86
choc frontal E 7
choc moléculaire M 147
choc oblique de compression O 4
choc par électrons E 93
choc triple T 87
chromage noir B 125
chromatographie de gaz A 80
chromatographie gazeuse G 18
chromatographie sous vide V 23
chromé C 136
chromer C 137
chute de concentration C 251
chute de pression P 158
ciment à support de tube T 179
ciment à vide V 21
cinétique des gaz G 50
circlips R 152
circuit d'accumulation S 370
circuit de référence R 73
circuit en pont B 201
circuit flip-flop F 106
circuit intégré des couches minces T 76
circuit microminiature M 115
circulation d'eau C 167
circulation d'huile O 23
cire adhésive A 67
cire d'étanchéité S 50
cire-laque S 50
cire pour joint étanche S 50
cire pour joints S 40
claie à sublimation C 117
clapet V 193
clapet à air A 103
clapet de dilatation E 213
clapet d'expansion E 213
clapet élastique de vanne S 2
claque sous vide V 17
classification hydrolytique H 150
clé K 4
clef dynamométrique T 128
clivage C 160
clivage des couches P 91
cliver C 163
cloche à double paroi D 218
cloche à vide B 90, V 14
cloche basculable à vide T 100
cloche de dégazage D 40
cloche de verre G 81
cloche du récipient B 90
cloche en acier S 348
cloche pulvérisateuse S 308
coefficient d'accommodation A 32
coefficient d'adhérence S 359
coefficient d'adhésion S 359
coefficient d'arrêt C 31
coefficient d'auto-capture S 74
coefficient d'auto-collage S 74
coefficient d'auto-fixation S 74
coefficient de blocage C 31
coefficient de condensation A 33

coefficient d'écoulement
    thermique F 51
coefficient de diffusion
    D 130
coefficient de dilatation
    C 180
coefficient de dispersion
    S 300
coefficient d'émission
    spectrale S 264
coefficient d'entrée
    anodique P 27
coefficient de puissance
    C 181
coefficient de réflexion
    R 83
coefficient de sortie D 159
coefficient de transmission
    T 155
coefficient de transmission
    thermique F 51, H 50
coefficient d'évaporation
    E 171
cœfficient du passage de
    chaleur O 72
coefficient du vide V 63
coefficient Ho H 100
coefficient moyen linéaire
    de dilatation A 220
coefficient proportionnel de
    la transpiration thermique
    T 47
coefficient thermique
    de dilatation T 37
coffre congélateur D 26
coffret d'alimentation P 128
cohésion C 183
ccin élastique F 19
col de diffuseur D 127,
    N 35
col de tuyère N 35
collage B 164
colle A 63
collecteur T 133
collecteur de tubes pour la
    dessiccation des ampoules
    D 260
collecteur d'huile O 37
collecteur d'impuretés
    D 156
collecteur ionique I 133
collecteur pour la dessicca-
    tion D 257
collet F 72
collet d'égouttage ondula-
    toire S 99
collier d'accès E 40
collier de serrage C 150
collier pour tuyaux H 119
collision murale W 1
collodion C 224
colloïdal C 225
colonne à corps de
    remplissage P 1
colonne de dégazage D 42
colonne de dégazage de
    couches minces T 70
colonne de distillation sous
    vide poussé H 83
colonne de fractionnement
    F 179
colonne de mercure M 76
colonne de rectification
    à vide V 134
colonne de rondelle
    élastique D 176
colonne de séparation
    F 179
colonne de séparation d'air
    A 115
colonne de sous-ensemble
    U 18
colonne de vapeur J 13
colonne disposée en couches
    S 381
colonne positive P 110
colonne stratiforme S 381
colophane C 226

combinaison de pompage
    P 216
combinaison de soupape
    V 196
combustion B 230
commande P 197
commande à distance
    R 107
commande de robinet
    V 194
commande d'étuvage B 28
commande de vanne V 194
commande du bâti de
    pompage P 220
commande électro-
    magnétique / à E 48
commande électro-
    pneumatique / à E 114
commande manuelle / à
    M 22
commandé par un ressort
    S 303
commande thermique du
    côté sensible du film
    S 407
commandeur à pro-
    gramme C 424
commutateur à temps
    T 111
commutateur de contrôle à
    l'eau de refroidissement
    C 331
commutateur par noyau
    de fuite F 29
commutateur pour air
    liquide L 77
commutation des zones
    de mesure R 22
compatibilité à la vapeur
    d'eau M 42
compensation barométrique
    B 52
compensation de pression
    P 151
comportement de fluage
    C 364
composant semiconducteur
    S 77
composé aliphatique A 120
composition de couche
    F 52
composition de gaz G 19
composition de gaz
    résiduel R 121
compound P 118
compresseur C 250
compresseur à piston
    R 50
compresseur à piston
    rotatif R 172
compresseur axial A 226
compresseur centrifuge
    C 88
compresseur de déplace-
    ment P 111
compression C 238
compression à chaud
    sous vide V 80
compression finie de sortie
    C 242
compteur d'après
    Geiger-Müller G 70
compteur de temps T 110
compteur lumineux P 43
concasseur de pont B 200
concentrateur C 254
concentration de rayon
    E 19
concentration du faisceau
    F 146
concentration du vide
    V 29
concentration ionique
    C 252
concentration molaire
    M 140
concentration par congé-
    lation F 191

concrétion préliminaire
    P 140
condensateur C 260
condensateur à glace I 1
condensateur à jet d'eau
    W 15
condensateur à mélange
    C 284
condensateur à plusieurs
    couches M 191
condensateur à plusieurs
    couches minces M 191
condensateur à surface
    S 433
condensateur extérieur
    E 218
condensateur intermédiaire
    I 109
condensateur principal
    M 11
condensation C 238,
    C 256, L 74
condensation capillaire
    C 24
condensation retardée
    D 57
condensation sous vide
    V 32
condenseur L 76
condenseur à jet I 67
condenseur à mélange
    C 284
condenseur à surface S 433
condenseur à vide F 81
condenseur en cascade
    C 44
condenseur intermédiaire
    I 109
condenseur interposé I 109
condenseur par injection
    I 67
condenseur rotatif à mé-
    lange R 192
condition de bord B 178
condition extérieure B 178
conditionnement sous vide
    V 111
conditions normales S 321
condition sous vide V 33
condition superficielle
    S 447
conductance C 262
conductance d'écoulement
    moléculaire M 148
conductance du clapet
    ouvert O 45
conductance d'une
    combinaison S 38
conductance du robinet
    fermé R 124
conductance residuelle
    R 117
conductibilité de masse
    B 219
conductibilité de
    température T 33
conductibilité de volume
    B 219
conductibilité ionique
    I 141
conductibilité thermique
    H 24, H 26, T 28
conduction de chaleur
    H 23
conduction thermique
    H 23, H 49
conductivité calorifique
    H 26
conductivité moléculaire
    M 149
conduite D 266
conduite à reflux d'huile
    O 32
conduite à vide V 97
conduite by-pass B 239
conduite d'aspiration I 87
conduite d'échappement
    D 165

conduite de circulation par
    pompage C 143
conduite de dérivation
    B 239, P 168
conduite de prévide R 197
conduite de raccordement
    T 182
conduite de soupape V 194
conduite d'évacuation
    préliminaire R 197
conduite sous pression
    D 164
conduit vidé de trans-
    mission D 120
cône mâle M 14
configuration croisée
    H 113
configuration en nid
    d'abeille H 113
congélateur à basse
    température D 26
congélateur par
    jaillissement S 295
congélation F 199
congélation à basse
    température D 27
congélation à grande
    vitesse Q 16
congélation à rotation
    R 194
congélation à rotation
    verticale rapide S 277
congélation de petite
    vitesse S 194
congélation par contact
    C 289
congélation sous vide V 70
congeler F 190
connexion J 17
connexion circulaire de
    couches minces T 68
connexion de protection
    P 199
connexion détachable D 62
connexion de verre G 95
connexion en cascade C 45
connexion en pont B 201
connexion flexible F 99
connexion métallique
    M 96
consolidation H 13
consommation d'azote
    liquide L 86
consommation de courant
    de chauffage H 31
consommation de
    réfrigérant C 318
constance C 275
constance du zéro Z 6
constante de Boltzmann
    B 161
constante de conversion de
    la pression P 153
constante de désintégration
    D 13
constante de grille L 17
constante de la dépression
    D 69
constante de manomètre
    G 67
constante de Planck P 76
constante de Poisson P 97
constante de réseau molécu-
    laire L 17
constante des ions de gaz
    G 41
constante des photo-
    électrons P 45
constante de Sutherland
    S 451
constante de temps de
    pompage P 219
constante diélectrique
    D 102
constante radioactive D 13
constante solaire S 216
constante universelle
    des gaz B 161, I 7

constitution de la surface
F 65
construction de soudure
W 32
construction entièrement
métallique A 123
construction tout-métal
A 123
contact bimétallique B 115
contact de gaz protecteur
(protectif) P 203
contact de ligne L 64
contact d'étanchéité J 19
contact élastique S 302
continu du courant C 297
continuité de couches F 53
contourner B 238
contraction S 134
contrainte S 378
contre-bride C 357
contre-courant C 358
contrôle de vide V 168
contrôleur automatique
du niveau L 53
contrôleur à vide à haute
fréquence H 75
contrôleur de la tempé-
rature T 16
convection C 314
convection forcée F 152
convection libre N 1
conversion cathodique C 63
convertisseur de mesure
M 55
convoyeur B 100
copeau C 131
copie R 112
copie superficielle S 444
cordonnet d'amiante A 196
corindon fritté S 166
corne polaire P 101
corps d'activité optique
O 53
corps de pompe B 54,
P 210
corps d'équilibrage C 231
corps de soupape B 153
corps gris G 120
corps noir B 124
corpuscule C 346
corrosion ionique E 153
corrosion superficielle S 436
cortège électronique
planétaire E 107
cosinus de direction D 152
côté d'aspiration C 220
côté de vide T 9
côté du vide préliminaire
F 159
côté prévide F 159
côté vide C 220
couche antiréfléchissante
A 176
couche d'adsorption A 76
couche d'arrêt B 55
couche d'eau W 8
couche de barrage B 55
couche de couverture
R 140
couche de fermeture R 140
couche de feuilles d'étain
L 28
couche de fond P 186
couche de fond de vernis
B 56
couche de gaz G 33
couche de graphite A 181
couche de laque L 3
couche de laque
d'accrochage B 56
couche de métallisation sous
vide C 176
couche de protection P 200
couche de recouvrement
O 73
couche de travail bas de
sortie L 120

couche d'évaporation C 176
couche d'interférence I 100
couche électronique E 108
couche étrangère F 157
couche fluorescente F 137
couche-getter C 175
couche graphitique G 113
couche hydrofuge D 252
couche insulaire I 173
couche intermédiaire I 99
couche limite B 180
couche magnétique à
mémoire M 7
couche magnétique
d'emmagasinage M 7
couche métallique par évaporation V 216
couche monoatomique
M 168
couche multiple M 192
couche protectrice P 200
couche résistant à la
corrosion C 349
couche spéculaire M 124
couche support de laque
B 57
coude B 102
coulage d'induction sous
vide V 85
coulage par injection sous
vide V 44
coulage par pression P 149
coulage sous vide S 410
coulée centrifuge C 87
coulée de précision sous
vide poussé H 91
coulée du sol B 177
coulée liquide F 131
coulée sous vide S 410,
V 59
coulisse S 179
coulisse à expansion E 212
couloir de transport C 316
coup de bélier W 13
coupe de coulée C 53
couplage en pont B 201
couple de décrochage P 71
couple de rotation T 127
couple moteur de dé-
marrage T 98
coupler en série C 272
couple thermoélectrique
T 56
coupure par rayon
électronique E 60
coupure par rayons
d'électrons E 60
courant de couche limite
B 181
courant de décalage D 186
courant de décharge D 160
courant de déplacement
D 186
courant de faisceau
électronique E 74
courant de fond B 5
courant de fuite L 37,
R 118
courant de gaz F 123
courant de glissement S 182
courant de perte L 37
courant descendant D 223
courant de soudure W 27
courant d'ions E 163
courant d'une fuite L 39
courant gazeux F 123
courant gazeux
dynamique D 275
courant instationnaire U 23
courant inverse U 21
courant ionique G 42
courant laminaire L 8
courant moléculaire M 155
courant montant U 25
courant photo-électrique
par rayonnement X doux
S 215
courant résiduel R 118

courant sans friction E 142
courant sans frottement
E 142
courant stationnaire S 332
courant superficiel S 438
courant turbulent T 197
courant ultrasonique S 424
courbe de débit Y 1
courbe de la vitesse d'aspiration et de la pression
S 268
courbe de pompage P 211
courbe de pression en fonction du temps P 176
courbe de rendement Y 1
courbe d'étalonnage C 14
courbe enchevêtrée S 356
courbe en paliers S 356
courroie B 99
courroie conique V 231
courroie en forme de coin
V 231
courroie trapézoïdale V 231
course de soupape V 198
couvercle de la soupape
V 189
couvercle superficiel S 434
couverture finale de grille
G 122
couverture monoatomique
M 173
couverture superficielle
S 434
couvrir C 173
creuset C 383
creuset à bouton B 234
creuset à épancher de fond
B 174
creuset basculable T 99
creuset basculant T 99
creuset céramique C 97
creuset de cuivre C 336
creuset d'évaporation E 191
creuset graphitique G 114
crevasse C 361
crible moléculaire M 159
cristal capillaire C 412
cristal de contrôle M 166
cristal idéal I 5
cristallerie de cuivre C 337
cristallisation sous vide V 37
cristal réel I 19
croisillon de centrage S 276
croissance de gouttelettes
D 246
croissance des grains G 108
cryodessiccation C 391
cryofixation C 395
cryopompage C 394
cryostat de vide V 36
cryosurface C 397
cryotrapping C 395
cubique-plaine centrée F 3
cucurbite C 197
cuirasse E 131
cuivrage par immersion
C 340
cuivre GFPH G 37
cuivrer C 334
cuivre sans gaz et de grande
pureté G 37
cuivre très pur O 83
culasse M 8
culasse de l'aimant Y 4
culot à baïonnette B 68
cuvette de coulée C 53
cuvette de décongélation
D 242
cuvette d'égouttage D 242
cycle d'eau C 167
cycle de Carnot C 38
cycle de Linde L 61
cycle de travail A 40
cycle d'étuvage B 27
cyclotron C 427
cylindre secondaire S 450
cylindre-support de
traversées E 40

D

débit E 16
débit de concentration
C 253
débit de masses M 32
débit de passage M 31
débit de pompage P 215
débit d'ions I 167
débit du courant R 34
débit d'une fuite I 68
débit d'une pompe P 215
débit d'une pompe à air
A 116
débit d'une pompe à lest
d'air B 37
débit d'une pompe sans lest
d'air U 14
débit effectif E 15, I 118
débit effectif de pompage
P 215
débit global T 139
débit horaire H 142
débit intrinsèque I 119
débit massique C 20
débit massique d'une pompe
E 199
débit massique intrinsèque
I 120
débit massique inverse
mesuré M 50
débit massique maximum
d'une pompe C 109
débit massique théorique
I 120
débitmètre F 122
débit nominal R 33
débit spécifique S 263
débit total d'une pompe
T 139
débit utile E 15
débit utile du froid U 27
débit volumétrique S 270,
V 249
déblocage U 20
décalage de fréquence
F 209
décalage de phases P 38
décalaminer D 71
décantation D 10
décanter D 9
décapage P 52
décapage final F 62
décarbonisation sous vide
V 39
décarboniser le filament de
tungstène D 11
décarburation sous vide
V 39
décarburer D 12
décharge à l'arc A 183
décharge automatique
S 70
décharge dans un gaz
G 26
décharge des cathodes
froides dans un gaz C 194
décharge de Townsend
D 4
décharge en aigrette B 225
décharge glissante S 181
décharge luminescente
G 99
décharge luminescente
anormale A 1
décharge luminescente
négative N 3
décharge luminescente
normale N 27
déchargement spontané
S 70
décharge noire B 126
décharge non indépendante
N 24
décharge non visible B 126
décharge obscure B 126,
D 4
décharge par effluves D 158

déchet W 4
déclenchement de pression
  de sécurité S 6
déclenchement thermique
  T 43
décompositeur d'air A 115
décomposition D 16
décompositions apparais-
  santes aux électrodes D 18
décongélation naturelle O 6
découvrir M 24
défaut de grille L 18
défaut de la descente
  hélicoïdale H 63
défaut d'étanchéité I 68
déflecteur B 24
déformation F 169
déformation à froid C 219
déformation sous vide
  V 67
dégagement de gaz G 54
dégagement de gaz à plu-
  sieurs étages M 206
dégagement de l'acier
  sous vide V 159
dégagement de poche L 5
dégagement de poussière
  F 168
dégagement gazeux G 30,
  O 68
dégagement gazeux à air
  comprimé sous vide
  V 160
dégagement gazeux par
  ruissellement O 49
dégagement instantané
  de gaz G 17
dégagement par circulation
  C 144
dégagement par siphon
  sous vide P 64
dégagement spontané de gaz
  G 17
dégager P 182
dégager / se F 124
dégazage D 39, G 54, O 68
dégazage à plusieurs étages
  M 206
dégazage continu de poêle à
  poêle L 7
dégazage dans la poche de
  coulée T 5
dégazage de l'acier S 349
dégazage par bombarde-
  ment D 41
dégazage par circulation
  C 144
dégazage par ruissellement
  O 49
dégazage par ruissellement
  du jet S 291
dégazage pendant la coulée
  S 376
dégazation D 39
dégazer D 37
dégazeur de courant dans le
  vide V 161
dégel D 36
dégeler D 34
dégel naturel O 6
dégel par eau fine W 12
dégel par gaz chaud H 131
dégivrage D 36
dégivrer D 34
déglacer D 56
dégraissage de vapeur
  V 212
dégraisser D 47
degré Celsius D 48
degré centésimal D 48
degré d'adsorption C 360
degré de couverture C 360
degré de dissociation D 49
degré de liberté de la translation T 151
degré de pression dynamique
  D 280
degré de pureté D 50

degré de sensibilité S 81
degré d'humidité M 139
degré d'usinage D 51
degré du vide V 63
degré Rankine D 52
délit L 10
demi-vie H 2
densité brute G 126
densité de courant
  d'émission E 124
densité de courant du faisceau électronique E 59
densité de déchargement
  B 220
densité de garniture P 3
densité de gaz D 63
densité de la dislocation
  D 178
densité d'énergie P 123
densité de particules P 19
densité de puissance P 124
densité de revêtement
  C 359
densité des endroits dé-
  fectueux d'un réseau
  moléculaire L 19
densité des porteurs de
  charge C 112
densité de vapeur V 214
densité du jet B 75
densité du rayon B 75
densité finale U 1
densité moléculaire N 45
densité nette N 5
densité spécifique S 260
dépendant de la pression
  P 154
déphasage P 38
déplacement à étages E 6
dépoli intérieurement I 75
dépolir F 218
dépolissage R 196
déposé par vaporisation
  V 215
déposer / se S 97
déposer en couches
  successives P 55
déposition par faisceau
  ionique I 127
dépôt de couches
  réfléchissantes M 123
dépôt de gaz G 60
dépôt gazeux G 53
dépôt sous vide V 42
dépresseur radial R 170
dépresseur Roots R 170
dépression L 112, S 165a
dépression capillaire C 26
dérangement / sans T 176
dérangement cristallin
  C 405
dérivation A 145
dériver B 199
désaération A 7
désagrégation W 23
désaimantation D 60
descente cathodique C 64
déshydratation D 54
déshydratation sous vide
  V 41
désintégration C 160
désintégration inverse B 111
désorption D 73
désorption de gaz O 68
désorption électronique
  E 90
désorption thermique
  T 30
désorption thermique
  pulsée F 90
désoxydant D 64
dessèchement D 55
desséchant avec noyau de
  glace D 258
dessèchement D 72
dessiccateur D 55
dessiccateur à vide V 43,
  V 53

dessiccation D 54
dessiccation centrifuge
  C 90
dessiccation diélectrique
  D 103
dessiccation finale A 90
dessiccation par congélation
  C 391
dessiccation par congélation
  de tissu T 114
dessiccation par courant
  de vapeur V 180
dessiccation par ruissellement F 105
dessiccation par sublimation
  D 255
dessiccation principale
  M 12
dessiccation secondaire
  A 90
dessiccation sous vide V 54
dessouder U 22
détecteur de fuites L 43
détecteur de fuites à filtres
  de palladium P 13
détecteur de fuites à haute
  fréquence S 255
détecteur de fuites à palladium P 13
détecteur de fuites au platine-
  halogène H 133
détecteur de fuites aux
  halogènes H 4, H 7
détecteur de fuites de diodes
  halogènes H 6
détecteur de fuites d'un récipient par palladium et
  hydrogène P 10
détecteur de fuites d'un récipient par rayons infra-
  rouges I 56
détecteur de fuites par gel silice S 142
détecteur de fuites par
  hélium H 59
détecteur de fuites par
  oxygène O 84
détecteur de fuites par
  pompe ionique à getter
  par pulvérisation
  cathodique I 159
détecteur de fuites par
  spectrométrie de masse
  M 39
détecteur de fuites sensible
  aux halogènes H 7
détecteur différentiel de
  fuites D 113
détecteur d'ionisation I 145
détection de fuites L 38
détection de fuites avec indicateur de vide à hautes
  fréquences S 256
détection de fuites avec indicateur radio-actif R 19
détection de fuites avec le
  tube de Geissler G 21
détection de fuites par
  ammoniaque A 141
détection dynamique de
  fuites D 278
détection par l'espace primaire B 16
détender E 206
détendeur de pression P 167
détermination de la fré-
  quence des isotopes I 182
détourner B 238
détremper A 150
deuxième électrode A 26
déverrouillage U 20
déviation D 79
déviation de rayons
  cathodiques C 69
déviation maxima V 185
déviation totale V 185
dévitrification D 83
dévitrifier D 84

diagramme de phases P 37
diagramme polaire P 98
diagramme polaire du choc
  S 123
dialyser D 91
diamètre du cercle des trous
  P 70
diamètre extérieur E 219
diamètre intérieur I 73
diamètre nominal N 15
diaphragme D 92
diaphragme automatique
  A 214
diaphragme circulaire
  C 139
diaphragme de sortie E 203
diaphragme iris I 169
diaphragme ondulé C 350
différence de pression P 155
diffraction de réflexion
  R 84
diffraction des électrons
  à basse énergie D 123
diffraction des électrons
  à grande vitesse H 72,
  T 157
diffraction des électrons à
  grande vitesse par
  réflexion R 86
diffraction des électrons
  lents D 123
diffraction des électrons
  par réflexion R 85
diffraction des électrons par
  transmission T 156
diffraction des rayons X
  X 2
diffraction électronique
  d'exploration S 22
diffractomètre à rayons X
  à vide V 114
diffus D 125
diffuseur A 157, D 126,
  D 134, E 21, M 129, N 31
diffuseur-déflecteur A 157
diffusion D 128, D 184
diffusion de retour B 2
diffusion miroitante S 267
diffusion séparée S 161
diffusion sous pression
  P 156
diffusion spéculaire S 267
diffusion superficielle S 435
diffusion thermique T 32
diffusion volumétrique
  B 221
dilatabilité M 15
dilatation de rupture B 193
dilatomètre S 375
dilatomètre à vide V 46
dilution successive S 408
dimensionnement de fuite
  calibrée C 12
diminution de concen-
  tration d'ions D 21
diminution du vide
  d'un tube D 22
diode d'amortissement
  D 3
diode de récupération
  B 169
diode halogène H 5
diode-pentode D 140
diode-triode D 143
direction de propagation
  D 154
discontinu B 64
discontinuation de contact
  C 285
discontinuité d'énergie
  E 138
dislocation D 179
dislocation de cristal C 407
dislocation de vis S 32
disperser D 183
dispersion D 184
dispersion d'angle
  miniature S 196

dispersion d'angle petit S 196
dispersion des limites de grains G 107
dispersion inverse B 111
dispersion miroitante S 267
dispersion multiple M 199
dispersion séparée S 161
dispersion spéculaire S 267
dispositif à soudure froide sous pression et à déserrage C 218
dispositif d'alimentation C 115
dispositif d'arrêt R 29
dispositif de centrage C 95
dispositif d'échappement de lingots I 63
dispositif d'éclusage L 96
dispositif de commande R 30
dispositif de décharge D 161
dispositif de fixation J 16
dispositif de fractionnement automatique S 67
dispositif de mise à zéro Z 4
dispositif d'enregistrement P 29
dispositif d'épreuve de friture S 28
dispositif de protection P 201, S 1
dispositif de régénération d'huile O 31
dispositif de sécurité P 201, S 1
dispositif de sécurité de soupape V 195
dispositif de serrage G 125, I 16
dispositif de surveillance de pression P 163
dispositif d'ionisation I 146
dispositif pour ajouter R 79
dispositif protecteur P 201
dispositif rotatif R 174
disposition de l'appareillage des taux de vaporisation avec électrodes chauffées H 30
disposition de soupapes V 206
disque céramique C 98
disque de tungstène T 186
disque de vanne V 190
disque élastique de vanne S 2
disque en lisière de tissu B 213
disque intermédiaire C 421
dissipation anodique A 163
dissipation de chaleur H 28
dissolvant S 233
distance anodique A 164
distance de décharge D 163
distance de plaine de réseau I 116
distance explosive sous vide V 151
distillateur D 194
distillation à court trajet O 47
distillation à libre parcours F 189
distillation dans le vide V 47
distillation de vapeur d'eau S 341
distillation fractionnée F 176
distillation fractionnée sous vide poussé H 87
distillation moléculaire M 150
distillation pelliculaire T 71

distillation pelliculaire à court trajet T 79
distillation pelliculaire de surfaces mixtes M 131
distillation sous vide V 47
distillerie D 196
distillerie à court trajet O 48
distillerie des balais B 210
distillerie moléculaire M 162
distillerie moléculaire à film tombant F 9
distributeur D 199
distribution angulaire A 148
distribution cosinusoïdale de Lambert C 353
distribution de chaleur H 28
distribution de charge C 113
distribution de potentiel P 114
distribution de vitesse V 233
distribution statistique R 21
divergence A 149
diviseur de rayon B 83
doigt de refroidissement à plonger I 12
doigtier en caoutchouc R 202
doigt refroidisseur C 201, R 114
doigt refroidisseur d'immersion I 12
domaine de coulée M 67
domaine de pression P 165
domaine transitoire I 110
dôme à vapeur S 340
dôme de mesure M 54
donneur D 200
dorure à froid C 203
dosage gazeux G 22
double fusion D 216
double liaison conjuguée C 268
double surface électrique E 30
douille S 177
douille à baïonnette B 70
douille du robinet K 5
douille glissante L 62
drapage D 234
druse de cristal C 170
ductilité D 267, M 15
Dumet-fusion D 269
durcissement H 13
durcissement à froid W 20
durcissement de vieillissement A 92
durcissement électrique E 29
durée A 61
durée de fonctionnement T 150
durée de fusion F 236
durée de non-contamination H 102, S 337
durée de parcours T 150
durée de réponse R 148
durée de service de filament F 42
durée d'évacuation P 212
durée de vie de filament F 42
dureté H 16
dureté Brinell B 205
dureté de pénétration I 38
dureté de polissage G 124
dureté de rebondissement R 47
dureté de refonte R 106
dureté Rockwell R 165
dureté sclérométrique A 4
dureté Shore S 128
dureté Vickers V 240
dyne D 282

E

eau condensée C 259
eau de condensation C 259
eau de conduite T 10
eau distillée D 195
eau réfrigérée C 128
eau saline de refroidissement C 321
ébauche pour presse à filet E 229
ébonite E 3
ébullition B 157
ébullition instable B 224
écaillement de mica M 102
écartement des plaques C 158
échangeable I 92
échange de rayonnement R 15
échange thermique H 33
échangeur à contre-courant C 356
échangeur ionique I 136
échangeur thermique H 34
échantillon S 8
échantillonnage S 10
échantillonnage intermédiaire I 111
échappée de gaz G 17
échappement D 166, E 120
échappement de lingot I 60
échappement de vapeur V 218
échapper / s' F 124
échauffement diélectrique à haute fréquence D 105
échauffement par courants de Foucault E 5
échauffement par résistance R 134
échelle Beaufort B 88
échelle Celsius C 83
échelle centigrade C 83
échelle de la température absolue A 5
échelle de température absolue A 9
échelle Fahrenheit F 5
échelle graduée D 53
échelle Réaumur R 44
éclatement F 89
éclateur sous vide V 151
éclat sous vide V 17
écluse S 195
écluse à étage de pression D 114
écluse à gaz L 96
écluse à lingot fusible I 58
écluse à vide V 98
écluse d'entrée E 144
écluse du radiateur [électronique] G 141
écluse par d'avance à barre B 47
écoulement E 120, F 118
écoulement d'après Knudsen K 12
écoulement d'augmentation de pression R 35
écoulement de couche limite B 181
écoulement de Poiseuille P 93
écoulement dynamique des gaz D 275
écoulement en régime intermédiaire K 12
écoulement en régime laminaire L 8
écoulement froid C 202
écoulement laminaire L 8
écoulement moléculaire M 155
écoulement stationnaire S 332
écoulement thermique H 49

écoulement tourbillonnaire V 252
écoulement tourbillonnant V 252
écoulement turbulent T 197
écoulement ultrasonique S 424
écoulement ultrasonique dans pompes T 159
écoulement visqueux S 182
écran B 24, O 62
écran à éjecteur A 180
écran à injecteur A 180
écran à tuyère A 180
écran automatique A 214
écran contre la radiation H 46
écran de la source d'évaporation E 183
écran de protection contre les aspersions S 259
écran fluorescent F 138
écran protecteur à moulage S 259
écran protecteur contre le rayonnement H 46
écraser P 61
écrou à chapeau C 29
écrou à raccord C 29
écrou-chapeau C 29
écuelle de coulée C 53
écume F 144
écumer F 143
effervescence E 4
effet capillaire C 22
effet d'arrêt B 142
effet d'avertissement M 74
effet de Blears B 133
effet de blocage B 142
effet de cryofixation C 401
effet de cryotrapping C 401
effet de démixtion S 84
effet de pompage P 213
effet de sens O 60
effet de séparation S 85
effet de soudage à basse température D 28
effet des rayons X X 3
effet d'incrustation S 19
effet directif D 151
effet d'orientation O 60
effet du faisceau lancé B 79
effet getter C 154
effet getter d'évaporation D 182
effet Kerr K 2
effet pelliculaire S 172
effet photoélectrique externe P 44
effet photoélectrique intérieur P 41
effet tunnel T 191
effet tunnel des électrons E 111
effet utile du froid U 27
effeuiller C 132
efficacité de fusion M 65
efficacité d'émission E 125
efficacité du piège refroidi C 214
efflorescence W 23
effort annulaire H 116
effusion E 17
effusion moléculaire M 153
éjecteur E 20, E 21, N 31
éjecteur à diffusion D 131
éjecteur à gaz G 44
éjecteur à tuyère annulaire A 158
éjecteur à vapeur E 20
éjecteur à vapeur d'eau J 9
éjecteur d'huile O 22
éjecteur diffuseur A 158
élastomère E 26
électrode auto-consommateur S 62
électrode auxiliaire A 217

électrode conductrice
G 138
électrode consommable
C 280
électrode d'accélération
A 26
électrode de balayage R 164
électrode de frein S 429
électrode de réseau N 7
électrode en graphite G 115
électrode en toile
métallique N 7
électrode fusible C 280
électrode fusible pour
soudage à l'arc C 280
électrode inverse pour
électrons E 106
électrode inverse pour ions
I 160
électrode permanente N 19
électrode réticulaire N 7
électrode scellée étanche à
l'air H 66
électro-érosion E 42
électrolyse E 45
électromètre à condensateur
oscillant V 237
électron de cascade C 47
électron de grille L 20
électronique de feuilles
minces P 24
électronique de gaz G 27
électronique moléculaire
M 154
électron périphérique S 108
électrovalence E 118
élément chauffant en forme
de barre C 43
élément combustible F 220
élément compensateur
C 231
élément de construction
C 279
élément de masse C 346
élément de mémoire de
couches minces T 75
élément d'étanchéité G 45
élément fusible F 235
élément mono M 170
élément sensible S 79
élément thermoélectrique à
couches minces T 81
élément tubulaire de
chauffage T 180
élève sous vide poussé H 89
élimination de couches
métallisées D 15
éliminer par filtration F 60
émanation E 120
emballage sous vide V 111
embouchure de diffuseur
N 34
embouchure de tuyère N 34
embouti à capuchon C 21
embouti de joint S 41
embranchement A 145
émission de gaz G 54
émission de gaz de filets
minces T 69
émission d'incandescence
T 52
émission électronique E 87
émission exoélectronique
E 204
émission ionique de champ
F 38
émission photoélectrique
P 44
émission thermoélectrique
T 52
émission totale T 134
émissivité totale T 134
emplacement de sorption
S 241
empoisonnement P 96
empoisonnement de la
cathode C 68
empoisonner P 95

émulseur à air comprimé
G 16
émulsionner E 129
enceinte à vide V 34
enchaînement L 70
encliquetage R 30
encoche à brides F 76
encoche en queue d'aronde
D 221
encoche trapézoïdale T 165
encre à verre E 154
endothermique E 135
endroit d'impuretés D 30
enduit adhésif A 64
enduit aquadag A 181
endurance à la flexion B 104
énergie d'activation A 44
énergie de combinaison
B 122
énergie de liaison B 122
énergie de seuil T 90
énergie de surface limite
I 97
énergie de translation T 152
énergie d'ionisation I 147
énergie interfaciale I 97
énergie libre F 184
énergie moyenne
d'excitation A 219
énergie thermique des
électrons T 34
enfoncement d'un rayon
B 80
enfumage sous vide V 71
engrenage à crémaillère R 1
engrenage différentiel D 107
enregistrement par rayon
électronique E 75
enregistrement rapide S 272
enregistreur P 29
enregistreur à point
lumineux F 142
enregistreur en pointillé
D 204
enregistreur en trait continu
C 298
enregistreur multicourbe par
points M 204
enroulement bifilaire B 112
ensemble base M 203
ensemble de scellement
monofilaire S 165
ensemble diffuseur J 4
ensemble mécanique de
vannes V 188
enthalpie E 141
entonnoir filtre V 64
entraînement P 197
entrée d'air A 89
entrée d'air accidentelle
A 106
entrée d'eau W 14
entrée de l'air A 98
entrée glissante L 62
entretien / sans M 13
entretoise S 336
entropie E 143
enveloppe à l'eau de
refroidissement C 332
enveloppe à vide V 92
enveloppe de tuyère S 173
enveloppe de verre G 86
enveloppe électronique
E 107
enveloppe protectrice E 217
enveloppe réfrigérante
C 325
epitaxie E 146
éponge de caoutchouc
S 284
éponge de titane T 118
épreuve à blanc B 131
épreuve à l'air comprimé
P 181
épreuve d'étanchéité C 105,
L 45
épreuve d'étanchéité à
surpression O 79

épreuve enveloppante
E 145
épreuve pour le contrôle du
vide à haute fréquence
S 257
éprouvé F 7
éprouver à l'air comprimé
P 61
épurateur d'huile O 14,
O 31
épuration C 153
épuration à l'argon
par pulvérisation A 190
épuration automatique / d'
S 68
épuration cathodique
C 72
épuration des gaz G 55
équation de Bernoulli
B 105
équation de Clausius-
Clapeyron C 151
équation de continuité
C 294
équation des gaz idéaux
E 149
équation du gaz parfait
(idéal) G 29
équation universelle des gaz
E 149
équilibrage de faisceau
électronique E 55
équilibrage de rayon
électronique E 55
équilibre thermique T 35
équivalent calorifique T 36
équivalent d'azote N 9
équivalent manométrique
M 21
erg E 151
érosion d'étincelles S 258
érosion supersonique U 11
érosion ultrasonore U 11
erreur admissible A 125
erreur fortuite A 31
espace à étincelle sous vide
V 151
espace d'ionisation I 151
espace évacué intermédiaire
G 136
espace intérieur I 106
espace mort D 6
espace noir de Faraday F 12
espace nuisible D 6
espace obscur anodique
A 162
espace sous vide primaire
B 15
espace vide intermédiaire
G 136
essai S 8
essai à blanc B 131
essai de bruit N 14
essai de dureté H 17
essai de flexion au choc
I 17
essai de fuite L 40
essai de résilience N 28
essai d'étanchéité L 45
essai permanent E 137
essai pour le contrôle du
vide à haute fréquence
S 257
essorage V 65
essoreuse C 89
essoreuse à vide S 413
estampage à froid C 211
estamper S 320
étage à basse pression L 114
étage de diffusion D 139
étage d'éjecteur E 22
étage d'éjection de vapeur
E 22
étage de pompage
différentiel D 121
étage de pression P 172
étage de pression dynamique
D 280

étage d'injecteur à vapeur
S 138
étage du jet à vapeur S 138
étain à souder S 219
étanche L 44
étanche au vide V 173
étanche aux gaz G 61
étanchéité aux gaz G 62
étanchéité du vide V 96
étanchement G 45
étanché par métal M 93
étancher S 36
état de surface S 447
état métastable M 100
état permanent S 338
état stationnaire S 333,
S 338
étendre S 299
étendue d'ajustage S 95
étendue de réglage S 95
étendue transitoire I 110,
T 148
étirage brillant B 203
étoupé hermétiquement
H 65
étranglement T 93
étrier en U S 363
étuvable B 26
étuve B 29, D 256
étuve à vide V 55
évacuation D 229, V 127
évacuation de l'air D 7
évacuation grossière C 171
évacuation préliminaire
C 171
évacuer B 134, D 237,
E 158
évaporateur E 190
évaporateur à chute libre
F 186
évaporateur de circulation
à vide V 24
évaporateur de titane T 114a
évaporateur interchangeable
I 93
évaporateur par bombarde-
ment électronique E 65
évaporateur par rayons
d'électrons E 65
évaporateur par tuyauteries
verticales L 103
évaporateur rotatif à vide
V 141
évaporation E 165
évaporation à arc sous vide
avec électrode consomma-
ble C 281
évaporation à l'arc de
charbon C 32
évaporation de champ F 37
évaporation de charbon
C 34
évaporation de creuset
C 384
évaporation de la surface
S 437
évaporation de pointes de
charbon C 34
évaporation de poussée
F 84
évaporation des couches
monomoléculaires M 174
évaporation en suspension
L 55
évaporation oblique O 3
évaporation par lévitation
L 55
évaporation par recuit F 87
évaporation sous vide V 60
évaporer / s' E 164
exactitude d'ajustage A 36
excitation E 194
excitation électronique E 91
exothermique E 205
expansibilité M 15

expansion libre F 185
exsiccateur C 1
extensibilité M 15
extracteur E 223
extraction de métal M 89
extraction sous vide V 61
extraire E 222
extraire par pression P 61
extrapoler E 225
extrudeuse E 227
extrudeuse à deux cylindres
  T 210
extrusion E 228
extrusion métallique M 95
extrusion sous vide V 62

F

fabrication des tubes
  électroniques E 110
face de fente de diffusion
  T 92
face d'émission E 126
facette C 403
façonnage F 169, S 103
facteur binaire de la thermodiffusion B 120
facteur d'absorption A 21
facteur de Clausing C 152
facteur de compression
  C 240
facteur de conversion
  C 315
facteur de forme S 102
facteur de gaz G 31
facteur de pertes
  diélectriques D 191,
  L 106
facteur de puissance C 181
facteur de séparation S 83
facteur de transmission
  E 171
facteur du vide V 63
facteur séparateur S 83
Fahrenheit-graduation de
  température F 1
faire circuler par pompage
  C 141
faire le vide E 158
faire le vide primaire F 160
faire sortir par pression
  P 61
faisceau cathodique C 70
faisceau électronique E 52
faisceau électronique de
  haute pervéance H 76
fatigue F 14
fausse bride B 128
fêlure C 361
fenêtre à la pression P 180
fenêtre de contrôle I 79
fenêtre de contrôle à niveau
  d'huile I 79
fenêtre de coup d'œil I 78
fenêtre de diffusion D 133
fenêtre de glucinium B 106
fenêtre de lamelles F 148
fenêtre de mica M 108
fenêtre d'observation I 78
fente C 361
fente capillaire C 25
fente d'aspiration I 88
fente par contraction C 120
fente thermique H 27
ferailles S 26
fer à souder S 223
fer en U C 108
fer forgé L 108
fermé hermétiquement
  H 65
fermer S 366
fermer par une fausse bride
  B 129
fermeture à baïonnette B 69
fermeture à vide V 144
fermeture de vide poussé
  H 85

fermeture hermétique H 67
fermeture instantanée Q 17
fermeture rapide Q 17
feuille d'aluminium A 134
feuille mince d'argent fin
  laminé T 82
feuille polaroïdale P 100
fibre de verre G 87
ficelle d'amiante A 196
fiche de guidage G 139
fil à lier T 97
filament F 40
filament à double bobinage
  C 185
filament de carbide de
  tungstène C 35
filament de réserve S 252
filament de tungstène
  T 185
filament d'évaporation
  hélicoïdal H 56
filament d'iridium I 168
filament spiral S 153
fil chaud F 40
fil d'âme C 343
fil d'attache T 97
fil d'entrée L 32
fil d'enveloppe de cuivre
  C 232
fil enrobé de cuivre C 335
filer E 226
filet E 221, F 40
filet additionnel A 58
filetage E 221
filetage à pas multiple
  M 201
filetage plat S 315
fil étiré à froid C 212
filet plat S 315
filet produit par des ions
  I 157
filet supplémentaire A 58
fil fusible F 235
filière D 100
film de surface limite
  I 98
film interfacial I 98
fil monocristallin M 171
fil porteur C 42
fil scellé S 51
fil sous tube I 84
fil thermique F 40
filtrage F 59
filtrage par succion V 65
filtration F 61
filtration de la lumière par
  couches minces T 73
filtration sous vide V 65
filtre à air A 102
filtre à huile O 25
filtre à membrane
  métallique M 87
filtre antibrouillard d'huile
  O 29
filtre à poussière D 272
filtre à tambour à vide
  V 52
filtre d'aspiration S 413
filtre debout E 10
filtre d'échappement D 162
filtre de masse E 31
filtre de nuage d'huile
  O 29
filtre de verre G 89
filtre d'interférence I 102
filtre d'interférence à
  réflexion de chaleur H 44
filtre électrique de masse
  E 31
filtre gris G 121
filtre neutre G 121
filtre pour l'épuration de
  l'huile O 30
filtre pour l'huile O 25
filtrer F 57
filtrer par aspiration F 58
filtre tournant à vide R 187
fini de la surface F 65

fiole V 236
fissilité C 160
fission du noyau N 36
fission nucléaire N 36
fissure C 361
fissure capillaire C 25
fissure de contraction C 120
fixateur C 80
flambage F 125
flash sous vide V 17
floculation F 112
flottation F 117
fluage M 116
fluage d'huile O 16
fluage froid C 202
fluctuation F 115
fluctuation de la tension de
  secteur L 67
fluide actif P 197
fluide moteur P 197
fluide moteur d'une pompe
  O 50
fluidité F 134
flux d'air comprimé A 97
flux de cisaillement S 105
flux de dégazage O 68
flux de gaz F 123
flux de matière en fusion
  C 232
flux des lacunes V 2
flux de verre V 244
flux dynamique des gaz
  D 275
flux gazeux F 123
flux inverse des gaz G 9
flux ionique E 163
flux massique C 20
flux moléculaire M 155
flux total F 135
flux volumétrique S 270
focalisation F 146
focalisation des gaz G 36
fonction de distribution de
  Maxwell-Boltzmann
  M 46
fonctionnement sûr / de
  F 7
fond à déplacement vers le
  bas L 110
fondage F 237
fondage bout à bout B 236
fondage de cheville P 63
fondage de tuyau T 181
fondage sous vide poussé
  H 88
fondant à braser H 18,
  S 222
fondant à souder S 222
fondant à souder à l'étain
  S 214
fond de poche S 174
fond mobile S 180
fonte C 55
fonte brute P 54
fonte de lingots sous vide
  V 89
fonte de moulage C 55
fonte malléable M 16
fonte par moulage système
  Croning C 119
fonte primaire F 67
fonte secondaire S 55
foration à évacuation de
  l'air A 113
foration de désaération
  A 113
force de liaison B 165
force moléculaire M 156
force répulsive de Coulomb
  C 355
forgeabilité F 165, M 15
formage F 169
formation dans les tuyères
  J 10
formation d'eau de condensation S 452
formation de bulles B 139,
  P 207

formation de bulles d'air
  P 207
formation de givre F 219
formation de glace I 2
formation de poussière
  F 168
formation d'un courant
  vagabond S 440
formation d'un noyau N 38
formation d'un nucléus
  N 38
formation par forgeage,
  laminage, pression et
  tirage F 1
forme d'aspiration S 414
forme de coulée C 50
former un boudin E 226
formule barométrique [de
  Boltzmann] B 53
four à arc A 184
four à arc à vide V 11
four à arc sous vide V 11
four à arc sous vide avec
  une électrode consom-
  mable C 206
four à bombardement
  électronique E 66
four à bombardement
  électronique à plusieurs
  chambres E 72
four à chambres C 104
four à chambres horizontales
  H 117
four à cornue R 154
four à cornue à vide V 139
four à creuset C 386
four à cuve P 72
four à étirer des cristaux
  C 404
four à induction à vide V 86
four à moufle M 186
four à pots P 116
four à recuire A 152
four à recuire à vide V 8
four à recuire à vide poussé
  H 94
four à recuire en forme de
  pot P 112
four à recuire pendant la
  coulée à bandes sous vide
  C 303
four à réduction R 68
four à résistance R 132
four à résistance à vide
  V 138
four à résistance de vide
  V 138
four à réverbère A 104
four à tunnel T 193
four à une seule direction
  U 16
four à usage multiple M 205
four à vide V 72
four à vide à pression chaude
  et trempe V 79
four basculant T 104
four basculant de fusion à
  vide T 105
four continu T 95, T 193
four continu à recuire C 295
four de brasure à vide V 16
four de chauffe B 29
four de coulée à bombardement électronique E 57
four de dégazage sous vide
  V 40
four de forme rectangulaire
  B 185
four de frittage à vide
  poussé H 94
four de frittage de passage
  continu C 302
four de frittage sous pression
  P 145
four de fusion M 66
four de fusion à bombardement électronique avec
  canon à distance R 109

four de fusion à canons
  multiples par faisceaux
  électroniques E 73
four de fusion à coquille
  S 175
four de fusion à induction
  sous vide V 88
four de fusion à vide V 103
four de fusion à vide à
  basculant creuset fusible
  T 103
four de fusion par induction
  I 43
four de gaz de protection
  C 309
four de gaz protecteur
  C 309
four de paroi froide à vide
  V 28
four de paroi surchauffée à
  vide V 81
four de rayonnement R 12
four de soudure sous vide
  à haute température de
  paroi froide avec chauffeur intérieur C 216
four de tirage de cristaux
  C 404
four d'étuvage B 29
four Martin O 46
fournée F 230
four tubulaire R 154
four tunnel C 301
foyer S 286
foyer linéaire-anticathode
  S 373
fraction de moles M 140
fraction intercristalline I 95
fragilité B 207
fragilité à chaud H 136
frein à friction F 211
fréquence à l'excitation
  E 195
fréquence de collision
  C 222, C 223, I 21
fréquence de cyclotron
  C 428
fréquence de phase B 87
fréquence de résonance
  R 143
fréquence de résonance de
  cyclotron C 430
fréquence du choc de
  surface I 21
fréquence propre F 225
friction F 210
friction à sec D 254
friction de l'interface B 179
friction de surface limite
  B 179
friction interne I 113
frigorigène C 317
frittage par bombardement
  électronique E 76
frittage sous pression sous
  vide V 124
frittage sous vide V 148
fritte F 216
front de choc S 122
frottement F 210
frottement adhérent S 327
frottement d'écoulement
  F 212
frottement de glissement
  S 183
frottement statique S 327
fuel-gaz F 221
fugacité F 222
fuite L 36
fuite apparente V 241
fuite argent-oxygène S 148
fuite calibrée C 11
fuite d'angle miniature
  S 196
fuite de dosage de gaz G 23
fuite de membrane M 73
fuite d'énergie E 140
fuite de paroi mince T 83

fuite moléculaire M 157
fuite palladium-hydrogène
  P 12
fuite réglable A 71, D 203
fuite totale I 89
fuite virtuelle V 241
fuite visqueuse V 243
fusée ionique I 163
fusibilité F 233
fusible F 234
fusion F 231, F 237
fusion à cathode liquide
  C 282
fusion à coquille S 176
fusion à fond de moule
  refroidi S 176
fusion à l'arc A 185
fusion à l'arc sous vide
  V 12
fusion à zone Z 15
fusion à zone flottante
  F 108
fusion à zones sous vide
  V 182
fusion d'acier inoxydable et
  verre dur H 15
fusion de barre R 166
fusion d'égouttage D 241
fusion d'égouttage sous vide
  V 50
fusion de zone Z 15
fusion de zone sous vide
  V 182
fusion de zone verticale
  F 108
fusion d'induction sous vide
  V 87
fusion en zones sans creuset
  C 385
fusion nucléaire N 37
fusion par bombardement
  électronique E 71
fusion par extrusion E 230
fusion par induction I 45
fusion par lévitation F 188
fusion sous vide V 102
fusion sous vide poussé
  H 88
fusion verre-métal M 97

G

gabarit du microcircuit
  M 113
gaine en résine synthétique
  C 178
galvaniser au zinc Z 10
galvanomètre à cadre
  mobile M 184
gamme d'ajustage S 95
gamme de réglage S 95
garde du courant d'eau
  P 206
garniture cylindrique S 41
garniture de joint F 75
garniture de presse-étoupe
  G 79
garniture en cannelures L 1
garniture étanche à anneau
  glissant F 4
garniture étanche à étage de
  pression P 173
garniture étanche d'arbre
  R 5
garniture étanche de fil
  W 48
gaz absorbé S 236
gaz adsorbé A 75
gaz condensable C 255
gaz de balayage B 136
gaz d'éclairage T 140
gaz de plasma P 80
gaz d'épreuve P 190
gaz de rinçage B 136
gaz de ville T 140
gaz électronique E 88
gaz en bouteille B 163

gaz étalon C 15
gazeux G 24
gaz idéal I 6
gaz incondensable N 18
gaz inerte I 47
gaz mixte F 170
gaz parfait I 6
gaz pauvre P 126
gaz permanent G 8, N 18
gaz porteur C 41
gaz rare I 47
gaz raréfié R 25
gaz réel I 20
gaz résiduel R 119
gaz-sonde P 190
gaz sorbé S 236
gaz témoin P 190
gaz traceur P 190
gel de silice S 141
générateur des jets
  d'électrons E 67
générateur de vapeur à vide
  V 156
génération de gaz G 38
génération de vapeur S 342
germanium G 72
germe cristallin C 406
getter G 73
getter actif A 49
getter à étrier S 364
getter à évaporation V 246
getter annulaire R 162
getter de dispersion D 181
getter de pores P 107
getter de sorption N 26
gettérisation par immersion
  D 146
getter massif B 222
getter réactif R 42
getter sélectif S 59
getter superficiel C 175
glace sèche C 33
glaçure G 98
glissement S 182, S 187
glissière S 188
gorge rectangulaire R 59
goujon S 390
goupille d'ajustage A 119
gradient de concentration
  C 251
graduation logarithmique de
  pression L 99
graduel G 104
grain abrasif A 3
graissage L 123
graissage central sous
  pression F 116
graissage forcé F 153
graissage limite T 74
graissage par barbotage
  S 281
graissage par circulation
  sous pression F 116
graissage par flux de masse
  fondue M 71
graissage par immersion
  S 281
graissage sous pression
  F 153
graisse à vide P 5, V 77
graisse d'étanchéité P 5
graisse pour robinets S 365
graisse silicone S 144
graisseur de gouttes D 240
grandeur du foyer S 287
granulation G 109
granule G 112
grattoir S 27
gratture sur le joint G 49
gravité G 116
gravure cathodique C 73
gravure de congélation
  F 197
grenaille S 26
grid final E 134
griffe de raccordement
  C 271
grillage B 230

grille à couche F 54
grille anodique A 165
grille d'arrêt C 266
grille de protection S 30
grille de séchage S 116
grille-écran S 30
grille finale E 134
grille spatiale S 248
grippage S 58
grosseur des grains G 109
groupe de pompage G 130
groupe de pompage à vide
  V 128
groupe frigorifique R 98
groupe frigorifique de
  compression C 248
guide de tige à soupape
  V 204
guide de vis à soupape
  V 204
guidon G 140
gutta-percha G 142

H

halogène H 3
hauteur de pression P 161
hélice de refroidissement
  C 323
hélice de tungstène T 185
hélice double D 212
hélicoïde H 55
hermétique A 110
hermétiquement clos H 65
hétérogène I 64
homéopolaire H 111
homogénéisation H 109
homogénéiser H 110
homogénéité H 108
homogénéité de densité
  U 17
hotte de la soupape V 189
hotte d'évacuation D 224
hublot I 78
huile à pompe O 50
huile à pompe primaire
  R 184
huile de pompe à diffusion
  D 136
huile de silicone S 145
huile d'une pompe à vapeur
  V 225
humidité H 144
humidité relative R 101
humidité résiduelle R 123
hydrate d'aluminium A 135
hydrocarbure H 147
hydrogénation H 146

I

iconoscope I 4
ignition I 9
image d'optique électro-
  nique E 97
immersion D 145
impact d'ions I 142
impédance F 120
imperfection de réseau
  moléculaire L 21
imperméabilité aux gaz G 62
imperméable aux gaz G 39,
  G 61
imploder I 22
implosion I 23
imprégnation I 28, S 202
imprégnation sous vide
  V 83
imprégné de résine
  synthétique R 128
imprégner I 25
imprégné sous vide V 82
impression enveloppante
  E 145
impression superficielle
  S 444

impulsion E 194, I 29
impulsion électronique
  E 91
impureté I 32
Inbus-clé I 33
Inbus-clef I 33
incandescence d'acier
  inoxydable A 153
incandescence intermédiaire
  P 192
incandescence par bom-
  bardement électronique
  E 53
incidence touchante G 118
incrustation I 37
indépendant de la tension
  I 39
indicateur T 142
indicateur à ionisation de
  haute pression H 77
indicateur acoustique de
  fuite A 209
indicateur de fuite par tube
  à décharge D 170
indicateur de fuites L 42
indicateur de niveau d'huile
  O 35
indicateur de position
  P 109
indicateur de vide V 84
indicateur du niveau L 52,
  L 82
indicateur du niveau liquide
  L 80
indicateur radioactif R 18
indication à distance R 110
indice adiabatique A 69
indice de réfraction R 90
indice d'usure W 21
induction magnétique
  M 6
infiltration de volume
  S 270
infiltration thermique T 26
inflammation I 9
injecteur à vapeur V 217
injecteur condenseur I 67
injecteur de vapeur S 346
injecteur enveloppe C 28
injecteur latéral S 138
injection à flammes F 70
injection d'air G 10
injection de gaz G 40
injection de plasma P 82
inoxydable C 348
insensibilité à la vapeur
  d'eau M 42
insérer E 121
instabilité d'argon A 189
installation I 81
installation à coller C 82
installation à décirer D 86
installation à dégraissage de
  vapeur V 213
installation à deux salles
  T 208
installation à étirer des
  cristaux C 404
installation à incandescence
  à bombardement électro-
  nique E 54
installation à récupération
  pour l'hélium H 62
installation à séparer des gaz
  G 56
installation à vide V 119
installation à vide poussé à
  imprégnation de cire
  H 98
installation continue A 117
installation d'agglomération
  S 171
installation d'agglomération
  à barres R 167
installation d'aspiration
  E 202
installation d'échappement à
  filer D 81

installation de cloche B 91
installation de combustion
  lente G 100
installation de concentration
  à vide V 30
installation de congélation à
  basse température S 104
installation de congélation à
  claies S 117
installation de congélation
  en forme de plaques P 89
installation de congélation
  par contact C 288
installation de coulée à vide
  V 20
installation de dégagement
  métallique M 86
installation de dégazage
  D 44
installation de dégazage de
  l'acier S 350
installation de dégazage de
  métaux M 86
installation de dégivrage
  D 35
installation de dessiccation
  D 259
installation de dessiccation
  sous vide V 56
installation de distillation
  D 194, D 196
installation de distillation à
  court trajet O 48
installation de distillation à
  court trajet avec purification murale mécanique
  S 130
installation de distillation
  centrifuge moléculaire
  C 93
installation de distillation
  moléculaire à brosse
  fractionnante W 45
installation de distillation
  moléculaire à brosse
  tournante W 45
installation de distillation
  sous vide V 49
installation de frittage sous
  vide V 149
installation de fusion à
  rayons électroniques à
  vide V 58
installation de fusion à zones
  Z 16
installation de lyophilisation
  F 195
installation de lyophilisation
  à tambour D 249
installation de lyophilisation
  de vide V 69
installation de métallisation
  au déroulé S 385
installation de métallisation
  sous vide H 82
installation de métallisation
  sous vide V 26, V 105
installation démontable à
  vide poussé D 61
installation de nettoyage par
  bombardement ionique
  G 102
installation de passage
  linéaire L 63
installation de pompage
  avec une pompe à sorp-
  tion D 263
installation de pompage et
  installation d'évaporation
  sous vide V 166
installation de pompes à
  grand débit H 80
installation de préparation
  des minerais O 59
installation de préparation
  d'huile O 15
installation de pression
  P 184

installation de pression à
  vide V 123
installation d'épuration
  P 225
installation de purification à
  zones Z 18
installation de rectification à
  vide V 133
installation de refroidissement L 115
installation de résine mou-
  lable à vide V 137
installation de séchage
  D 259
installation de séchage à
  couche fine T 72
installation de séchage à
  vide V 56
installation de séchage en
  poche de coulée L 6
installation de séchage par
  courant F 83
installation de séchage par
  ruissellement C 46
installation de séchage par
  ruissellement du jet S 294
installation d'essai P 56
installation de sublimation
  S 396
installation de tirage de cristaux C 404
installation d'évaporation
  sous vide V 26
installation de vaporisation
  de rayon électronique
  E 64
installation de vide V 118
installation de vide poussé
  H 90
installation d'imprégnation
  I 27
installation d'imprégnation
  à la cire W 18
installation d'imprégnation
  au vernis V 230
installation d'ionisation
  I 146
installation d'irradiation
  I 171
installation en attente
  S 324
installation frigorifique
  L 115
installation frigorifique
  à air froid C 189
installation frigorifique
  à jet de vapeur S 345
installation frigorifique
  de compression C 244
installation frigorifique par
  immersion I 13
installation par frittage
  S 171
installation poste de métallisation sous vide H 82
installation pour la coulée de
  résines synthétiques R 129
installation pour le chargement dosé sous vide de
  poudre V 121
installation scellée de vide
  S 39
instrument de mesure à
  épaisseur de couche F 56
instrument de mesure
  d'épaisseur de couche F 56
instrument de mesure du
  taux d'évaporation E 182
instrument de mesure uni-
  versel M 194
instrument en forme de secteur S 57
intensité du champ magné-
  tique M 5
interconnexion L 70
interférométrie à plusieurs
  rayons M 196
intermittent B 64

intermoléculaire I 112
interrupteur à minuterie
  T 110, T 111
interrupteur à pression
  P 174
interrupteur à vide V 91,
  V 164
interrupteur bimétallique
  B 117
interrupteur bimétallique
  à pression B 119
interrupteur de diaphragme
  S 137
interrupteur de fin de course
  L 59
interrupteur de flux F 129
interrupteur de sûreté d'eau
  de refroidissement C 331
interrupteur flotteur F 109
interrupteur limiteur L 59
interrupteur manométrique
  à membrane D 96
intervalle de masses H 35
inverseur de masque à
  substrat S 406
ion du gaz G 28
ionique I 140
ionisation des électrons par
  choc E 95
ionisation par choc I 14,
  I 143
ionisation partielle D 110
ionisation photoélectrique
  P 47
ionisation spécifique
  D 111, S 262
ionisation superficielle
  S 439
ionisation thermique T 39
ioniser I 154
ionomètre I 138
isentropique I 172
isobare I 175
isolant I 83
isolation calorifique du vide
  V 169
isolation calorifuge H 36
isolation de mica M 103
isolation thermique H 36
isolation thermique de verre
  H 43
isolement à vide V 90
isolement pulvérulent
  à vide V 122
isoplèthe I 177
isotherme d'adsorption
  A 81
isotherme de soufflage
  S 453
isothermique I 178

J

jaillissement F 89
jauge G 65
jauge à huile O 18
jauge à ionisation I 137,
  I 152
jauge à vide à frottement
  D 23
jauge à vide thermoélec-
  trique T 57
jauge de Bayard-Alpert
  B 66
jauge du diaphragme D 95
jauge Pirani P 66
jauge thermique H 25
jet atomique A 204
jet creux H 105
jet de gaz G 43
jet de vapeur V 210
jet moléculaire ultrasonique
  S 425
jet plat d'électrons E 109
jet supersonique S 423,
  S 426
jeu du palier B 85

joint G 45
joint à anneaux glissants F 4
joint à bille B 36
joint à brides F 73
joint à chapeau C 21
joint à cisaillement S 106
joint à compression
  C 245, C 342
joint à compression en verre
  C 247
joint à couteau K 11·
joint à double couteau
  D 213
joint à double siège B 20
joint à écrasement C 245
joint à fente C 159
joint à labyrinthe L 1
joint à lèvre L 71
joint à membrane
  métallique M 88
joint annulaire A 156,
  R 161
joint à rotule B 36
joint avec un anneau à angle
  droit C 188
joint avec un anneau rectangulaire C 188
joint brasé S 225
joint céramique-métal
  C 103
joint conique C 263
joint d'angle C 345, F 47
joint d'arbre R 5, S 100
joint d'arbre à section angulaire C 21
joint d'arbre étanche au vide
  R 189
joint de bride F 73, F 75
joint de chapeau et de corps
  de la vanne B 166
joint de coulée C 56
joint de couvercle de soupape B 167
joint de diffusion D 138
joint de dilatation E 209
joint de fente D 94
joint de feuilles minces de
  cuivre C 338
joint de fil N 38
joint de fil en or C 344,
  G 103
joint de lamelles F 147
joint de l'obturateur d'un
  robinet D 174, V 191
joint de plomb L 34
joint de rodage en verre
  G 127
joint d'étanchéité G 45
joint d'huile O 33
joint du disque de vanne
  D 174, V 191
joint du siège d'une sou-
  pape V 203
joint du siège d'un robinet
  V 203
joint élastomérique E 27
joint en caoutchouc R 203
joint en cascade S 357
joint en cuir L 48
joint en I P 74
joint en tige S 354
joint en V S 164
joint en X D 217
joint étanche G 45
joint liquide L 88, S 43
joint métallique M 90
joint métallique à soufflet
  M 83
joint métallique d'indium
  I 41
joint mica-métal M 107
joint par feuilles minces
  F 147
joint par fusion S 47
joint par presse-étoupe
  S 392
joint plat F 95
joint plein B 128

joint profilé P 195
joint rabattu F 78
joint rodé C 129
joint rotatif à vide R 189
joint scellé G 45, S 47
joint sec D 264
joint soudé S 225
joint torique D 214, O 64
joint torique à section circulaire D 214
joint torique trapézoïdal
  V 232
joint verre-cuivre C 341
joint verre-Kovar K 17
joint verre-métal G 96
joint Wilson W 43
jonction B 164, J 17
jonction adaptée M 40
jonction à pénétration
  K 11
jonction de serrage P 59
jonction de verre G 95
jonction en caoutchouc
  R 201
jonction étanche G 45
jonction fixe F 68

K

kénotron V 176

L

lâcher B 134
lacune V 1
lacune de miscibilité M 125
Lafferty-vacuomètre à
  ionisation H 123
laisser échapper B 134
laitier de hauts fourneaux
  I 170
lame de mica M 106
lame fusible F 235
lamelle adhésive A 65
lamelle collante A 65
lamelle de soufflage B 148
lamelle plate F 94
lamelle respiratoire B 198
laminage à chaud H 135
laminage à froid C 209
laminage d'enveloppe
  S 109
laminé à froid C 208
laminer R 168
lampe à deux filaments
  B 114
lampe à incandescence
  I 34
lampe à vapeur de mercure
  à basse pression L 117
lampe de recherche H 8
lancé B 76
laque d'accrochage B 57
laque de couverture M 28
laque de protection P 204
laquer L 2
laques à étanchéifier S 46
laques d'étanchéité S 46
laques de savon S 79
largeur de clivage de diffusion J 7
largeur de fente R 3
largeur de joint de soudure
  W 33
largeur de maille d'un tamis
  métallique W 46
largeu du joint soudé
  W 33
laser-miroir L 16
lavage F 141
Laval-injecteur L 26
laveur d'air A 114
L-cathode L 29
lecture de pression P 166
lentille électronique E 96

lest d'air G 10
lest de gaz G 10
levée de soupape V 198
liaison B 164, L 70
liaison par sorption chimique
  C 124
liant C 80
liberté d'oscillations F 183
libre de parasites T 176
libre parcours moyen M 48
ligne de fuite G 366
ligne de glissement S 184
ligne de masse M 34
ligne fantôme G 77
ligne focal-anticathode
  S 373
limitant le courant C 418
limite d'allongement
  C 368, M 44, Y 2
limité de charge d'espace
  S 246
limite de décèlement D 77
limite de dilatation O 7
limite de grain G 106
limite de la sensibilité
  décelable D 78
limite de résistance E 24
limite de rupture
  thermique H 29
limite de tolérance des impuretés P 35
limite de vaporisation V 208
limite d'indication D 77
limite inférieure de mesure
  de la pression L 113
lingot B 113
lingot fusible I 57
lingotière I 59
lingotière refroidie par eau
  W 10
liquation E 119
liquéfacteur I 76
liquéfaction L 74
liquéfaction de gaz L 75
liquéfaction de l'air A 107
liquide à rendre étanche
  S 43
liquide avec une tension de
  vapeur peu considérable
  L 118
liquide obturateur C 267
liquidus dans le plan L 90
lit S 402
litharge-glycérine L 92
lit intercalé I 99
loi cosinusoïdale de
  Lambert C 354
loi d'action des masses
  L 27
loi d'Avogadro A 223
loi de Boyle-Mariotte
  B 186
loi de Dalton D 1
loi d'effusion E 18
loi de Gay-Lussac C 119
loi de Kirchhoff K 9
loi de Knudsen K 16
loi de Lambert C 354
loi de Paschen P 20
loi de Poiseuille P 94
loi de Stefan-Boltzmann
  S 352
loi de translation de Wien
  W 42
loi de Wiedemann-Franz
  W 41
loi du rayonnement de
  Planck P 77
longueur de cohérence
  C 122
longueur d'onde des rayons
  X X 6
longueur du trajet des électrons E 102
longueur limite d'onde
  C 374
lubrifiant L 121
lubrifiant à vide L 122

lubrifiant de contact C 149
lubrifiant sec D 253
lubrification L 123
lubrification de surfaces limites B 182
lubrification limite T 74
lubrification par matière
  fusée M 71
luminescence anodique
  A 166
luminescence cathodique
  C 66
lyophilisation C 391, T 114
lyophilisation chancelante
  T 183
lyophiliser F 193

M

machine à air froid A 111
machine à expansion D 19
machine à froid de compression C 248
machine à percer de faisceau
  électronique E 62
machine à piston de fusion
  B 217
machine à piston de souf-
  flage G 83
machine à tréfiler de monocristaux S 155
machine de suintement
  B 72
machine frigorifique
  C 329, R 97, R 98
machine frigorifique à gaz
  C 392
machine frigorifique de
  compression C 248
machine pressante de pied
  S 353
machine réfrigérante C 329
machine tournante à disque
  F 79
macle T 203
magasin à support d'émul-
  sion S 405
magnétron M 9
magnétron à fonctionne-
  ment continu C 305
magnétron à ondes pro-
  gressives T 167
magnétron pompe ionique à
  getter par pulvérisation
  M 10
magnétron-vacuomètre à
  ionisation avec cathode
  chauffée H 123
magnétron-vacuomètre
  inverse C 195
malaxeur K 10
malléabilisation T 19
malléabilité M 15
manchon de contraction
  S 135
manchon de glissement
  S 185
manchon d'entrée I 70
manchon de protection
  contre les aspersions
  S 297
manchon de quartz S 143
manchon de soudage
  W 26
manchon d'étanchéité S 41
manchon d'étuvage B 30
manchon douille collé
  C 81
manchon pour la protection
  contre les jaillissements
  S 297
mandrin M 17
manège de pompage A 213
manège tournant
  d'évaporation E 184
manomètre M 19
manomètre à aiguille D 89

manomètre à amortisse-
ment D 23
manomètre à amortissement
de Dushman D 271
manomètre absolu A 6
manomètre à cadran D 89
manomètre à ionisation
I 137, I 152
manomètre à ionisation à cathode chaude H 122,
T 51
manomètre à ionisation à cathode froide C 196,
S 282
manomètre à ionisation avec
électrode de frein S 430
manomètre à ionisation avec
extractor E 224
manomètre à ionisation de
Houston H 143
manomètre à ionisation par
radiation nucléaire R 16
manomètre à ionisation sans
aimant N 21
manomètre à membrane
D 95, D 98
manomètre à mercure en U
U 29
manomètre à niveau liquide
L 81
manomètre à vide V 74
manomètre à viscosité D 23
manomètre à viscosité de
Dushman D 271
manomètre bimétallique
B 116
manomètre clos à mercure
C 166
manomètre d'électro-
luminescence E 44
manomètre de Pirani P 66
manomètre déplaceur
D 187
manomètre de pression
partielle P 14
manomètre différentiel
D 115
manomètre différentiel à
soufflet avec lecture de
zéro B 97
manomètre d'ionisation à
cathode chaude de Bayard
et Alpert B 67
manomètre enregistreur
M 18
manomètre en U tronqué
M 80
manomètre magnétron
C 199
manomètre magnétron à ca-
thode chaude et suppres-
sion du courant photo-
électrique H 124
manomètre rotatif à com-
pression T 190
manomètre thermique
H 25
manomètre total T 137
manostat bimétallique
B 119
marchandise séchée par
congélation F 192
marche continue C 299
marche continue de masse
M 36
marmite à vide C 190
marque de calibrage R 74
masque à microcircuit
M 112
masque de cuivre corrodé
E 152
masque de protection S 98
masque d'évaporation D 66
masque du vaporisateur
E 179
masse de damage T 4
masse de remplissage F 44
masse isolante I 85, P 118

masse plastique P 118
mastic à culot de tube
T 179
mastic à fusion M 64
mastic à prendre S 92
mastic à verre G 84
mastic de constitution
C 278
mastic de glycérine et
litharge L 33
mastic de réaction R 38
mastic d'étanchéité S 40
mastic marbre M 23
mastic pour le vide V 21
mastic superficiel S 442
matériel d'étanchéité S 37
matériel d'étanchéité pour
fuites L 46
matériel gomme élastique
R 205
matière à faire le plein F 45
matière collante A 63
matière de mousse A 88
matière de remplissage
F 44
matière d'évaporation
E 162
matière fusée C 232
matière isolante I 83
matière massive B 223
matière plastique P 85
matière plastique empilée
L 9
matière thermoplastique
T 64
mécanisme à mettre en
circuit R 30
mécanisme moteur à injec-
tion J 15
mélange d'air et de gaz
A 105
mélange explosif E 215
mélange frigorifique F 202
mélangeur M 127
mélangeur avec un agitateur
M 126
mélangeur à vide V 109
membrane D 92
membrane cylindrique
C 432
membrane ondulée C 350
mercure M 75
mesure à court temps du gaz
résiduel S 131
mesure de précision H 68
mesure de pression différentielle D 119
mesure du vide V 99
mesure finale E 132
mesure finie F 116
mesureur de temps T 110
mesureur du courant gazeux
G 35
métal antifriction B 1
métal brut C 390
métal de bain B 65
métal de palier lisse A 172
métal de transition T 147
métal de Wood W 50
métal étiré D 182
métallisation M 85
métallisation adhésive A 64
métallisation à l'arc sous vide
V 10
métallisation au déroulé
B 42
métallisation au pistolet
M 95
métallisation conique C 264
métallisation de céramique
S 169
métallisation d'une matière
plastique P 86
métallisation extérieure
O 66
métallisation par pulvérisation M 95
métallisation par trou T 94

métallisation sous vide
E 170, V 104
métallisé sous vide V 215
métallurgie des poudres
P 122
métallurgie du vide V 106
métal mixte C 234
métal transitoire T 147
métastable M 99
métaux difficilement
fusibles R 93
métaux fondus sous vide
V 101
méthode à éprouver à l'air
comprimé S 203
méthode à faire l'épreuve de
pression S 203
méthode d'accroissement de
pression I 176
méthode à l'hydrure de
titane T 116
méthode de bulles de savon
S 204
méthode de burette V 250
méthode de compensation
B 34
méthode de concentration
B 215
méthode de conductance
C 10, T 209
méthode de conservation
P 141
méthode de détection de
fuites par différence de
pression D 112
méthode de diaphragme
O 63
méthode de double cloche
D 208
méthode de la double enceinte
ceinte D 207
méthode de la pression
constante C 276
méthode d'enrichissement
B 215
méthode d'enveloppe H 114
méthode de perméation
G 52
méthode de pipette P 65
méthode de rayon de coulée
à vide V 162
méthode de surpression
O 78
méthode de temps de vol
T 109
méthode du filament
chauffé F 85
méthode du gaz de balayage
W 3
méthode du volume cons-
tant C 277
méthode molybdène-man-
ganèse M 165
méthode par décroissance de
pression P 159
méthode reniflante S 11
mettre à l'épreuve de pression P 182
mettre à zéro A 72
mica M 101
mica des Indes M 2
mica en feuilles S 113
mica pressé M 104
microbalance M 109
microbalance à vide V 108
microbalance de torsion
T 133
microbalance en quartz-
vide Q 10
microcircuit M 111
microdosage M 114
micro-électronique des corps
solides T 92
micromanomètre différentiel
D 116
micromètre T 66

microscope à émission de
champ F 36
microscope à émission
électronique F 34
microscope à émission
ionique de champ F 35
microscope électronique à
miroir E 98
microscope électronique
d'exploration S 23
migration M 116
migration de la surface
S 441
migration en retour B 18
migration superficielle
S 441
miroir à plusieurs couches
M 193
miroir de surface F 217
miroir froid C 205
miroir réfléchissant à la
surface F 217
miroir superficiel F 217
miroitant S 266
mise à la terre E 1
mise au point de direction
D 153
mise au point focal F 146
mise à zéro Z 7
mitraille S 26
mobilité des électrons E 99
mobilité du point zéro Z 5
mode d'aspiration S 414
modèle de sphère rigide
H 19
modulateur M 134
module d'élasticité M 135
module de rigidité M 136
Mohs-dureté M 137
mole G 111
moléculaire brut M 158
moment de rappel R 150
moment de rotation T 127
moniteur de quote-part
d'ionisation I 139
monocellule M 170
monochromateur à vide
V 110
monocouche M 172
monomoléculaire M 176
montage en cascade C 45
montage en pont B 201
montage flip-flop F 106
montée de grille G 123
monter en série C 272
morceau collecteur H 21
morceau final T 2
mouillabilité W 36
mouillure dans le vide V 181
moulage C 51, S 101, S 316
moulage du vide F 182
moulage par injection
D 101
moulage par injection sous
vide V 44
moulage par pression P 149
moulage permanent P 31
moule M 181
mousse F 144
mousser F 143
mouvement brownien
B 209
moyen réfrigérant C 317
multimètre M 194
multivibrateur F 106

N

nacelle de molybdène
M 163
nacelle d'évaporation E 191
neige carbonique C 33
nettoyage C 153
nettoyage par bombarde-
ment ionique D 158
nipple N 8
niveau de baromètre B 51

niveau de bruit N 12
niveau de Fermi F 28
niveau de saturation S 18
niveau d'huile O 27
niveau liquide L 79
niveau-stabilisateur L 51
nombre d'Avogadro
  A 224, L 105
nombre de collision C 223
nombre de collisions avec
  une paroi I 21
nombre d'écoulement
  D 159
nombre de Knudsen K 14
nombre d'élasticité E 23
nombre de la transmission
  thermique F 51
nombre de Mach M 1
nombre de mollesse S 213
nombre de Nusselt N 47
nombre des plots N 46
nombre de Prandtl P 129
nombre de Reynolds
  R 155
nombre de sortie D 159
nombre de viscosité I 121
nombre d'usure W 21
nombre ordinal A 205
nombre ordinal effectif
  E 11
non corrosif C 348
non destructif N 20
non homogène I 64
non interchangeable F 150
non volatil N 25
noyau de glace I 3
noyer E 121
NTC-résistance N 4
nuage d'électrons C 168,
  E 85
nucléon N 39
nucléus N 40
nucléus de glace I 3

O

obstacle à décharge D 168
obturateur d'un robinet
  V 190
obturateur élastique d'un
  robinet S 2
obturation instantanée Q 17
obturation rapide Q 17
obturer par bride B 129
occlusion O 5
OFHC-cuivre O 83
olive H 120
ombrage O 3
ombragement conique
  C 265
ombrage métallique M 94
omégatron O 39
omégatron-spectromètre
  de masse incorporé N 42
omégatron-vacuomètre
  O 40
onde de choc S 126
onde de spin S 278
opacité O 42
opaque O 43
opération de charge B 63
opération erronée F 18
opéré à la main M 22
optique électronique E 100
optiquement étanche O 43
orbite O 57
orbite d'électrons E 101
orbite électronique E 107
orbitron-pompe à vide
  O 56
orbitron-pompe ionique
  O 56
orifice O 62
orifice d'échappement
  D 166
orifice de refoulement
  D 166

orifice d'expulsion D 166
orifice du brûleur B 228
oscillateur à cristal R 144
oscillateur commandé par
  quartz R 144
oscillateur de quartz R 144
oscillation V 238
outre H 118
ouverture d'aspiration I 88
ouverture du jet B 73
oxydation anodique A 167
oxydation thermique T 40
oxyde d'aluminium A 133
oxyde d'aluminium activé
  A 41
oxyde de baryum B 49
oxyder électrolytiquement
  A 168

P

palette S 179
palette du getter G 75
palier à billes B 39
palier à roulement A 171
palladium noir P 11
panier rotatif R 171
papier héliographique
  « Ozalid » O 85
papillon d'obturation
  B 233
paroi de Bloch B 140
partage de pression P 157
particule C 346
particule alpha A 129
particule bêta B 107
particule de vapeur V 222
particule nucléaire N 39
partie constituante du vide
  préliminaire F 163
partie d'usure W 22
passage L 30
passage à courant fort H 53
passage d'arbre R 185
passage de chaleur H 51
passage de courant C 417
passage pour un système de
  vacuomètre d'ionisation
  d'immersion N 41
passage rapide R 24
passage tournant R 178
pâte de résine synthétique
  S 456
peau d'eau W 8
peau de verre G 82
peau métallique M 82
pechblende P 23
pellicule d'eau W 8
pendule à deux fils D 215
pénétration P 34
pénétration d'air A 106
pénétration d'hélium H 60
pénétromètre P 28
penning-pompe C 197
perçage par rayons d'électrons E 61
percement de tension S 449
percussion moléculaire
  M 147
période de dégazage B 27
période de travail A 40
perle de verre G 80
permanent vaporiseur
  métallique à vide C 304
perméabilité P 33
perméabilité à rayons R 13
perméabilité aux gaz G 51
perméabilité d'hélium de
  verre H 61
persistance de forme F 172
persorption P 36
perte de pression P 162
perte de pression statique
  L 107
perte d'évaporation E 178
perte par évaporation E 178
perturbation / sans T 176

pesanteur G 116
petit détecteur de fuites aux
  halogènes H 9
petite bride Q 15
pétrisseuse K 10
phase de travail A 40
photocorrosion P 46
photo-luminescence P 48
photo-masque de vernis
  L 4
photomètre avec rayon
  modulé M 133
photon P 49
physique de couches minces
  T 77
physique de métaux M 91
physique des corps solides
  S 228
physique du vide V 115
physisorption P 51
pièce coulée C 52
pièce d'adaptation R 64
pièce de raccord de
  réduction R 65
pièce de raccordement
  C 274
pièce de rechange S 253
pièce de réduction R 64
pièce détachée du vide
  préliminaire F 163
pièce d'usure W 22
pièce en croix C 377
pièce en croix avec bride de
  réduction R 67
pièce en T T 141
pièce intermédiaire A 53,
  S 249
pièce polaire P 101
piège T 164, V 174
piège à absorption A 19
piège à adsorption A 86
piège à basse température à
  haut rendement H 71
piège à décomposition
  thermique T 29
piège à effet Peltier P 25
piège à feuille de cuivre
  C 339
piège à froid C 213
piège à huile O 37
piège à impuretés D 156
piège à sorption A 86
piège à sorption avec charbon actif A 48
piège à sorption de charbon
  C 110
piège à vapeur V 221
piège à vapeur avec toile annulaire M 198
piège à vapeur opaque O 52
piège à vapeur optiquement
  étanche O 52
piège avec perte petite de réfrigérant L 111
piège à vide primaire B 11
piège à zéolite Z 3
piège cryogénique C 213
piège de refroidissement
  C 213
piège ionique I 126
piège pour le vide primaire
  B 11
piège réfrigérant C 213
piège refroidi C 213
piège refroidi sphérique
  S 275
piège sphérique S 275
piège thermoélectrique
  P 25
piézoélectrique P 53
pignon D 243
pile atomique à vide V 131
pile thermoélectrique T 63
pilon P 226
pince à socle pressant P 144
pince à tige P 144
pince au filament H 140
pince pour tuyaux H 119

pince pressante C 164
pincette T 202
pinch-effet P 60
Pirani-détecteur différentiel
  de fuites D 117
pisé réfractaire damé de céramique oxydée P 82
pistolet pulvérisateur S 296
piston de déplacement
  D 188
piston déplaceur D 188
piston de soupape V 199
piston de vanne V 200
piston moteur W 54
piston rotatif I 18
placage électrolytique
  E 113
place de la surface S 446
place non occupée du réseau
  moléculaire L 23
plaine de base B 58
plaine de diffraction D 124
plaine de glissement S 188
plaine de grille L 22
plaine de réglage S 114
plaine de réseau C 408,
  L 22
plaine focale F 145
planche d'égouttage D 230
plan de clivage C 161
plan de clivage de diffusion
  A 74
plan de réglage S 114
plan lumineux de montage
  M 119
planning S 25
plaque chauffante J 2
plaque de base B 59
plaque de chauffage J 2
plaque de montage B 59
plaque de rupture B 232
plaque du récipient B 92
plaque en mica M 105
plaque-support B 59
plaquette cristalline C 411
plastifiant P 84
plateau de remplissage
  C 116
plateau de soupape V 190
plate tige pressante F 97
platine élastique de vanne
  S 2
platine en verre fritté
  P 143
pleine soudure bilatérale
  F 93
plein joint bilatéral F 93
plier F 149
plusieurs étages / à C 235
plus petit changement détectable (décelable) de pression M 122
plus petit débit détectable
  (décelable) d'une fuite
  S 197
plus petite fuite détectable
  (décelable) M 121
poche à quenouille B 176
poche de coulée C 53, T 8
poche de coulée avec bec de
  coulée L 73
poche de coulée refroidie
  par eau W 11
poche intermédiaire M 60
poêle de four F 228
poids du lingot I 62
point à reculer S 94
point critique C 371
point d'amollissement
  S 210
point d'annulation molé-
  culaire M 160
point d'attache par soudure
  S 224
point d'ébullition B 159
point de compression S 317
point de congélation F 203
point de Curie C 415

point de décongélation P 120
point de floculation F 113
point de fusion F 239, L 91
point de goutte D 247
point de pliage de la température I 51
point de référence R 74
point de rosée D 87
point de scellement G 45
point de solidification P 120, S 231
point de tension thermique S 243
point de transformation T 144
point de trouble C 169
point d'inflammation F 91
pointe de chute B 197
point eutectique E 157
point fixe F 69
point non occupé du réseau moléculaire L 23
point thermique d'arrêt T 25
point triple T 173
polarisation de vide V 120
polariseur de feuille polaroïdale P 99
pôle négatif S 165a
polir électrolytiquement E 115
polissage électrique E 116
polissage électrolytique E 46
polissure électrolytique E 46
polymérisation P 103
polymérisation à bombardement électronique E 74
polymérisation d'addition P 102
polymorphisation P 104
polytropique P 105
pompage V 127
pompage cryogénique C 394
pompage d'emboutissage D 25
pompage primaire C 171
pompe P 209
pompe à absorption A 22
pompe à adsorption A 83
pompe à air humide R 54
pompe à anneau d'eau W 17
pompe à anneau de liquide F 136
pompe à chaleur H 42
pompe à condensation C 257
pompe à cryo-getter C 396
pompe à diffusion D 135
pompe à diffusion à haut débit H 79
pompe à diffusion à huile O 17
pompe à diffusion à (de) mercure M 78
pompe à diffusion de vapeur d'huile O 17
pompe à diffusion d'huile à ultra-vide U 6
pompe à diffusion du type booster B 171
pompe à éjecteur E 20
pompe à éjecteur de mercure M 79
pompe à éjecteur d'huile O 22
pompe à engrenages G 69
pompe à évaporation E 192
pompe à évaporation de titane T 115
pompe à flux axial A 227
pompe à flux de vapeur V 220
pompe à flux de vapeur d'huile O 38

pompe à flux radial R 2
pompe à getter G 76
pompe à getter par évaporation de titane T 115
pompe à injection d'air G 11
pompe à injection d'air de Gaede G 1
pompe à jet d'eau A 197
pompe à jet de liquide F 135
pompe à jet de vapeur E 20
pompe alimentaire F 23
pompe à membrane D 97
pompe annulaire à eau W 17
pompe à palettes B 184, S 186, V 207
pompe à palettes multiples M 200
pompe à piston P 67
pompe à piston axiale A 229
pompe à piston élévatoire R 51
pompe à piston rotatif R 182
pompe à piston tournant R 182
pompe à plusieurs étages M 207
pompe à prévide B 14
pompe à rectification F 180
pompe à reflux R 89
pompe à sorption S 240
pompe à sorption par getter G 76
pompe à sorption par zéolite Z 2
pompe aspirante D 235
pompe aspirante et refoulante D 205
pompe à tiroir S 178
pompe à tiroirs rotatifs R 176
pompe à vide V 126
pompe à vide à vapeur d'eau J 9
pompe à vide poussé H 92
pompe à vide préliminaire B 14, R 198
pompe ballast à gaz G 11
pompe centrifuge C 94
pompe cryogénique C 393
pompe cryostatique C 393
pompe cryostatique à sorption C 398
pompe d'absorption de gaz C 155
pompe d'alimentation F 23, F 24
pompe d'eau de refroidissement C 333
pompe de booster B 170
pompe de chaleur H 42
pompe de circulation C 145
pompe d'écoulement F 130
pompe de distillat D 76
pompe de dosage D 202
pompe d'effet getter C 155
pompe de getterage C 155
pompe de gettérisation C 155
pompe de maintien H 101
pompe de maintien avec dispositif de lest d'air G 12
pompe d'entretien H 101, P 194
pompe d'entretien avec dispositif de lest d'air G 12
pompe déplaceur D 189
pompe de prévidage R 198
pompe de retenue P 194
pompe de sublimation S 397
pompe de sublimation de titane T 119
pompe de Toepler T 121

pompe différentielle à getter D 108
pompe élévatoire D 235
pompe fractionnante F 180
pompe Gaede G 2
pompe getter G 76
pompe getter à évaporation de titane T 115
pompe getter-ionique G 74
pompe giratoire C 94
pompe glissante S 178
pompe idéale I 8
pompe intermédiaire B 170
pompe ionique I 150
pompe ionique à cathode froide C 197
pompe ionique à getter G 74
pompe ionique à getter avec vaporisation de fil W 47
pompe ionique à getter par pulvérisation cathodique D 142
pompe ionique à getter par pulvérisation cathodique du type de manomètre magnétron renversé I 125
pompe ionique à sorption I 165
pompe ionique de pulvérisation C 197
pompe ionique de titane T 117
pompe ionique par évaporation E 177
pompe ionique par vaporisation E 177
pompe mammouth G 16
pompe mécanique M 58
pompe mécanique à étages C 237
pompe mécanique étanchéifiée par liquide L 89
pompe mécanique non volumétrique N 22
pompe moléculaire M 143, M 152
pompe monobloc M 169
pompe multiétagée M 207
pompe multi M 200
pompe pour maintenir le vide H 101
pompe pour maintenir le vide avec dispositif de lest d'air P 13
pompe pour vide primaire B 14
pompe primaire B 14
pomper D 237, P 208
pompe Roots R 170
pompe rotative R 183
pompe rotative à huile R 179
pompe rotative à jet d'eau R 193
pompe rotative à joint d'huile R 180
pompe rotative à mercure R 177
pompe rotative à palettes S 186
pompe rotative à vide R 188
pompe semi-fractionnante S 72
pompe silencieuse W 40
pompe Sprengel S 301
pompe thermomoléculaire T 48
pompe tournante à joint d'huile R 180
pompe turbo-moléculaire T 196

pont de Wheatstone W 38
porosité P 106
portatif P 108
porte-creuset C 388
porte-électrode E 41
porte-objets O 1
porte-outil W 53
porteur de charge C 111
poste de coulée à vide V 20
poste de distillation à cycle intermittent C 423
poste de pompage C 229
poste roulant de pompage T 179
potentiel de contact C 291
potentiel de la paroi de tube B 216
potentiel de terre E 2
potentiel d'ionisation I 148
potentiel oscillant F 107
potentiel zéro Z 8
poudre à dépolir E 155
poussée L 56, T 96
poussée d'Archimède L 56
poussée verticale L 56
poussoir P 226
pouvoir absorbant A 18
pouvoir adsorbant A 82
pouvoir d'absorption A 18
pouvoir de déformation W 51
pouvoir d'émission thermique R 11
pouvoir d'émission totale T 134
pouvoir de résolution R 142
pouvoir dissolvant R 142
pouvoir émissif d'électrons T 53
pouvoir frigorifique R 95
pouvoir isolant I 86
pouvoir lubrifiant L 124
pouvoir réfléchissant R 82
pouvoir réflecteur R 82
pouvoir séparateur R 142
préchauffage C 172, P 137
préchauffer P 134
préchauffeur P 136
précision A 35
précision de mesure M 52
précompression S 415
précomprimer S 415
précondensation S 416
précondenser S 415
première couche P 186
préparation mécanique P 139
prérefroidissement P 131
prérefroidisseur P 130
préséchage P 138
présécher P 132
pressage d'écoulement froid C 200
presse à extrusion E 227
presse de vis sans fin S 35
presse-étoupe P 4
presse par vis sans fin S 162
pression P 146, S 316
pression absolue A 7
pression ambiante A 139
pression à rayonnement R 14
pression atmosphérique A 202
pression atmosphérique normale A 200
pression continue Y 3
pression critique C 372
pression d'admission F 63
pression d'aspiration F 63
pression d'échappement D 167
pression de charge F 48
pression de comparaison R 75
pression de compensation B 35

pression de dissociation
D 192
pression de dissolution
D 192
pression de distribution
S 455
pression de gaz résiduel
R 122
pression de joint S 48
pression de mise en marche
S 326
pression d'enclenchement
S 326
pression de référence R 75
pression de refoulement
D 167
pression de remplissage F 48
pression des électrons E 103
pression de service O 51
pression de sortie D 167
pression de travail O 51
pression de vapeur V 223
pression de vapeur résiduelle
R 125
pression de vapeur saturante
S 17
pression de vapeur saturée
S 17
pression différentielle D 118
pression du jet B 82
pression du rayon B 82
pression du vide préliminaire
B 13, C 382, D 167,
D 277
pression dynamique D 279
pression effective O 77
pression finale B 130,
D 167
pression initiale I 65
pression limite d'amorçage
C 370
pression nominale N 16
pression normale du débit
du vide préliminaire
D 276
pression normale et température normale S 321
pression partielle P 15
pression primaire de rupture
B 194
pression sous 10⁻³ Torr
B 127
pression statique S 328,
S 329
pression tolérée du vide
préliminaire F 158
pression totale T 136
pressostat à membrane D 96
prévidage C 171
prévider P 133
principe de boîte de
construction B 214
principe de cryostat C 399
principe de la cloche double
D 207
principe de la focalisation
intense C 389
principe des unités de montage pour l'agencement
rapide B 214
principe schématisé de
câblage B 60
prise de courant S 205
prise de résistance de
contrôle M 117
prise d'essai S 10
probabilité d'adhérence
C 31
probabilité d'adhésion C 31
probabilité d'association
P 188
probabilité de captage d'ions
I 132
probabilité de changement
T 158
probabilité de collage C 31
probabilité de fixation
C 31

probabilité de passage
T 158
probabilité d'ionisation
I 149
probabilité du choc I 15
probabilité du pouvoir
adhérent C 31
procédé « Aircomatic » I 50
procédé à l'hydrure de
titane T 116
procédé à masque M 27
procédé à pas de pèlerin
R 52
procédé avec masse catalyti-
que C 292
procédé à vide V 107
procédé capillaire C 9
procédé d'aspiration sous
vide S 374
procédé de concentration
A 34
procédé de coulée par le
fond B 175
procédé de creuset
horizontal Z 13
procédé de Czochralski
C 433
procédé de drapage sous
vide D 233
procédé de fusion à zones
par tirage des cristaux
Z 14
procédé de fusion, d'alliage
et de coulage F 1
procédé de masque de
moulage S 13
procédé de pompage E 160,
P 217
procédé de réflexion de Laue
L 25
procédé de soudure
argonarc A 188
procédé de Telefunken
S 168
procédé de vernis de
déchirure S 379
procédé d'extraction à
chaude H 130
procédé du métal actif A 47
procédé fondamental U 19
procédé Heliarc H 54
procédé molybdène-manga-
nèse M 165
procédé par le sac en
caoutchouc D 31
production de gaz G 38
production de jet B 77
production des rayons
d'électrons E 58
production de vapeur sous
dépression V 158
production de vapeur sous
vide V 158
production de vide C 362
produit de clivage C 162
produit de fission C 162
produit demi-fini S 78
produit lyophilisé F 192
produit sec avec noyau de
glace F 258
produit seché par
congélation F 192
produit synthétique P 85
profil du rayon très étroit
C 381
profondeur de pénétration
D 70
programme S 25
proportion de dilatation
E 210
proportion de la limite
d'allongement E 25
proportion de résistance à la
fatigue F 17
proportion de résistance
continue F 17
proportion de résistances
R 136

proportion d'expansion
E 210
proportion d'isotopes I 180
propulsion à réaction J 12
protection O 3
protection anti-rayonnante
R 37
protection conique C 265
protection contre le
rayonnement R 37
protection contre les
surcharges O 76
protection contre l'implosion I 24
protection de migration
M 117
protection électrique P 199
protection électrostatique
E 117
protection thermique de
verre H 43
puissance à assurer
l'étanchéité S 44
puissance à rendre étanche
S 44
puissance calorifique R 32
puissance continue C 300
puissance d'admission
E 199
puissance d'admission
nominale N 17
puissance d'aspiration
nominale N 17
puissance de chauffage R 32
puissance de chauffage
absorbé H 31
puissance de congélation
F 200
puissance de raccordement
P 127
puissance de rayon B 81
puissance d'une pompe
D 185, E 199
puissance effective d'une
pompe N 6
puissance frigorifique R 95
puissance maxima d'une
pompe C 109
puits de potentiel P 115
pulvérisateur S 298, S 309
pulvérisation A 207, S 307
pulvérisation à haute
fréquence R 20
pulvérisation cathodique
C 71
pulvérisation de diodes
D 141
pulvérisation de triode
T 172
pulvérisation inverse B 111
pulvérisation monocristaux
S 156
pulvérisation par tétrodes
T 23
pulvériser A 208, S 306
pupitre de commande
C 307
pupitre mobile M 130
pupitre roulant M 130
pur en état atomique A 203
pureté de couche F 55
purgeur P 224
purificateur d'huile O 31
purification C 153
purification automatique / à
S 68
purification des zones Z 17
purification du vide V 129
purification supersonique
(ultrasonore) U 10
pyrolyse P 227
pyromètre à filament
D 157
pyromètre à radiation totale
T 138
pyromètre à rayonnement
total A 187
pyromètre optique O 54

Q

quadrupôle-vacuomètre
d'ionisation Q 3
qualité de surface F 65
quantité de gaz Q 4
quantité fondamentale
F 224
quantité mésurée de reflux
M 50
quantummécanique Q 5
quartz de résonance C 410
quartz oscillatoire C 410
quenouille S 368
queue-soudure T 1
queusot E 201
quote-part de dégazage
D 45
quote-part de flux inverse
B 22
quote-part de reflux [par
migration] B 22
quote-part effective
d'évaporation E 14
quote-part normal
d'écoulement S 322
quote-part normal de fuite
S 22

R

raccord C 273, C 274
raccord à bride F 73
raccord à écrasement C 245
raccord angulaire B 102
raccord à petites brides
S 198
raccord à rodage G 129
raccord au vide primaire
F 155
raccord à vis S 31
raccord à vis des tuyaux
T 85
raccord céramique-métal
C 100
raccord collé C 81
raccord coudé B 102
raccord de l'eau de refroidissement C 330
raccord de réduction T 86
raccord de verre G 95
raccord de vide préliminaire
B 10
raccordement J 17
raccordement à vide poussé
H 84
raccordement en caoutchouc
R 201
raccordement fixe F 68
raccordement par collier en
U U 28
raccord en caoutchouc
R 201
raccord fileté N 8
raccord intermédiaire A 53
raccord plié L 14
raccord pour le vide E 161
raccord pour tuyaux H 120
raccord tubulaire T 182
radiateur tubulaire C 184
radiation R 10
raffinage R 81
raffinage de surface S 443
raffinage en zones progressi-
ves sous vide V 183
raffinage par immersion
H 128
rainure pour joint étanche
G 46
rainure rectangulaire R 59
ralentissement D 14
ramification A 145
rapidité de vaporisation
E 180
rapport de compression
C 243
rapport signal-bruit S 139

rapport signal sur bruit du signal de sortie O 69
raréfier R 26
râteau sec D 257
rayon atomique A 204
rayon cathodique E 51
rayon corpusculaire C 347
rayon de courbure B 103
rayon d'inversion I 123
rayon électronique E 52
rayon électronique de haute pervéance H 76
rayon hydraulique H 145
rayon moléculaire M 145
rayonnement R 10
rayonnement thermique T 41
rayonnement X limite X 4
rayonner R 9
rayons γ G 4
rayons canaux C 18
rayons cathodiques aberrants V 184
rayons de nettoyage B 132
rayons de Röntgen X 5
rayons gamma G 4
rayons X X 5
Re R 155
réactif R 40
réaction de creuset C 389
réaction réciproque I 91
réaction sous vide V 130
réaction thermonucléaire N 37
réactivation R 39
réagir R 146
réalisation de surfaces miroitantes M 123
rebondissement élastique R 127
rebord F 72
rebord à guidage de gaz G 34
rebouilleur R 46
recalibrage R 48
recarbonisation R 49
recarburation R 49
recharge B 3
réchaud sur tiroir H 32
réchauffeur de substratum S 403
recherche de fuites L 38
recherche de fuites avec du pétrole K 1
recherche d'espace S 250
recherche fondamentale B 61
récipient C 293
récipient à vide V 34
récipient de dégagement D 8
récipient du vide préliminaire F 164
récipient étuvable H 22
recombinaison R 56
recombinaison électronique E 105
recongélation R 94
recouvercle S 63
recouvert d'huile O 26
recouverture S 63
recouvrement métallique de protection P 205
recristallisation R 61
rectification C 153
rectification des zones Z 17
rectification de vide V 132
rectifier R 60
recuire A 150
recuire par électricité A 151
recuit F 223
recuit brillant B 202
recuit brillant sous vide V 7
récupération R 55
redresseur à vide poussé H 93
réducteur R 64
réducteur avec raccord unilatéral S 377

reducteur de sorption N 26
réémission R 70
réétalonnage R 48
réévaporateur R 45
réévaporiser R 72
réflecteur thermique H 45
refluer F 119
reflux R 87
reflux des gaz G 9
reflux d'huile B 4, O 8
reflux par migration B 18
réfondage à socle pressant F 96
réfondage à tige F 96
refondage de disque D 175
refondre R 105
refoulement à froid C 204
refoulement thermique H 48
refraîchir C 127
réfrigérant C 317
réfrigéranten cascade C 44
réfrigérant préliminaire P 130
réfrigérant tubulaire C 184
réfrigérateur C 329
réfrigérateur à absorption A 23
réfrigérateur à basse température D 26
réfrigérateur à gaz d'hélium H 58
réfrigérateur de compression C 248
réfrigération R 99
réfrigération thermo-électrique T 62
réfrigéré par ébullition V 211
réfrigéré par liquide L 78
réfrigérer C 127
refroidi par air A 100
refroidi par l'hydrogène C 320
refroidir C 127
refroidissement C 129
refroidissement à basse température D 24
refroidis ement à l'ébullition H 127
refroidissement de grille C 327
refroidissement mécanique M 59
refroidissement par circulation C 165
refroidissement par circulation d'eau C 322
refroidissement par évaporation E 188
refroidissement par expansion E 211
refroidissement par vaporisation E 174
refroidissement rapide Q 13
refroidissement sous vide V 35
regard I 78
regard de contrôle I 78
regard de niveau d'huile O 35
région de seuil T 91
réglage S 90
réglage à distance R 107
réglage d'avance F 21
réglage de pression P 171
réglage de sensibilité S 81
réglage de température T 15
réglage d'étuvage B 28
réglage du bâti de pompage P 220
réglage du point zéro R 115
réglage du zéro Z 7
réglage par vaporisation E 172
règle de phase P 39

régulateur à étrier mobile H 115
régulateur à niveau L 53
régulateur de dégazage D 38
régulateur de pression P 152
régulateur de vaporisation E 173
régulateur thermique T 16
régulation de pression P 171
relais à excès de courant C 419
relais à retardement T 107
relais à vide V 135
relais de surintensité C 419
relais temporisé T 107
relation de fréquence des isotopes I 179
relation de la profondeur de soudure et de la largeur R 36
relation réelle de compression E 12
reluminescence A 91
remouillage R 100
remplissage de l'huile O 11
rendement E 16
rendement calorifique H 31
rendement d'aspiration E 199
rendement de Carnot C 39
rendement d'élasticité R 127
rendement d'émission E 127
rendement de Rankine R 23
rendement de transport T 162
rendement du rayon J 8
rendement électronique C 65
rendement ionique I 135
rendement quantique Q 6
rendement volumétrique V 251
rendre étanche S 36
renforcé par fibre de verre G 88
reniflar par oxygène O 84
renifleur P 164
renifleur par oxygène O 84
renouvellement R 111
renouvellement d'huile O 10
répartir D 183
répartiteur D 199
répartition de vitesse V 233
repousser P 182
reproduction enveloppante E 145
réseau de résistance R 141
réserve de vide primaire F 164
réserve gazeuse G 60
réservoir S 371
réservoir à vide V 136
réservoir de transport M 132
réservoir distillatoire D 197
réservoir intermédiaire B 38, F 164
réservoir-piège à refroidissement R 114
réservoir stationnaire S 334
résidu R 126
résilience I 16, N 29
résine à couler C 54
résine à mouler C 54
résine époxydique E 147
résines alkylphénoliques A 121
résines de cétone K 3
résine synthétique C 54
résistance à changement de température T 44

résistance à choc de température H 47
résistance à la fatigue F 16, R 31
résistance à la flexion F 103
résistance à la pression C 249
résistance à la raideur F 102
résistance à la rayure A 4
résistance à la rupture B 195, T 11
résistance à la rupture chaude H 138
résistance à la torsion T 131
résistance à la traction B 195, T 20
résistance à la vapeur d'eau M 42
résistance à l'eau W 16
résistance à l'écoulement F 120
résistance à l'entaille N 29
résistance à l'incandescence I 10
résistance à l'usure R 138
résistance au choc I 16
résistance au cisaillement S 107
résistance au claquage B 190
résistance au flambage B 212
résistance au fluage C 367
résistance au flux d'un orifice A 179
résistance au percement disruptif B 190
résistance au rouge H 137
résistance continue à la flexion B 104
résistance de conturnement F 90
résistance de jaillissement F 90
résistance de l'air D 226
résistance de seuil I 117
résistance de trempe H 47
résistance électrique au percement E 36
résistance limite L 58
résistance thermique d'effort T 46
résistant à la corrosion C 348
résistant aux acides A 37
résistant aux fluorures F 139
résistivité électrique E 33
résonance des ions I 161
résonance du cyclotron C 429
respiration D 7
respirer A 73, S 409
ressort en disque C 413
ressuage tacheté F 181
retard D 14
retard à l'ébullition B 158
retardation D 14
retard de temps T 106
retard d'indication R 147
retassure C 76
réticulation C 380
retour d'huile B 4
rétrodiffusion B 2
revenu T 19
réverbération R 113
revêtement L 69
revêtement à vide V 145
revêtement de surfaces optiques L 49
revêtement d'ornement D 20
revêtement gazeux G 53
revêtement ornemental D 20
revêtement pyrolytique C 174
revêtement réfléchissant M 124
revêtement sous vide V 42

révolver de source d'évaporation V 227
rhéologie R 156
richesse de la pulvérisation S 314
rigidité diélectrique B 190
rigole de coulée P 119
rincer R 163
robinet G 64
robinet à aiguilles N 2
robinet à boisseau P 90
robinet à clapet articulé F 77
robinet à colonne de mercure M 77
robinet à coulisse oscillante P 26
robinet à deux voies T 211
robinet à fermeture rapide F 13
robinet à flux libre F 187
robinet à injection d'air G 13
robinet à membrane D 99
robinet à programme P 196
robinet à sas L 98
robinet à siège oblique B 110
robinet à tiroir G 64
robinet à trois voies T 89
robinet à vide V 38
robinet compensateur E 148
robinet d'arrêt P 90
robinet d'arrêt à mercure M 81
robinet de contrôle C 121
robinet de fermeture S 136
robinet de lest d'air G 13
robinet d'entrée d'air A 95
robinet d'entrée d'air de la pompe préliminaire P 221
robinet de prévidage R 199
robinet de remplissage des gaz G 32
robinet de rinçage à gaz G 15
robinet d'étranglement B 233
robinet d'évacuation D 228
robinet de vidage D 228, B 135
robinet-flotteur F 110
robinet obturateur C 420
robinet piant expansible à vide poussé B 94
robinet pour l'évacuation préliminaire R 199
robinet pour vide poussé H 95
robinet sans presse-étoupe P 6, P 7
robinet solénoïde M 3
robinet sphérique de retenue B 40
robinet tournant à trois lumières T 89
robinet tout-métal A 124
robinet-vanne S 369
robinet-vanne à écluse L 98
robinet-vanne à soufflet B 95/6
rodage en verre G 93
rodage mâle M 14
rodage sphérique G 128
roder L 12
rondelle à section circulaire O 24
rondelle axiale A 228
rondelle de joint G 48
rondelle d'épaisseur S 249
rondelle d'étanchéité S 42
rondelle élastique C 413
rondelle en cuir L 48
rondelle joint S 42
rondelle joint au vide V 143

rotor I 18, M 187, R 195
roue de roulement M 187
roue mobile M 187
roulant P 108
rouleau moteur D 244
roulement à billes B 39
rugosité de la surface S 445
rugosité superficielle S 445
rupture C 361
rupture de fatigue F 15
rythme de travail A 40

S

sablage S 12
sablage à jet de sable d'acier S 132
sans fluide F 132
sans fuites L 44
sans joint G 47
sas L 96
sas à vide V 98
sas d'air A 108
satiné intérieurement I 75
saturer S 15
saumon fusible I 57
saumure frigorigène C 321
scellement G 45
scellement annulaire A 161
scellement céramique C 102
scellement céramique-métal C 103
scellement de barre simple S 160
scellement de verre à pénétration E 8
scellement de verre extérieur O 71
scellement de verre intérieur I 77
scellement de Wilson W 43
scellement métal-céramique C 103
scellement métal-céramique d'après la méthode molybdène-manganèse M 164
scellement métal-saphir S 14
scellement mica-métal M 107
scellement verre-cuivre C 341
scellement verre-Kovar K 17
scellement verre-métal G 96, M 97
sceller S 366
schéma de termes E 139
schéma lumineux des connexions M 119
science du vide V 142
scrubber d'air A 114
séchage D 54
séchage à vide V 54
séchage de congélation à rotation C 92
séchage de contact C 287
séchage final A 90
séchage par courant de vapeur V 180
séchage par ébullition E 175
séchage par évaporation E 189
séchage par ruissellement du jet S 293
séchage par vaporisation E 175
séchage primaire P 138
séchage principal M 12
séchage secondaire A 90
sécher par congélation F 193
sécheur D 251
sécheur à ailettes P 8
sécheur à bande B 101

sécheur à bande de filtrage S 29
sécheur à boucles F 30
sécheur à couche fine T 72
sécheur à disque P 88
sécheur à fourrage vert G 119
sécheur à palettes P 8
sécheur à vide V 53
sécheur à vide avec agitateur mécanique V 113
sécheur à vide par bandes rotatives M 197
sécheur chancelant T 101
sécheur chancelant à vide V 57
sécheur continu T 190
sécheur cylindrique C 431
sécheur cylindrique à vide V 51
sécheur de congélation à rotation C 91
sécheur de contact C 286
sécheur de gaz pendant A 99
sécheur de lyophilisation de vide V 68
sécheur en armoire C 1
sécheur en auge S 33
sécheur en lés S 112
sécheur en pale à vide V 112
sécheur monocylindrique S 157
sécheur par action rapide F 82
sécheur par couches tourbillonnaires T 198
sécheur par plaque de chauffage J 3
sécheur par pulvérisation S 292
sécheur par tuyau F 82
sécheur suspendu F 30
sécheur tournant S 431
sécheur tubulaire C 431
sécheur tunnel T 190
séchoir D 251
séchoir à vide V 55
séchoir électrique D 43
section d'aspiration I 69
section de soupape V 187
section droite différentielle de choc D 106
section droite du choc C 221
sécurité de marche F 151
sécurité de service F 151
sécurité d'exploitation F 151
segment de piston P 68
ségrégation E 119
sel à ternir E 156
self-dispersion S 73
sens de blocage I 122
sens de non-conduction I 122
sensibilisateur S 82
sensibilité S 80
sensibilité à la pression partielle P 17
sensibilité à l'encoche N 30
sensibilité à l'entaille N 30
sensibilité de mise en action R 149
sensibilité de pression P 170
sensibilité d'un tube électronique G 68
sensible au choc S 124
séparateur S 86
séparateur d'air D 8
séparateur de condensat C 260
séparateur de poussière D 273
séparateur de sublimat S 394
séparateur de sublimé S 394
séparateur d'huile O 34
séparateur d'huile d'échappement O 36

séparateur sur le côté prévide B 9
séparation des couches P 91
séparation des isotopes I 181
séparation des masses M 37
séparation par fusion T 113
séparer par fusion S 36
serpentin condenseur C 187
serpentin réfrigérant C 187
serpentin refroidisseur C 187
serrage chaud H 134
serrer C 147
seuil de cristallisation R 63
shellac S 118
shunter B 199
siccatif D 239
siège de soupape V 202
signal sonore A 210
silencieux N 13
silicagel S 141
silicate de soude S 209
silice fondue O 44
silice vitreuse O 44
simple-cellule anodique S 152
simple effet / à S 150/1
simulation d'espace S 247
simulation d'espace de l'univers 247
simulateur cosmique A 132
siphon P 25
siphon avec enveloppe J 1
siphon enveloppé J 1
socle à compression en verre G 94
socle pressant P 58
socle pressant en croix C 140
solarisation S 217
solidification H 13, S 91
solidus S 230
solubilité S 232
solubilité des gaz G 58
solution solide S 227
solvant S 233
sonde P 189
sonde de mesure S 79
sonde de perçage T 7
sonde de refroidissement C 201, R 114
sonde de température T 18
sonde d'un détecteur de fuites L 42
sonde électronique E 104
sonde reniflante P 164
sonde thermique T 18
sorbant S 237
sorbat S 236
sorber S 235
sorption S 238
sorption à l'aide de couches-getter C 290
sorption chimique C 123
sorption ionique I 164
sorption par getter de décharge ionisante E 35
sortie d'arbre tournant W 49
sortie de courant C 417
soudabilité W 24
soudage à basse tension L 119
soudage à froid par pression C 207, C 217
soudage à l'étain S 221
soudage à rapprochement sous pression P 148
soudage avec tungstène et gaz inerte T 187
soudage d'agitation T 206
soudage d'arêtes E 9
soudage de cylindre extérieur O 70
soudage gaz inerte-métal I 48

soudage par bombardement électronique E 56, E 79
soudage par flammes F 71
soudage par fonte F 92
soudage par fusion F 240
soudage par points S 288
soudage par résistance R 139
soudage ultrasonore U 12
souder à l'étain S 218
soudeur W 31
soudure S 219
soudure à basse température D 29
soudure à électro-percussion E 112
soudure à haute tension H 99
soudure à l'arc A 186
soudure à l'étain par flamme T 126
soudure à l'hydrogène atomique A 182
soudure arcatome A 182
soudure autogène oxyacétylénique G 37a
soudure avec compensation flexible F 101
soudure bout à bout B 237
soudure de cylindre extérieur O 70
soudure de cylindre intérieur I 76
soudure de diffusion sous vide V 45
soudure de jonction J 18
soudure de verre S 220
soudure électrique bout à bout F 80
soudure en atmosphère gazeuse protectrice C 308
soudure en filet S 53
soudure en I P 74
soudure en ligne continue L 68
soudure en V S 164
soudure en X D 217
soudure forte sous hydrogène H 149
soudure par bombardement électronique E 56
soudure par coulage C 57
soudure par couture H 141, S 53
soudure par frottement F 213
soudure par goujon S 391
soudure par immersion D 148
soudure par impulsion thermique T 38
soudure par pointe S 391
soudure par rapprochement F 80
soudure sous gaz inerte I 49
soufflage de feuilles S 110
soufflage de lamelles S 110
soufflante B 147
soufflante à commande manuelle H 11
soufflante à jet de vapeur S 344
soufflante centrifuge C 86
souffler B 134
soufflerie W 44
soufflerie à pistons rotatifs R 181
soufflerie axiale A 225
soufflerie de courant transversal C 379
soufflerie radiale R 173
soufflet B 93, C 432
soufflure D 138, P 62
soupape V 186
soupape / sans V 197
soupape à admettre de l'air A 95
soupape à bille B 41

soupape à boulet B 41
soupape à cathode incandescente H 126
soupape à cloche B 98
soupape à commande d'air comprimé A 109
soupape à coulisse oscillante P 26
soupape à disque D 177
soupape à fermeture instantanée F 13
soupape à lames L 35
soupape à languette R 69
soupape à main à fermeture instantanée (rapide) Q 12
soupape à passage angulaire A 146
soupape à passage direct à vide G 64
soupape à piston S 200
soupape à programme P 196
soupape à ressort F 20
soupape à rodage sphérique S 274
soupape à rupture B 196
soupape à ultravide U 8
soupape à vanne à vide V 150
soupape à vide V 179
soupape à vide de métal liquide L 84
soupape à vide poussé H 97
soupape à vide préliminaire B 12
soupape à vide ultra-poussé avec une garniture de joint de métal liquide L 83
soupape d'admission à gaz G 6
soupape dans la conduite de dérivation B 240
soupape d'arrêt C 420
soupape d'échappement B 149, D 171
soupape d'échappement commandée par un ressort S 304
soupape de chargement C 118
soupape de commande S 88
soupape de commutation P 57
soupape de contrôle C 121
soupape de fermeture C 420
soupape de lest d'air G 13
soupape de lest de gaz réglable C 310
soupape de pompe à vide préliminaire B 12
soupape de purge A 101, B 149
soupape d'équilibrage de pression P 160
soupape de rebondissement B 19
soupape de refoulement D 171
soupape de réglage C 313
soupape de remplissage d'huile O 12
soupape de retenue B 19
soupape de rinçage à gaz G 15
soupape de sécurité S 7
soupape de service S 88
soupape de sortie d'air A 101
soupape de surpression R 104
soupape de traversée I 72
soupape d'évacuation B 149
soupape de vidange D 232, P 223
soupape de vidange d'huile O 21
soupape droite I 72

soupape flotteur F 110
soupape magnétique à rodage sphérique M 4
soupape réducteur de pression P 167
soupape réglable D 203
soupape reniflante S 201
soupape sphérique B 41
soupape thermostatique à expansion T 65
source de choc d'électrons en forme de cathode annulaire R 160
source de flash évaporation P 121
source de gaz G 59
source de vapeur V 226
source d'évaporation E 193
source de vaporisation à choc d'électrons E 82
source de vaporisation bombardée par électrons E 80
source d'ions I 166
source d'ions à cathode froide C 198
source d'ions à haute fréquence H 73
source d'ions de duoplasmatron D 270
source d'ions du plasma P 81
source d'ions par choc électronique E 83
source du rayon moléculaire M 146
source ionique d'immersion I 74
source ponctuelle P 92
source radioactive R 17
sourdine N 13
souspression L 112
sous-refroidissement S 419
sous-sol gaz B 6
sous-sonore S 401
spatule S 27
spectre cannelé B 43
spectre continu C 306
spectre d'absorption A 24
spectre de lignes L 66
spectrographe à vide V 153
spectrographe de masse à double focalisation D 211
spectromètre à adsorption A 84
spectromètre à désorption D 74
spectromètre de masse M 38
spectromètre de masse à déflexion magnétique S 56
spectromètre de masse à détection de fuites L 47
spectromètre de masse à haute fréquence H 74
spectromètre de masse à quadrupôle de Paul P 21
spectromètre de masse à secteur magnétique S 56
spectromètre de masse cycloïdale C 426
spectromètre de masse de respiration R 145
spectromètre de masse de temps de propagation T 108
spectromètre de masse par micro-sondes ioniques I 155
spectromètre de masse quadripolaire à haute fréquence Q 2
spectromètre ionique à résonance I 162
spectromètre monopole M 179
spectromètre quadrupôle de Paul P 21
spectrométrie de désorption D 75

spectrométrie par adsorption d'emission de champ F 33
spectroscopie d'adsorption infrarouge I 54
spectroscopique S 265
sphère à pouvoir couvrant P 198
spirale chauffante H 57
spirale du filament H 57
spirale pour lampes à vide V 27
spot de rayons B 84
stabilisateur de courant de source S 242
stabilisateur du niveau L 51
stabilité d'argon A 191
stabilité de longue durée L 102
stabilité du vide / vec V 154
stabilité du zéro Z 6
stabilité permanente P 32
stabilité temporaire C 369
stand de pompage C 229
station de commande R 30
station roulante de pompage T 175
stator S 335
stéatite naturelle B 143
stériliser S 358
stratification L 10
stries S 380
structure croisée H 113
structure de colonne C 228
structure de flux de dispersion F 215
structure d'île I 174
structure double T 204
structure en nid d'abeille H 113
structure spécifique S 75
sublimation S 395, S 400
sublimation de lit sous vide E 159
sublimation sous vide V 163
sublimer S 393
submergé d'huile O 26
substance absorbée A 16
succion I 80, P 214
succion mesurée M 51
sucer P 208
sulfate de cadmium et zinc Z 12
support d'électrode E 41
support de lingotière I 61
support de séchage D 231
support de substratum S 404
support-guide G 137
suppression du zéro Z 9
supraconducteur S 418
supraconductibilité S 417
supraconduction S 431
sûr F 7
surcharge O 75
surchauffage S 421
surchauffe de vapeur du fluide moteur V 228
surchauffeur S 420
surexpansion O 74
surface centrée F 2
surface cryogénique C 397
surface de contact du joint J 19
surface de joint J 19
surface des corps solides S 229
surface d'étanchage J 19
surface d'étanchéité J 19
surface d'obturation de soupape V 192
surface limite I 96
surpression O 77
surpression atmosphérique A 201
sursaturation S 422
surtension impulsionnelle I 31

surveillance / sans M 13
susceptibilité d'aspiration sans baffle U 13
susceptibilité de décomposition D 17
susceptibilité initiale I 66
susceptibilité mesurée d'aspiration M 51
suspension de cardan G 78
synthèse de vaporisation E 185
système à double paroi D 209
système à immersion N 43
système à ultra-vide U 7
système à vide V 118
système de calibration volumétrique V 248
système de choc de compression S 127
système de déviation D 80
système de foyer à distance T 14
système d'électrodes avec anneau protecteur G 135
système de levier M 57
système de pompage P 218
système de rayon J 14
système de rayon annulaire R 158
système de réglage du taux d'évaporation E 181
système des couches d'interférence I 101
système des couches interférentielles I 101
système d'étage de pression D 122
système de verrouillage protecteur S 5
système de vide V 165
système d'immersion B 46
système dynamique à vide D 281
système dynamique du vide D 281
système en cascade C 48
système frigorifique R 98
système frigorifique d'absorption A 25
système noyé F 114
système séparé par fusion S 39
système statique S 330

T

table à étuver B 31
table à sécher en chauffant B 31
table pivotante T 200
taillant C 422
talc B 143
tambour à double rotation P 78
tambour du support d'objet O 2
tambour sécheur à vide V 140
tambour sécheur en position inclinée T 102
tamis moléculaire M 159
taraudage [de vis intérieur] F 26
tâteur S 79
taux de collision C 223, I 15
taux de condensation C 258
taux de déposition par vaporisation D 67
taux de fuite I 68
taux de la sublimation S 398
taux de métallisation sous vide D 67
taux d'entrée I 52

taux de pression C 243
taux de pression en pompes P 179
taux de pulvérisation S 310
taux de récupération R 71
taux d'évaporation D 67, E 180
taux d'incidence I 21
taux maximum d'évaporation K 15
technique T 12
technique à l'hydrure de titane T 116
technique d'adsorption et de désorption A 79
technique de cryodécapage F 198
technique de détection de fuites par un jet de gaz-témoin P 191
technique de faisceau ionique I 128
technique de jonction I 94
technique de mesures du vide V 100
technique de métallisation C 177
technique d'enveloppement H 114
technique de raccordement I 94
technique des couches minces T 80
technique des matières premières T 13
technique des procédés sous vide V 125
technique de sublimation S 399
technique d'évaporation E 186
technique du filament chauffé F 86
technique du froid R 96
technique du vide V 167
technique du vide préliminaire B 17
technique fondamentale U 19
technologie de déposition par vaporisation D 68
technologie de métallisation sous vide D 68
technologie des procédés industriels P 193
technologie des procédés sous vide V 125
technologie des rayons d'électrons E 78
télécommande R 107
télécontrôle R 107
température absolue A 8
température ambiante A 140
température ambiante normale S 323
température cassante à froid B 208
température continue F 126
température critique C 373
température d'amollissement S 211
température de chauffage B 32
température de couleur C 227
température de Curie C 416
température de déformation D 33
température de détente instantanée I 82
température de façonnement W 55
température de filament F 43
température de l'air ambiant A 140

température d'entretien S 318
température de recristallisation R 62
température de référence R 76
température de tambour à disques S 115
température de transformation T 149
température d'évaporation E 187
température limite M 45
température lumineuse B 204
température moyenne A 222
température supérieure de desserrage A 154
température ultrabasse U 9
température véritable T 178
temporisation D 14
temps d'adsorption A 85
temps d'aération V 234
temps de chauffage W 2
temps de chauffage de la cathode W 2
temps de clean-up C 156
temps de congélation P 206
temps de contact A 61
temps de cycle C 425
temps de dessiccation D 262
temps de formation d'une couche monomoléculaire M 175
temps de maintien H 102, S 337
temps d'enregistrement R 58
temps d'entrée d'air V 234
temps de pompage P 212
temps de pompage primaire F 161
temps de prévidage F 161
temps de pulvérisation S 311
temps de réaction R 148
temps de réadsorption M 175
temps de recouvrement M 175
temps de réfrigération C 130
temps de refroidissement C 319
temps de régénérabilité C 157
temps de réglage A 38, S 96
temps de relaxation R 102
temps de répétition C 425
temps de réponse R 148, S 96
temps de retardement D 58
temps de séchage D 262
temps d'étranglement C 135
temps de transit T 150
temps d'exposition A 61
temps d'exposition à la lumière E 216
temps d'irradiation E 216
temps mort D 225
temps refroidissant C 130
temps régénérateur C 157
temps tombant R 103
tendance à être cassant E 123
teneur en eau W 9
teneur en gaz G 20
teneur en humidité M 139
teneur gazeuse G 20
tension S 378
tension / sous U 15

tension accélératrice A 28
tension activée F 171
tension au percement disruptif B 191
tension auxiliaire A 218
tension d'accélération A 28
tension d'allumage I 11
tension d'amorçage I 11
tension de choc I 31
tension de polarisation A 218
tension de pulvérisation S 313
tension de référence R 78
tension de repos E 38
tension de vapeur V 223
tension de vapeur de fluide moteur dans la chaudière B 156
tension de vapeur saturée S 17
tension d'ignition I 11
tension formée F 171
tension / sous U 15
tension superficielle S 448
ternir F 218
test de déchets E 150
test de fuite L 40
test d'enveloppe C 105
test de pression P 175
test d'étanchéité L 45
tête de mesure G 66
tête de pompe H 20
tête de tuyère N 32
tête fusible M 62
tête manométrique G 66
tête-révolver T 201
tête soupape V 190
tétrode-vacuomètre d'ionisation T 22
texture fibreuse F 32
théorie cinétique des gaz K 8
théorie de signalisation petite S 199
thermistor T 54
thermistor-vacuomètre T 55
thermobalance à vide V 170
thermocouple à couches minces T 81
thermocouple du vide V 171
thermodynamique T 59
thermo-élément du vide V 171
thermo-élément enrobé S 120
thermogravimétrie de vide V 172
thermomètre à aiguille D 90
thermomètre à contacts électriques E 34
thermomètre à pression de vapeur V 224
thermomètre à résistance R 137
thermomètre à résistance à couches minces T 78
thermomètre bimétallique B 118
thermosonde T 18
thermostabilité R 92
thermostat T 16
thermostat chauffant par circulation C 142
tige T 58
tige de guidage G 140
tige de verre G 94
tige en croix C 140
tirage de cristaux C 409
tirant S 336
tiroir S 179
tiroir de blocage H 107
tissu métallique M 92
titrer T 120

toile de fibre de verre F 31
tombac T 122, T 123
torr T 129
torsadé T 205
tourbillon W 39
tour de dégazage D 46
tourelle révolver T 201
tournant K 4
tournure C 131
traction de bouchage B 45
train à feuillards S 387
train de pompage I 71, T 143
traitement à chaud au four à vide V 66
traitement adhésif par métallisation A 64
traitement à l'argon A 192
traitement d'agent d'imprégnation T 168
traitement mécanique préalable P 139
traitement métallique M 98
traitement par film F 50
traitement par métallisation à l'arc sous vide V 10
traitement préalable P 185
traitement préparatoire P 185
traitement sous vide V 175
traitement thermique H 52
traitement thermique au four à vide V 66
traitement thermique sous vide V 78
trajet du courant F 125
tranchant C 422
transfert d'impulsion I 30
transformateur de mesure M 56
transformateur de pression P 177
transition de l'état solide à l'état liquide C 106
translucide T 153
translucidité T 160
transmission T 154
transmission calorifique H 49
transmission de chaleur H 49
transmission de courant I 134
transmission de courant ionique I 134
transmission de flux ionique I 134
transmission d'ions I 134
transmission thermique H 49
transparence T 160
transpiration T 161
transpiration thermique T 31
transporteur à courroie B 100
transporteur oscillatoire V 239
transvaser par pompage C 141
trap de fibre de verre F 31
trappe T 164
trappe angulaire E 28
trappe à vide V 235
trappe à zéolite Z 3
trappe cryogène de sublimation C 400
trappe d'adsorption A 86
trappe de Meißner M 61
travail des métaux M 98
traversée L 30
traversée à courant de haute intensité H 53
traversée coaxiale C 179

traversée coulissante R 53
traversée d'arbre R 185
traversée de courant C 417
traversée glissante L 62, R 53
traversée rotative R 178
trempage D 145
trempe C 414
trempe Q 11
trempé antiréflexe B 145
tremper H 12
tremper à froid C 210
trempe superficielle L 50
trempeur de résine R 130
triatomique T 169
tripolaire T 174
trois pôles / à T 174
trompe à vide F 135
trompe rotative à mercure de Gaede G 2
trou C 76
trouble / sans T 176
trou borgne B 137
trou de coulée T 6
tube à cathode chaude H 125
tube accumulateur S 372
tube à décharge D 169
tube à détection de fuites L 41
tube à gaz inerte R 27
tube à l'interférence I 103
tube à mémoire S 372
tube analyseur A 142
tube à rayons X à foyer linéaire L 65
tube à vide V 176
tube capillaire C 23
tube circulaire à vide poussé D 220
tube compteur de lumière P 43
tube compteur Geiger-Müller G 70
tube d'amenée de vapeur J 13
tube d'ascension à double paroi J 1
tube d'ascension à double paroi à vide V 93
tube de Bourdon B 183
tube d'échappement D 165
tube de décharge lumineuse G 101
tube déclenché de mesure à décharge T 170
tube de Crookes C 376
tube de Geissler G 71
tube de longévité L 101
tube de manomètre T 166
tube de mesure G 66
tube de quartz Q 9
tube de sorption ionique I 156
tube détecteur de fuites L 41
tube de thérapie sous vide poussé H 96
tube de vanne à plusieurs parties M 202
tube de Venturi N 31, V 235
tube de verre G 97
tube diagnostique sous vide poussé H 86
tube directeur du jet B 78
tube directeur du rayon B 78
tube d'orbitron-vacuomètre à ionisation O 58
tube électrométrique E 49
tube électronique E 92
tube en verre de transition G 105
tube final E 136
tube gradué à viscosité D 23

tube gradué en miniature M 120
tube microcapillaire M 110
tube ondulé C 351
tube photoélectrique P 40
tube thermionique H 125
tube unipolaire S 159
tubulaire d'évacuation E 161
tubulaire de vide E 161
tubulure de pompage E 201
tungstate de cadmium H 3
tungstate de calcium C 7
tungstène T 184
tunnel T 192
tunnel aérodynamique W 44
tunnel de congélation F 207
tunnel effet T 191
turbine d'échappement E 198
turbocompresseur T 194
turbopompe N 22
turbosécheur T 195
turbosurpresseur T 194
tuyau D 266
tuyau à manchon S 207
tuyau à onde de choc S 125
tuyau à vide V 177
tuyau collecteur D 199
tuyau d'aspiration T 3
tuyau de caoutchouc R 204
tuyau de clivage S 193
tuyau de mélange M 129
tuyau de pompage J 87
tuyau de pompe à vide V 117
tuyau de raccordement T 182
tuyau de séparation T 58
tuyau de vapeur J 13
tuyau en croix C 377
tuyau flexible H 118
tuyau flexible en métal F 100
tuyau métallique flexible F 100
tuyau rabattu F 74
tuyau raccordé par vis T 85
tuyau séparateur T 58
tuyère E 21, N 31
tuyère à gouttes D 248
tuyère à jet inversé A 157, A 159
tuyère annulaire A 159
tuyère cuirassée C 28
tuyère de Laval L 26
tuyère de vapeur S 346
tuyère latérale S 138
tuyère mélangeuse M 127
tuyère multiple M 195
type de curseur rotatif R 186
type de palettes rotatives R 186

U

UHV U 4
UISTAV I 115
ultracentrifugateur U 2
ultravide U 4
Union Internationale pour la Science, la Technique et les Applications de Vide I 115
unique conducteur / à S 154
unité Angström A 147
unité de pression P 178
unité de pulvérisation S 312
unité de réserve S 325
unité du pulvérisateur S 312
uranine P 23

uranium U 26
usinage de faisceau électronique E 70
usinage des métaux M 98
usinage électro-érosif E 43
usure W 19
usure des électrodes E 39
utilisés en technique du vide V 6
UV-source de vide V 178

V

vacuologie V 142
vacuomètre B 116, V 74
vacuomètre à adsorption A 87
vacuomètre à amortissement O 65
vacuomètre à amortissement à fil de quartz Q 8
vacuomètre absolu A 11
vacuomètre à capsule C 30
vacuomètre à compression C 241
vacuomètre à conductibilité thermique H 25
vacuomètre adhérent à fil de quartz Q 8
vacuomètre à émission ionique de champ F 39
vacuomètre à frottement O 65
vacuomètre à membrane D 98
vacuomètre à tube courbé S 285
vacuomètre à tube élastique B 183
vacuomètre à vibration rotative T 132
vacuomètre d'adhérence D 23
vacuomètre de Bourdon B 183
vacuomètre de frottement D 23
vacuomètre de Mc Leod M 47
vacuomètre de pression partielle P 18
vacuomètre de type d'un magnétron inverse I 124
vacuomètre différentiel d'interférence D 109
vacuomètre d'ionisation I 152
vacuomètre d'ionisation à deux chambres T 207
vacuomètre d'ionisation à haute pression H 77
vacuomètre d'ionisation avec produit radio-actif R 16
vacuomètre d'ionisation palladium-hydrogène P 9
vacuomètre interférentiel I 104
vacuomètre interférométrique à huile I 105
vacuomètre ionique à cathode chaude T 51
vacuomètre moléculaire M 151
vacuomètre moléculaire de Gaede G 3
vacuomètre moléculaire de Langmuir-Dushman L 11
vacuomètre radiométrique K 13
vacuomètre radiométrique de Knudsen K 13
vacuomètre réfractométrique R 91

vacustat V 4
valeur de conductivité du clapet fermé R 124
valeur de la rugosité P 22
valve V 186
valve à commande manuelle H 10
valve actionnée à la main H 10
valve à lames L 35
valve à languette R 69
valve à vapeur S 347
valve de rebondissement B 19
valve de ventilation de la pompe préliminaire P 221
vanne S 179
vanne à baffle B 25
vanne à charnière F 77
vanne à clapet articulé F 77
vanne à commande magnétique M 3, S 226
vanne à coulisse G 64
vanne à coulisse oscillante P 26
vanne à écran B 25
vanne à flux libre F 187
vanne à levier L 54
vanne annulaire à piston A 160
vanne à opercule K 4
vanne à passage droit I 72
vanne à pointeau N 2
vanne à presse-étoupe P 2
vanne à programme P 196
vanne à rupture B 189
vanne à siège oblique B 110
vanne à soufflet B 95/6
vanne à tiroir G 64
vanne dans la canalisation de prévidage R 199
vanne d'arrêt C 420, S 136
vanne d'arrêt à vide V 147
vanne de contrôle C 121
vanne de dosage D 203
vanne de lest de gaz réglable C 310
vanne d'équerre A 146
vanne de réglage C 313
vanne-écluse L 98
vanne entièrement métallique A 124
vanne étuvable à passage direct de grand diamètre nominal L 15
vanne magnétique M 3
vanne obturatrice C 420
vanne sphérique de retenue B 40
vapeur V 209
vapeur d'eau S 339
vapeur d'échappement E 200
vapeur du fluide moteur M 128
vapeur épuisée E 200
vapeur humide W 37
vapeur moteur M 128
vapeur non saturée U 21
vapeur saturée S 16
vapeur sous dépression V 155
vapeur sous vide V 155
vapeur surchauffée D 265
vapeur sursaturée O 80
vaporisateur E 190
vaporisateur à couche fine F 8
vaporisateur de puissance L 93
vaporisateur en fin S 158
vaporisation E 165
vaporisation à bande B 42

vaporisation ascendante U 24
vaporisation d'aluminium A 136
vaporisation de diffusion D 132
vaporisation descendante D 222
vaporisation explosive E 214
vaporisation directe D 149
vaporisation par bombardement électronique E 63
vaporisation par faisceau d'électrons E 63
vaporisation propre S 66
vaporisation sous vide V 60
vaporiser / se E 164
variation de la tension de secteur L 67
vase de dégazage D 8
vase Dewar D 85
véhicule spatial S 251
ventilateur F 10
ventilateur aspirant E 197
ventilateur d'extraction E 197
ventilation D 7
vérification de l'étalonnage R 48
vernir L 2
vernir électrolytiquement E 115
vernis V 229
vernis à enlever S 384
vernis à tirer S 384
vernis givré B 206
verre à couches multiples C 236
verre à refondre S 45
verre à vitre collé C 236
verre calcaire-natron L 57a
verre compound C 236
verre court S 129
verre de borosilicate B 172
verre de jonction C 270
verre de platine G 91
verre de protection à rayonnement de chaleur T 42
verre de raccordement C 270
verre de soude S 208
verre de transition T 146
verre fritté S 167
verre intermédiaire A 54
verre long L 100
verre molybdique G 90
verre mou S 212
verre mousse C 78
verre pour assemblage C 270
verre quartzeux F 232
verres cristallisés G 85
verre semblable à silice H 78
verre se soudant bien au tungstène G 92
verre sintérisé S 167
verrou d'eau W 7
verrouillage I 108
verrouillage de sécurité S 4
vibration V 238
vibration de cisaillement d'épaisseur T 67
vibration du réseau moléculaire L 24
vibration du treillis L 24
vidange D 229
vidange d'huile O 19
vidange inférieure B 173
vidange rapide Q 14
vide V 5
vide / à V 3
vide / dans le V 3

vide / sous V 3
vide absolu A 10
vide adhésif C 27
vide à maintenir H 103
vide auxiliaire V 136
vide bas L 116
vide collant C 27
vide d'adhésion du mercure C 27
vide de comparaison R 77
vide d'entretien H 103
vide de protection G 136
vide de Torricelli T 130
vide garanti G 133
vide grossier R 200
vide intermédiaire F 64
vide moyen F 64
vide poussé H 81
vide préliminaire F 162
vide primaire R 200
vider E 158
vide sans agent de propulsion F 133
vide sans fluide F 133
vide sans hydrocarbures H 148
vide sans poudre propulsive F 133
vide ultrapoussé U 4
vide ultraviolet V 152
vie de filament F 42
vieillissement cathodique C 62
vieillissement des filaments thoriés A 93
vis à cheville S 390
vis à repousser F 154
vis calante S 89
viscosité V 242
viscosité cinématique K 7
viscosité dynamique A 12
vis de décharge D 59
vis de purge d'huile O 20
vis de réglage S 89
vis de remplissage d'huile O 24
vis de sortie d'air A 112
vis d'évacuation d'air A 112
vis de vidange d'huile O 20
vis de vidange d'huile en champ de gravitation G 117
vis du niveau d'huile O 28
vis pointeau G 132
vis régulatrice S 89
vis sans tête S 132
vitesse d'aspiration E 16, P 215, V 249
vitesse de balayage S 24
vitesse de congélation F 205
vitesse d'écoulement F 127, R 34
vitesse de croissance G 131
vitesse de déposition par vaporisation D 67
vitesse de désintégration D 172
vitesse de dessiccation D 261
vitesse de diffusion D 137
vitesse de dilatation C 365
vitesse de faire le vide S 269
vitesse de fusion M 68
vitesse de glissement S 191
vitesse de migration M 118
vitesse de pompage P 215, S 269
vitesse de pompage ionique I 158
vitesse de pompage mesurée M 51
vitesse de réaction S 271
vitesse de refroidissement C 328

vitesse de réponse S 271
vitesse de séchage D 261
vitesse de soudage W 30
vitesse d'évacuation S 269
vitesse de vaporisation E 180
vitesse d'exploration S 24
vitesse du courant R 34
vitesse du séchage par congélation F 196
vitesse élévatoire du piston S 388
vitesse la plus probable M 180
vitesse moléculaire M 161
vitesse moyenne M 49
vitesse moyenne des particules A 221
vitesse moyenne moléculaire A 221
vitesse moyenne quadratique R 169
vitesse périphérique C 146
vitesse thermique T 49
vitrer V 245
vitrifier V 245
voie à bande S 387
volet d'aération A 103
voltage de formation F 171
voltage de pulvérisation S 313
voltage d'excitation E 196
volume à prévide B 15
volume aspiré E 199
volume atomique A 206
volume de l'appareillage P 79
volume de l'installation P 79
volume d'ionisation I 153
volume du grammemolécule d'un gaz V 247
volume effectif E 13
volume engendré S 454
volume molaire M 141
volumètre à gaz G 63
volume utile E 13
voûte de véhicule du froid C 191
vulcanisation à chaud H 139
vulcanisation à froid C 215

W

wolfram T 184

X

xénon X 1

Z

zéolite Z 1
zéolite artificielle A 193
zéro absolu A 13
zéro absolu de la température A 14
zincifère Z 11
zinquer Z 10
zone de fusion F 111
zone de pénombre P 30
zone de solidification F 208
zone de soudage W 34
zone de température de la décharge thermique T 27
zone de température du percement disruptif D 190
zone d'usinage W 56
zone transitoire T 148

РУССКИЙ

СОДЕРЖАНИЕ

1.	Вакуумная физика
1.1.	Материя в газообразном состоянии
1.2.	Кинетическая теория газов
1.3.	Течение газа
2.	Вакуумная техника
2.1.	Величины и единицы
2.2.	Вакуумметры
2.2.1.	Манометры для измерения полного давления
2.2.2.	Измерители парциального давления
2.3.	Вакуумные насосы и геттеры
2.3.1.	Объемные насосы
2.3.2.	Молекулярные насосы
2.3.3.	Эжекторные насосы
2.3.4.	Диффузионные насосы
2.3.5.	Газопоглотительные насосы
2.3.6.	Криогенные насосы
2.3.7.	Принадлежности насосов
2.4.	Отражатели
2.5.	Вентили и стопорные краны
2.5.1.	Запорные клапаны
2.5.2.	Вентили пуска воздуха
2.5.3.	Вентили для впуска газа
2.5.4.	Комбинации вентилей
2.5.5.	Стопорные краны
2.6.	Вакуумные коммуникации
2.7.	Соединения
2.7.1.	Неразъемные соединения
2.7.2.	Разборные соединения
2.8.	Вакуумные камеры
2.9.	Вводы
2.10.	Окна
2.11.	Течеискание
2.11.1.	Течеискатели
2.11.2.	Методы течеискания
2.12.	Вакуумные системы
2.12.1.	Запаянная вакуумная система
2.12.2.	Динамические вакуумные системы
2.13.	Техника сверхвысокого вакуума
3.	Применения вакуумной техники
3.1.	Вакуумная сушка
3.2.	Вакуумное обезгаживание
3.3.	Вакуумная пропитка
3.4.	Вакуумная пайка твердым припоем
3.5.	Упаковка в вакууме
3.6.	Вакуумная дистилляция
3.7.	Вакуумная сублимация
3.8.	Вакуумная конденсация
3.9.	Вакуумное фильтрование
3.10.	Техника напыления в вакууме
3.11.	Осветляющее покрытие
3.12.	Защитные оболочки
3.13.	Металлизация
3.14.	Технология тонких пленок
3.15.	Вакуумная плавка
3.16.	Вакуумное обезгаживание
3.17.	Вакуумное литье
3.18.	Вакуумный отжиг
3.19.	Спекание в вакууме

А

абразивное зерно А 3
абсолютная активность F 222
абсолютная скорость испарения K 15
абсолютная температура А 8
абсолютная температурная шкала А 9
абсолютная температурная шкала Фаренгейта А 5
абсолютное давление А 7
абсолютный вакуум А 10
абсолютный вакуумметр А 11
абсолютный манометр А 6
абсолютный нуль А 13
абсолютный нуль температуры А 14
абсорбент А 17
абсорбированное вещество А 16
абсорбировать А 15
абсорбтив А 16
абсорбционная ловушка А 19
абсорбционная способность А 18
абсорбционный насос А 22
абсорбционный спектр А 24
абсорбционный холодильный агрегат А 25
абсорбционный электрохолодильник А 23
абсорбция А 20
абсорбция инфракрасных лучей J 53
автоблокировка S 69
автоионизационный вакуумметр F 39
автомат А 212
автоматическая диафрагма А 214
автоматический затвор А 214
автоматический насос А 213
автоматический прибор для контроля уровня L 15
авторадиография А 215
автоэмиссионный микроскоп F 36
адатом А 56
адгезив А 63
адгезивная прочность А 60
адгезивный слой А 65
адгезионная способность А 66
адиабатический А 68
адсорбат А 75
адсорбент А 77
адсорбирующее вещество А 77
адсорбирующий слой А 76
адсорбционная газовая хроматография А 80
адсорбционная ловушка А 86
адсорбционная спектрометрия, основанная на электростатической эмиссии F 33
адсорбционная спектроскопия в инфракрасных лучах J 54
адсорбционная способность А 82
адсорбционно-десорбционная техника А 79

адсорбционный вакуумметр А 87
адсорбционный насос А 83
адсорбционный спектрометр А 84
адсорбция А 78
адсорбция газа G 25
адсорбция порами Р 36
азотный эквивалент N 9
активатор А 46
активационный анализ А 43
активированная окись алюминия А 41
активированный уголь А 42
активировка катода С 63
активировка оксидных катодов А 45
активность А 50
активный газопоглотитель А 49
активный геттер А 49
актиний А 39
акустический сигнал А 210
акцептор А 29
алифатическое соединение А 120
алкилфенольные смолы А 121
альфа-излучатель А 128
альфатрон А 130
альфа-частица А 129
алюминиевая фольга А 134
алюминирование А 136
амальгамация А 138
амальгамировать А 137
амортизационное крепление А 178
ампула V 236
анализаторная трубка А 142
анализатор остаточного газа R 120
анализатор парциального давления Р 16
анализ газов способом фракционного испарения F 177
анализ газов способом фракционной адсорбции F 173
анализ газов способом фракционной десорбции F 175
анализ газов способом фракционной конденсации F 174
ангстрем А 147
анкерный болт S 336
анодирование А 167
анодировать А 168
анодная сетка А 165
анодная темная область А 162
анодное расстояние А 164
анодное свечение А 166
анодное темное пространство А 162
аномальный тлеющий разряд А 1
антидетонационный А173
антикатод А 169
антикатод со штриховым фокусным пятном S 373
антикоррозионное покрытие С 349
антимиграционный барьер А 170
антиобледенитель D 35
антиотражательное покрытие А 176
антиотражательный В 145
антифрикционный сплав А 172, В 1
апериодический D 5

апертура пучка В 73
аппарат для молекулярной перегонки М 162
аппарат для пополнения R 79
аргонная нестабильность А 189
аргонная стабильность А 191
аргоновая обработка А 192
армированный стекловолокном G 88
асбест А 194
асбестовый картон А 195
асбестовый корд А 196
астоново темное пространство А 198
атм А 200
атмосферное давление А 202
атомизация А 207
атомно-водородная дуговая сварка А 182
атомно-дуговая сварка А 182
атомно-чистый А 203
атомный номер А 205
атомный объем А 206
атомный пучок А 204
аустенитная нержавеющая сталь А 211
аэрация А 89
аэрировать А 73
аэродинамическая труба W 44

Б

баббит В 1
байонетный запор В 69
байпасная магистраль В 239
байпасный вентиль В 240
байпасный трубопровод В 239
бак для кабеля С 2
балластный бак В 38
баллонный газ В 163
бар В 44
барабан для закладки объекта О 2
барабанная сушилка С 431
барабанный вакуумфильтр V 52
барокамера А 131
барометр В 50
барометрическая компенсация В 52
барометрическая формула [Больцмана] В 53
барометрическое давление В 51
барьер для защиты от проникновения масла по поверхности А 170
барьерная сетка С 266
батарея термоэлементов Т 61
безаварийный Т 176
безвентильный V 197
безмасляный вакум F 133
безопасность в работе F 151
безопасный F 7
бериллиевое окно В 106
бескислородная медь О 83
бесклапанный V 197
беспорядочное распределение R 21
беспрокладочный G 47
бессальниковый вентиль Р 7
бестигельная зонная плавка С 385
бесшумно-работающий насос W 40

бетатрон В 108
бета-частица В 107
бивалентный В 123
биметаллический выключатель В 117
биметаллический контакт В 115
биметаллический регулятор давления В 119
биметаллический термометр В 118
биспиральная нить накала С 185
биспиральный подогреватель С 186
бифилярная навивка В 112
бифилярная намотка В112
благородный газ I 47
блокировать I 107
блокировка I 108
блокирующее действие В 141
блок-катод В 141
блок питания Р 128
блок питания манометра W 75
блуждающие электронные лучи V 184
болванка В 113, В 144
болт-катод В 160
болтовое соединение S 31
бомбардировка В 162
боросиликатное стекло В 172
бронированный провод I 84
бросок давления Р 147
броуновское движение В 209
брутто-плотность G 126
бугельный газопоглотитель S 364
бурное кипение Е 4
буртик вала, смазываемый капельной масленкой S 99
буртик кокиля М 183
буртик уплотнения S 49
бустерно-диффузионный насос В 171
бустерный диод В 169
бустерный диффузионный насос В 168
бустерный насос В 170
быстродействующий вентиль F 13
быстродействующий вентиль с ручным приводом Q 12
быстродействующий клапан F 13
быстродействующий регистратор S 272
быстрое замораживание Q 16
быстрое опорожнение Q 14
быстрое охлаждение Q 13
быстрое разрежение Q 14
быстрое сканирование R 24
быстроразборное соединение Q 17
быстрота действия ионного насоса I 158
быстрота действия насоса Р 215
быстрота действия насоса без подачи балласта U 14
быстрота действия насоса с подачей балласта В 37
быстрота откачки V 249
быстрота разрежения объекта Е 15, N 6
БЭТ-изотерма В 109

В

вакансия V 1
вакансия в кристаллической решетке L 23
вакантное место в решетке L 23
вакустат V 4
вакуум V 5
вакуум без остаточного давления углеводородов H 148
вакууме /в V 3
вакуумирование E 160
вакуумированный бетон V 31
вакуумметр V 74
вакуумметр парциального давления P 18
вакуумная арматура V 119
вакуумная вальцовая сушилка V 51
вакуумная возгонка V 163
вакуумная давильная установка V 123
вакуумная дистилляционная камера V 48
вакуумная дистилляция V 47
вакуумная диффузионная сварка V 45
вакуумная дуговая печь V 11
вакуумная дуговая печь с расходуемым электродом C 206
вакуумная дуговая печь с холодным кристаллизатором C 206
вакуумная дуговая плавка V 12
вакуумная задвижка G 64, V 150
вакуумная замазка S 50
вакуумная зонная очистка V 183
вакуумная зонная плавка V 182
вакуумная изоляция V 90
вакуумная индукционная печь V 86
вакуумная индукционная плавка V 87
вакуумная инкапсуляция V 59
вакуумная камера B 90, V 22, V 34
вакуумная камера с двойными стенками D 219
вакуумная капельная плавка V 50
вакуумная качающаяся сушилка V 57
вакуумная коммуникация V 97
вакуумная конвейерная печь C 303
вакуумная конденсация V 32
вакуумная концентрация V 29
вакуумная кристаллизация V 37
вакуумная лампа V 176
вакуумная лампа для терапии H 96
вакуумная ловушка V 174
вакуумная металлизация V 104
вакуумная металлургия V 106
вакуумная мешалка V 109
вакуумная насосная система V 128
вакуумная оболочка V 145, V 146
вакуумная обработка V 175

вакуумная отливка V 89
вакуумная очистка V 129
вакуумная пайка твердым припоем V 15
вакуумная перегонка V 47
вакуумная печь V 72
вакуумная печь для горячей прессовки и закалки V 79
вакуумная печь для пайки V 16
вакуумная печь для плавки V 103
вакуумная печь с нагретыми стенками V 81
вакуумная печь с опрокидывающимся тиглем T 103
вакуумная печь сопротивления V 138
вакуумная печь с холодными стенками V 28
вакуумная плавка V 102
вакуумная плавка и разливка S 176
вакуумная плавка слитка V 89
вакуумная полочная сушилка V 55
вакуумная порошковая изоляция V 122
вакуумная пропитка V 83
вакуумная реакция V 130
вакуумная ректификационная колонна V 134
вакуумная ректификационная установка V 133
вакуумная ректификация V 132
вакуумная реторная печь V 139
вакуумная рубашка V 92
вакуумная система с двойным колпаком D 210
вакуумная система V 118, V 165
вакуумная смазка V 77
вакуумная сублимационная сушилка V 68
вакуумная сублимация V 163
вакуумная сублимация в движущемся слое E 159
вакуумная сублимация в псевдоожиженном слое E 159
вакуумная сушилка V 53
вакуумная сушилка с вращающимся барабаном V 140
вакуумная сушилка с лопастной мешалкой V 113
вакуумная сушилка с перелопачивающей мешалкой V 112
вакуумная сушка V 54
вакуумная сушка струей пара V 180
вакуумная термогравиметрия V 172
вакуумная термопара V 171
вакуумная техника V 167
вакуумная установка V 118
вакуумная установка для литья смолы V 137
вакуумная установка для наполнения порошком V 121
вакуумная физика V 115
вакуумная формовка V 67

вакуумная формовка в резиновую форму D 31
вакуумная формовка с глубокой вытяжкой D 25
вакуумная хроматография V 23
вакуумная экстракция V 61
вакуумная экструзия V 62
вакуумная электроннолучевая плавильная печь V 58
вакуумно-дуговое испарение с расходуемым электродом C 281
вакуумное замораживание V 70
вакуумное индукционное литье V 85
вакуумное испарение V 60
вакуумное кальцинирование V 19
вакуумное капельно-струйное рафинирование V 162
вакуумное литье S 410
вакуумное масло O 50
вакуумное нанесение V 42
вакуумное напыление E 170
вакуумное обезвоживание V 41
вакуумное обезгаживание с непрерывным переходом от одной позиции к другой L 7
вакуумное обезгаживание стали V 159
вакуумное обезуглероживание V 39
вакуумное обкуривание V 71
вакуумное обогащение V 29
вакуумное охлаждение V 35
вакуумное покрытие V 42
вакуумное реле V 135
вакуумное спекание под давлением V 124
вакуумное струйное обезгаживание V 160
вакуумное трубчатое соединение V 116
вакуумное фильтрование V 65
вакуумное формование F 182
вакуумно импрегнированный V 82
вакуумно-паровой нагрев V 157
вакуумно-плотный V 173
вакуумно-плотный ввод вала R 189
вакуумно-тепловая изоляция V 169
вакуумноустойчивый V 154
вакуумные весы V 13
вакуумные испытания V 168
вакуумные микровесы V 108
вакуумные микровесы с кварцевой нитью Q 10
вакуумные принадлежности V 6
вакуумные термовесы V 170
вакуумные условия V 33
вакуумный L 44, V 173
вакуумный бойлер C 190

вакуумный ввод с осевым перемещением R 53
вакуумный вентиль V 179
вакуумный вентиль без сальника P 6
вакуумный выключатель V 164
вакуумный дилатометр V 46
вакуумный дистиллятор V 49
вакуумный искровой промежуток V 151
вакуумный искровой разрядник V 151
вакуумный испаритель H 82
вакуумный кипятильник C 190
вакуумный кожух V 146
вакуумный колпак B 90, V 14
вакуумный колпак с двойными стенками D 218
вакуумный концентратор V 30
вакуумный кран V 38
вакуумный криостат V 36
вакуумный курбель W 49
вакуумный лак S 46
вакуумный монохроматор V 110
вакуумный насос V 126
вакуумный обжиг V 19
вакуумный обогатитель V 30
вакуумный отжиг V 7
вакуумный отсечной клапан V 147
вакуумный охладитель F 81
вакуумный пар V 155
вакуумный парогенератор V 156
вакуумный перегонный аппарат V 49
вакуумный прерыватель V 91
вакуумный прием V 107
вакуумный пробой V 17
вакуумный процесс V 107
вакуумный реактор V 131
вакуумный резервуар V 136
вакуумный рентгеновский дифрактометр V 114
вакуумный сосуд V 34
вакуумный спектрограф V 153
вакуумный струйный аспиратор V 94
вакуумный сушильный шкаф V 55
вакуумный трубопровод V 117
вакуумный ультрафиолет V 152
вакуумный цемент V 21
вакуумный шкаф V 18
вакуумный шланг V 177
вакуумный шлюз V 98
вакуумный эксикатор V 43
вакуум-паровая установка V 158
вакуумплотность V 96
вакуумплотный V 173
вакуум прилипания C 27
вакуумпровод V 97, V 117
вакуум-сушилка V 56
вакуум Торричелли T 130
вакуум-упаковка V 111
вакуумфактор V 63
вакуумфильтр S 413, V 64
вальцовая сушилка C 431
ванна с расплавом M 63
вверх по течению U 25

ввод L 30
ввод воды W 14
ввод вращающегося вала R 185
ввод вращения R 178
ввод для открытого ионизационного манометра с накаленным катодом N 41
ввод для передачи линейного перемещения L 62
вводить E 121
вводить каплю за каплей D 245
вводный кабель L 31
вводный провод L 32
ведущая шестерня D 243
ведущий шкив D 244
величина неровности P 22
величина шероховатости P 22
вентиль V 186
вентиль в коммуникации предварительной откачки R 199
вентиль для впуска газа G 6
вентиль для наполнения газом G 32
вентиль для пуска атмосферы в форвакуумный насос P 221
вентильная коробка V 188
вентильное распределение V 194
вентильный фотоэлемент P 50
вентиль пуска воздуха A 95
вентиль с наклонным шпинделем B 110
вентиль со сферическим клапаном B 41
вентиль со сферическим шлифом S 274
вентиль со сферическим шлифом и магнитным приводом M 4
вентиль с качающейся заслонкой P 26
вентиль с ручным приводом H 10
вентиль с сильфонным уплотнением B 95/6
вентиль с соленоидным приводом S 226
вентиль с уплотнением жидким металлом L 84
вентиль с электромагнитным приводом S 226
вентилятор F 10
вентилятор с поперечным потоком C 379
вентиляционный туннель W 44
вероятность ассоциации P 188
вероятность захвата C 31
вероятность ионизации I 149
вероятность ионного захвата I 132
вероятность перехода T 158
вероятность прилипания C 31
вероятность соударения I 15
вероятность столкновения I 15
вертикальное испарение U 24
вертикальный перегонный куб P 117

вертикальный способ зонной плавки F 108
весовой гигрометр M 138
вес слитка I 62
весы для измерения лобового сопротивления D 227
вжигать B 229
взаимодействие I 91
взаимодействие газа с твердым телом G 57
взаимозаменяемый I 92
взрываться внутрь I 22
взрыв, вызванный наружным давлением I 23
взрывное испарение E 214
взрывной испаритель с порошковым питанием P 121
взрывчатая смесь E 215
вибрационный желоб D 198
вибрация V 238
вилсоновское уплотнение W 43
винт без головки G 132
винт для выпуска воздуха A 112
винт для заливки масла O 24
винтовая дислокация S 32
винтовая пробка для заливки масла O 24
винтовая пробка для регулировки уровня масла O 28
винтовое соединение S 31
винтовой H 55
винтовой инжектор с одним витком S 162
винт с граненным отверстием в головке S 206
вискер C 412
витой T 205
вихревое течение V 252
вихрь W 39
включать последовательно C 272
включающий механизм R 30
влажность H 144
влажный пар W 37
влияние собственного рентгеновского излучения на показания X 3
вместилище C 293
внешнее поле E 220
внешний фотоэффект P 44
внутреннее паяное цилиндрическое соединение I 76
внутреннее пространство I 106
внутреннее трение I 113
внутренний диаметр I 73
внутренний нагреватель I 114
внутренний подогреватель I 114
внутренний трубчатый спай I 77
внутренний фотоэффект P 41
внутренняя резьба F 26
водокольцевой насос W 17
водопоглощение W 6
водопроводная вода T 10
водосодержание W 9
водоструйный конденсатор W 15
водоструйный насос A 197
водянаяпленка W 8
всдяная рубашка C 332

водяное защитное реле P 206
водяной затвор W 7
водяной пар S 339
возбуждение E 194
возвратно-поступательный насос R 51
возвратный клапан B 19
возгонка S 395
возгонная технология S 399
возгонять S 393
воздуходувка B 147
воздухоотделитель D 8
воздухоочиститель A 114
воздушная холодильная машина A 111
воздушный удар A 106
воздушный клапан A 103
воздушный фильтр A 102
воздушный шлюз A 108
возникновение дуги S 383
возрастание давления P 169
волнистость C 352
волокнистая текстура F 32
волосяная трещина C 25
волосяной кристалл C 412
волочение на оправке B 45
вольфрам T 184
вольфрамат кадмия C 3
вольфрамат кальция C 7
вольфрамовая спираль T 185
вольфрамовое стекло G 92
вольфрамовый диск T 186
вольфрамовый наконечник T 188
волюметрическая калибровочная система V 248
восприимчивый к удару S 124
воспринимающий элемент S 79
восстанавливающий момент R 150
восстановитель D 64
восстановительная печь R 68
впайка G 45
впускать воздух A 73
впуск воздуха A 89
впускное давление F 63
впускное окно I 88
впускной патрубок E 161, I 88
впускной трубопровод I 87
впускной фланец I 70
вращательный водоструйный насос R 193
вращательный масляный воздушный насос R 179
вращательный момент T 127
вращательный насос R 183, R 188
вращательный ртутный насос Геде R 177
вращающаяся барабанная сушилка C 431
вращающаяся клетка R 171
вращающееся вакуумное уплотнение R 189
вращающееся поле R 191
вращающийся анод R 190
вращающийся вакуумный испаритель V 141
вращающийся вакуумный фильтр R 187
вращающийся ввод R 178

вращающийся смесительный конденсатор R 192
вращающийся фланец L 104
вредное пространство D 6
временная задержка T 106
временной ход повышения давления R 35
время восстановления C 157
время для восстановления C 157
время до начала повышения давления в сверхвысоковакуумной системе S 337
время достижения среднего значения H 2
время дросселирования C 135
время газопоглощения C 156
время задержки D 58
время запаздывания D 58, R 148
время облучения E 216
время образования мономолекулярного слоя M 175
время откачки P 212
время охлаждения C 319
время повышения давления до атмосферного V 234
время предварительной откачки F 161
время пролета T 150
времяпролетный масс-спектрометр T 108
времяпролетный метод T 109
время простоя D 225
время разогрева W 2
время разогрева катода C 67
время распыления S 311
время регистрации R 58
время регулирования A 38, S 96
время релаксации R 102
время спада R 103
время срабатывания R 148
время сушки D 262
время удерживания A 61
время установления S 96
всасывание I 80
всасывать P 208
всасывать обратно S 409
всасывающая труба T 3
всасывающее действие P 214
всасывающее окно I 88
всасывающее устройство S 412
всасывающий и нагнетательный насос D 205
всасывающий насос D 235
всасывающий трубопровод I 87
всасывающий щуп P 164
вскипание E 4
вспомогательный диффузионный насос B 168
вспомогательный катод A 216
вспомогательный насос B 170
вспомогательный электрод A 117
втекать F 121
вторая плавка S 55
вторичная осушка A 90
вторичная плавка S 55
вторичный воздух S 54

вторичный цилиндр S 450
второе катодное свечение N 3
второе катодное темное пространство C 375
втулка S 177
входная мощность P 127
входной патрубок I 70
входной поток I 120
входной шлюз I 144
выброс газа G 17
выветривание W 23
выгорание электрода E 39
выдавливание C 200, E 228
выдавливать E 226
выдвижная лопатка S 179
выделение E 120
выделение газа G 30
выдержка на солнце S 217
выдувной клапан S 201
выдутая пленка B 148
выжигать F 140
выключатель, срабатывающий от давления P 174
выключающее давление S 455
вымывать F 140
выпадение хлопьями F 112
выпарной аппарат E 190
выпускать B 134
выпуск воздуха A 98, D 7
выпуск газа G 54
выпускная диафрагма E 203
выпускная сторона T 9
выпускная труба D 165
выпускное давление B 13, D 167, D 277
выпускное давление срыва B 194
выпускное отверстие T 6
выпускной вентиль D 232
выпускной воздушный вентиль A 101
выпускной канал D 224
выпускной клапан D 171
выпускной клапан, удерживаемый пружиной S 304
выпускной патрубок D 165, D 166
выпускной трубопровод D 164
выпуск пара V 218
выработка с помощью электронных лучей E 58
выравнивание давления P 151
выравнивать F 98
выравнивающий вентиль E 148
выращивание в высоком вакууме H 89
высокий вакуум H 81
высоковакуумная диагностическая лампа H 86
высоковакуумная отсечка H 85
высоковакуумная перегонная колонна H 83
высоковакуумная печь для спекания H 94
высоковакуумная установка H 90
высоковакуумная установка для пропитки воском H 98
высоковакуумное прецизионное литье H 91
высоковакуумное соединение H 84

высоковакуумный вентиль H 97
высоковакуумный вентиль с сильфонным уплотнением B 94
высоковакуумный кран H 95
высоковакуумный насос H 92
высоковакуумный патрубок H 90
высоковольтная сварка H 99
высокоглиноземистая керамика H 69
высокопроизводительный диффузионный насос H 79
высокопроизводительный насосный агрегат H 80
высокотемпературная вакуумная печь с холодными стенками и внутренним нагревателем C 216
высокочастотное распыление R 20
высокочастотный диэлектрический нагрев H 105
высокочастотный ионный источник H 73
высокочастотный масс-спектрометр H 74
высокочастотный нагрев I 44
высота давления P 161
выступ уплотнения S 49
высушенный сублимацией продукт F 192
высушивание D 72
вытекать F 124
вытесняющий поршень D 188
вытягивание кристалла C 409
вытягивающее поле D 238
выхлопная труба D 165
выхлопное окно D 166
выхлопной пар E 200
выхлопной фильтр D 162
выход ионов I 167
выходная диафрагма E 203
выходная лампа E 136
выход распыления S 314
вязкостное течение S 182
вязкостный манометр D 23
вязкостный манометр с кварцевой нитью Q 8
вязкость V 242

Г

газ в состоянии плазмы P 80
газ для продувки B 136
газ-носитель C 41
газобалластное устройство G 13
газобалластный клапан G 13
газобалластный насос G 11
газобалластный насос Геде G 1
газобалластный насос для создания и поддержания предварительного разрежения G 12
газовая кинетика K 50
газовая пайка T 124
газовая пленка G 33
газовая сварка F 71, G 37a
газовая струя G 43

газовая сушилка суспендированного вещества A 99
газовая фокусировка G 36
газовая холодильная машина C 189
газовая хроматография G 18
газовая электроника G 27
газовое содержание G 20
газовоздушная смесь A 105
газовый анализ G 7
газовый анализ методом экстракции из расплава T 112
газовый балласт G 10
газовый волюметр G 63
газовый ион G 28
газовый поток F 123
газовый разряд G 26
газовый разряд с холодным катодом C 194
газовый расходомер G 35
газовый состав G 19
газовый эжектор G 44
газовый эжекторный насос G 44
газовыпускной вентиль G 15
газодинамическое течение D 275
газонепроницаемый **G** 39
газообразный G 24
газоотдача O 68
газопламенная металлизация F 70
газопламенное напыление F 70
газоплотность G 62
газоплотный G 61
газопоглотитель G 73
газопоглотительный насос C 155, G 76
газопоглощение C 154
газопроницаемость G 51
газопузырьковый насос G 16
газосодержание G 20
галоид H 3
галоидная горелка H 8
галоидная лампа H 8
галоидный детектор H 4
галоидный диод H 5
галоидный диодный детектор H 6
галоидный индикатор H 4
галоидный течеискатель H 7
галоидный течеискатель с накаленным платиновым анодом H 133
гальванометр с подвижной катушкой M 184
гальванопокрытие E 113
гальваностегия E 113
гамма-лучи G 4
гарантированный вакуум G 133
гарниссажная плавка S 176
гашение Q 11
гелиевая рекуперационная установка H 62
гелиевая холодильная установка H 58
гелиевый течеискатель H 59
генераторный газ P 126
генерация газа G 38
генерация пучка H 77
генерирующий кварц C 410
генерирующий кварцевый кристалл C 410
германий G 72

герметизировать S 36
герметичное уплотнение H 67
герметично заделанный H 65
герметично запаянный электрод H 66
герметичный A 110
геттер G 73
геттеро-ионный насос G 74
геттеро-ионный насос с проволочным испарителем W 13
гибкое соединение F 99
гидравлический радиус H 145
гидравлический удар W 13
гидравлическое сопротивление W 16
гидридный процесс H 146
гидродинамическое сопротивление W 16
гидролитическая классификация H 150
гидроокись алюминия A 135
гидростатический напор P 161
гидрофобный слой D 252
глазурь G 98
глинозем A 133
ицерино свинцовый глет L 92
глубина проникновения D 70
глубина проникновения луча B 80
глубокое замораживание D 27
глубокое охлаждение D 24
глухое отверстие B 137
глухой фланец B 128
глушитель N 13
глушитель звука N 13
глушитель шума N 13
головка насоса H 20
головка сопла N 32
гомеополярный H 111
гомогенизация H 109
гомогенизировать H 110
гомогенность H 108
горелка B 227
горелка со скрещивающимися огнями C 378
горелка со смесительным соплом N 33
горение B 230
горизонтальная камерная печь H 117
горловина диффузионного насоса T 92
горловина диффузора D 127
горловина сопла N 35
горшковая печь P 116
горючий газ F 221
горячая вулканизация H 139
горячая прессовка H 134
горячая прессовка в вакууме V 80
горячая прокатка H 135
гофрированная мембрана C 350
гофрированная трубка C 351
гравитационная маслоспускная пробка G 117
гравитация G 116
градиент концентрации C 251
градуированная шкала D 53
градуированный G 104
градуировка C 13

градуировочная кривая C 14
градус Ренкина D 52
градус Цельсия D 48
грамм-атом G 110
грамм-молекула G 111
граница доменов B 140
граница зерна G 106
граница Ферми F 27
граничная пленка I 98
граничная поверхность I 96
граничная сетка E 134
граничный слой B 180
гранула G 112
грань кристалла C 403
графитовое покрытие G 113
графитовый тигель G 114
графитовый электрод G 115
гремучая смесь E 215
грунтовка P 186
грунтовочный лак B 57
грязеуловитель D 156
губчатая резина S 284
губчатый газопоглотитель P 107
губчатый титан T 118
гуттаперча G 142

Д

давление P 146
давление диссоциации D 192
давление излучения R 14
давление истечения Y 3
давление на выхлопе D 167
давление наполняющего газа F 48
давление насыщенных паров S 17
давление ниже атмосферного L 141
давление окружающей среды A 139
давление остаточных газов R 122
давление остаточных паров R 125
давление паров V 223
давление пучка B 82
давление рабочего пара в кипятильнике L 112
давление сжатия D 167
давление, созданное поддерживающим насосом H 103
давление струи B 82
датчик вакуумметра V 76
датчик манометра G 66
датчик течеискателя L 41
датчик уровня жидкости L 82
двойная конструкция T 204
двойная спираль со встречной намоткой D 212
двойниковый кристалл T 203
двойное уплотнение с круглой кольцевой прокладкой D 214
двойной спай D 216
двухвальцовая шприцмашина T 210
двухкамерная установка T 208
двухкамерный ионизационный манометр T 207
двухроторный насос R 170
двухседельное уплотнение B 20

двухходовой кран T 211
деаэрационная камера D 8
деаэрация D 7
деаэрирование D 7
деблокирование U 20
дегазация капель в процессе протекания S 291
дегазация при разливке T 5
дегазировать D 37
дегидратация D 54
дежурная установка S 324
действие направленного луча B 79
декантация D 10
декантировать D 9
декарбонировать вольфрамовый катод D 11
декоративное покрытие D 20
деление ядра N 36
демпфирующий диод D 3
демпфирующий капилляр D 2
держатель заготовки W 53
держатель ключа K 5
держатель объекта O 1
держатель подложки S 404
держатель тигля C 388
держатель электрода E 41
десорбционная спектрометрия D 75
десорбционный спектрометр D 74
десорбция D 73
десорбция со действием электронов E 90
детандер D 19
детектор S 79
дефект кристаллической решетки L 18
деформация F 169
диаграмма в полярных координатах P 98
диаграмма состояния P 37
диаграмма удара в полярных координатах S 123
диаграмма энергетических уровней E 139
диазотипная бумага O 85
диализировать D 91
диаметр окружности центров отверстий P 70
диаметр условного прохода N 15
диапазон давлений P 165
диапазон масс M 35
диапазон переработки W 56
диапазон регулировки S 95
диапазон температур затвердевания S 208
диапазон установки S 95
диафрагма D 92
диафрагменный сальник D 94
дилатометр S 375
дина D 282
динамическая вакуумная система D 281
динамическая вязкость A 12
динамическая проверка на течи D 278
динамический манометр D 23
динамический манометр Дэшмана D 271
динамический манометр с кварцевой нитью Q 8

динамический манометр с колеблющейся пластиной O 65
динамический манометр с крутильными колебаниями T 132
динамическое давление D 279
динамометрический ключ T 128
диодное распыление D 141
диодный магнитно-электроразрядный насос D 142
диод-пентод D 140
диод-триод D 143
диск вентиля V 190
дисковая пружина C 413
дисковый впай D 175
дисковый катод D 173
дислокация D 179
дислокация в кристалле C 407
диспенсерный катод D 180
диспергировать D 183
дисперсионный газопоглотитель D 181
дисперсия D 184
дистанционная индикация R 110
дистанционное управление R 107
дистиллированная вода D 195
дистиллятный насос D 76
дистиллятор D 194
дистилляционная установка D 196
дистилляционная установка со щетками W 45
дистилляционный аппарат D 194
дистилляция со свободным пробегом F 189
дистилляция с подвижной пленкой M 131
дифракция быстрых электронов H 72
дифракция быстрых электронов в отраженном пучке R 86
дифракция быстрых электронов в проходящем пучке T 157
дифракция в отраженных лучах R 84
дифракция медленных электронов D 123
дифракция рентгеновских лучей X 2
дифракция электронов в отраженном пучке R 85
дифракция электронов в проходящем пучке T 156
дифференциальная ионизация D 110
дифференциальная передача D 107
дифференциальное давление P 155
дифференциальное эффективное сечение соударений D 106
дифференциально откачиваемая канавка D 120
дифференциальный интерферометрический манометр D 109
дифференциальный манометр D 115
дифференциальный микроманометр D 116
дифференциальный сорбционный насос D 108

дифференциальный течеискатель D 113
дифференциальный течеискатель Пирани D 117
диффузионная активность D 129
диффузионная камера S 290
диффузионная мембрана D 133
диффузионная сварка C 217
диффузионная ступень D 139
диффузионное испарение D 132
диффузионное соединение D 138
диффузионное сопло D 134
диффузионный насос D 135
диффузионный эжекторный насос D 131
диффузия D 128
диффузия под давлением P 156
диффузор D 125
диффузор D 126
диэлектрическая постоянная D 102
диэлектрическая сушка D 103
диэлектрический нагрев D 104
длина волны рентгеновских лучей X 6
длина когерентности C 182
длина пробега электрона E 102
«длинное» стекло L 100
длиннофокусная система T 14
длительная работа C 299
длительное испытание E 137
добавка A 57
дозвуковой S 401
дозирование газа G 22
дозировочный вентиль D 203
дозировочный насос D 202
дозирующее устройство D 201, F 49
долговечная лампа L 101
долговечность нити накала F 42
долговременная стабильность L 102
доменный чугун P 54
донная разгрузка B 173
донная разливка B 175
донор D 200
дополнительный испаритель R 45
дополнительный кипятильник R 46
дополнительный слой A 58
допустимая ошибка A 125
допустимый предел примесей P 35
дорн M 17
доступ газа G 5
досушивание A 90
досушка A 90
дрейф D 79
дрейф нуля Z 5
дробеструйная обработка S 132
дробная перегонка F 176
дробовой шум S 133
дросселирование T 93
дроссельный вентиль B 233

дублированная конструкция T 204
дуга в вакууме V 9
дуговая печь A 184
дуговая сварка A 186
дуговая сварка вольфрамовым электродом в среде инертного газа T 187
дуговая сварка в среде инертного газа I 49
дуговая сварка в среде инертного газа с расходуемым электродом I 48
дуговой разряд A 183
дуктильность D 267
дуоплазматронный ионный источник D 270
дутье A 97
дыра O 62
дырка C 76, O 62

Е

единица давления P 178
емкостный измеритель толщины пленок C 19
емкость печи F 227
естественная конвекция N 1
естественное оттаивание O 6

Ж

«железная дорога» T 143
железный шлак I 170
желоб для транспортировки C 316
жесткость при изгибе F 102
жестчение E 29
жидкий азот L 85
жидкостное уплотнение L 88
жидкостнокольцевой насос F 136
жидкостный манометр L 81
жидкостным охлаждением /с L 78
жидкоструйный насос F 135
жидкость для [перекрытия] затвора S 267
жидкость с малой упругостью паров L 118

З

зависимость скорости откачки от давления S 268
зависящий от давления P 154
завихрение W 39
заглушить B 129
заглушка B 128
загораживать M 24
заготовка B 113
заготовка для выдавливания E 229
загрузка C 114
загрузочная решетка для сублимации C 117
загрузочное приспособление L 94
загрузочное распределительное устройство D 199
загрузочное устройство C 115
загрузочный клапан C 118

загрузочный лоток C 116
загрязнение I 32
задвижка G 64
заделка E 122
заделывать E 121, S 36
задержанная конденсация D 57
задерживающее поле R 153
задержка D 14
задержка кипения B 158
задержка показаний R 147
заедание S 58
зажигание I 9
зажигание дуги S 383
зажигание погружением D 147
зажимать S 147
зажим для шланга C 164, H 119
зажимное болтовое соединение S 428
зажимное винтовое соединение S 428
зажимное приспособление G 125
зажимное соединение P 59
зажимное фланцевое соединение S 427
заземление E 1
зазор в подшипнике B 85
закалка Q 11
закапывать D 245
заключатьв капсулу E 130
заключение в капсулу E 131
закон Авогадро A 223
закон Бойля-Мариотта B 186
закон Вилемана-Франца W 41
закон Гей-Люссака C 119
закон Дальтона D 1
закон действующих масс L 27
закон излучения Планка P 77
закон Кирхгофа K 9
закон Кнудсена K 16
закон Ламберта C 353, C 354
закон Пашена P 20
закон Пуазейля P 94
закон смещения Вина W 42
закон Стефана-Больцмана S 352
закон эффузии E 18
закрывать S 366
заливка под вакуумом V 59
заливочная масса P 118
замазка C 80
замазка на основе глицерина и свинцового глета L 33
замкомое кольцо R 152
замораживание F 199
замораживать F 190
замораживающее устройство F 201
замуровывание ионов I 130
запаечная машина B 217
запаздывание D 14
запайка G 45
запайка плоской ножки F 96
запас газа G 60
запасная часть S 253
запасной агрегат S 325
запаянный капилляр S 52
запирание I 108
запирать S 366
запирающее действие B 142

запись электронным лучом E 75
заполненная система F 114
запоминающая трубка S 372
запорная заслонка S 369
запорная часть вентиля V 200
запорный вентиль C 420, S 136, S 369
запорный клапан C 420, S 136
запорный слой B 55
запотевание S 452
зародыш кристалла C 406
засасывание I 80
засасывать P 208
заслонка G 64, S 137
заслонка вентиля V 193
заслонка для источника испарения E 183
заслонка клапана V 193
застойная зона S 317
затвердевание S 91
затвор L 98
затенение металлом M 94
затопление F 115
затравочный кристалл C 406
затухающий D 5
затягивающий момент T 98
захват дыркой H 104
захват электрона E 84
защита от излучения R 37
защита от имплозии I 24
защита от перегрузок O 76
защитная атмосфера P 202
защитная облицовка P 200
защитная пленка P 200
защитная способность охлаждаемой ловушки C 214
защитная сфера P 198
защитное металлическое покрытие P 205
защитное покрытие E 217
защитное приспособление P 201
защитное реле водяного потока P 206
защитное устройство P 201, S 1
защитный вакуум G 136
защитный лак P 204
защитный слой P 200
защитный электрод S 429
защищенный манометр T 166
защищенный от неосторожного обращения F 150
звездочка S 276
звуковой индикатор течи A 209
звуковой сигнал A 210
звукоизоляция N 11
зейгерование E 119
зеркальная поверхность F 217
зеркальное покрытие M 124
зеркальное рассеяние S 267
зеркальный S 266
зеркальный электронный микроскоп E 98
змеевиковый конденсатор C 184
змеевиковый охладитель C 184
золотниковый привод V 194
золотниковый насос S 178
зона плавления F 111
зона сварки W 34
зонд P 189

зонная очистка Z 17
зонная плавка Z 15
зонное выращивание Z 13
зонтичная ступень A 157
зонтичное сопло A 157
зубчато-реечная передача R 1
зубчатый насос G 69

И

игольчатый вентиль N 2
идеальный вакуум A 10
идеальный газ I 6
идеальный кристалл I 5
идеальный насос I 8
избирательный газопоглотитель S 59
избыточное атмосферное давление A 201
избыточное давление O 77
известковое стекло L 57a
извлекать E 222
извлечение металла M 89
изготовление корковых литейных форм S 119
изентропический I 172
изложница I 59
изложница с водяным охлаждением W 10
излучательный нагреватель R 7
излучательный обмен R 15
излучать R 9
излучающая газовая горелка R 6
излучение R 10
изменение давления P 150
изменение состояния C 107
измерение вакуума V 99
измерение дифференциального давления D 119
измеренная скорость откачки M 51
измеритель G 65
измерительная бюретка M 53
измерительная схема ионизационного манометра I 138
измерительный кристалл M 166
измерительный преобразователь M 55
измерительный прибор G 65
измерительный трансформатор M 56
измеритель объема газа G 63
измеритель параметров полупроводниковых диодов C 402
измеритель парциального давления P 14
измеритель проникающей способности излучения P 28
измеритель скорости испарения E 182
измеритель тока ионизационного манометра I 139
измеритель толщины пленки F 56
изнашивающаяся деталь W 22
изнашивающийся узел W 22
износ W 19
износоустойчивость R 138
износ электрода E 39

изобарический I 175
изолировать S 36
изолирующий состав
  I 85
изоляционная способ-
  ность I 86
изоляционный вентиль
  C 420
изоляционный материал
  I 83
изоплета I 177
изотерма адсорбции A 81
изотерма Брунауера-
  Эммета-Теллера B 109
изотерма набухания S 453
изотермический I 178
изотопное отношение
  I 180
изотопный индикатор
  T 142
изотопный состав I 180
иконоскоп I 4
имитатор космического
  пространства A 132
иммерсионная слюда M 2
иммерсионный датчик
  манометра D 150
иммерсионный манометр
  D 150
имплозия I 23
импрегнирование I 28,
  S 202
импрегнированный катод
  I 26
импрегнировать I 25
импульс I 29
импульсная десорбция
  F 86
импульсное напряжение
  I 31
инверсно-магнетронный
  манометр I 124
инверсно-магнетронный
  манометр с холодным
  катодом C 195
инверсно-магнетронный
  электроразрядный на-
  сос I 125
индиевое проволочное
  уплотнение I 41
индийская слюда M 2
индикатор вакуума V 84
индикатор уровня L 52
индикатор уровня жид-
  кости L 80, L 82
индукционная печь ва-
  куумной плавки V 88
индукционная плавильная
  печь I 43
индукционная плавка I 45
индукционная плазменная
  горелка I 42
индукционное перемеши-
  вание I 46
индукционный нагрев
  I 44
инертный газ I 47
инжекция газа G 40
интегральная электроника
  I 90
интегральный поток T 135
интенсивность охлажде-
  ния C 328
интерференционная лампа
  I 103
интерференционная плен-
  ка I 100
интерференционный
  фильтр I 102
интерферометрический
  манометр I 104
интерферометрический
  масляный манометр
  I 105
инфразвуковой S 401
инфракрасный нагрев
  I 55
инфракрасный течеиска-
  тель I 56

ионизационная камера
  I 144
ионизационная установка
  I 146
ионизационное простран-
  ство I 151
ионизационный вакуум-
  метр I 152
ионизационный детектор
  I 145
ионизационный манометр
  I 137, I 152
ионизационный манометр
  Байярда Альперта
  B 67
ионизационный манометр
  без оболочки B 46
ионизационный манометр
  для высоких давлений
  H 77
ионизационный манометр
  с автоэлектронным ка-
  тодом F 39
ионизационный манометр
  с накаленным катодом
  H 122, T 51
ионизационный манометр
  с палладиевой перего-
  родкой P 9
ионизационный манометр
  с экстрактором E 224
ионизационный манометр
  с электродом для по-
  давления фототока
  S 430
ионизационный манометр
  Хаустона H 143
ионизационный объем
  I 153
ионизация под действием
  электрического поля
  высокой напряженно-
  сти F 38
ионизация при соударении
  I 143
ионизация электронным
  ударом E 95
ионизировать I 154
ионит I 136
ионная бомбардировка
  I 129
ионная ловушка I 126
ионная проводимость
  I 141
ионная очистка D 158
ионная ракета I 163
ионная сорбция I 164
ионная эффективность
  I 135
ионно I 140
ионное травление E 153
ионной форме / в I 140
ионнолучевая техника
  I 128
ионнолучевое напыление
  I 127
ионно-распылительный
  насос C 197
ионно-резонансный спек-
  трометр I 162
ионно-сорбционная трубка
  I 156
ионно-сорбционный насос
  I 165
ионный захват I 131
ионный источник I 166
ионный источник с холод-
  ным катодом C 198
ионный коллектор I 133
ионный насос I 150
ионный насос с холодным
  катодом C 197
ионный ракетный двига-
  тель I 163
ионный резонанс I 161
ионный ток, обусловлен-
  ный остаточными газа-
  ми G 42

ионный удар I 142
ионообменник I 136
иридиевая нить накала
  I 168
ирисовая диафрагма I 169
искровая эрозия S 258
искровой течеискатель
  S 255, H 75
искусственное охлажде-
  ние R 99
искусственный цеолит
  A 193
испарение E 165
испарение вверх U 24
испарение в дуге с уголь-
  ными электродами C 32
испарение вниз D 222
испарение во взвешенном
  состоянии L 55
испарение в поле F 37
испарение вспышкой
  F 84
испарение выступов угле-
  рода C 34
испарение мономоле-
  кулярного слоя M 174
испарение поверхности
  S 437
испарение рабочей жид-
  кости в систему B 18
испаритель B 155, E 190
испаритель мощности
  L 93
испаритель, нагреваемый
  электронной бомбарди-
  ровкой E 80
испарительная камера
  E 168
испарительная лодочка
  E 191
испарительная сушка
  E 189
испарительное газопогло-
  щение D 182
испарительное охлажде-
  ние E 188
испарительный анализ
  E 166
испарительный ионный
  источник E 163
испарительный ионный
  насос E 177
испарительный нагрев
  F 87
испарительный насос
  E 192
испарительный синтез
  E 185
испаритель с длинной вер-
  тикальной трубкой
  L 103
испаритель со свободно
  падающей пленкой
  F 186
испаритель с падающей
  пленкой F 8
испаритель титана T 114a
испаряемое вещество
  E 162
испаряемый газопоглоти-
  тель V 246
испарять E 164
использованное тепло
  W 5
испытание на герметич-
  ность L 45
испытание на истирание
  E 150
испытание на твердость
  H 17
испытание на удар N 28
испытание на удар при
  изгибе I 17
испытание на усталость
  E 137
испытание на шумы N 14
испытание под давлением
  P 175

испытательный колпак
  M 54
исследование космиче-
  ского пространства S 250
истечение E 120
истинная скорость откачки
  I 118
истинная температура
  T 178
истирание A 2, W 19
источник вакуумного
  ультрафиолета V 178
источник газа G 59
источник для создания
  молекулярного пучка
  M 146
источник для электрон-
  ной бомбардировки
  с кольцевым катодом
  R 160
источник ионов, созда-
  ваемых электронной
  бомбардировкой E 83
источник испаряющегося
  вещества E 193
источник пара V 226
исходная точка R 74

К

кабельный бокс E 133
кавитация S 75
кажущаяся течь V 241
каландрировать C 5
калибратор C 16/7
калиброванная течь C 11
калиброванное натекание
  C 12
калибровать C 8
калибровка C 13
калибровочная кривая
  C 14
калибровочный газ C 15
кальцинировать C 6
камера для имитации усло-
  вий космического прос-
  транства O 67
камера для напыления
  D 65
камера для сублимацион-
  ной сушки F 194
камера, моделирующая
  космическое простран-
  ство S 244
камера, моделирующая
  пространство S 244
камера расширения E 208
камера сжатия C 239
камерная печь C 104
камерная поршневая
  задвижка A 160
канавка в форме ласточ-
  киного хвоста D 221
канавка для уплотнения
  G 46
каналовые лучи C 18
канифоль C 226
капельная масленка D 240
капельная плавка D 241
капельно-струйное обезга-
  живание S 291
капельный рост D 246
капилляр C 23
капиллярная депрессия
  C 26
капиллярная конденсация
  C 24
капиллярная трещина
  C 25
капиллярная трубка C 23
капиллярное притяжение
  C 22
капиллярность C 22
капиллярный метод C 9
капсульный насос D 184
капсульный пружинный
  вакуумметр C 30

капсуляция E 131
карбонат бария B 48
карбонизированный вольфрамовый подогреватель C 35
карбюрация C 37
карданный подвес G 78
карусель C 40
карусельный откачной автомат R 175
карусель с испарителями E 184
касательное падение G 118
каскадная сушилка C 46
каскадная установка C 48
каскадное соединение C 45
каскадный конденсатор C 44
каскадный охладитель C 44
каскадный электрон C 47
кассета для подложек S 405
катализатор C 58
катафорез C 59
катетометр C 60
L-катод L 29
катод из рафинированного золота R 80
катодная активировка C 61
катодная очистка C 72
катодное падение C 64
катодное распыление C 71
катодное свечение C 66
катодное тлеющее свечение N 3
катодное травление C 73
катодолюминесценция C 74
катушка Гельмгольца H 64
катушка Тесла S 254
качать P 208
качающийся конвейер V 239
качающийся транспортер C 316
качество поверхности F 65
квадратная резьба S 315
квадрупольное поле Q 1
квадрупольный ионизационный манометр Q 3
квадрупольный масс-спектрометр Q 2
квадрупольный масс-спектрометр Паули P 21
квантовомеханический Q 5
квантовый выход Q 6
кварцевая лампа Q 9
кварцевая порода Q 44
кварцевая футеровка S 143
кварцевое стекло F 232
кварцевый генератор R 144
кенотрон H 126
кенотронный выпрямитель H 93
керамика C 101
керамическая шайба C 98
керамический диск C 98
керамический изоляционный материал C 99
керамический спай C 102
керамический тигель C 97
керосиновый течеискатель K 1
кетоновые смолы K 3
кинематическая вязкость K 7
кинетическая теория газов K 8

кипение B 157
кипение рывками B 224
кипятильник B 155
кипящая сталь R 157
кипящий при высокой температуре H 70
кипящий при низкой температуре L 109
кислородный течеискатель O 84
кислотостойкий A 37
кислотоупорное покрытие R 140
кислотоустойчивый A 37
кистевой разряд B 225
клапан V 186
клапан для напуска балластного газа G 13
клапанное распределение V 194
клапанный вентиль F 77
клапанный поршень V 199
клапанный привод V 194
клапанный упор V 195
клапан одноразового действия B 196
клапан с пневматическим приводом A 109
клапан управления P 57
клей A 63
клейкое вещество A 63
клетка Фарадея F 11
клещи для горячей проволоки H 140
клиновидный ремень V 231
ключ Инбуса I 33
Кнудсеновское течение K 12
коаксиальный ввод C 179
кованый фланец F 167
ковкая сталь L 108
ковкий сплав F 166
ковкий чугун M 16
ковкость M 15
ковшовая сушилка L 6
ковшовое обезгаживание L 5
ковш печи F 228
ковш с водяным охлаждением W 11
когезия B 183
кожаное уплотнение L 48
кокиль C 53
колебание решетки L 24
колебания напряжения сети L 57
колебания толщины среза T 67
колено L 57
колено манометра M 20
колено трубы B 102
количество газа Q 4
коллектор-сборник H 21
коллодий C 224
коллоидный C 225
колонка для обезгаживания D 42
колонка с дисковыми пружинами D 176
колонна для обезгаживания D 46
колонна для тонкопленочного обезгаживания T 70
колонная конструкция C 228
колонная структура C 228
колпак для обезгаживания D 40
колпачковая манжета C 21
колпачок вентиля V 189
колпачок сопла J 5
кольцевая горелка A 155
кольцевая диафрагма C 139

кольцевая прокладка R 161
кольцевое напряжение H 116
кольцевое сопло A 159
кольцевое уплотнение R 161
кольцевой газопоглотитель R 162
кольцевой зажим C 150
кольцевой катод R 159
кольцевой спай A 161
кольцо круглого сечения O 64
кольцо Рашига R 28
комбинация вентилей V 186
комбинированная выпрямительная лампа M 202
коммуникация для выравнивания давления P 168
компенсационное соединение E 209
компенсационный метод B 34
компенсирующий элемент C 231
компонентная камера C 233
компонент форвакуумной системы F 163
компрессионный вакуумметр Мак-Леода M 47
компрессионный капилляр C 246
компрессионный манометр C 241
компрессия C 238
компрессор C 250
компрессорная холодильная машина C 248
компрессорная холодильная установка C 244
компрессор с вращающимся ротором M 172
компрессор с поперечным потоком C 379
конвейер S 387
конвейерная сушилка на ситах S 29
конвекция C 314
конденсатор C 260
конденсатор поверхностного типа S 433
конденсатор смешения с охлаждением впрыском I 67
конденсатор со льдом I 1
конденсационная вода C 259
конденсационный насос C 257
конденсация C 256
конденсируемый газ C 255
конденсирующийся газ C 255
конечная плотность U 1
конечная сушка A 90
коническое уплотнение C 263
контакт в защитном газе P 203
контактная паста C 149
контактная стыковая сварка F 80
контактная сушилка C 286
контактная сушка C 287
контактное газопоглощение C 290
контактное замораживание C 269
контактный морозильный аппарат C 288
контактный потенциал C 291
контактный способ C 292

контактный термометр E 34
контейнер C 293
континуум C 306
контроль испарения E 172
контрольная плитка E 132
контрольное окно для наблюдения за уровнем масла I 79
контрольное стекло I 78
контрольный клапан C 121
контрфланец C 357
конусное затенение C 265
конусное экранирование C 264
концевой выключатель L 59
концентратор C 254
концентрация ионов C 252
концентрирование вымораживанием F 191
корзиночная катушка B 62
коробка для перчаток D 250
коробка планетарной передачи P 78
короткая печь B 185
«короткое» стекло S 129
корпус вентиля B 153
корпускула C 346
корпускулярный пучок C 347
корпус насоса P 210
корпус печи F 229
косвенным накалом / c I 40
косвенным подогревом / c I 40
косинус направления D 152
косинусоидальный закон распределения C 353
косинус угла направления D 152
космический корабль S 251
котел B 154
коэффициент аккомодации A 32
коэффициент аккомодации при абсорбции A 21
коэффициент аккомодации при конденсации A 33
коэффициент внутреннего трения I 121
коэффициент диффузии D 130
коэффициент диэлектрических потерь D 191
коэффициент захвата C 31
коэффициент износа W 21
коэффициент испарения E 171
коэффициент истечения D 159
коэффициент компрессии C 240
коэффициент мощности C 181
коэффициент отражения R 83
коэффициент пересчета C 315
коэффициент полезного действия E 16
коэффициент полезного действия цикла Карно C 39
коэффициент полезного действия цикла Ренкина R 23

коэффициент потерь D 191, L 106
коэффициент предела выносливости (усталости) F 17
коэффициент предела текучести E 25
коэффициент прилипания S 359
коэффициент проницаемости P 33
коэффициент проницаемости анода P 27
коэффициент пропускания T 155
коэффициент Пуассона P 97
коэффициент рассеяния S 300
коэффициент расхода D 159
коэффициент расширения C 180, E 210
коэффициент реэмиссии R 71
коэффициент самоприлипания S 74
коэффициент сепарации S 83
коэффициент сцепления S 359
коэффициент температуропроводности T 33
коэффициент теплового расширения T 37
коэффициент теплоотдачи F 51
коэффициент теплопередачи H 50, O 72, F 51
коэффициент теплопроводности H 26
коэффициент термической транспирации T 47
коэффициент упругости E 23
коэффициент, учитывающий род газа G 31
коэффициент Xo H 100
краевая дислокация E 6
краевое условие B 178
краевой поток F 215
краевой угол C 283
кран P 90
кран для продувки P 224
кран с ртутным уплотнителем M 81
красноломкость H 136
кремнекаучук S 140
кремнийорганическое масло S 145
кремнистая сталь S 146
крестовина C 377, S 276
крестовина с переходным фланцем R 67
крестообразная ножка C 140
кривая выхода Y 1
кривая изменения давления во времени P 176
кривая откачки P 211
криогенная откачка C 394
криогенная установка C 392
криогенный насос C 393
криогеттерный насос C 396
криозахват C 395
криопанель C 397
криосорбционный насос C 396, C 398
криосублимационная ловушка C 400
кристаллизация S 91
кристаллическая друза C 170
кристаллическая пластина C 411

кристалл кварцевого генератора C 410
критическая длина волны C 374
критическая температура C 373, M 45
критическая точка C 371
критическое давление C 372
кричная сталь B 146
кромкогибочный станок B 72
кроссовер C 381
круговой конвейер C 40
круксово темное пространство C 375
крутящий момент T 127
крышка вентиля V 189
ксенон X 1
кубически-гранецентрированный F 3
кулоновская сила отталкивания C 355

Л

лабиринтное уплотнение L 1
лабораторное сито T 21
лазерное зеркало L 16
лак для защиты при обработке M 28
лакировать L 2
лаковая основа B 56
лаковое покрытие L 3
ламинарное течение L 8
лампа двойного света B 114
лампа накаливания I 34
лампа, наполненная инертным газом R 27
лампа с двойной нитью B 114
лампа с односторонней проводимостью S 159
лампа тлеющего разряда G 101
легирующая добавка A 127
ледяное ядро I 3
лезвенный спай E 8
лезвие C 422
ленточная конвейерная сушилка B 101
ленточная сталь S 351
ленточный конвейер B 100
ленточный нагрев S 386
ленточный самописец C 298
ленточный электронный поток E 109
летка T 6
летучесть F 222
ликвация E 119
ликвидус L 90
линейный контакт L 64
линейный коэффициент ионизации D 111
линейный ускоритель L 60
линейчатый спектр L 66
линия скольжения S 184
линия фантома G 77
лиофилизация C 391
листовая слюда S 113
листовая сушилка S 112
листовой слиток B 144
литейная печь с электроннолучевым нагревом E 57
литейная форма M 181
литейный ковш C 53
литье C 51
литьевая смола C 54
литье в вакууме S 176, S 410

литье в оболочковые формы S 13
литье в почвенные формы B 177
литье под давлением D 101, P 149
литье под давлением в вакууме V 44
лобовое сопротивление D 226
ловушка T 164
ловушка для защиты от выплесков масла O 36
ловушка для конденсата на форвакуумной стороне B 9
ловушка Майснера M 61
ловушка с активированным углем A 48, C 110
ловушка с астероидальным профилем жалюзи A 199
ловушка с высокой пропускной способностью, охлаждаемая жидким азотом H 71
ловушка с коленом E 28
ловушка с медной фольгой C 339
ловушка с множественными кольцами M 198
ловушка с термоэлектрическим охлаждением T 60
логарифмическая шкала давления L 99
ложная течь V 241
ломкий лак B 206
ломкость B 207
лоток для капель D 242
лотошная сушилка S 33
луночный тигель B 234
лучевод B 78
лучевой B 76
лучеиспускание R 10
лучеиспускательный нагреватель R 7
лучистый нагрев R 8
лучистый радиатор R 7
люминесцентный манометр E 44
люминесцентный экран F 138

М

магистраль низкого вакуума R 197
магнетрон M 9
магнетрон непрерывного режима C 305
магнетронный вакуумметр C 199
магнетронный ионизационный манометр с накаленным катодом H 123
магнетронный ионизационный манометр с накаленным катодом с подавлением фототока H 124
магнетронный электроразрядный насос M 10
магнетрон типа бегущей волны T 167
магнитная индукция M 6
магнитная мешалка с нагретой пластиной H 132
магнитная пленка для запоминания M 5
магнитный электроразрядный манометр с холодным катодом C 196
магнитный электроразрядный насос C 197

магнитооптический эффект Керра K 2
максимальная производительность C 109
максимально-допустимое выпускное давление F 158
максимальное допустимое впускное давление паров воды M 42
максимальное реле C 419
малый фланец Q 15
малый фланец подготовленный к сварке W 29
малый щуп галоидного течеискателя H 9
маммут-насос G 16
манжетное уплотнение L 71
манометр M 19
манометр Байярда-Альперта B 66
манометр Бурдона B 183
манометр для измерения полного давления T 137
манометрический эквивалент M 137
манометр Кнудсена K 13
манометр Пеннинга C 196
манометр Пирани P 66
манометр с биметаллической полоской B 116
манометр с кварцевой нитью Q 8
манометр сопротивления P 66
манометр с охлаждаемой ловушкой T 166
манометр с перемещающимся чувствительным элементом D 187
манометр с трубчатой пружиной S 285
манодетандер P 167
мартеновская печь O 46
маска для испарения E 179
маска для нанесения микросхемы M 112
маска из светочувствительного лака L 4
маска из фотолака L 4
маскирование M 27
маскировать M 24
масковая карусель M 30
масковое накопление M 29
масло для вращательного насоса R 184
масло для высоковакуумного насоса O 50, V 225
масло для диффузионного насоса D 136
маслоизмерительный стержень O 18
маслоналивной вентиль O 12
маслоналивной клапан O 12
маслоотделитель O 34, O 36
маслоочиститель O 14
маслосборник B 11
маслосливная пробка O 20
маслосливной винт O 20
маслоспускной клапан O 21
маслоспускной вентиль O 21
маслоуловитель O 34
маслоуловительный экран O 9
маслощуп O 18
масляная ловушка O 37
масляное уплотнение O 33

масляный вращательный насос R 180
масляный диффузионный насос O 17
масляный рециркуляционный фильтр O 30
масляный фильтр O 25
масляный эжекторный насос O 22
масса для набивки T 4
масса для прессования оксидной керамики P 82
массовая производительность M 32
массовая сепарация M 37
массовый пик M 34
массовый расход M 31
масс-спектрограф с двойной фокусировкой D 211
масс-спектрометр M 38
масс-спектрометр для анализа выхлопных газов R 145
масс-спектрометрический течеискатель L 47, M 39
масс-спектрометр с ионным микрозондом I 155
масс-спектрометр с квадрупольным анализатором Q 2
мастика для цоколевки T 179
материал для замазки течей L 46
матирование R 196
матированный изнутри I 75
матировать F 218
матовое серебро D 268
матричный катод M 41
машина для выдувания стеклянных колб G 83
машина для штамповки ножек S 353
машина для электроннолучевого сверления E 62
машинное охлаждение M 59
маятник с двойной кварцевой нитью D 215
мгновенное испарение F 88
медленное замораживание S 194
медный кристаллизатор C 337
медный тигель C 336
медь GFPH G 37
медь OFHC O 83
медь особо чистая с низким содержанием газов G 37
Международный вакуумный научно-технический союз I 115
межкристаллическая трещина I 95
межкристаллический излом I 95
межмолекулярный I 112
межплоскостное расстояние решетки I 116
мембрана D 92
мембранная коробка D 93
мембранная течь M 73
мембранный вакуумметр D 98
мембранный вентиль D 99
мембранный манометр D 95
мембранный насос D 97
мембранный регулятор давления D 96
менделеевское число A 205

«мертвое» время D 225
мертвое пространство D 6
место дефектов D 30
место пайки S 224
место припайки S 47
место спая G 45, S 47
металл в ванне B 65
металлизация M 85
металлизация керамики S 169
металлизация распылением M 95
металлическая пленка, нанесенная испарением V 216
металлическая прокладка M 90
металлическая сетка для экранировки M 92
металлический мембранный фильтр M 87
металлический наконечник M 96
металлический сильфон M 82
металлический уплотнитель M 90
металлический шланг F 100
металлическое мембранное уплотнение M 88
металлическое сильфонное уплотнение M 83
металлокерамика M 84
металлокерамическая композиция C 234
металлокерамический ввод C 100
металлокерамическое соединение C 103
металлургические способы производства F 1
металлы вакуумной плавки V 101
метастабильное состояние M 100
метастабильный M 99
метастабильный уровень M 100
метод T 12
метод «бюретки» V 250
метод всасывающего щупа S 11
метод вспышки F 85
метод двойного колпака D 208
метод двух манометров C 10
метод диафрагмы O 63
метод измерения быстроты возрастания давления I 176
метод изоляции I 176
метод мыльных пузырей S 203, S 204
метод накопления A 34
метод непрерывной разливки C 296
метод обнаружения течи искровым течеискателем S 256
метод обнаружения течи по свечению газового разряда G 21
метод обогащения B 215
метод обратного отражения Лауэ L 25
метод определения быстроты действия насоса по известной пропускной способности C 10
метод определения скорости откачки по пропускной способности T 209
метод опрессовки с индикацией мыльными пузырями S 203

метод, основанный на разнице в газопроницаемости G 52
метод откачки сосуда под вакуумным колпаком D 207
метод падения давления P 159
метод «пипетки» P 65
метод поиска течей опрессовкой O 79
метод постоянного давления C 276
метод постоянного объема C 277
метод проверки под избыточным давлением O 78
метод проверки помещением объекта в оболочку H 114
метод промывки газом W 3
метод течеискания по разности давлений D 112
метод течеискания с помощью разрядной трубки D 170
метод течеискания с присоединением течеискателя к форвакуумной магистрали B 17
метод трескающегося покрытия S 379
механизм T 111
механизм блокировки обратного хода R 88
механизм для опрокидывания тигля D 82
механизм для смены масок-подложек S 406
механическая ловушка B 23
механическая ловушка, в которой исключена миграция N 10
механический насос M 58
механический насос с жидкостным уплотнением L 89
меченый атом R 18, T 142
мешалка A 94
меш проволочной сетки W 46
миграция M 116
миканит M 104
микровесы M 109
микроиспаритель S 158
микрокапилляр M 110
микроминиатюрная схема M 115
микроопределение M 114
микросхема M 111
микроэлектроника I 90
миниатюрный датчик манометра M 120
минимально обнаруживаемое изменение давления M 122
минимально обнаруживаемое натекание M 121
мнемоническая схема M 119
многозаходная резьба M 201
многокамерная электронно-лучевая печь E 72
многократное рассеяние M 199
многоленточная конвейерная сушилка M 197
многолопастная крыльчатка M 187
многолучевая интерферометрия M 196
многониточная резьба M 201

многопластинчатый насос M 200
многослойное зеркало M 193
многослойное покрытие M 192
многослойный тонкопленочный конденсатор M 191
многоструйная горелка M 189
многоступенный насос M 207
многоступенчатое обезгаживание M 206
многоступенчатый C 235
многоступенчатый механический насос C 237
многоступенчатый насос M 207
многоходовая резьба M 201
многоцелевая печь M 205
моделирование космического пространства S 247
модель абсолютно упругой сферы H 19
модель твердой сферы H 19
модуль сдвига M 136
модуль упругости при растяжении M 135
модуль Юнга M 135
модулятор M 134
мокровоздушный насос R 54
молектроника M 154
молекулярная адгезия M 142
молекулярная дистилляция M 150
молекулярная перегонка O 47
молекулярная плотность N 45
молекулярная проводимость M 149
молекулярная сила M 156
молекулярная течь M 157
молекулярная электроника M 154
молекулярная эффузия M 153
молекулярное притяжение M 144
молекулярное сито M 159
молекулярное соударение M 147
молекулярное течение M 155
молекулярный воздушный насос M 143
молекулярный дистиллятор с падающей пленкой F 9
молекулярный манометр M 151
молекулярный насос M 152
молекулярный объем M 141
молекулярный пучок M 145
молекулярный сборник M 160
молибденовая лодочка M 163
молибденовое стекло G 90
молибдено-марганцевый процесс M 165
моль G 111
мольвакуумметр Геде G 3
мольвакуумметр Ленгмюра-Дэшмана L 11
мольманометр M 151

мольная доля M 140
мольная проводимость M 149
молярная долевая концентрация M 140
молярный объем M 141
моноатомная пленка M 168
моноатомный слой M 168
моноблочный насос M 169
монокристаллическая проволока M 171
монокристаллическое распыление S 156
мономолекулярная адсорбция M 177
мономолекулярное покрытие M 173
мономолекулярный M 176
монополярный измеритель парциальных давлений M 178
монополярный спектрометр M 179
монослой F 144
монослойное покрытие M 173
моноспираль S 153
монохроматический коэффициент излучения S 264
моноячейка M 170
морозильный аппарат для замораживания орошением S 295
морозильный аппарат с замораживанием на стеллажах S 117
морозильный туннель F 207
мостовая схема B 201
мост Уитстона W 38
мощность луча B 81
мощность накала H 31
мощность подогревателя H 31
мощность при непрерывной работе C 300
мраморная замазка M 23
мраморный цемент M 23
мультивибратор F 106
мутная сердцевина в блоке льда I 3
муфельная печь M 186
муфта S 177
муфта горячей посадки S 135
муфта для защиты от разбрызгивания S 297
муфта с резьбой T 85
муфтовое соединение M 185
мягкая сталь L 108
мягкое стекло S 212
мягкость S 213
мятый пар E 200

Н

набивка сальника G 79
набивная колонна P 1
набивная масса T 4
набивочная манжета S 41
набивочный материал S 37
наблюдательное окно I 78
наводить «мороз» F 218
нагнетатель A 226
нагнетать S 415
нагревательная рубашка B 30
нагреватель подложки S 403
нагрев вихревыми токами E 5

нагрев инфракрасными лучами I 55
нагрев паром S 343
нагрев сопротивления R 134
нагрев циркуляцией масла O 13
нагрев электронной бомбардировкой E 94
наддув S 416
надежный F 7
нажимное уплотнение C 245
наиболее вероятная скорость M 180
наибольшая производительность по воде M 43
наибольшее выпускное давление C 370
наибольшее форвакуумное давление C 370
наименьшая обнаруживаемая скорость натекания S 197
накаленный катод H 121
накидная гайка C 29
накип F 144
наклеп H 13
наклонное напыление O 3
наклонный удар O 4
наконечник T 2
наконечник для шланга H 120
накопительное кольцо S 370
накопительный баллон S 371
накопительный бачок S 371
накопление тепла H 48
нанесение в вакууме E 170
нанесение газопоглотителя маканием D 146
нанесение покрытия на ленту B 42
нанесение слоев под вакуумом V 95
нанесение смазки L 123
нанесенный испарением V 215
наплавка твердым слоем H 14
наполнитель F 44
направление распространения D 154
направлении течения / в D 223
направленное наращивание O 61
направленный B 76
направляющая опора G 137
направляющая шпонка F 19
направляющий стержень G 140
направляющий штифт G 139
направляющий шток вентиля V 204
направляющий электрод G 138
напряжение возникновения разряда I 11
напряжение зажигания I 11
напряжением / под U 15
напряжение смещения A 218
напряжение смещения на электроде E 38
напряжение формовки F 171
напряженное состояние S 378
напряженность магнитного поля M 5

напряженный спай стекла с металлом C 247
напуск балластного газа A 96
напускной вентиль A 95
напыление в вакуумной дуге V 10
напыление сквозь отверстие T 94
напыленная маска D 66
напыленный катод E 167
напыленный слой O 176
нарезной редукторный переход T 86
наружная металлизация O 66
наружная резьба E 221
наружное паянное цилиндрическое соединение O 70
наружный диаметр E 219
наружный конденсатор E 218
наружный слой O 73
наружный трубчатый спай O 71
нарушение в кристалле C 405
нарушение контакта C 285
нарушение кристаллической решетки D 179, L 21
насадка P 1
насос P 209
насос, в котором используется поток (течение) вещества F 130
насос двойного действия D 205
насос для получения грубого вакуума R 198
насосная система P 218
насосное соединение P 216
насосный агрегат G 130
насос Пеннинга C 197
насос предварительного разрежения B 14
насос Рутса R 170
насос с вращающимся поршнем R 182
насос с частичным фракционированием S 72
насос Теплера T 121
насос Шпренгеля S 301
настыль в ковше S 174
насыщать S 15
насыщенный пар S 16
натекание I 68
натриевое жидкое стекло S 209
натриевое стекло S 208
натяжение S 378
наука о вакууме V 142
начальная точка отсчета R 74
начальная чувствительность I 66
начальное давление I 65, S 326
негативное вакуумное формование S 374
независимость от напряжения I 39
неидеальный газ I 20
нейтральный фильтр G 121
неконденсируемый газ N 18
нелетучий N 25
немагнитный ионизационный манометр N 21
ненасыщенный пар U 21
необработанный металл C 390
неоднородный I 64
неоспаряемый газопоглотитель N 26
неохлаждаемая ловушка N 23

неподвижный газ G 8
непрерывное прессование E 228
непрерывность пленки F 53
непрерывность потока C 297
непроводящее направление I 122
непроводящее покрытие R 140
непрозрачность O 42
непрозрачный O 43
непрозрачный кварц O 44
неразъемное соединение F 68
неразрушающий N 20
нераспыляемый газопоглотитель B 222, N 26
нерасходуемый электрод N 19
нержавеющая сталь S 319
нержавеющий C 348
несамостоятельный разряд N 24
несовершенный кристалл I 19
несовершенство кристалла C 405
несовершенство кристаллической решетки L 21
несущий провод C 150
не требующий квалифицированного обслуживания F 150
не требующий обслуживания M 13
нетто-плотность N 5
неуспокоенная сталь R 157
неустановившееся течение U 23
нижний предел измерения давления L 113
низкий вакуум L 116, R 200
низковольтная сварка L 119
низкотемпературная установка L 115
низкотемпературное замораживание D 27
низкотемпературное охлаждение D 24
низкоуглеродистая сталь L 108
ниппель N 8
нитроцеллюлозный лак C 79
нить накала F 40
ножка-цоколь с остеклованными выводами M 203
ножничное уплотнение S 106
номинальная мощность C 300
номинальная мощность накала R 32
номинальная производительность N 17
номинальная скорость откачки R 33
номинальное давление N 16
нормальная комнатная температура S 323
нормальное выпускное давление срыва D 276
нормальные условия температуры и давления S 321
нормальный тлеющий разряд N 27
носитель заряда C 111
нуклон N 39
нулевой потенциал Z 8
нуль-индикатор N 44

О

обдувание А 89
обдувать S 289
обеднение ионами D 21
обезвоживание D 54
обезвоживатель D 239
обезгаживание D 39, Т 5
обезгаживание во время откачки с помощью газовой горелки Т 125
обезгаживание в разливочной струе S 376
обезгаживание в расплавленной струе S 291
обезгаживание путем бомбардировки D 41
обезгаживание стали S 349
обезжиривать D 47
обезуглероживать D 12
обжигать В 229
обкладка L 69
область давлений Р 165
область несмесимости М 125
область полусвета Р 30
область полутени Р 30
облицовка L 69
обмен лучеиспусканием R 15
обнаружение течей L 38
обновление R 111
обогащение руды О 59
обрабатываемость W 51
обработка давлением S 103
обработка импрегнирующим составом Т 168
обработка металла М 98
образец S 8
U-образный манометр М 80
U-образный ртутный манометр U 29
образование водяного пара S 342
образование инея F 219
образование конденсата S 452
образование корки I 37
образование льда I 2
образование налета I 37
образование пузырей Р 207
образование пузырьков В 139
образование раковин Р 73
образование следа от утечки S 440
образование струи в сопле J 10
образование центров N 38
образование ядер N 38
образовывать эмульсию Е 129
обратная диффузия В 2
обратная струя I 123
обратное зажигание В 7
обратное излучение R 113
обратное направление I 122
обратное проникновение масла О 8
обратное течение R 87
обратное течение масла В 4
обратный клапан В 19
обратный поток газа G 9
обратный поток масла В 4
обратный поток паров рабочей жидкости В 21
обращенное сопло Лаваля А 157
обтекатель сопла S 173
обходить В 238
общее натекание I 89

объем грамм-молекулы газа V 247
объемная диффузия В 221
объемная плотность В 220
объемная проводимость В 219
объемная скорость течения S 270
объемное поглощение В 218
объемно-центрированный В 152
объемный заряд S 245
объемный компрессор Р 111
объемный коэффициент ионизации S 262
объемный коэффициент полезного действия V 251
объемный насос D 189
объемный расход S 270
объем рабочей камеры S 454
объем установки Р 79
ограниченный пространственным зарядом S 246
ограничивающая сетка С 266
ограничивающее сопротивление L 58
ограничительное кольцо В 86
ограничительное уплотнительное кольцо R 151
ограничитель хода клапана V 195
одновальцовая сушилка S 157
одноввводная ножка S 165
одноввводный проходник S 165
одножильный S 154
однозерновой испаритель S 158
однократного действия S 150/1
однократно действующий S 150/1
однократное рассеяние S 161
однородность Н 108
однородность по плотности U 17
однорычажный блок клапанов О 41
одношнековый инжектор S 162
одноячеистый анод S 152
ожижение L 74
окись алюминия А 133
окись бария В 49
окклюзия О 5
окно из фольги F 148
окно, находящееся под давлением Р 180
оконечное покрытие сетки G 122
окончательная сушка А 90
окончательный размер F 66
окружающая температура А 140
окружная скорость С 146
окружность центров отверстий Р 69
оксидный катод О 81
оливка Н 120
олифа V 229
омегатрон О 39
омегатронный манометр О 40
омеднение погружением С 340
омеднять С 334
описываемый объем S 454

опорная плита В 59
опорное кольцо В 86
опорное напряжение R 78
опорный кольцевой уплотнитель R 151
опорный потенциал R 78
оправка М 17
опрокидывающаяся вакуумная камера Т 100
опрокидывающаяся печь Т 104
опрокидывающий момент Р 71
опрокидывающийся тигель Т 99
оптически активное тело О 53
оптический измеритель толщины пленки О 55
оптический монохроматический пирометр D 157
оптический пирометр О 54
оптически непрозрачная ловушка О 52
оптически непросматриваемая ловушка О 52
опускаемое дно L 110
опытная установка Р 56
орбита О 57
орбитронный ионизационный манометр О 58
орбитронный ионный насос О 56
ориентированное наращивание О 61
оросительная сушилка F 105
осадок R 126
осаждаться S 97
осаждение под углом О 3
осветляющее покрытие L 50
осевой вентилятор А 225
осевой компрессор А 226
осевой насос А 227
осевой поршневой насос А 229
осколок С 131
ослабление звука N 11
ослабление отражения А 175
основание изложницы I 61
основная величина F 224
основная гармоническая F 225
основная постоянная F 224
основная сушка М 12
основная частота F 225
основной конденсатор М 11
основной слой Р 186
основной метод обработки U 19
остаток R 126
остаточная активность R 116
остаточная влажность R 123
остаточная проводимость R 117
остаточная пропускная способность вентиля R 124
остаточное давление В 130
остаточный газ R 119
остаточный ток R 118
остеклованный V 245
островная пленка I 173
островная структура I 174
осушитель D 55
отбор проб S 10
отбор промежуточных проб I 111

отбортовывать В 71
отвердитель смолы R 130
отверждение С 414
отверстие О 62
отверстие горелки В 228
отверстие для откачки воздуха А 113
отвод от измерительного контрольного сопротивления М 167
отводящий канал D 224
отдавливать Р 61
отделительный диск С 421
отделка F 65
отделять сжатием Р 61
отжатие Р 181
отжать Р 182
отжигать А 150
отжиг для снятия внутренних натяжений Т 19
отжиг нержавеющей стали А 153
отжимать Р 61
отжимный винт F 154
отзывчивость R 149
откачивать Е 158
откачивающее действие Р 213
откачка Е 160, V 127
откачка до предварительного разрежения С 171
откачная система Е 202
откачная система с дифференциальной откачкой D 122
откачная установка с сорбционным насосом D 263
откачной автомат S 331
откачной агрегат G 130
откачной патрубок Е 201
откачной стенд С 229
откидное дно S 180
откидной вентиль F 77
отклонение шага намотки Н 63
отклонение электронного луча С 69
отклоняющая система D 80
отклоняющий магнит D 32
отключающая способность В 192
открытая система N 43
открытый ионизационный манометр В 46
открытый ионный источник I 74
открытый омегатронный масс-спектрометр N 42
отливка С 51, С 52
отметчик времени Т 110
относительная влажность R 101
относительная концентрация С 253
относительная распространенность изотопа I 179
отношение выходного сигнала к шуму О 69
отношение глубины провара к ширине R 36
отношение сигнал — шум О 69, S 139
отношение сопротивлений R 136
отношения давлений в насосах Р 179
отпадать С 132
отпаивание Т 113
отпаивать S 36
отпаянная вакуумная система S 39
отпаянный вакуумный прибор S 39

отпаянный капилляр S 52
отпечаток оболочки E 145
отпуск T 19
отработавший пар E 200
отравление P 96
отравление катода C 68
отравлять P 95
отражатель B 23
отражатель ионов I 126, I 160
отражательная печь A 104, R 12
отражательная пластина B 24
отражательная способность R 82
отражатель электронов E 106
отрицательное тлеющее свечение N 3
отсасывать D 237
отсекающий клапан E 213
отслаивание слюды M 102
отслаиваться C 132
отсоединение C 342
отсоединять пережатием P 61
отсос паров V 218
отсчет давления P 166
отсчетное давление R 75
оттаивание D 36
оттаивание во время нерабочей части цикла O 6
оттаивание горячим газом H 131
оттаивать D 34
оттиск R 112
оттиск поверхностных микронеровностей S 444
отфильтровывать F 60
отходы W 4
отходящее тепло W 5
охладитель C 317
охлаждаемая коническая насадка C 192
охлаждаемая ловушка C 213
охлаждаемая ловушка с малым расходом хладоагента L 111
охлаждаемая ловушка с резервуаром для хладоагента R 114
охлаждаемый водородом C 320
охлаждаемый жидкостью L 78
охлаждаемый колпачковый отражатель C 192
охлаждаемый паром V 211
охлаждать C 127
охлаждающая рубашка C 325
охлаждающая смесь F 202
охлаждающая среда C 317
охлаждающее ребро C 324
охлаждающий агент C 317
охлаждающий водяной насос C 333
охлаждающий змеевик C 187, C 323
охлаждение C 129
охлаждение водяной циркуляцией C 322
охлаждение испарением E 174
охлаждение кипением H 127
охлаждение непосредственным испарением D 149
охлаждение расширением E 211

охлаждение сетки C 327
охлажденная вода C 128
охлажденный воздухом A 100
охранное кольцо G 134
охрупчивание E 123
оцинковывать Z 10
очистительная установка P 225
очистительная установка с парообразным растворителем S 234
очистка C 153, R 81
очистка газа G 55
очистка пара от масел V 212
очистка погружением в нагретом состоянии H 128
очистка путем распыления в аргоне A 190
очистка разрядом D 158
очистка тлеющим разрядом D 158
очистка ультразвуком U 10
очищающее поле D 238
ошибочное управление F 18

П

падение давления P 158
пайка в защитной атмосфере C 308
пайка в печи P 226
пайка горелкой T 126
пайка мягким припоем S 221
пайка погружением D 144, D 148
пайка с гибкими переходными частями F 101
пайка с металлизацией молибдено-марганцевым порошком M 165
пайка твердым припоем B 188
пайка твердым припоем в водороде H 149
пайка ультразвуком U 12
палец S 390
палладиевая чернь P 11
палладиевый водородный натекатель P 12
палладиевый течеискатель P 10, P 13
пальцевой холодильник C 201
пар V 209
паровое сопло S 346
паровой вентиль S 347
паровой затвор V 221
паровой колпак S 340, V 219
паровой эжектор J 9
паровой эжекторный насос J 9
парогазовый термометр V 224
паромасляный насос O 38
пароотводящий колпак V 219
пароперегреватель S 420
паропровод J 13
парортутный диффузионный насос M 78
параструйная холодильная установка S 345
пароструйный инжектор S 344
пароструйный насос E 20, V 220
пароструйный насос с кольцевым соплом A 158

пароструйный эжектор J 9
пароструйный эжекторный вакуумный насос J 9
пароулавливающий колпак V 219
пароэжекторная холодильная установка S 345
пароэжекторный насос J 9, V 217
парциальное давление P 15
паста из синтетической смолы S 456
патронный нагреватель C 43
патрон Свана B 70
паяльная горелка B 150/1
паяльник S 223
паять мягким припоем S 218
паять твердым припоем B 187
пена F 144
пенетрометр P 28
пениться F 143
пенопласт A 88
пеностекло C 78
первая катодная темная область A 198
первая плавка F 67
первичная плавка F 67
первичная сушка M 12
переводный множитель C 315
перегонка с водяным паром S 341
перегонная колба A 118
перегонная установка с циклической загрузкой C 423
перегонный аппарат D 194
перегонный сосуд D 197
перегорание B 231
перегорание нити накала F 41
перегрев S 421
перегрев паров V 228
перегрев рабочей жидкости V 228
перегретый пар D 265
перегрузка O 75
передача I 154
передача импульса I 30
передвижная откачная позиция T 175
передвижной пульт M 130
передвижной резервуар M 132
пережог B 231
перекачивать C 141
переключение диапазона R 22
переливание D 10
перемешивание расплавленного металла S 362
перемешивающая катушка S 361
перемещающийся поршень D 188
перенаполнение B 3
перенапряжение S 449
перенасыщение S 422
переносный P 108
перенос тепла H 49
переохлаждение S 419
перепад давлений P 155
перепад концентрации C 251
переплавлять R 105
перепускной вентиль L 98
перерасширение O 74
пересыщение S 422
пересыщенный пар O 80

переход в стекловидное состояние V 244
переход из твердого в жидкое состояние C 106
переходная муфта R 66
переходная область T 148
переходная область течения I 110
переходная соединительная деталь R 65
переходник R 66, A 53
переходное стекло A 54, T 146
переходный металл T 147
переходный фланец A 55, T 145
период восстановления C 157
периодическое электронное группирование B 226
период повторения C 425
период полураспада H 2
персорбция P 36
пескоструйная очистка B 132, S 12
печь вакуумного отжига V 8
печь для вакуумного обезгаживания V 40
печь для вакуумного отжига (термообработки) V 8
печь для выращивания кристаллов C 404
печь для гарниссажной плавки S 175
печь для нагрева B 29
печь для отжига A 152
печь для спекания под давлением P 145
печь непрерывного действия T 193
печь отжига непрерывного действия C 295
печь периодического действия C 104
печь с защитной атмосферой C 309
печь с нагревом электронной бомбардировкой E 66
печь сопротивления R 132
печь с проталкиванием деталей C 301
пинцет T 202
пинч-эффект P 60
пипетка D 248
пирокерамы G 85
пиролиз P 227
пиролитическое нанесение C 174
пирометр суммарного излучения A 187, T 138
пистолет для распыления S 296
питание C 114
питательный клапан C 118
питательный насос F 23
питающее устройство C 115
питающее устройство с вакуумной рубашкой V 93
плавающий потенциал F 107
плавильная мощность M 65
плавильная печь M 66
плавильный котел M 70
плавильный пруток M 69
плавильный штаб M 69
плавка F 232, F 237
плавка в дуговой печи A 185

плавка во взвешенном состоянии F 188
плавка прутка R 166
плавка с вытягиванием слитка E 230
плавка с переплавляемым электродом C 282
плавкая вставка F 235
плавкая замазка M 64
плавкая мастика M 64
плавкая пробка F 235
плавкий F 234
плавкость F 233
плавление F 237
плазменная горелка P 83
плазменное напыление P 82
плазменный газ P 80
плазменный ионный источник P 81
плакированный металл C 234
пламенная печь A 104
пластина S 179
пластина газопоглотителя G 75
пластинчато-роторная конструкция R 186
пластинчато-роторный насос R 176, S 186
пластинчато-статорный насос R 182
пластинчатый насос V 207
пластинчатый отражатель P 87
пластификатор P 84
пластичность D 267
пластичный сплав F 166
пластмассы P 85
платиновое стекло G 91
пленка, полученная ионным осаждением I 157
пленка примеси F 157
пленочная дегазация O 49
пленочная сушилка T 72
плита вакуумного колпака B 92
плита с обогревом J 2
плиточная морозильная установка F 89
плоская ножка F 97
плоская пленка F 94
плоская прокладка F 95
плоская тонкая пластина F 94
плоский слиток B 144
плоский соединитель P 75
плоский стыковой сварной шов P 74
плоское уплотнительное кольцо F 95
плоскость базы B 58
плоскость дифракции D 124
плоскость кливажа C 161
плоскость кристаллической решетки C 408, L 22
плоскость основания B 58
плоскость решетки C 408
плоскость скольжения S 188
плоскость спайности C 161
плоскость уплотнения J 19
плотность S 260
плотность газа D 63
плотность дефектов кристаллической решетки L 19
плотность дислокаций D 178
плотность набивки P 3
плотность носителей заряда C 112
плотность паров V 214

плотность покрытия C 359
плотность пучка B 75
плотность струи B 75
плотность тока электронного луча E 59
плотность упаковки P 3
плотность частиц P 19
плотность эмиссионного тока E 124
плотность энергии P 123
площадка на кривой изменения температуры T 25
площадка поверхности S 446
площадка регулировки S 114
площадь кольцевого зазора A 74
площадь поверхности диффузии A 74
площадь сечения горловины сопла I 69
плунжерный вакуумный насос с золотниковым распределением S 178
плунжерный насос R 182
плющить F 98
поверхностная диффузия S 435
поверхностная емкость S 111
поверхностная закалка C 49
поверхностная ионизация S 439
поверхностная миграция S 441
поверхностная очистка S 443
поверхностная реплика S 444
поверхностная утечка масла O 16
поверхностная шероховатость S 445
поверхностная электрическая прочность F 90
поверхностная энергия I 97
поверхностное зеркало F 217
поверхностное натяжение S 448
поверхностное покрытие S 434
поверхностное твердение C 49
поверхностное течение S 438
поверхностное травление S 436
поверхностное трение B 179
поверхностно-центрированный F 2
поверхностный атом S 432
поверхностный разряд S 181
поверхность раздела I 96
поверхность соединения J 19
поверхность твердого тела S 229
поворотная печь для вакуумной плавки T 105
поворотный барабан R 174
поворотный компрессионный манометр V 199
поворотный плунжер I 18
поворотный стол T 200
повторная гидратация R 100

повторное замораживание R 94
повторное наполнение B 3
повторное науглероживание R 49
повторно испаряться R 72
повышение давления газа I 35
поглотитель A 17
поглотительная способность A 18
поглощающее вещество A 17
пограничный слой B 180
погружаемый пальцевый холодильник I 12
погруженный в масло O 26
подавление нулевого значения измеряемой величины Z 9
подача масла O 23
подача материала F 22
подача электрода E 37
подающий насос F 24
подвесная сушилка F 30
подвижная откачная система I 71
подвижное соединение F 99
подвижность электронов E 99
подводимая мощность P 127
подгонять A 52
поддерживающий вакуум H 103
поддерживающий насос P 194, H 101
подложка S 402
подогрев P 137
подогреватель P 136
подогреватель колпачкового отражателя J 6
подогревательная вставка H 32
подогреватель тигля C 387
подогреть P 134
подпружиненный вентиль F 20
подстраиваемость T 171
подсушивание P 138
подшипник качения A 171
подшипниковый сплав B 1
подъемная сила L 56
подъемная тележка T 177
позитивно-вакуумное формование D 233
позитивное формование D 234
позиционер P 109
поиски течеискателем, присоединенным к форвакууму B 16
показатель адиабаты A 69
показатель преломления R 90
показатель степени в уравнении адиабаты A 69
покоящийся газ G 8
покрывать медью C 334
покрытие аквадагом A 181
покрытие отражательным слоем M 123
покрытие пластмассой P 86
покрытие пленкой F 50
покрытие погружением D 145
покрытие синтетической смолой C 178

покрытие с низкой работой выхода L 120
покрытие твердым слоем H 14
покрыть C 173
полезная хладопроизводительность U 27
поле рассеяния F 214
полимеризация P 103
полимеризация электронным лучом E 74
полимолекулярная адсорбция M 190
полиморфизм P 104
полировальный круг B 213
политропный P 105
полная быстрота разрежения T 139
полная излучательная способность T 134
полная лучеиспускательная способность T 134
полная скорость откачки T 139
полное давление T 136
полный отжиг F 223
полный отсчет шкалы V 185
полный поток T 135
положительный столб P 110
полосатый спектр B 43
полосовая сталь S 351
полость C 77
полочная сушилка S 116
полупроводниковый катод S 76
полупроводниковый компонент S 77
полуфабрикат S 78
получение газа G 38
получение пленок выдуванием S 110
полый катод H 106
полый плунжер H 107
полый пучок H 105
полюсный наконечник P 101
поляризационная фольга P 100
поляризация вакуума V 120
полярная диаграмма P 98
полярная диаграмма удара S 123
поляроидный анализатор A 143
поляроидный поляризатор P 99
понижение давления P 158
понижение давления в единицу времени S 269
понижение точки (температуры) замерзания F 204
поперечное сечение соударений C 221
поперечное течение S 105
поплавковый вентиль F 110
поплавковый выключатель F 109
поплавковый клапан F 110
поправочный коэффициент Клаузинга C 152
пора P 62
пористость P 106
пористый газопоглотитель P 107
порог кристаллизации R 63
пороговая область T 91
пороговая энергия T 90
порог рекристаллизации R 63

порог чувствительности обнаружения D 78

порошковая металлургия P 122

порошок для матирования E 155

порция залитого масла O 11

поршневое кольцо P 68

поршневой вакуумный насос с золотниковым распределением S 178

поршневой вентиль S 200

поршневой компрессор R 50

поршневой насос P 67, R 51

порядковый номер элемента A 205

посаженный на мастике цоколь C 81

последовательное разбавление S 408

последовательно наносить слои P 55

послесвечение A 91

постороннее поле E 220

постоянная Больцмана B 161

постоянная времени откачки P 219

постоянная газовых ионов G 41

постоянная депрессии D 69

постоянная литейная форма P 31

постоянная манометра G 67

постоянная пересчета давления P 153

постоянная Планка P 76

постоянная решетки L 17

постоянная Сезерленда S 451

постоянство C 275

постоянство формы F 172

поступательная степень свободы T 151

потайной винт G 132

потенциал возбуждения E 196

потенциал земли E 2

потенциал ионизации I 148

потенциал стенки B 216

потенциальная яма P 115

потенциальный барьер P 113

потери W 4

потери на испарение E 178

потеря давления P 162

потеря напора L 107

поток F 118

поток газа F 123

поток дырок V 2

поток жидкости F 131

поток утечки F 4

правило фаз P 39

преграда против миграции C 363

предварительная механическая обработка P 139

предварительная обработка P 185

предварительное охлаждение P 131

предварительное спекание P 140

предварительно откачивать F 160

предварительно сушить P 132

предварительно удалять P 133

предварительный вакуум F 162

предварительный нагрев P 135, P 137

предварительный охладитель P 130

предварительный подогрев C 172

предел выносливости F 16, R 31

предел долговременной прочности C 369

предел излома B 195

предел обнаружения D 77

предел парообразования V 208

предел ползучести C 367, C 368

предел прочности на разрыв T 11

предел прочности при знакопостоянной периодической нагрузке I 117

предел прочности при знакопостоянном положительном цикле I 117

предел прочности при изгибе F 103

предел прочности при растяжении T 20

предел прочности при сдвиге S 107

предел прочности при сжатии C 249

предел текучести M 44, Y 2

предел термостойкости H 29

предел упругости E 24

предел усталости F 16

предел усталости при переменном изгибе B 104

предельная плотность U 1

предельное давление B 130

предельное давление сжатия C 242

предельный вакуум B 130

предохранительная диафрагма B 232

предохранительная мембрана B 232

предохранительное выключение от давления S 6

предохранительное подъемное устройство S 3

предохранительное устройство S 1

предохранительный вентиль S 7

предохранительный выключатель водяного охлаждения C 331

предохранительный клапан S 7

преобразователь давления P 177

прерыватель мостика B 200

прерывистый B 64

прессование S 316

прессованная слюда M 104

прессованный катод M 182

прессованный янтарь P 142

прецизионное измерение H 68

прибор для измерений пленочных сопротивлений R 135

прибор для измерения массового расхода M 33

прибор для измерения толщины тонкой пленки по омическому сопротивлению R 131

прибор для обнаружения течей L 42

прибор с кварцевым датчиком для измерения толщин пленок Q 7

прибор с установкой на нуль Z 4

прибыль M 62

приваренный фланец W 25

призматическая шпонка F 19

приклеенное покрытие A 64

приклеенный слой A 65

приклеенный цоколь C 81

прилеплять A 59

прилипаемость A 60

прилипать A 59

прилипающий воск A 67

прилипший A 62

примесный атом F 156

примесь A 57, I 32

принадлежности A 30,

принудительная конвекция F 152

принудительная смазка F 153

принципиальная схема B 60

принцип «криостата» C 399

принцип сборки из готовых узлов B 214

принцип сильной фокусировки B 389

припой S 219

природный стеатит B 143

присадка A 57

присадочный металл F 45

присадочный пруток F 46

присоединение J 17

присоединительная коробка E 133

приспосабливать A 52

приспособление для вытягивания слитка I 60

приспособление для регулировки S 93

притертое соединение G 129

приток газа G 5

пришлифованное соединение G 129

пришлифовывать L 12

проба S 8

проба, отлитая в виде лепешки B 235

пробивная прочность B 190

пробивное напряжение B 191

пробка M 14, S 367

пробка вентиля V 200

пробка для выпуска воздуха A 112

пробка крана K 4

пробный газ P 190

пробой F 89

проверка искровым течеискателем S 257

проверка на герметичность L 45

проверка на течи L 40

проверка помещением объекта в оболочку C 105

проверочная калибровка R 48

проводимость C 262

провод накала F 40

провод сердечника C 343

проволока для впая S 51

проволока для связки T 97

проволочное уплотнение W 48

программный вентиль P 196

программный выключатель T 111

программный переключатель C 424

программный регулятор C 424

прогреваемый B 26

прогреваемый контейнер H 22

прогреваемый проходной вентиль с большим диаметром условного прохода L 15

прогрев в печи B 33

прогрев паром S 343

продолжительная работа C 299

продолжительность адсорбции A 85

продолжительность замораживания F 206

продолжительность охлаждения C 130, H 102

продолжительность переходного процесса A 38

продолжительность плавки F 236

продолжительность распыления S 311

продувать A 73

продувочный вентиль B 149

продувка A 98, D 229

продукт обжига F 216

продукт расщепления C 162

прозрачность T 160

прозрачный для излучения T 153

производительность конденсации C 261

производительность морозильного аппарата F 200

производительность насоса D 185

производительность печи F 230

производительность плавильной печи M 65

производительность по всасыванию E 199

производительность по вымораживанию F 200

производство электронных ламп E 110

прокаливать A 150

прокатка в контейнере S 109

прокатка труб на пилигримовом (пильгерном) станке R 52

прокатывать C 5, R 168

прокладка G 48

прокладка для уплотнения вала S 100

прокладка из медной фольги C 338

прокладка из фольги F 147

прокладочное кольцо G 48

прокладочный материал S 37

прокованный сплав F 166

промежуточный ковш M 60

промежуточный конденсатор I 109

промежуточный отжиг P 192
промежуточный элемент A 53
промывать F 140, R 163
промывка газом F 141
промывочный газ B 136
промывочный электрод R 164
проникновение P 34
проникновение гелия H 60
проникновение гелия через стекло H 61
проницаемость P 33
проницаемость анода P 27
проницаемость для лучей R 13
пропитанный под вакуумом V 82
пропитанный синтетической смолой R 128
пропитка I 28
пропиточная установка I 27
пропитывание S 202
пропускная способность закрытого вентиля R 124
пропускная способность открытого вентиля O 45
пропускная способность печи F 230
пропускная способность при молекулярном течении M 148
пропускная способность соединения S 38
прорыв воздуха A 106
просветление оптических линз L 49
просветление оптических стекол L 49
просветленный B 145
просвечивающий T 153
проскальзывание S 187
прослойка I 99
пространственная решетка S 248
пространственный заряд S 245
противодавление полного срыва C 382
противоразрежение A 177
противоток C 358
профильное кольцо P 195
профильный уплотнитель P 195
проходная втулка E 40
проходная печь T 95
проходное кольцо E 40
проходное сечение вентиля V 187
проходной вентиль I 72
прохождение тепла H 51
прохождение туннеля T 192
процесс горячей экстракции H 130
процесс металлизации керамики по способу фирмы Телефункен S 168
процесс откачки P 217
процесс плавки в высоком вакууме H 88
прочно приставший A 62
прочность на излом B 195
прочность на износ R 138
прочность по отношению к термическим напряжениям T 46
прочность при изгибе F 103
прочность при продольном изгибе B 212

прочность при растяжении T 20
прочность при сдвиге S 107
прочность сцепления A 60, B 165
пружинная сталь S 305
пружинный контакт S 302
пружинящий контакт S 302
прямонакальный D 155
прямопролетный вентиль F 187
прямоточная печь U 16
прямоугольная резьба S 315
прямоугольный паз R 59
пузырек V 236
пузырь в металле B 138
пузырьковая камера B 211
пульверизатор S 296
пульт управления S 307
пусковое давление S 326
путь потока F 125
путь утечки C 366
пылеобразование F 168
пылесос V 25
пылеуловитель D 273
пьезоэлектрический P 53
пятнистая ликвация F 181
пятнистая сегрегация F 181
пятно от луча B 84

Р

рабочая камера S 411
рабочее давление O 51
рабочей жидкости / без F 132
рабочий измерительный прибор S 87
рабочий пар M 128
рабочий поршень W 54
рабочий такт A 40
радиальное сальниковое кольцо с манжетой для уплотнения вала R 4
радиальное уплотнение вращающегося вала R 5
радиальный компрессор R 173
радиационный нагрев R 8
радиация R 10
радиоавтография A 215
радиоактивная постоянная D 13
радиоактивный индикатор R 18
радиоактивный ионизационный манометр A 130, R 16
радиоактивный источник R 17
радиоизотопный газовый анализ D 182
радиоизотопный ионизационный манометр A 130
радиолампа E 92
радиометрический манометр K 13
радиометрический манометр Кнудсена K 13
радиус кривизны B 103
разбиваемый кончик B 197
разборная вакуумная система D 61
разборное соединение D 62
разброс границ зерен G 107

разброс скоростей V 233
разброс энергий E 140
развертка спектра масс M 36
развертывать S 21
разгоночный насос F 180
раздавливание I 23
разделение давления P 157
разделение изотопов I 181
разделение слоев P 91
разделительная колонна F 179
разделительная прокладка S 249
разделительное кольцо S 283
разделительный диск C 421
разливка C 51, C 52
разливочный желоб P 119
разливочный ковш C 53, T 8
разливочный ковш с донным стаканом B 176
разливочный ковш со сливным носком L 73
разложение D 16
разложение, происходящее на электродах D 18
размагничивание D 60
размер зерна G 109
размер после доводки F 66
размер фокусного пятна S 287
размещение вентилей V 206
размораживание орошением водой W 12
размораживатель D 35
размораживать D 34
размыкание U 20
разность давлений D 118
разрежать E 158, R 26
разрежение L 112
разреженный газ R 25
разреженный пар V 155
разрешающая способность R 142
разрушаемая перегородка B 196
разрушаемый стеклянный спай B 189
разрыв в энергетической кривой E 138
разрывная мощность B 192
разрядная лампа D 169
разрядная трубка D 169
разрядный промежуток D 163
разрядный ток D 160
раковина B 138, C 77
рамка маски M 26
раскалываемость C 160
раскалывать C 163
раскисленная сталь K 6
раскислитель D 64
распаивать U 22
расплав C 232
расплавление F 237
расплавленный металл M 63
расплавленный металл в ванне B 65
расплавляемая замазка M 64
расплавляемый катод M 72
распорный болт S 336
распределение зарядов C 113
распределение по скоростям V 233
распределение потенциала P 114

распределитель D 199
распределительная магистраль D 199
распределительный вал A 51
распределительный желоб D 198
распределительный катод D 180
распределительный щит C 312
распределять по поверхности S 299
распыление S 307
распыление жидкости A 207
распыление под напряжением смещения B 111
распылительная сушилка S 292
распылительная сушильная установка S 294
распылительный газопоглощение D 182
распылительное сопло S 298
распылительное устройство S 309
распылительный колпак S 308
распыляемый газопоглотитель V 246
распылять S 306
распылять жидкость A 208
распыляющее напряжение S 313
рассеяние D 184
рассеяние мощности на аноде A 163
рассеяние на аноде A 163
рассеяние под малым углом S 196
рассеяние тепла H 28
рассеянно D 125
расслоение I 10
расстекловывание D 83
расстекловываться D 84
расстояние между пластинами C 158
раствор для предварительной обработки S 382
растворимость S 232
растворимость газа G 58
растворитель S 233
растрескивающийся лак B 206
растровый электронный микроскоп S 23
растягивать E 206
растяжение при разрыве B 193
расход R 34
расход жидкого азота L 86
расходомер F 122
расходуемый электрод C 280
расходуемый электрод плавильной печи C 280
расход хладоагента C 318
расход холода C 326
расширительная машина D 19
расширительный золотник E 212
расширительный клапан E 213
расширять E 206
расширяющееся сопло E 207
расщепление L 10
расщепление ядра N 36
расщепляемость C 160
реагировать P 146
реагирующий R 40
реактивация R 39

реактивное движение J 12
реактивное испарение R 41
реактивное распыление R 43
реактивное топливо P 197
реактивный газопоглотитель N 4
реакционный цемент R 38
реакция в вакууме V 130
реакция в тигле C 389
реальный газ I 20
реальный кристалл I 19
ребристый фильтр E 10
ребро уплотнения S 49
револьверная головка T 201
револьверная головка с испарителями V 227
регенерация R 55
регидратация R 100
регистрирующее устройство R 57
регистрирующий прибор P 29
регистрирующий прибор для записи нескольких величин M 204
регулирование загрузки F 21
регулирование подачи F 21
регулирование температуры подложки S 407
регулировка S 90
регулировка испарения E 172
регулировка процесса нагрева B 28
регулировка чувствительности S 81
регулировочный вентиль S 88
регулируемая течь A 71
регулируемый газобалластный вентиль C 310
регулирующий вентиль C 313
регулирующий клапан C 121
регулятор давления P 152
регулятор испарения E 173
регулятор обезгаживания D 38
регулятор с падающей дужкой H 115
регулятор уровня L 51
редкий газ I 47
редуктор R 64
редукторный переход R 64
редуктор с двухсторонней резьбой T 86
редуктор с односторонней муфтой S 377
редукционный клапан P 167, R 104
режим текучести C 364
режущая кромка C 422
резервная спираль накала S 252
резервная установка S 324
резервуар V 34
резиновая прокладка R 203
резиновая трубка R 204
резиновый напальчник R 202
резиновый соединитель R 201
резиновый уплотнитель R 203
резиновый шланг R 204
резиноподобный эластичный материал R 205

резистивная схема R 141
резистивная цепь R 141
резистивный нагрев R 133
резистор с отрицательным температурным эффициентом N 4
резонансная частота R 143
резьбовое соединение T 85
рекарбонизация R 49
рекомбинация R 56
реконструирование R 111
рекристаллизация R 61
ректифицировать R 60
рекуперация R 55
реле времени T 107
реле давления P 163
реле жидкого воздуха L 77
реле максимального тока C 419
реле перегрузки C 419
реле потока F 129
реле расхода F 129
ремень B 99
рентгеновская граница X 4
рентгеновская трубка с штриховым фокусом L 65
рентгеновские лучи X 5
рентгеновский фототок S 215
реология R 156
реперная точка температурной шкалы F 69
реплика R 112
респираторная плёнка B 198
реторная печь R 154
рефрактометрический вакуумметр R 91
рефрижератор R 97
рециркуляционный насос C 145, R 89
решетка для сушки ампул D 260
решетчатая сушилка S 116
реэмиссия R 70
риска на прокладке G 49
рифление C 352
роликовая сварка L 68, S 53
роликовая сварка через проволочную прокладку H 141
рост зерна G 108
ротационная воздуходувка R 181
ротационное замораживание R 194
ротационный компрессор R 170, R 172
ротор I 18, R 195
роторная сублимационная сушилка D 249
ртутная лампа низкого давления L 117
ртутный вращательный насос Геде G 2
ртутный затвор M 77
ртутный манометр с запаянным концом C 166
ртутный пароструйный насос M 79
ртутный столб M 76
ртуть M 75
рубашка сопла S 173
ручная горелка H 11
ручной теплоэлектровентилятор D 43
рычажная система M 57
рычажный вентиль L 54

C

сальник P 4
сальниковая коробка P 4
сальниковое уплотнение S 392
сальниковый вентиль P 2
самовсасывающий S 71
самодиффузия S 64
саможестчение [рентгеновской] трубки S 65
самозагрязнение S 63
самозаполняющийся S 71
самоиспарение S 66
самоочистка S 61
самоочищающийся S 68
самописец P 29
самопишущий манометр со световым лучом F 142
самопишущий манометр M 18
саморассеяние S 73
саморассеивающийся электрод S 62
самостоятельный разряд S 70
сатинировать C 5
сброс на нуль R 115
свариваемость W 24
сварка внахлестку L 13
сварка встык E 9
сварка заливкой расплавленным металлом C 57
сварка кромок E 9
сварка плавлением F 240
сварка плавящимся электродом в аргоновой среде I 50
сварка прихваточными швами T 1
сварка прямолинейным швом L 68
сварка с глубоким проплавлением D 29
сварка сопротивлением R 139
сварка с перемешиванием T 206
сварка трением T 213
сварная конструкция W 32
сварная муфта W 26
сварное соединение встык с двухсторонним скосом кромок D 217
сварное соединение со скосом двух кромок S 164
сварной стыковой шов в форме X D 217
сварочная горелка W 31
сварочный пруток F 46
сварочный ток W 27
сверхвысокий вакуум U 4
сверхвысоковакуумная система U 7
сверхвысоковакуумный вентиль U 8
сверхвысоковакуумный вентиль с уплотнением жидким металлом L 83
сверхвысоковакуумный масляный диффузионный насос U 6
сверхвысоковакуумный фланец U 5
сверхзвуковая молекулярная струя S 425
сверхзвуковая паровая струя S 426
сверхзвуковая струя S 423
сверхзвуковое течение S 424
сверхзвуковое течение в насосах T 159
сверхзвуковой молекулярный пучок S 425

сверхнизкая температура U 9
сверхпроводимость S 431, S 423
сверхпроводник S 418
сверхрасширение O 74
сверхтвердая сталь U 3
светильный газ T 140
светлая протяжка B 203
светлый отжиг B 202
светочувствительная бумага O 85
свили S 380
свинцовое соединение L 34
свинцовое уплотнение L 34
свобода от вибрации F 183
свобода от колебаний F 183
свободная конвекция N 1
свободная энергия F 184
свободное расширение F 185
свободный фланец L 104
связывающее вещество C 80
связывание S 121
связь B 164, L 70
сгибать F 149
сгорание B 230
сдвиг T 96
сдвиг по частоте F 209
сдвиг фаз F 38
сегрегация E 119
седло вентиля V 202
седло клапана V 202
секторный масс-спектрометр S 56
секторный прибор S 57
селективный газопоглотитель S 59
сенсибилизатор S 82
сепаратор S 86
сердечник шлифа M 14
серебрение S 149
серебрить S 147
серебрянокислородный натекатель S 148
серое тело G 120
серый фильтр G 121
сжатие S 134, S 316
сжатие струи E 19
сжатый воздух P 183
сжигание B 230
сжижение L 74
сжижение воздуха A 107
сжижение газов L 75
сжижитель L 76
сиккатив D 239
сила сцепления B 165
сила тяжести G 116
силикагелевый течеискатель S 142
силикагель S 141
силиконовая смазка S 144
силиконовое масло S 145
силоконовый каучук S 140
сильноточный ввод H 53
сильфон B 93, C 432
сильфон из томпака T 123
сильфонный вентиль B 96
сильфонный дифференциальный манометр с нулевым отсчетом B 97
синтез ядер N 37
синтерированное стекло S 167
система без оболочки N 43
система вакуумпроводов V 119
система для вакуумного струйного обезгаживания V 161

система защитной блоки-
ровки S 5
система с интерферен-
ционными пленками
I 101
система с кольцевым
лучом R 158
система сопел J 4
система управления ско-
ростью испарения
E 181
ситаллы G 85
сифон F 25
сифон, заключенный в
рубашку J 1
сифонное обезгаживание
C 144
сифонное обезгаживание
в вакууме P 64
скалывание T 96
сканированная электрон-
ная дифракция S 22
сканировать S 21
скачок давления P 147
скачок давления S 121
скин-эффект S 172
склеивать A 59
склеивающее вещество
A 63
склерометр S 28
склерометрическая
твердость A 4
склонность к разложению
D 17
скольжение T 96
скользящая лопатка
S 179
скользящее дно S 180
скользящее кольцевое
уплотнение F 4
скользящее кольцо S 189
скользящий ввод R 53
скользящий разряд
S 181
скопление тепла H 48
скоростной напор D 279
скорость возгонки S 398
скорость всасывания S 52
скорость диффузии D 137
скорость замораживания
F 205
скорость испарения E 180
скорость конденсации
C 258
скорость миграции M 118
скорость молекул M 161
скорость напыления D 67
скорость натекания I 68
скорость обезгаживания
D 45
скорость обратного
течения B 22
скорость осаждения D 67
скорость откачки C 20,
P 215
скорость откачки воздуха
A 116
скорость откачки в отсут-
ствии ловушки U 13
скорость охлаждения
C 328
скорость парообразова-
ния E 180
скорость потока F 127
скорость развертки S 24
скорость распада M 172
скорость расплавления
M 68
скорость распыления
S 310
скорость растяжения
C 365
скорость реагирования
R 149, S 271
скорость роста G 131
скорость сварки W 30
скорость скольжения
S 191

скорость сублимации
S 398
скорость сублимационной
сушки F 196
скорость сушки D 261
скорость течения F 127
скорость хода поршня
S 388
скрап S 26
скребок S 27
скрепление формы C 50
скрученный T 205
слепая проба B 131
слив масла O 19
сливной кран B 135
сливной носок L 72
слиток B 113, I 57
слияние ядер N 37
слоистая пластмасса L 9
слоистая решетка F 54
слоистый столб S 381
слой краски D 274
слойный газопоглотитель
C 175
слойный геттер C 175
случайная ошибка A 31
слюда M 101
слюдяная изоляция
M 103
слюдяная пластина M 105
слюдяная прокладка
M 106
слюдяное окно M 108
сляб B 144
смазка L 121
смазка граничных поверх-
ностей B 182
смазка для кранов S 365
смазка для применения в
вакууме L 122
смазка под давлением
F 153
смазка разбрызгиванием
S 281
смазка расплавлением
M 71
смазка тонким слоем T 74
смазывание L 123
смазывающая способность
L 124
смачиваемость W 36
смачиваемость в вакууме
V 181
смачивающий агент W 35
смена масла O 10
сменный испаритель I 93
смесительная камера
S 290
смесительная машина
K 10
смесительная трубка
M 129
смесительное сопло
M 127
смесительный конденса-
тор C 284
смеситель с мешалкой
M 126
смеситель с наклонным
барабаном T 102
смешивающий конденса-
тор C 284
смещение на электроде
E 38
смещение фаз P 38
смола для заливки C 54
смола для литья C 54
смотровое окно I 78
смотровое отверстие
I 78
смотровое стекло I 78
смывать F 140
снег из твердой углеки-
слоты C 33
снижение давления P 162
снимаемое покрытие
S 384
собирающая сторона C 220

собственная полимериза-
ция P 102
собственная структура
S 75
согласованный спай
M 40
содержание влаги M 139
содержащий цинк Z 11
соединение B 164, C 273,
G 45, J 17, L 70
соединение внахлестку
L 13, L 14
соединение металла с ке-
рамикой C 103
соединение металла со
слюдой M 107
соединение металл-
сапфир S 14
соединение насоса с фор-
вакуумом B 10
соединение насосов P 216
соединение охлаждающей
воды C 330
соединение пайкой S 225
соединение раструбом
B 89
соединение с малыми
фланцами S 198
соединение с U-образным
хомутом U 28
соединение со скользящей
муфтой S 185
соединение, состоящее из
нескольких переходных
стекол G 105
соединение с отбортовкой
F 78
соединение с отбортовкой
кромок F 73
соединение с форвакуу-
мом B 10
соединительная пайка
J 18
соединительная сварка
J 18
соединительная трубка с
резьбой N 8
соединительное стекло
C 270
соединительные части
труб C 274
соединительный зажим
C 271
соединительный трубо-
провод T 182
соединительный фланец
D 269
создавать избыточное
давление S 415
создание вакуума C 362,
V 127
создание избыточного
давления S 416
создание с помощью элек-
тронных лучей E 58
солидус S 230
солнечная постоянная
S 216
соль для матирования
E 156
соляризация S 217
сопло N 31
сопло Вентури V 235
сопло Лаваля L 26
сопло с многими отвер-
стиями M 195
сопло с оболочкой C 28
сопло с обтекателем C 28
сопротивление апертуры
A 179
сопротивление вдавли-
ванию I 38
сопротивление воздуха
D 226
сопротивление ползуче-
сти C 367
сопротивление при кру-
чении T 131

сопротивление разрыву в
нагретом состоянии
H 138
сопротивление термиче-
ским напряжениям
T 46
сопротивление течению
газа F 120
сопротивление удару I 16
сопротивление шлифова-
нию G 124
сопряженное двойное сое-
динение C 268
«сопутствующая»
откачка C 395
сорбат S 236
сорбент S 237
сорбированный газ S 236
сорбировать S 235
сорбирующее вещество
S 237
сорбционная петля S 239
сорбционная площадка
S 241
сорбционное вещество
S 237
сорбционно-ионный
насос G 74
сорбционно-ионный насос
с проволочным испари-
телем W 47
сорбционный насос
S 240
сорбция S 238
сортированный G 104
составление графика тех-
нологического процесса
S 25
составная колонка U 18
составное колено L 95
составное стекло C 236
состав остаточного газа
R 121
состав пленки F 52
состояние поверхности
F 65, S 447
сосуд Дьюара D 85
сотовая структура H 113
соударение со стенкой
W 1
соударение с электроном
E 86
спай F 231, G 45
спай металла с керамикой
C 103
спай металла с керамикой
с металлизацией молиб-
дено-марганцевым
составом M 164
спай металла со стеклом
M 97
спай нержавеющей стали
с тугоплавким стеклом
H 15
спай с внутренней трубкой
I 77
спай с наружной трубкой
O 71
спай стекла с коваром
K 17
спай стекла с медью C 341
спай стекла с металлом
G 96
спай стекла с проволокой
«Думет» D 269
спейсер S 249
спекание в вакууме V 148
спекательная печь не-
прерывного действия
C 302
спекать обжигом S 170
спектральный коэффи-
циент излучения S 264
спектроскопический
S 265
спектр поглощения A 24
спеченный корунд S 166
спиновая волна S 278

спираль для вакуумной лампы V 27
спираль накала H 57
спиральная мешалка S 279
спиральный H 55
спиральный испаритель H 56
сплав A 126
сплав Вуда W 50
сплав для заливки подшипников B 1
сплошной материал B 223
спокойная сталь K 6
споласкивать S 4
способ Айркоматик I 50
способ аргоно-дуговой сварки A 188
способ гелиево-дуговой сварки H 54
способ дуговой сварки в гелиевой среде H 54
способ дуговой сварки плавящимся электродом в среде аргона A 188
способ наращивания G 53
способность подстраиваться T 171
способ хранения P 141
спуск масла O 19
спускная пробка D 59
спускной вентиль D 232, P 223
спускной винт D 59
спускной кран B 135, D 228
«спутная» откачка C 395
сравнительный капилляр C 230
средний вакуум F 64
средний коэффициент линейного расширения A 220
средний свободный пробег M 48
средняя квадратичная скорость R 169
средняя скорость M 49
средняя скорость молекул A 221
средняя температура A 222
средняя энергия возбуждения A 219
срез T 96
стабилизатор тока источника S 242
стабильность C 275
стабильность нулевой точки Z 6
сталь корытного профиля C 108
стальная лента S 351
стальная полоса S 351
стальной вакуумный колпак S 348
стандартная течь C 11
стандартное натекание S 322
станиолевое покрытие L 28
станок с вращающимся диском (кругом) F 79
старение катода C 62
старение торированных катодов A 93
статистическое распределение R 21
статическая вакуумная система S 330
статическое давление S 328, S 329
статор S 335
стационарное состояние S 333
стационарное течение S 332

стационарный насосный автомат S 331
стационарный резервуар S 334
створчатый вентиль F 77, L 35
стекловолокно G 87
стекло для запайки S 45
стекло для защиты от теплоизлучения G 84
стекло для наблюдения за уровнем масла O 35
стекло с большим содержанием двуокиси кремния H 78
стеклоцемент G 84
стеклянная бусинка G 80
стеклянная ножка G 94
стеклянная оболочка G 86
стеклянная трубка G 97
стеклянное шлифовое соединение G 127
стеклянный ионизационный манометр A 122
стеклянный колпак G 81
стеклянный наконечник G 95
стеклянный переход G 105
стеклянный сильфон G 82
стеклянный пробка G 89
стеклянный шлиф G 93
стекольный припой S 220
степень влажности M 139
степень диссоциации D 49
степень компрессии C 243
степень обработки D 51
степень покрытия C 360
степень сжатия C 243
степень чистоты D 50
стержень для крепления и центровки C 85
стержень для пробивания T 7
стержень для разбивания T 7
стержень толкателя P 226
стержень электрода S 360
стержневое уплотнение S 354
стержневой впай S 160
стерилизовать S 358
стойкий по отношению к солям фтора F 139
стойкость к пережогу I 10
стойкость к температурным колебаниям T 44
сток S 165a
столб со стратами S 381
стол для прогрева B 31
стопорное устройство R 29
стопорный кран P 90
стопорный стержень S 368
сторона впуска C 220
сточная доска D 230
стрелочный барометр D 88
стрелочный вакуумметр D 89
стрелочный манометр D 89
стрелочный термометр D 90
стружка C 131
струйная система J 14
струйное зажигание J 11
струйное напыление P 82
струйное сопло E 21
структурное соответствие E 142
структурный цемент C 278
структурный элемент C 279

струя воздуха A 97
струя пара V 210
ступенчатая кривая S 356
ступенчатое уплотнение S 357
ступенчатый фланец S 355
ступень давления P 172
ступень низкого давления L 114
ступень с динамическим давлением D 280
ступень с дифференциальной откачкой D 121
стыковая припайка B 236
стыковая сварка B 237
стыковая сварка давлением P 148
стыковая сварка оплавлением F 92
стыковое V-образное сварное соединение S 164
стыковой сварной шов без скоса кромок F 93
стыковой спай B 236
стягивание S 134
стяжная гайка C 29
сублимационная очистка F 197
сублимационная сушка D 255
сублимационная сушка из замороженного состояния C 391
сублимационная технология S 399
сублимационная установка S 396
сублимационный насос S 397
сублимационный сепаратор S 394
сублимация S 395, S 400
сурьма A 174
сухая смазка D 253
сухое трение D 254
сухое уплотнение D 264
сухой лед C 33
сухой пар D 265
сухой продукт с ледовым ядром D 258
сухопарник S 340
сушилка D 251
сушилка для высушивания в тонком слое T 72
сушилка для зеленого корма C 119
сушилка с качающимся цилиндром T 101
сушилка с перелопачивающей мешалкой P 8
сушилка с подогревными полками J 3
сушилка с псевдоожиженным слоем T 198
сушилка типа «пьяной бочки» T 101
сужильная доска D 230
сушильная камера D 256
сушильная решетка D 231, D 257
сушильная установка D 259
сушильный барабан C 431
сушильный коллектор D 257
сушильный шкаф C 1
сушить сублимационным способом F 193
сушка в вакууме V 54
сушка выпораживанием в качающейся сушилке T 183
сушка замораживанием тканей T 114
сушка испарением E 175
сушка распылением S 293

сферическая аберрация S 273
схватывание B 121, B 164
схватывающая замазка S 92
схема защитной блокировки S 4
схема защиты P 199
схема питания манометра V 75
сцеживанье D 10
сцепление B 164, L 70
счетчик Гейгера-Мюллера G 70
счетчик фотонов P 43
сшивание C 380

T

таймер T 110
тальк B 143
тангенциальное напряжение H 116
тарелка вентиля V 190
тарелка клапана V 190
тарельчатая пружина C 413
тарельчатая сушилка P 88
тарельчатый вентиль B 25, D 177
таунсендовский разряд D 4
твердая резина E 3
твердая углекислота C 33
твердение C 414
твердение при старении A 92
твердость H 16
твердость на вдавливание I 38
твердость на истирание A 4
твердость по Бринелю B 205
твердость по Виккерсу V 240
твердость по методу отскока R 47
твердость по Моосу M 137
твердость по Роквеллу R 165
твердость после переплавки R 106
твердость по царапанию A 4
твердость по Шору R 47, S 128
твердый раствор S 227
текучесть F 134
телеуправление R 107
«темный» вакуум B 127
темный разряд B 126, D 4
температура быстрого отжига I 82
температура воспламенения F 91
температура выпадения капель D 247
температура задержки S 318
температура замерзания F 203
температура застывания c P 120
температура испарения E 187
температура Кюри C 416
температура ликвидуса L 91
температура нити накала F 43
температура обработки W 55

температура окружающей среды A 140
температура отжига A 154
температура полочного барабана S 115
температура помутнения F 113
температура превращения T 149
температура прогрева B 32
температура размягчения D 33, F 126, S 211
температура рекристаллизации R 62
температура солидуса S 231
температура текучести F 126
температура хрупкости B 208
температурная область разрушающего пробоя D 190
температурная транспирация T 31
температурная шкала Фаренгейта F 6
температурная шкала Цельсия C 84
температурное колебание T 17
температурный датчик T 18
температурный диапазон плавления M 67
температурный диапазон теплового пробоя T 27
температурный шов E 209
температуропроводность T 33
теневая маска S 98
тензодатчик S 375
тензодатчик омического сопротивления E 32
тензометр S 375
теоретическая быстрота откачки I 119
теоретический поток I 120
теория малых возмущений S 199
теория малых сигналов S 199
тепловая десорбция T 30
тепловая диффузия T 32
тепловая камера для имитации космического пространства T 45
тепловая рубашка H 35
тепловая скорость T 49
тепловая транспирация T 31
тепловая энергия электронов T 31
тепловое излучение T 41
тепловой манометр H 25
тепловой насос H 42
тепловой отражатель H 45
тепловой пробой T 26
тепловой разрыв H 27
тепловой шум T 24
тепловой эквивалент T 36
тепловой экран H 46
теплозащитное стекло H 43
теплоизоляция H 36
теплообмен H 33
теплообменник H 34
теплообменник с встречным током C 356
теплоотражающее стекло H 43
теплоотражающий интерференционный фильтр H 44

теплопередача H 23, H 49
теплопроводность H 24, H 26, T 28
теплосодержание E 141
теплостойкость H 137, R 92
теплота десорбции H 37
теплота испарения E 176
теплота плавления H 38
теплота растворения H 40
теплота смешения H 39
теплота сублимации H 41
теплоэлектрический манометр H 25
термистор T 54
термисторный манометр T 55
термическая десорбция T 30
термическая ионизация T 39
термическая ловушка T 29
термическая обработка в вакууме V 78
термическая трещина H 27
термическая эмиссионная способность R 11
термический отпуск T 43
термическое окисление T 40
термическое равновесие T 35
термобатарея T 61
термодинамика T 59
термодиффузионная разделительная колонна T 58
термодиффузионный коэффициент для бинарной смеси B 120
термодиффузия T 32
термоимпульсная сварка T 38
термокатод H 121, T 50
термометр сопротивления R 137
термомолекулярный насос T 48
термообработка H 52
термообработка в вакуумной печи V 66
термопара T 56
термопара в кожухе S 120
термопарный манометр T 57
термопластичные пластмассы T 64
термопласты T 64
терморегулировка T 15
терморегулятор T 16
термостатированный расширительный вентиль T 65
термостат с циркуляционным подогревом C 142
термостойкость H 137, R 92
термостолбик T 63
термоэлектрическая ловушка P 25
термоэлектрическое охлаждение T 62
термоэлектронная эмиссия T 50
термоэлектронный катод H 121
тетродное распыление T 23
тетродный ионизационный манометр T 22
техника T 12
техника вакуумных измерений V 100
техника зонной плавки для выращивания кристалла Z 14

техника соединений I 94
техника течеискания путем обдувания пробным газом P 191
технологическая проработка S 25
технологический процесс P 193
технология T 12
технология вакуумной обработки V 125
технология испарения E 186
технология материалов T 13
технология нанесения пленки испарением D 68
технология напыления C 177
технология с применением гидридов титана или циркония A 47
технология сублимационной очистки поверхности F 198
технология тонких пленок T 80
технология Чохральского C 433
течеискание L 38
течеискание с помощью аммиака A 141
течеискатель L 43
течеискатель с ионным насосом I 159
течеискатель с палладиевым барьером P 10
течеискатель с радиоактивным индикатором R 19
течение F 118
течение без трения E 142
течение в пограничном слое B 181
течение жидкости F 131
течение, определяемое энтальпией E 142
течение при наличии трения T 212
течение Пуазейля P 93
течение со скольжением S 182
течение с поперечным градиентом скорости S 105
течь L 36
течь в тонкой стенке T 83
течь для дозировки газа G 23
течь обратно F 119
течь с вязкостным потоком V 243
тигель C 383
тигельная печь C 386
тигельное испарение C 384
тигель с донным сливом B 174
титановый геттерно-ионный насос T 117
титановый испарительный насос T 115
титановый сублимационный насос T 119
титано-гидридная пайка T 116
титровать T 120
тихий разряд G 99
ткань из стекловолокна F 31
тлеющий разряд G 99
токозвод C 417
токоограничивающий C 418
ток пучка B 74
ток смещения D 186
ток утечки L 37

толкатель P 226
толщиномер T 66
томпак T 122
тонкопленочная дистилляция T 71
тонкопленочная интегральная схема T 76
тонкопленочная схема T 68
тонкопленочная термопара T 81
тонкопленочная электроника P 24
тонкопленочное обезгаживание T 69
тонкопленочный дистиллятор с коротким путем T 79
тонкопленочный запоминающий элемент T 75
тонкопленочный светофильтр T 73
тонкопленочный термометр сопротивления T 78
тонкопленочный элемент памяти T 75
тонко прокатанный лист чистого серебра T 82
топливный газ F 221
топливный элемент F 220
топливо P 197
тор T 129
торированный катод T 84
тороидальная вакуумная камера D 220
тороидальная вакуумная камера бетатрона D 220
торсионные микровесы T 133
торцевое соединение E 7
точечная коррозия P 73
точечная сварка S 391
точечная электросварка S 288
точечный источник P 92
точечный самописец D 204
точка воспламенения F 91
точка деформации D 33
точка замерзания F 203
точка застоя S 317
точка кипения B 159
точка Кюри C 415
точка натяжения S 94, S 243
точка обработки W 55
точка отжига A 154
точка отсчета R 74
точка пайки S 224
точка перегиба температурной кривой I 51
точка перехода T 144
точка плавления F 239
точка помутнения C 169
точка превращения T 144
точка размягчения S 210
точка росы D 87
точка текучести F 126
точность A 35
точность измерения M 52
точность настройки A 36
травленая медная маска E 152
травление накипи S 20
травление окалины S 20
травление поверхности P 52
траектория потока F 125
трансмиссия ионного тока I 134
транспирация T 161
транспортабельный P 108
трапецеидальное кольцо V 233
трапецеидальный паз T 165

трение F 210
трение покоя S 327
трение скольжения S 183
трехатомный T 169
трехполюсный T 174
трехходовой кран T 89
трещина C 361
трещина при усадке
C 120
триодное распыление
T 172
триодный магнитно-
электроразрядный
насос T 88
тройная точка T 173
тройник T 141
тройник в виде буквы
«У» A 145
тройное соударение T 87
труба с бортиком F 74
труба с раструбом S 207
труба с фланцем F 74
трубка Вентури V 235
трубка Гейслера G 71
трубка для возврата масла
O 32
трубка Крукса C 376
трубопровод D 266
трубореактивный
двигатель J 15
трубчатое резьбовое сое-
динение J 15
трубчатый манометр
S 285
трубчатый нагреватель-
ный элемент T 180
трубчатый провод I 84
трубчатый спай T 181
трупного жира / из A 70
тугоплавкие металлы
R 93
туннельная печь C 301,
T 193
туннельная сушилка
T 190
туннельное действие
T 189
туннельный морозильный
аппарат F 207
туннельный эффект T 191
туннельный эффект для
электронов E 111
турбина, работающая на
отработавших газах
E 198
турбокомпрессор T 194
турбомолекулярный
насос T 196
турбонасос N 22
турбосушилка T 195
турбулентное течение
T 197
тусклое серебрение D 268
тяга T 96
тянутый металл D 236

У

углеводород H 147
углекислый барий B 48
углеродистая сталь C 36
угловая ловушка E 28
угловое распределение
A 148
угловое соединение
C 345
угловое уплотнение
с кольцевой золотой
прокладкой C 344
угловой вентиль A 146
угловой раствор пучка
B 73
угловой сварной шов
F 47
угловой сектор A 149
угол диэлектрических по-
терь P 125

угол падения A 144
угол смачивания C 283
удаление воздуха A 98,
D 7
удаление газа E 160
удаление напыленного
слоя D 15
удаленная электронная
пушка R 108
удаленный катод D 193
удалять газы D 37
удалять лед D 56
удалять окалину D 71
удалять смазку D 47
удар S 121
ударная волна S 126
ударная вязкость I 16
ударная вязкость образца
с надрезом N 29
ударная ионизация I 14,
I 143
ударная сварка E 112
ударная труба S 125
ударноволновая система
S 127
ударное испарение F 84
удельная быстрота откач-
ки S 263
удельная мощность
P 124
удельная теплоемкость
S 261
удельное электрическое
сопротивление E 33
удерживаемый пружиной
S 303
удлинение при разрыве
B 193
уловитель для конденсата
на форвакуумной сто-
роне B 9
улучшение вакуума в лам-
пе I 36
ультразвуковая очистка
U 10
ультразвуковая эрозия
U 11
ультрацентрифуга U 2
уменьшение числа ионов
D 21
универсальная газовая
постоянная I 7
универсальное газовое
уравнение E 149
универсальный измери-
тельный прибор M 194
упаковка в вакууме V 111
уплотнение G 45, I 40
уплотнение вала R 5
уплотнение зазора C 159
уплотнение заливкой
C 56
уплотнение кольцевого
зазора A 156
уплотнение крышки вен-
тиля B 167
уплотнение между крыш-
кой и корпусом вентиля
B 166
уплотнение острой кром-
кой K 11
уплотнение пайкой S 225
уплотнение с двойными
ножевидными высту-
пами D 213
уплотнение с двухсторон-
ними острыми кромками
D 213
уплотнение с золотой про-
волокой G 103
уплотнение с ножевид-
ными выступами K 11
уплотнение со скользя-
щим кольцом F 4
уплотнение с рельефной
прокладкой C 188
уплотнение ступени давле-
ния P 173

уплотнение тарелки вен-
тиля D 174
уплотнение тарелки зат-
вора D 174
уплотненный металлом
M 93
уплотнительная поверх-
ность вентиля V 192
уплотнительная шайба
V 143
уплотнительное кольцо
G 48
уплотнительное кольцо
для оси A 228
уплотнитель седла вен-
тиля V 203
уплотнитель тарелки вен-
тиля V 191
уплотнять S 36
уплотняющая смазка P 5
уплотняющая жидкость
C 267, S 43
уплотняющая замазка
S 40
уплотняющая манжета
S 41
уплотняющая поверх-
ность J 19
уплотняющая сила S 44
уплотняющая шайба S 42
уплотняющее давление
S 48
уплотняющее кольцо
G 48
употняющее кольцо с
образным поперечным
сечением V 232
уплотняющее кольцо с
L-образным сечением
C 21
уплотняющее кольцо с
X-образным сечением
X 7
уплотняющий воск S 50
уплотняющий материал
S 37
уплотняющий элемент
G 45
управление откачным
стендом P 220
управляемый вручную
M 22
управляющее устройство
C 311
упрочнение H 13
упрочнять H 12
упругая твердость R 47
упругий элемент вентиля
S 2
упругость R 127
упругость диссоциации
D 192
уравнение Бернулли
B 105
уравнение идеального газа
G 29
уравнение Клапейрона
C 151
уравнение непрерывности
C 294
уравнение состояния
идеального газа E 149
уравнительный вентиль
P 160
уравновешивание элек-
тронным лучом E 55
уравновешивающее дав-
ление B 35
уран U 26
уранинит P 23
урановая смолка P 23
уровень жидкости L 79
уровень масла O 27
уровень насыщения S 18
уровень Ферми F 28
уровень шума N 12
усадочная трещина C 120
усилитель расхода F 128

ускоряющее напряжение
A 28
ускоряющее поле A 27
ускоряющий электрод
A 26
условный предел текуче-
сти O 7
успокоенная сталь K 6
усталостная прочность
P 32, R 31
усталостное растрескива-
ние F 15
усталость F 14
устанавливать на нуль
A 72
установившееся состоя-
ние S 338
установившееся течение
S 332
установка I 81, S 90
установка вакуумного на-
пыления V 26
установка давления P 171
установка для анализа ме-
тодом горячей экстрак-
ции F 238
установка для безмасля-
ной откачки D 263
установка для быстрого
замораживания S 104
установка для вакуумного
синтерирования V 149
установка для вакуумного
спекания V 149
установка для вакуумной
металлизации V 105
установка для вакуумной
сублимационной сушки
V 69
установка для выращива-
ния кристаллов C 404
установка для горячей
экстракции V 213
установка для дистилля-
ции при минимально-
коротком расстоянии
с механической очист-
кой стенки S 130
установка для заморажи-
вания погружением I 13
установка для зонной
очистки Z 18
установка для зонной
плавки Z 16
установка для линейной
протяжки L 63
установка для литья под
вакуумом V 26
установка для литья
смолы R 129
установка для молекуляр-
ной перегонки O 48
установка для нанесения
покрытия на ленту
S 385
установка для непрерыв-
ной металлизации испа-
рением в вакууме C 304
установка для обезгажи-
вания D 44
установка для обезгажи-
вания металла M 86
установка для обезгажи-
вания стали S 350
установка для обезжири-
вания паров V 213
установка для облучения
I 171
установка для однократ-
ного - двойного покры-
тия S 163
установка для откачки и
вакуумного испарения
V 166
установка для приготов-
ления масла O 15
установка для приклеива-
ния C 82

установка для пропитки
I 27
установка для пропитки
воском W 18
установка для пропитки
лаком V 230
установка для разделения
воздуха A 115
установка для разделения
газов G 56
установка для разливки
жидкого азота L 87
установка для распыления
S 312
установка для регенера-
ции масла O 31
установка для синтери-
вания S 171
установка для спекания
S 171
установка для спекания
штабика R 167
установка для сублима-
ционной сушки F 195
установка для сушки
мгновенным распыле-
нием F 83
установка для сушки рас-
пылением (методом
сброса давления) F 82
установка для электрон-
нолучевого испарения
E 64
установка для электрон-
нолучевого напыления
на стальную ленту E 77
установка для электрон-
нолучевого нагрева
(отжига) E 54
установка, использующая
тлеющий разряд G 100
установка на нуль Z 7
установка непрерывного
действия A 117
установка очистки тлею-
щим разрядом G 102
установка под давлением
P 184
установка с вакуумным
колпаком B 91
установка с двойным кол-
паком D 206
установление поперечной
связи C 380
установочный винт S 89
устойчивость C 275
устойчивость к термо-
удару H 72
устройство I 81
устройство для вакуум-
ного уплотнения
V 144
устройство для выгрузки
D 161
устройство для вытяжки
монокристалла S 155
устройство для депарафи-
низации D 86
устройство для защиты от
миграции M 117
устройство для извлече-
ния слитка I 63
устройство для измерения
скорости испарения
с подогревными
электродами H 30
устройство для наполне-
ния F 49
устройство для отпайки
методом холодной
сварки C 218
устройство для поддержа-
ния уровня жидкости
L 51
устройство для пополне-
ния R 79
устройство для расщепле-
ния пучка B 83

устройство для смены ма-
сок M 25
устройство для смены об-
разцов S 9
устройство для удаления
металлического прутка
D 81
устройство для шлюзова-
ния D 161
устройство с автоматиче-
ским фракционирова-
нием S 67
устройство со скользящим
кольцом S 190
устье сопла N 34
ухудшение вакуума в лам-
пе D 277
участок поверхности S 446

Ф

фазовая диаграмма P 37
фальцевать F 149
фарадеева темная область
F 12
фарадеево темное про-
странство F 12
фасонное литье S 101
ферритовый сердечник
с прямоугольной петлей
F 29
фестонная сушилка F 30
физика металлов M 91
физика твердого тела
S 228
физика тонких пленок
T 77
физическая атмосфера
A 200
физическая сорбция
P 51
фиксирующая шпилька
A 119
фиксирующее приспособ-
ление J 16
фиксирующий зажим J 16
фильера D 100
фильтрация F 61
фильтр для улавливания
паров масла O 29
фильтр для улавливания
пыли D 272
фильтрование F 59, F 61
фильтровать F 57
фильтровать с помощью
вакуума F 58
фланец F 72
фланец для подсоедине-
ния промывочного газа
G 14
фланец для подвода
потока газа G 34
фланец подготовленный
к сварке W 28
фланец, прижимаемый
скобами C 148
фланец с угловым уплот-
нителем из двух кон-
центрических золотых
проволок D 210
фланцевое соединение
F 73
фланцевое уплотнение
F 75
флоккуляция F 112
флотация F 117
флюоресцентное покры-
тие F 137
флуоресцирующий экран
F 138
флюс C 232
флюс для пайки S 222
флюс для пайки мягким
припоем S 214
флюс для пайки твердым
припоем H 18
фовмовка S 103

фокальная плоскость
F 145
фокусировка F 146
фокусировка по направле-
нию D 153
фокусное пятно S 286
фоновый газ B 6
фоновый ток B 5
форвакуум F 162
форвакуумная ловушка
B 11
форвакуумная магистраль
B 8
форвакуумная сторона
F 159
форвакуумное давление
D 277
форвакуумное пространс-
тво B 15
форвакуумный бак B 38
форвакуумный баллон
F 164
форвакуумный вентиль
B 12, R 199
форвакуумный насос
B 14
форвакуумный патрубок
F 155
форма C 53
форма с отсосом S 414
формиргаз F 170
формула Планка P 77
формфактор S 102
форсированное расшире-
ние O 74
фотоионизация P 47
фотокатод P 42
фотолюминесценция P 48
фотометр с модулирован-
ным световым пучком
M 133
фотон P 49
фототравление P 46
фотоэлектрический счет-
чик P 43
фотоэлектронная постоян-
ная P 45
фотоэлектронная эмиссия
P 44
фотоэлектронный катод
P 42
фотоэлемент P 40
фотоэлемент с запорным
слоем P 50
фракционированная дис-
тилляция F 176
фракционированная пере-
гонка под высоким
вакуумом H 87
фракционирующая щетка
F 178
фракционирующий насос
F 180
фрикционный тормоз
F 211
фритта F 216
фронт ударной волны
S 122
фугитивность F 222
фундаментальное исследо-
вание B 61
функция распределения
Максвелла-Больцмана
M 46
футеровка L 69
фыркающий клапан S 201

Х

хаотическое распреде-
ление R 21
характеристика испарения
E 169
хемосорбционная связь
C 124
хемосорбция C 123
химически чистый C 122

хладоагент C 317
хладотекучесть C 202
хлорид кобальта C 133
хлорид платины C 134
ход клапана V 198
холодильная машина
R 97, R 98
холодильная система
C 329, R 98
холодильная техника
R 96
холодильная установка
L 115
холодильный агент C 317
холодильный рассол
C 321
холодильный шкаф для
глубокого заморажи-
вания D 26
холодная вулканизация
C 215
холодная обжимка C 204
холодная обработка C 219
холодная прессовка
C 200
холодная прокатка C 209
холодная сварка C 216
холодная сварка давле-
нием C 207
холодное зеркало C 205
холодное золочение C 203
холодное осаживание
C 204
холодное схватывание
C 210
холодное упрочнение
W 20
холодное штампование
C 211
холоднокатанный C 208
холоднотянутая прово-
лока C 212
холодный катод C 193
холодный колпак C 191
холодный наклеп W 20
холодопроизводитель-
ность R 95
хомут C 150, S 363
хомутообразный поглоти-
тель S 364
хромированный C 136
хромировать C 137
хронотрон T 108
хрупкость B 207

Ц

цапоновый лак C 79
цапфа S 390
цветовая температура
C 227
цезий C 4
целиковое вещество
B 223
целостность пленки F 53
цельнометаллическая
конструкция A 123
цельнометаллический
вентиль A 124
цемент на основе глице-
рина и свинцового глета
L 33
центр N 40
централизованная смазка
под давлением F 116
центрирующая шпилька
A 119
центрирующее кольцо
C 96
центрирующее приспособ-
ление C 95
центробежная воздухо-
дувка C 86
центробежная отливка
C 87
центробежная сублима-
ционная сушилка C 91

центробежная сублимационная сушка C 92
центробежная сушилка C 89
центробежная сушка C 90
центробежное замораживание тонких слоев S 277
центробежное литье C 87
центробежный компрессор C 88
центробежный молекулярный дистиллятор C 93
центробежный насос C 94, R 2
цеолит Z 1
цеолитовая ловушка Z 3
цеолитовый адсорбционный насос Z 2
циклический режим B 63
циклический ускоритель C 138
цикл Карно C 38
цикл Линде L 61
циклоидальный масс-спектрометр C 426
циклотрон C 427
циклотронная резонансная частота C 430
циклотронная частота C 428
циклотронный резонанс C 429
цикл прогрева B 27
цилиндрическая мембрана C 432
цилиндрический вентиль S 200
цилиндр насоса B 54
цинко-кадмиевый сульфат Z 12
циркуляционная смазка под давлением F 116
циркуляционное обезгаживание C 144
циркуляционное охлаждение C 165
циркуляционный вакуумный испаритель V 24
циркуляционный насос C 145
циркуляционный трубопровод C 143
циркуляция воды C 167
циркуляция масла O 23
цоколь S 205
цоколь Свана B 68

Ч

часовая производительность H 142
частицы пара V 222
частота биений B 87
частота возбуждения E 195
частота повторения B 87
частота соударений C 222
частота соударений с поверхностью J 21
частотный сдвиг F 209
чашечный вентиль B 98
чернила для стекла E 154
черновой металл C 390
черное тело B 124
черное хромирование B 125
«черный» вакуум B 127
чертик B 189
число Авогадро A 224
число каскадов N 46
число Кнудсена K 14
число Лошмидта L 105

число Маха M 1
число Нуссельта N 47
число Прандтля P 129
число Рейнольдса R 155
число соударений C 223
число ступеней N 46
чистовое волочение B 203
чистовое травление F 62
чистометаллический катод P 222
чистота поверхности F 65
чиитота слоя F 55
чувствительность S 80
чувствительность манометра G 68
чувствительность к насечке N 30
чувствительность по давлению P 170
чувствительность по парциальному давлению P 17
чувствительный элемент S 79
чугун C 55
чушка B 113

Ш

шаблон для микросхемы M 113
шаг сетки G 123
шариковый подшипник B 39
шарнирный вентиль F 77
шаровая охлаждаемая ловушка S 275
шаровое шлифовое соединение G 128
шаровой запорный клапан B 40
шаровой клапан B 41
шаровой шлиф B 36, G 128
шахтная печь P 72
швеллерная сталь C 108
шевронное уплотнение C 126
шевронный отражатель C 125
шейка фланца F 76
шеллак S 118
шероховатый на молекулярном уровне M 158
шестеренчатый насос G 69
ширина диффузионной поверхности J 7
ширина зазора R 3
ширина зазора сопла J 7
ширина кромки W 33
ширина сварного шва W 33
шкала Бэфорта B 88
шкала настройки D 53
шкала Реомюра R 44
шкала Фаренгейта F 5
шкала Цельсия C 83
шланг H 118
шлюз L 96, S 195
шлюз для подачи прутка B 47
шлюз для [пропускания] электронного луча G 141
шлюз для слитка I 58
шлюз для электронного луча G 141
шлюзовая камера A 108, L 97
шлюз с дифференциальной откачкой D 114
шнековый пресс S 35
шпаклевка S 442
шпилька S 390
шпилькообразный катод H 1

штампованная ножка P 58
штампованная стеклянная пластина P 143
штамповать S 320
штанга толкателя P 226
штенгель E 201, T 182
штифтовая сварка S 391
штифтовой запор B 69
штифтовой патрон B 70
штифтовой спай P 63
штифтовой цоколь B 68
шток вентиля V 201
шум фликкер-эффекта F 104
шунт B 199
шунтировать B 238
штыковой патрон B 70

Щ

щелевая лампа S 193
щелевой катод S 192
щелевой магнетронный ионизационный манометр S 282
щель C 361
щеточный перегонный аппарат B 210
щипцы для прессовки ножек P 144

Э

эбонит E 3
эвтектическая точка E 157
эжектор E 20, E 21
эжекторная ступень E 22
эжекторная ступень с боковым расположением S 138
эжекторный насос E 20
эжекция E 19
экзотермический E 205
экзоэлектронная эмиссия E 204
экран для защиты от излучения H 46
экран для защиты от разбрызгивания S 259
экран для источника испарения E 183
экран для подавления разряда D 168
экран для предотвращения возникновения разряда D 168
экранирующая сетка S 30
экранная сетка S 30
экран-отражатель S 280
экран сопла A 180
эксгаустер E 197
эксплуатационная надежность F 151
эксплуатационное давление O 51
экспресс-анализ остаточных газов S 131
экстрагировать E 222
экстрагирующий агент E 223
экстракционный анализ газов путем вакуумного расплавления H 129
экстраполировать E 225
экструдер E 224
эластичность R 127
эластомер E 26
эластомерический уплотнитель E 27
электрическая валентность E 118
электрическая пробивная прочность E 36
электрическая прочность B 190

электрический двойной слой E 30
электрический пробой в вакууме V 17
электрический разряд в газе G 26
электрический ток в газе G 26
электрический фильтр масс E 31, Q 2
электрически отжигать A 151
электрически прокаливать A 151
электрическое поглощение газа E 29
электрод в медной оболочке C 335
электрод из проволочной сетки N 7
электродная система с охранными кольцами G 135
электродуговая сварка A 186
электролиз E 45
электролитическая полировка E 46
электролитически полировать E 115
электролитическое серебро E 47
электромагнитно управляемый E 48
электромагнитный клапан M 3
электромагнитным приводом / с E 48
электрометрическая лампа E 49
электрометр с динамическим конденсатором V 237
электронная бомбардировка E 81
электронная лавина E 50
электронная лампа E 92
электронная лампа с накаленным катодом H 125
электронная линза E 96
электронная оболочка E 108
электронная оболочка у поверхности электрода E 107
электронная оптика E 100
электронная орбита E 99
электронная пушка E 89
электронная пушка с наклонным пучком T 163
электронная пушка с независимым ускорением W 52
электронная пушка с поворотом пучка T 163
электронная пушка с самоускорением S 60
электронная рекомбинация E 105
электронная эмиссионная способность T 53
электронная эмиссия E 87
электронное возбуждение E 91
электронное давление E 103
электронное облако C 168, E 85
электронно-ионный микроскоп F 35
электроннолучевая машина E 69
электроннолучевая обработка E 70
электроннолучевая пайка E 56
электроннолучевая печь E 66

электроннолучевая пла-
вильная печь с несколь-
кими пушками E 73
электроннолучевая пла-
вильная печь с удален-
ной пушкой R 109
электроннолучевая
плавка E 71
электроннолучевая резка
E 60
электроннолучевая сварка
E 79
электроннолучевая техно-
логия E 78
электроннолучевое испа-
рение E 63
электроннолучевое свер-
ление E 61
электроннолучевое синте-
рирование E 76
электроннолучевое спека-
ние E 76
электроннолучевой генера-
ратор E 67
электроннолучевой зонд
E 104
электроннолучевой испа-
ритель E 65
электроннолучевой нагрев
E 68
электроннолучевой отжиг
E 53
электроннооптическое
изображение E 97
электронный газ E 88
электронный захват
E 84
электронный зонд E 104

электронный луч E 51,
E 52
электронный пучок C 70
электронный пучок с вы-
соким первеансом H 76
электронный удар E 93
электрон оболочки S 108
электрон решетки L 20
электропневматическим
приводом / с E 114
электропневматически
управляемый E 114
электрополировка E 116
электроразрядное газо-
поглощение E 35
электроразрядный мано-
метр T 170
электроразрядный мано-
метр с холодным като-
дом C 196
электроразрядный насос
C 197
электростатический экран
E 117
электростатический элек-
тронный микроскоп F 34
электроэрозионная обра-
ботка E 43
электроэрозия E 42
эманация E 120
эмиссионная способность
E 128
эмиссионный выход E 127
эмиттирующая поверх-
ность E 126
эмульгировать D 183,
F 129
эндотермический E 135

энергетическая зона E 138
энергия активации A 44
энергия ионизации I 147
энергия поступательного
движения T 152
энергия связи B 122
энтальпия E 141
энтропия E 143
эпитаксия E 146
эпоксидная смола
E 147
эрг E 151
эталонная температура
R 76
эталонная течь C 11
эталонная цепь R 73
эталонное давление R 75
эталонный вакуум R 77
эталонный контур R 73
эффект Блирса B 133
эффект глубокого про-
плавления D 28
эффект запоминания
M 74
эффективная скорость
испарения E 14
эффективная скорость
откачки E 15, N 6
эффективная способность
всасывания P 215
эффективная степень сжа-
тия E 12
эффективное простран-
ство E 13
эффективность катода
C 65
эффективность охлаждае-
мой ловушки C 214

эффективность переноса
T 162
эффективность сепарации
S 85
эффективность струи J 8
эффективность эмиссии
E 125
эффективный атомный
номер E 11
эффективный объем E 13
эффект криозахвата C 401
эффект направленности
D 151
эффект ориентации O 60
эффект осадка S 19
эффект отложения S 19
эффект «памяти» M 74
эффект разделения S 84
эффузия E 17

Я

ядро N 40
язычковый вентиль R 69
язычковый клапан R 69
язычок криозахвата V 205
яркостная температура
B 204
яркостный пирометр
D 157
ярмо Y 4
ярмо магнитопровода M 8
ячеистая структура H 113
ячеистый анод M 188
ячейка H 112
ящичная печь для отжига
P 112

FÜR NOTIZEN

FÜR NOTIZEN

FÜR NOTIZEN

FÜR NOTIZEN

FÜR NOTIZEN